7TH EDITION

INTRODUCTORY MATHEMATICAL ANALYSIS

FOR BUSINESS, ECONOMICS, AND THE LIFE AND SOCIAL SCIENCES

STUDENT SOLUTIONS MANUAL

Ernest F. Haeussler, Jr.
THE PENNSYLVANIA STATE UNIVERSITY

Richard S. Paul
THE PENNSYLVANIA STATE UNIVERSITY

PRENTICE HALL, Englewood Cliffs, New Jersey 07632

© 1994 by Prentice-Hall, Inc.
A Paramount Communications Company
Englewood Cliffs, New Jersey 07632

Printed in the United States of America

10 9 8 7 6 5

ISBN   0-13-491754-5

Prentice-Hall International (UK) Limited, *London*
Prentice-Hall of Australia Pty. Limited, *Sydney*
Prentice-Hall Canada Inc., *Toronto*
Prentice-Hall Hispanoamericana, S.A., *Mexico*
Prentice-Hall of India Private Limited, *New Delhi*
Prentice-Hall of Japan, Inc., *Tokyo*
Simon & Schuster Asia Pte. Ltd., *Singapore*
Editora Prentice-Hall do Brasil, Ltda., *Rio de Janeiro*

CONTENTS

Preface

This student supplement to *Introductory Mathematical Analysis:* FOR BUSINESS, ECONOMICS, AND THE LIFE AND SOCIAL SCIENCES, 7TH EDITION provides solutions to odd-numbered problems in the text. For drill-type problems that basically involve one step, just the answer may be given.

If you attempt to solve one of these problems and are not successful, this manual should be useful. It is designed specifically to show you how to set up the problem and to allow you to check any resulting manipulations. Every effort has been made to ensure that this supplement is helpful to you; sufficient details and intermediate steps are included so that you can readily see how the correct answer can be found. Although many problems can be solved in a variety of ways, usually the most direct and natural method, consistent with the text, is shown.

0

Algebra Refresher

1. true; -2 is a *negative integer*.

3. false, because the natural numbers are 1, 2, 3, and so on.

5. true, because $5 = \frac{5}{1}$.

7. false, because $\frac{4}{2} = 2$, which is a positive integer.

9. true, because 0 and 6 are integers and $6 \neq 0$. Note: $\frac{0}{6} = 0$.

11. true

1. false, because 0 does not have a reciprocal.

3. false; the additive inverse of 5 is -5 because 5 + (-5) = 0.

5. true; $-x + y = y + (-x) = y - x$.

7. true; $\dfrac{x + 2}{2} = \dfrac{x}{2} + \dfrac{2}{2} = \dfrac{x}{2} + 1$.

9. false; the left side is $x + y + 5$, but the right side is $x + y + x + 5$.

11. distributive

13. associative

15. commutative

17. definition of subtraction

19. distributive

21. $5a(x + 3) = (5a)x + (5a)(3) = 5ax + (3)(5a)$
$= 5ax + (3 \cdot 5)(a) = 5ax + 15a$

23. $(x + y)(2) = 2(x + y) = 2x + 2y$

25. $x[(2y + 1) + 3] = x[2y + (1 + 3)] = x[2y + 4]$
$= x(2y) + x(4) = (x \cdot 2)y + 4x$
$= (2x)y + 4x = 2xy + 4x$

27. $a(b + c + d) = a[(b + c) + d] = a(b + c) + a(d)$
$= ab + ac + ad$

EXERCISE 0.4

1. $-2 + (-4) = -6$

3. $6 + (-4) = 2$

5. $7 - (-4) = 7 + 4 = 11$

7. $-8 - (-6) = -8 + 6 = -2$

9. $7(-9) = -(7 \cdot 9) = -63$

11. $(-1)(6) = -(1 \cdot 6) = -6$

13. $-(-6 + x) = -(-6) - x = 6 - x$

15. $-12(x - y) = (-12)x - (-12)(y) = -12x + 12y$ (or $12y - 12x$)

17. $\frac{-2}{6} = -\frac{2}{6} = -\frac{1}{3}$

19. $\frac{-4}{-2} = -\frac{4}{2} = -2$

21. $3[-2(3) + 6(2)] = 3[-6 + 12] = 3[6] = 18$

23. $(-5)(-5) = 5 \cdot 5 = 25$

25. $3(x - 4) = 3(x) - 3(4) = 3x - 12$

27. $-(x - 2) = -x + 2$

29. $8\left(\frac{1}{11}\right) = \frac{8 \cdot 1}{11} = \frac{8}{11}$

31. $\frac{-5x}{7y} = \frac{-(5x)}{7y} = -\frac{5x}{7y}$

33. $\frac{2}{3} \cdot \frac{1}{x} = \frac{2 \cdot 1}{3 \cdot x} = \frac{2}{3x}$

35. $(2x)\left(\frac{3}{2x}\right) = 3$

37. $\frac{7}{y} \cdot \frac{1}{x} = \frac{7 \cdot 1}{y \cdot x} = \frac{7}{xy}$

39. $\frac{1}{2} + \frac{1}{3} = \frac{3}{6} + \frac{2}{6} = \frac{3 + 2}{6} = \frac{5}{6}$

41. $\frac{3}{10} - \frac{7}{15} = \frac{9}{30} - \frac{14}{30} = \frac{9 - 14}{30} = \frac{-5}{30} = -\frac{5 \cdot 1}{5 \cdot 6} = -\frac{1}{6}$

43. $\frac{x}{9} - \frac{y}{9} = \frac{x - y}{9}$

45. $\frac{2}{3} - \frac{5}{8} = \frac{16}{24} - \frac{15}{24} = \frac{16 - 15}{24} = \frac{1}{24}$

47. $\dfrac{\frac{x}{6}}{y} = \dfrac{x}{6} \div y = \dfrac{x}{6} \cdot \dfrac{1}{y} = \dfrac{x}{6y}$

49. not defined (we cannot divide by 0)

51. not defined (we cannot divide by 0)

EXERCISE 0.5

1. $(2^3)(2^2) = 2^{3+2} = 2^5 \ (= 32)$

3. $w^4 w^8 = w^{4+8} = w^{12}$

5. $\dfrac{x^2 x^6}{y^7 y^{10}} = \dfrac{x^{2+6}}{y^{7+10}} = \dfrac{x^8}{y^{17}}$

7. $\dfrac{(x^2)^5}{(y^5)^{10}} = \dfrac{x^{2 \cdot 5}}{y^{5 \cdot 10}} = \dfrac{x^{10}}{y^{50}}$

9. $(2x^2 y^3)^3 = 2^3 (x^2)^3 (y^3)^3 = 8x^{2 \cdot 3} y^{3 \cdot 3} = 8x^6 y^9$

11. $\dfrac{x^8}{x^2} = x^{8-2} = x^6$

13. $\dfrac{(x^3)^6}{x(x^3)} = \dfrac{x^{3 \cdot 6}}{x^{1+3}} = \dfrac{x^{18}}{x^4} = x^{18-4} = x^{14}$

15. $\sqrt{25} = 5$

17. $\sqrt[5]{-32} = -2$

19. $\sqrt[4]{\dfrac{1}{16}} = \dfrac{\sqrt[4]{1}}{\sqrt[4]{16}} = \dfrac{1}{2}$

21. $(100)^{1/2} = \sqrt{100} = 10$

23. $4^{3/2} = (\sqrt{4})^3 = (2)^3 = 8$

25. $(32)^{-2/5} = \dfrac{1}{(32)^{2/5}} = \dfrac{1}{(\sqrt[5]{32})^2} = \dfrac{1}{(2)^2} = \dfrac{1}{4}$

27. $\left(\dfrac{1}{16}\right)^{5/4} = \left(\sqrt[4]{\dfrac{1}{16}}\right)^5 = \left(\dfrac{1}{2}\right)^5 = \dfrac{1}{32}$

29. $\sqrt{32} = \sqrt{16 \cdot 2} = \sqrt{16}\sqrt{2} = 4\sqrt{2}$

31. $\sqrt[3]{2x^3} = \sqrt[3]{2}\sqrt[3]{x^3} = x\sqrt[3]{2}$

33. $\sqrt{16x^4} = \sqrt{16}\sqrt{x^4} = 4x^2$

35. $2\sqrt{75} - 4\sqrt{27} + \sqrt[3]{128} = 2\sqrt{25 \cdot 3} - 4\sqrt{9 \cdot 3} + \sqrt[3]{64 \cdot 2}$

$$= 2 \cdot 5\sqrt{3} - 4 \cdot 3\sqrt{3} + 4\sqrt[3]{2}$$

$$= 10\sqrt{3} - 12\sqrt{3} + 4\sqrt[3]{2}$$

$$= -2\sqrt{3} + 4\sqrt[3]{2}$$

37. $(9z^4)^{1/2} = \sqrt{9z^4} = \sqrt{3^2(z^2)^2} = \sqrt{3^2}\sqrt{(z^2)^2} = 3z^2$

39. $\left(\dfrac{27t^3}{8}\right)^{2/3} = \left(\left[\dfrac{3t}{2}\right]^3\right)^{2/3} = \left[\dfrac{3t}{2}\right]^2 = \dfrac{9t^2}{4}$

41. $\dfrac{x^3 y^{-2}}{z^2} = x^3 \cdot y^{-2} \cdot \dfrac{1}{z^2} = x^3 \cdot \dfrac{1}{y^2} \cdot \dfrac{1}{z^2} = \dfrac{x^3}{y^2 z^2}$

43. $2x^{-1}x^{-3} = 2x^{-1+(-3)} = 2x^{-4} = \dfrac{2}{x^4}$

45. $(3t)^{-2} = \dfrac{1}{(3t)^2} = \dfrac{1}{9t^2}$

47. $\sqrt[3]{7s^2} = (7s^2)^{1/3} = 7^{1/3}(s^2)^{1/3} = 7^{1/3}s^{2/3}$

49. $\sqrt{x} - \sqrt{y} = x^{1/2} - y^{1/2}$

51. $x^2\sqrt[4]{xy^{-2}z^3} = x^2(xy^{-2}z^3)^{1/4} = x^2x^{1/4}y^{-2/4}z^{3/4} = \dfrac{x^{9/4}z^{3/4}}{y^{1/2}}$

53. $(8x - y)^{4/5} = \sqrt[5]{(8x - y)^4}$

55. $x^{-4/5} = \dfrac{1}{x^{4/5}} = \dfrac{1}{\sqrt[5]{x^4}}$

57. $2x^{-2/5} - (2x)^{-2/5} = \dfrac{2}{x^{2/5}} - \dfrac{1}{(2x)^{2/5}} = \dfrac{2}{\sqrt[5]{x^2}} - \dfrac{1}{\sqrt[5]{(2x)^2}}$

$\qquad\qquad\qquad\qquad = \dfrac{2}{\sqrt[5]{x^2}} - \dfrac{1}{\sqrt[5]{4x^2}}$

59. $\dfrac{3}{\sqrt{7}} = \dfrac{3}{7^{1/2}} = \dfrac{3\cdot 7^{1/2}}{7^{1/2}\cdot 7^{1/2}} = \dfrac{3\sqrt{7}}{7}$

61. $\dfrac{4}{\sqrt{2x}} = \dfrac{4}{(2x)^{1/2}} = \dfrac{4(2x)^{1/2}}{(2x)^{1/2}(2x)^{1/2}} = \dfrac{4\sqrt{2x}}{2x} = \dfrac{2\sqrt{2x}}{x}$

63. $\dfrac{1}{\sqrt[3]{3x}} = \dfrac{1}{(3x)^{1/3}} = \dfrac{1(3x)^{2/3}}{(3x)^{1/3}(3x)^{2/3}} = \dfrac{\sqrt[3]{(3x)^2}}{3x} = \dfrac{\sqrt[3]{9x^2}}{3x}$

65. $\dfrac{\sqrt{32}}{\sqrt{2}} = \sqrt{\dfrac{32}{2}} = \sqrt{16} = 4$

67. $\dfrac{\sqrt[4]{2}}{\sqrt[3]{xy^2}} = \dfrac{\sqrt[4]{2}}{x^{1/3}y^{2/3}} = \dfrac{\sqrt[4]{2}\cdot x^{2/3}y^{1/3}}{x^{1/3}y^{2/3}\cdot x^{2/3}y^{1/3}} = \dfrac{2^{1/4}x^{2/3}y^{1/3}}{xy}$

$\qquad = \dfrac{2^{3/12}x^{8/12}y^{4/12}}{xy} = \dfrac{(2^3x^8y^4)^{1/12}}{xy} = \dfrac{\sqrt[12]{8x^8y^4}}{xy}$

69. $2x^2y^{-3}x^4 = 2x^6y^{-3} = \dfrac{2x^6}{y^3}$

71. $\sqrt{\sqrt[3]{t^4}} = \sqrt{t^{4/3}} = (t^{4/3})^{1/2} = t^{4/6} = t^{2/3}$

73. $\dfrac{2^0}{(2^{-2}x^{1/2}y^{-2})^3} = \dfrac{1}{2^{-6}x^{3/2}y^{-6}} = \dfrac{2^6 y^6}{x^{3/2}} = \dfrac{64y^6 \cdot x^{1/2}}{x^{3/2} \cdot x^{1/2}} = \dfrac{64y^6 x^{1/2}}{x^2}$

75. $\sqrt[3]{x^2 yz^3}\sqrt[3]{xy^2} = \sqrt[3]{(x^2yz^3)(xy^2)} = \sqrt[3]{x^3 y^3 z^3} = xyz$.

77. $3^2(27)^{-4/3} = 3^2(3^3)^{-4/3} = 3^2(3^{-4}) = 3^{-2} = \dfrac{1}{3^2} = \dfrac{1}{9}$

79. $(2x^{-1}y^2)^2 = 2^2 x^{-2} y^4 = \dfrac{4y^4}{x^2}$

81. $\sqrt{x}\sqrt{x^2 y^3}\sqrt{xy^2} = x^{1/2}(x^2 y^3)^{1/2}(xy^2)^{1/2}$

$\qquad = x^{1/2}(xy^{3/2})(x^{1/2}y) = x^2 y^{5/2}$

83. $\dfrac{(x^2 y^{-1}z)^{-2}}{(xy^2)^{-4}} = \dfrac{x^{-4}y^2 z^{-2}}{x^{-4}y^{-8}} = \dfrac{y^{10}}{z^2}$

85. $\dfrac{(x^2)^3}{x^4} \div \left[\dfrac{x^3}{(x^3)^2}\right]^{-2} = \dfrac{x^6}{x^4} \div \dfrac{(x^3)^{-2}}{(x^6)^{-2}} = x^2 \div \dfrac{x^{-6}}{x^{-12}}$

$\qquad = x^2 \div x^{-6-(-12)} = x^2 \div x^6 = \dfrac{x^2}{x^6} = \dfrac{1}{x^4}$

87. $-\dfrac{8s^{-2}}{2s^3} = -\dfrac{4}{s^3 s^2} = -\dfrac{4}{s^5}$

89. $\left(\dfrac{2x^2 y}{3y^3 z^{-2}}\right)^2 = \left(\dfrac{2x^2 z^2}{3y^2}\right)^2 = \dfrac{2^2(x^2)^2(z^2)^2}{3^2(y^2)^2} = \dfrac{4x^4 z^4}{9y^4}$

EXERCISE 0.6

1. $8x - 4y + 2 + 3x + 2y - 5 = 11x - 2y - 3$

3. $8t^2 - 6s^2 + 4s^2 - 2t^2 + 6 = 6t^2 - 2s^2 + 6$

5. $\sqrt{x} + \sqrt{2y} + \sqrt{x} + \sqrt{3z} = 2\sqrt{x} + \sqrt{2y} + \sqrt{3z}$

7. $6x^2 - 10xy + \sqrt{2} - 2z + xy - 4 = 6x^2 - 9xy - 2z + \sqrt{2} - 4$

9. $\sqrt{x} + \sqrt{2y} - \sqrt{x} - \sqrt{3z} = \sqrt{2y} - \sqrt{3z}$

11. $9x + 6y - 15 - 16x + 8y - 4 = -7x + 14y - 19$

13. $3(x^2 + y^2) - x(y + 2x) + 2y(x + 3y)$
$= 3x^2 + 3y^2 - xy - 2x^2 + 2xy + 6y^2 = x^2 + 9y^2 + xy$

15. $2\{3[3x^2 + 6 - 2x^2 + 10]\}$
$= 2\{3[x^2 + 16]\} = 2\{3x^2 + 48\} = 6x^2 + 96$

17. $-3\{4x(x + 2) - 2[x^2 - (3 - x)]\}$
$= -3\{4x^2 + 8x - 2[x^2 - 3 + x]\}$
$= -3\{4x^2 + 8x - 2x^2 + 6 - 2x\}$
$= -3\{2x^2 + 6x + 6\}$
$= -6x^2 - 18x - 18$

19. $x^2 + (4 + 5)x + 4(5) = x^2 + 9x + 20$

21. $(x + 3)(x - 2) = x^2 + (3 - 2)x + 3(-2) = x^2 + x - 6$

23. $(2x)(5x) + [(2)(2) + (3)(5)]x + 3(2) = 10x^2 + 19x + 6$

25. $(x + 3)^2 = x^2 + 2(3)x + 3^2 = x^2 + 6x + 9$

27. $x^2 - 2(5)x + 5^2 = x^2 - 10x + 25$

29. $(\sqrt{2y} + 3)^2 = (\sqrt{2y})^2 + 2(\sqrt{2y})(3) + 3^2 = 2y + 6\sqrt{2y} + 9$

31. $(2s)^2 - 1^2 = 4s^2 - 1$

33. $(x^2 - 3)(x + 4) = x^2(x + 4) - 3(x + 4)$
$= x^3 + 4x^2 - 3x - 12$

35. $x^2(2x^2 + 2x - 3) - 1(2x^2 + 2x - 3)$
$= 2x^4 + 2x^3 - 3x^2 - 2x^2 - 2x + 3$
$= 2x^4 + 2x^3 - 5x^2 - 2x + 3$

37. $x\{3(x - 1)(x - 2) + 2[x(x + 7)]\}$

    $= x\{3(x^2 - 3x + 2) + 2[x^2 + 7x]\}$

    $= x\{3x^2 - 9x + 6 + 2x^2 + 14x\}$

    $= x\{5x^2 + 5x + 6\} = 5x^3 + 5x^2 + 6x$

39. $x(3x + 2y - 4) + y(3x + 2y - 4) + 2(3x + 2y - 4)$

    $= 3x^2 + 2xy - 4x + 3xy + 2y^2 - 4y + 6x + 4y - 8$

    $= 3x^2 + 2y^2 + 5xy + 2x - 8$

41. $(x + 5)^3 = x^3 + 3x^2(5) + 3x(5)^2 + 5^3$

    $= x^3 + 15x^2 + 75x + 125$

43. $(2x)^3 - 3(2x)^2(3) + 3(2x)(3)^2 - 3^3 = 8x^3 - 36x^2 + 54x - 27$

45. $\dfrac{z^2 - 4z}{z} = \dfrac{z^2}{z} - \dfrac{4z}{z} = z - 4$

47. $\dfrac{6x^5}{2x^2} + \dfrac{4x^3}{2x^2} - \dfrac{1}{2x^2} = 3x^3 + 2x - \dfrac{1}{2x^2}$

49.

$$
\begin{array}{r}
x \phantom{{}+ 3\big)} \\
x + 3 \overline{\big)\, x^2 + 3x - 1} \\
\underline{x^2 + 3x} \phantom{{}- 1} \\
-1
\end{array}
$$

Ans. $\quad x + \dfrac{-1}{x + 3}$

51.

$$
\begin{array}{r}
3x^2 - 8x + 17 \phantom{{}- 3} \\
x + 2 \overline{\big)\, 3x^3 - 2x^2 + x - 3} \\
\underline{3x^3 + 6x^2} \phantom{{}+ x - 3} \\
- 8x^2 + x \phantom{{}- 3} \\
\underline{- 8x^2 - 16x} \phantom{{}- 3} \\
17x - 3 \\
\underline{17x + 34} \\
- 37
\end{array}
$$

Ans. $\quad 3x^2 - 8x + 17 + \dfrac{-37}{x + 2}$

53.

$$\require{enclose}\begin{array}{r}t + 8 \phantom{{}+ 000} \\ t - 8 \enclose{longdiv}{t^2 + 0t + 0} \\ \underline{t^2 - 8t\phantom{{}+ 000}} \\ 8t + 0 \\ \underline{8t - 64} \\ 64 \end{array}$$

Ans. $t + 8 + \dfrac{64}{t - 8}$

55.

$$\begin{array}{r}x - 2\phantom{{}+ 00} \\ 3x + 2 \enclose{longdiv}{3x^2 - 4x + 3} \\ \underline{3x^2 + 2x\phantom{{}+ 00}} \\ -6x + 3 \\ \underline{-6x - 4} \\ 7 \end{array}$$

Ans. $x - 2 + \dfrac{7}{3x + 2}$

EXERCISE 0.7

1. $6x + 4 = 2(3x + 2)$

3. $5x(2y + z)$

5. $8a^3bc - 12ab^3cd + 4b^4c^2d^2 = 4bc(2a^3 - 3ab^2d + b^3cd^2)$

7. $x^2 - 5^2 = (x + 5)(x - 5)$

9. $p^2 + 4p + 3 = (p + 3)(p + 1)$

11. $(4x)^2 - 3^2 = (4x + 3)(4x - 3)$

13. $z^2 + 6z + 8 = (z + 4)(z + 2)$

15. $x^2 + 2(\cdot3)x + 3^2 = (x + 3)^2$

17. $2x^2 + 12x + 16 = 2(x^2 + 6x + 8) = 2(x + 4)(x + 2)$

19. $3(x^2 - 1^2) = 3(x + 1)(x - 1)$

21. $6y^2 + 13y + 2 = (6y + 1)(y + 2)$

23. $2s(6s^2 + 5s - 4) = 2s(3s + 4)(2s - 1)$

25. $x^{2/3}y - 4x^{8/3}y^2 = x^{2/3}y(1 - 4x^2y^2) = x^{2/3}y[1^2 - (2xy)^2]$
$$= x^{2/3}y(1 + 2xy)(1 - 2xy)$$

27. $2x(x^2 + x - 6) = 2x(x + 3)(x - 2)$

29. $(4x + 2)^2 = [2(2x + 1)]^2 = 2^2(2x + 1)^2 = 4(2x + 1)^2$

31. $x(x^2y^2 - 10xy + 25) = x[(xy)^2 - 2(xy)(5) + 5^2] = x(xy - 5)^2$

33. $(x^3 - 4x) + (8 - 2x^2) = x(x^2 - 4) + 2(4 - x^2)$
$$= x(x^2 - 4) - 2(x^2 - 4)$$
$$= (x^2 - 4)(x - 2)$$
$$= (x + 2)(x - 2)(x - 2)$$
$$= (x - 2)^2(x + 2)$$

35. $y^2(y^8 + 8y^4 + 16) - (y^4 + 4)^2$
$$= y^2(y^4 + 4)^2 - (y^4 + 4^2) = (y^4 + 4)^2(y^2 - 1)$$
$$= (y^4 + 4)^2(y + 1)(y - 1)$$

37. $x^3 + 8 = x^3 + 2^3 = (x + 2)(x^2 - 2x + 4)$

39. $(x^3)^2 - 1^2 = (x^3 + 1)(x^3 - 1)$
$$= (x + 1)(x^2 - x + 1)(x - 1)(x^2 + x + 1)$$

41. $(x + 3)^3(x - 1) + (x + 3)^2(x - 1)^2$
$$= (x + 3)^2(x - 1)[(x + 3) + (x - 1)]$$
$$= (x + 3)^2(x - 1)[2x + 2]$$
$$= (x + 3)^2(x - 1)[2(x + 1)]$$
$$= 2(x + 3)^2(x - 1)(x + 1)$$

43. $[P(1 + r)] + [P(1 + r)]r = [P(1 + r)](1 + r) = P(1 + r)^2$

45. $x^4 - 16 = (x^2)^2 - 4^2$

$$= (x^2 + 4)(x^2 - 4)$$
$$= (x^2 + 4)(x + 2)(x - 2)$$

47. $(y^4)^2 - 1^2 = (y^4 + 1)(y^4 - 1) = (y^4 + 1)(y^2 + 1)(y^2 - 1)$

$$= (y^4 + 1)(y^2 + 1)(y + 1)(y - 1)$$

49. $x^4 + x^2 - 2 = (x^2 + 2)(x^2 - 1) = (x^2 + 2)(x + 1)(x - 1)$

51. $x(x^4 - 2x^2 + 1) = x(x^2 - 1)^2 = x[(x + 1)(x - 1)]^2$

$$= x(x + 1)^2(x - 1)^2$$

EXERCISE 0.8

1. $\dfrac{x^2 - 4}{x^2 - 2x} = \dfrac{(x + 2)(x - 2)}{x(x - 2)} = \dfrac{x + 2}{x}$

3. $\dfrac{x^2 - 9x + 20}{x^2 + x - 20} = \dfrac{(x - 5)(x - 4)}{(x + 5)(x - 4)} = \dfrac{x - 5}{x + 5}$

5. $\dfrac{6x^2 + x - 2}{2x^2 + 3x - 2} = \dfrac{(3x + 2)(2x - 1)}{(x + 2)(2x - 1)} = \dfrac{3x + 2}{x + 2}$

7. $\dfrac{y^2(-1)}{(y - 3)(y + 2)} = -\dfrac{y^2}{(y - 3)(y + 2)}$

9. $\dfrac{(2x - 3)(2 - x)}{(x - 2)(2x + 3)} = \dfrac{(2x - 3)(-1)(x - 2)}{(x - 2)(2x + 3)} = \dfrac{(2x - 3)(-1)}{2x + 3}$

$$= \dfrac{3 - 2x}{2x + 3}$$

11. $\dfrac{2(x - 1)}{(x - 4)(x + 2)} \cdot \dfrac{(x + 4)(x + 1)}{(x + 1)(x - 1)} = \dfrac{2(x - 1)(x + 4)(x + 1)}{(x - 4)(x + 2)(x + 1)(x - 1)}$

$$= \dfrac{2(x + 4)}{(x - 4)(x + 2)}$$

13. $\dfrac{x^2}{6} \div \dfrac{x}{3} = \dfrac{x^2}{6} \cdot \dfrac{3}{x} = \dfrac{3x^2}{6x} = \dfrac{x}{2}$

15. $\dfrac{2m}{n^3} \cdot \dfrac{n^2}{4m} = \dfrac{2mn^2}{4mn^3} = \dfrac{1}{2n}$

17. $\dfrac{4x}{3} \div 2x = \dfrac{4x}{3} \cdot \dfrac{1}{2x} = \dfrac{4x}{6x} = \dfrac{2}{3}$

19. $\dfrac{-9x^3}{1} \cdot \dfrac{3}{x} = \dfrac{-27x^3}{x} = -27x^2$

21. $\dfrac{x - 5}{\dfrac{x^2 - 7x + 10}{x - 2}} = (x - 5) \cdot \dfrac{x - 2}{x^2 - 7x + 10}$

$$= (x - 5) \cdot \dfrac{x - 2}{(x - 5)(x - 2)} = (x - 5) \cdot \dfrac{1}{x - 5} = 1$$

23. $\dfrac{10x^3}{(x + 1)(x - 1)} \cdot \dfrac{x + 1}{5x} = \dfrac{10x^3(x + 1)}{5x(x + 1)(x - 1)} = \dfrac{2x^2}{x - 1}$

25. $\dfrac{x^2 + 7x + 10}{x^2 - 2x - 8} \div \dfrac{x^2 + 6x + 5}{x^2 - 3x - 4}$

$= \dfrac{x^2 + 7x + 10}{x^2 - 2x - 8} \cdot \dfrac{x^2 - 3x - 4}{x^2 + 6x + 5}$

$= \dfrac{(x + 5)(x + 2)}{(x - 4)(x + 2)} \cdot \dfrac{(x - 4)(x + 1)}{(x + 5)(x + 1)} = 1$

27. $\dfrac{(2x + 3)(2x - 3)}{(x + 4)(x - 1)} \cdot \dfrac{(1 + x)(1 - x)}{2x - 3}$

$= \dfrac{(2x + 3)(1 + x)(1 - x)}{(x + 4)(x - 1)} = \dfrac{(2x + 3)(1 + x)(-1)(x - 1)}{(x + 4)(x - 1)}$

$= -\dfrac{(2x + 3)(1 + x)}{x + 4}$

29. $\dfrac{x^2}{x + 3} + \dfrac{5x + 6}{x + 3} = \dfrac{x^2 + (5x + 6)}{x + 3} = \dfrac{(x + 3)(x + 2)}{x + 3} = x + 2$

31. L.C.D. $= 3t$, so $\dfrac{1}{t} + \dfrac{2}{3t} = \dfrac{3}{3t} + \dfrac{2}{3t} = \dfrac{3 + 2}{3t} = \dfrac{5}{3t}$

33. L.C.D. $= p^2 - 1$

$1 - \dfrac{p^2}{p^2 - 1} = \dfrac{p^2 - 1}{p^2 - 1} - \dfrac{p^2}{p^2 - 1} = \dfrac{p^2 - 1 - p^2}{p^2 - 1} = \dfrac{-1}{p^2 - 1}$

$= \dfrac{1}{1 - p^2}$

35. L.C.D. = $(2x - 1)(x + 3)$

$$\frac{4(x + 3)}{(2x - 1)(x + 3)} + \frac{x(2x - 1)}{(x + 3)(2x - 1)} = \frac{4(x + 3) + x(2x - 1)}{(2x - 1)(x + 3)}$$

$$= \frac{2x^2 + 3x + 12}{(2x - 1)(x + 3)}$$

37. $x^2 - x - 2 = (x - 2)(x + 1)$ and $x^2 - 1 = (x + 1)(x - 1)$, so L.C.D. = $(x - 2)(x + 1)(x - 1)$.

$$\frac{1}{(x - 2)(x + 1)} + \frac{1}{(x + 1)(x - 1)}$$

$$= \frac{x - 1}{(x - 2)(x + 1)(x - 1)} + \frac{x - 2}{(x + 1)(x - 1)(x - 2)}$$

$$= \frac{(x - 1) + (x - 2)}{(x - 2)(x + 1)(x - 1)} = \frac{2x - 3}{(x - 2)(x + 1)(x - 1)}$$

39. L.C.D. = $(x - 1)(x + 5)$

$$\frac{4(x + 5)}{(x - 1)(x + 5)} - \frac{3(x - 1)(x + 5)}{(x - 1)(x + 5)} + \frac{3x^2}{(x - 1)(x + 5)}$$

$$= \frac{4x + 20 - 3(x^2 + 4x - 5) + 3x^2}{(x - 1)(x + 5)} = \frac{35 - 8x}{(x - 1)(x + 5)}$$

41. $(1 + x^{-1})^2 = \left(1 + \frac{1}{x}\right)^2 = \left(\frac{x}{x} + \frac{1}{x}\right)^2 = \left(\frac{x + 1}{x}\right)^2 = \frac{(x + 1)^2}{x^2}$

$$= \frac{x^2 + 2x + 1}{x^2}$$

43. $\left(\frac{1}{x} - y\right)^{-1} = \left(\frac{1}{x} - \frac{xy}{x}\right)^{-1} = \left(\frac{1 - xy}{x}\right)^{-1} = \frac{x}{1 - xy}$

45. Multiplying the numerator and denominator of the given fraction by x gives

$$\frac{1 + \frac{1}{x}}{3} = \frac{\left(1 + \frac{1}{x}\right)x}{3x} = \frac{x + 1}{3x}$$

47. Multiplying numerator and denominator by $2x(x + 2)$ gives

$$\frac{3(2x)(x + 2) - 1(x + 2)}{x(2x)(x + 2) + x(2x)} = \frac{(x + 2)[3(2x) - 1]}{2x^2[(x + 2) + 1]}$$

$$= \frac{(x + 2)(6x - 1)}{2x^2(x + 3)}$$

Exercise 0.8 -15-

49. L.C.D $= \sqrt{x + h} \cdot \sqrt{x}$.

$$\frac{2}{\sqrt{x + h}} - \frac{2}{\sqrt{x}} = \frac{2\sqrt{x}}{\sqrt{x + h}\sqrt{x}} - \frac{2\sqrt{x + h}}{\sqrt{x}\sqrt{x + h}} = \frac{2\sqrt{x} - 2\sqrt{x + h}}{\sqrt{x}\sqrt{x + h}}$$

51. $\dfrac{1}{2 + \sqrt{3}} \cdot \dfrac{2 - \sqrt{3}}{2 - \sqrt{3}} = \dfrac{2 - \sqrt{3}}{4 - 3} = 2 - \sqrt{3}$

53. $\dfrac{\sqrt{2}}{\sqrt{3} - \sqrt{6}} = \dfrac{\sqrt{2}}{\sqrt{3} - \sqrt{6}} \cdot \dfrac{\sqrt{3} + \sqrt{6}}{\sqrt{3} + \sqrt{6}}$

$$= \frac{\sqrt{2}(\sqrt{3} + \sqrt{6})}{3 - 6} = \frac{\sqrt{6} + \sqrt{12}}{-3} = - \frac{\sqrt{6} + 2\sqrt{3}}{3}$$

55. $\dfrac{2\sqrt{2}}{\sqrt{2} - \sqrt{3}} \cdot \dfrac{\sqrt{2} + \sqrt{3}}{\sqrt{2} + \sqrt{3}} = \dfrac{2\sqrt{2}(\sqrt{2} + \sqrt{3})}{2 - 3} = \dfrac{4 + 2\sqrt{6}}{-1}$

$$= -4 - 2\sqrt{6}$$

57. $\dfrac{1}{x + \sqrt{5}} = \dfrac{1}{x + \sqrt{5}} \cdot \dfrac{x - \sqrt{5}}{x - \sqrt{5}} = \dfrac{x - \sqrt{5}}{x^2 - 5}$

59. $\dfrac{5(1 - \sqrt{3})}{(1 + \sqrt{3})(1 - \sqrt{3})} - \dfrac{4(2 + \sqrt{2})}{(2 - \sqrt{2})(2 + \sqrt{2})}$

$$= \frac{5(1 - \sqrt{3})}{1 - 3} - \frac{4(2 + \sqrt{2})}{4 - 2} = \frac{5(1 - \sqrt{3})}{-2} - \frac{4(2 + \sqrt{2})}{2}$$

$$= \frac{5(\sqrt{3} - 1) - 4(2 + \sqrt{2})}{2} = \frac{5\sqrt{3} - 4\sqrt{2} - 13}{2}$$

1

Equations

1. $9x - x^2 = 0.$

   Set $x = 1$:

   $$9(1) - (1)^2 \overset{?}{=} 0$$
   $$9 - 1 \overset{?}{=} 0$$
   $$8 \neq 0$$

   Thus 1 does not satisfy the equation.

   Set $x = 0$:

   $$9(0) - (0)^2 \overset{?}{=} 0$$
   $$0 - 0 \overset{?}{=} 0$$
   $$0 = 0$$

   Thus 0 satisfies the equation.

3. $y + 2(y - 3) = 4.$

   Set $x = \frac{10}{3}$: $\frac{10}{3} + 2\left(\frac{10}{3} - 3\right) \overset{?}{=} 4$, $\frac{10}{3} + \frac{20}{3} - 6 \overset{?}{=} 4$, $4 = 4$.

   Set $x = 1$: $1 + 2(1 - 3) \overset{?}{=} 4$, $1 - 4 \overset{?}{=} 4$, $-3 \neq 4$.

   Thus $\frac{10}{3}$ satisfies the equation but 1 does not.

5. $x(7 + x) - 2(x + 1) - 3x = -2.$

   $x = -3: (-3)(7 - 3) - 2(-3 + 1) - 3(-3) \overset{?}{=} -2,$

   $-3(4) - 2(-2) + 9 \overset{?}{=} -2, \quad -12 + 4 + 9 \overset{?}{=} -2, \quad 1 \neq -2.$
   Thus -3 does not satisfy the equation.

   $x = 0: \quad 0(7) - 2(1) - 3(0) \overset{?}{=} -2, \quad -2 = -2.$
   Thus 0 satisfies the equation.

7. Adding 5 to both sides; equivalence guaranteed.

9. Squaring both sides; equivalence *not* guaranteed.

11. Dividing both sides by x; equivalence *not* guaranteed.

13. Multiplying both sides by x-1; equivalence *not* guaranteed.

15. Multiplying both sides by $\frac{x-5}{x}$; equivalence *not* guaranteed.

17. $4x = 10.$ Dividing both sides by 4 gives $x = \frac{10}{4} = \frac{5}{2}.$

19. $3y = 0.$ Dividing both sides by 3 gives $y = \frac{0}{3} = 0.$

21. $-5x = 10 - 15, \quad -5x = -5.$ Dividing both sides by 5 gives
    $x = \frac{-5}{-5} = 1.$

23. $5x - 3 = 9, \quad 5x = 12, \quad x = \frac{12}{5}$

25. $7x + 7 = 2(x + 1), \quad 7x + 7 = 2x + 2, \quad 5x + 7 = 2, \quad 5x = -5,$
    $x = \frac{-5}{5} = -1$

27. $2(p - 1) - 3(p - 4) = 4p, \quad 2p - 2 - 3p + 12 = 4p,$
    $-p + 10 = 4p, \quad 10 = 5p, \quad p = 2$

29. $\frac{x}{5} = 2x - 6, \quad x = 5(2x - 6), \quad x = 10x - 30, \quad 30 = 9x,$
    $x = \frac{30}{9} = \frac{10}{3}$

31. $5 + \frac{4x}{9} = \frac{x}{2}.$ Multiplying both sides by $9 \cdot 2$ gives
    $9 \cdot 2 \cdot 5 + 2(4x) = 9(x), \quad 90 + 8x = 9x, \quad x = 90.$

33. $q = \frac{3}{2}q - 4$. Multiplying both sides by 2 gives $2q = 3q-8$,
   $-q = -8$, $q = 8$.

35. $3x + \frac{x}{5} - 5 = \frac{1}{5} + 5x$. Multiplying both sides by 5 gives
   $15x + x - 25 = 1 + 25x$, $16x - 25 = 1 + 25x$, $-9x = 26$,
   $x = -\frac{26}{9}$.

37. $\frac{2y - 3}{4} = \frac{6y + 7}{3}$. Multiplying both sides by 12 gives
   $3(2y - 3) = 4(6y + 7)$, $6y - 9 = 24y + 28$, $-18y = 37$,
   $y = -\frac{37}{18}$.

39. $w + \frac{w}{2} - \frac{w}{3} + \frac{w}{4} = 5$. Multiplying both sides by 12 gives
   $12w + 6w - 4w + 3w = 60$, $17w = 60$, $\dot{w} = \frac{60}{17}$.

41. $\frac{x + 2}{3} - \frac{2 - x}{6} = x - 2$. Multiplying both sides by 6 gives
   $2(x + 2) - (2 - x) = 6(x - 2)$, $2x + 4 - 2 + x = 6x - 12$,
   $3x + 2 = 6x - 12$, $2 = 3x - 12$, $14 = 3x$, $x = \frac{14}{3}$.

43. $\frac{9}{5}(3 - x) = \frac{3}{4}(x - 3)$. Multiplying both sides by 20 gives
   $36(3 - x) = 15(x - 3)$, $108 - 36x = 15x - 45$,
   $153 = 51x$, $x = 3$.

45. $\frac{3}{2}(4x - 3) = 2[x - (4x - 3)]$, $3(4x - 3) = 4[x - 4x + 3]$,
   $12x - 9 = -12x + 12$, $24x = 21$, $x = \frac{21}{24} = \frac{7}{8}$.

47. $I = Prt$, $I = P(rt)$, $P = \frac{I}{rt}$

49. $p = 8q - 1$, $p + 1 = 8q$, $q = \frac{p + 1}{8}$

51. $S = P(1 + rt)$, $S = P + Prt$, $S - P = r(Pt)$, $r = \frac{S - P}{Pt}$

53. $S = \frac{n}{2}(a_1 + a_n)$, $2S = n(a_1 + a_n)$, $2S = na_1 + na_n$,
   $2S - na_n = na_1$, $a_1 = \frac{2S - na_n}{n}$

55. $P = 2\ell + 2w$ ⇒ $960 = 2(360) + 2w$, $960 = 720 + 2w$, $240 = 2w$, $w = 120$ m

57. $V = C\left(1 - \frac{n}{N}\right)$ ⇒ $1000 = 1600\left(1 - \frac{n}{8}\right)$, $1000 = 1600 - 200n$, $200n = 600$, $n = 3$

59. $\ell = \ell_0[1 + \alpha(T - T_0)]$ ⇒ $1.001 = 1[1 + \alpha(100 - 0)]$, $1.001 = 1 + 100\alpha$, $0.001 = 100\alpha$, $\alpha = \frac{0.001}{100} = 0.00001$

EXERCISE 1.2

1. $\frac{5}{x} = 25$. Multiplying both sides by x gives $5 = 25x$, $x = \frac{5}{25}$, $x = \frac{1}{5}$ (which checks)

3. Multiplying both sides by 7-x gives $3 = 0$, which is false. Thus there is no solution, so the solution set is ∅.

5. $\frac{4}{8 - x} = \frac{3}{4}$, $4(4) = 3(8 - x)$, $16 = 24 - 3x$, $3x = 8$, $x = \frac{8}{3}$

7. $\frac{q}{3q - 4} = 3$, $q = 3(3q - 4)$, $q = 9q - 12$, $12 = 8q$, $q = \frac{3}{2}$

9. $\frac{1}{p - 1} = \frac{2}{p - 2}$, $p - 2 = 2(p - 1)$, $p - 2 = 2p - 2$, $p = 0$

11. $\frac{1}{x} + \frac{1}{5} = \frac{4}{5}$, $\frac{1}{x} = \frac{3}{5}$, $5x\left(\frac{1}{x}\right) = 5x\left(\frac{3}{5}\right)$, $5 = 3x$, $x = \frac{5}{3}$

13. $\frac{3x - 2}{2x + 3} = \frac{3x - 1}{2x + 1}$, $(3x - 2)(2x + 1) = (3x - 1)(2x + 3)$, $6x^2 - x - 2 = 6x^2 + 7x - 3$, $1 = 8x$, $x = \frac{1}{8}$

15. $\frac{y - 6}{y} - \frac{6}{y} = \frac{y + 6}{y - 6}$, $y(y - 6)\left[\frac{y - 6}{y} - \frac{6}{y}\right] = y(y - 6)\left[\frac{y + 6}{y - 6}\right]$,

$(y - 6)^2 - 6(y - 6) = y(y + 6)$,

$y^2 - 12y + 36 - 6y + 36 = y^2 + 6y$,

$y^2 - 18y + 72 = y^2 + 6y$,

$72 = 24y$,

$y = 3$

17. $\frac{-4}{x-1} = \frac{7}{2-x} + \frac{3}{x+1}$.

Multiplying both sides by $(x-1)(2-x)(x+1)$ gives

$$-4(2-x)(x+1) = 7(x-1)(x+1) + 3(x-1)(2-x)$$
$$-4(-x^2 + x + 2) = 7(x^2 - 1) + 3(-x^2 + 3x - 2)$$
$$4x^2 - 4x - 8 = 4x^2 + 9x - 13$$
$$-13x = -5, \quad x = \frac{5}{13}$$

19. $\frac{9}{x-3} = \frac{3x}{x-3} \Rightarrow 9 = 3x$, $x = 3$. But the given equation is not defined for $x = 3$, so there is no solution. The solution set is $\emptyset$.

21. $\sqrt{x+6} = 3$, $(\sqrt{x+6})^2 = 3^2$, $x + 6 = 9$, $x = 3$ (checks)

23. $\sqrt{5x-6} - 16 = 0$, $\sqrt{5x-6} = 16$, $(\sqrt{5x-6})^2 = 16^2$, $5x - 6 = 256$, $5x = 262$, $x = \frac{262}{5}$, which checks.

25. $\sqrt{\frac{x}{2} + 1} = \frac{2}{3}$. Squaring both sides: $\frac{x}{2} + 1 = \frac{4}{9}$, $\frac{x}{2} = -\frac{5}{9}$, $x = 2\left(-\frac{5}{9}\right) = -\frac{10}{9}$, which checks.

27. $\sqrt{4x-6} = \sqrt{x}$, $(\sqrt{4x-6})^2 = (\sqrt{x})^2$, $4x - 6 = x$, $3x = 6$, $x = 2$, which checks.

29. $(x-3)^{3/2} = 8$, $[(x-3)^{3/2}]^{2/3} = 8^{2/3}$, $x - 3 = 4$, $x = 7$, which checks.

31. $\sqrt{y} + \sqrt{y+2} = 3$, $\sqrt{y+2} = 3 - \sqrt{y}$, $(\sqrt{y+2})^2 = (3 - \sqrt{y})^2$, $y + 2 = 9 - 6\sqrt{y} + y$, $6\sqrt{y} = 7$, $(6\sqrt{y})^2 = 7^2$, $36y = 49$, $y = \frac{49}{36}$, which checks.

33. $\sqrt{z^2 + 2z} = 3 + z$. Squaring both sides gives $z^2 + 2z = (3 + z)^2$, $z^2 + 2z = 9 + 6z + z^2$, $-9 = 4z$, $z = -9/4$, which checks.

35. $r = \frac{d}{1 - dt}$, $r(1 - dt) = d$, $r - rdt = d$, $r = d + rdt$, $r = d(1 + rt)$, $d = \frac{r}{1 + rt}$

37. $r = \frac{2mI}{B(n + 1)}$. Multiplying both sides by n + 1 gives

$r(n + 1) = \frac{2mI}{B}$, $n + 1 = \frac{2mI}{rB}$, $n = \frac{2mI}{rB} - 1$

39. $y = \frac{10x}{1 + 0.1x}$. With y = 50 the equation is $50 = \frac{10x}{1 + 0.1x}$.
Multiplying both sides by 1+0.1x gives 50(1 + 0.1x) = 10x,
50 + 5x = 10x,   50 = 5x,   x = 10, which checks.

EXERCISE 1.3

1. $x^2 - 4x + 4 = 0$, $(x - 2)^2 = 0$,   x - 2 = 0,   x = 2

3. (y - 4)(y - 3) = 0.
   y - 4 = 0   or   y - 3 = 0
       y = 4   or   y = 3

5. (x - 3)(x + 1) = 0.
   x - 3 = 0   or   x + 1 = 0
       x = 3   or   x = -1

7. $x^2 - 12x + 36 = 0$, $(x - 6)^2 = 0$,   x - 6 = 0,   x = 6

9. (x - 2)(x + 2) = 0.
   x - 2 = 0   or   x + 2 = 0
       x = 2   or   x = -2

11. $z^2 - 8z = 0$,   z(z - 8) = 0.
    z = 0   or   z - 8 = 0
    z = 0   or   z = 8

13. $4x^2 + 1 = 4x$, $4x^2 - 4x + 1 = 0$, $(2x - 1)^2 = 0$,
    2x - 1 = 0,   2x = 1,   x = 1/2

15. y(2y + 3) = 5, $2y^2 + 3y - 5 = 0$,   (y - 1)(2y + 5) = 0.
    y - 1 = 0   or   2y + 5 = 0
        y = 1   or   y = -5/2

17. $-x^2 + 3x + 10 = 0$, $x^2 - 3x - 10 = 0$, $(x - 5)(x + 2) = 0$.
    $x - 5 = 0$ or $x + 2 = 0$
       $x = 5$ or $x = -2$

19. $2p^2 - 3p = 0$, $p(2p - 3) = 0$.
    $p = 0$ or $2p - 3 = 0$
    $p = 0$ or $p = \frac{3}{2}$

21. $x(x - 1)(x + 2) = 0$.
    $x = 0$ or $x - 1 = 0$ or $x + 2 = 0$
    $x = 0$ or $x = 1$ or $x = -2$

23. $x(x^2 - 64) = 0$, $x(x - 8)(x + 8) = 0$.
    $x = 0$ or $x - 8 = 0$ or $x + 8 = 0$
    $x = 0$ or $x = 8$ or $x = -8$

25. $6x^3 + 5x^2 - 4x = 0$, $x(6x^2 + 5x - 4) = 0$,
    $x(2x - 1)(3x + 4) = 0$.
    $x = 0$ or $2x - 1 = 0$ or $3x + 4 = 0$
    $x = 0$ or $x = \frac{1}{2}$ or $x = -\frac{4}{3}$

27. $(x + 3)(x + 1)(x - 2) = 0$. Setting each factor equal to
    zero gives $x = -3$ or $x = -1$ or $x = 2$

29. $p(p - 3)^2 - 4(p - 3)^3 = 0$. Factoring out $(p - 3)^2$ gives
    $(p - 3)^2[p - 4(p - 3)] = 0$, $(p - 3)^2(12 - 3p) = 0$,
    $3(p - 3)^2(4 - p) = 0$. $p - 3 = 0$ or $4 - p = 0$
                                   $p = 3$ or $p = 4$

31. $x^2 + 2x - 24 = 0$. $a = 1$, $b = 2$, $c = -24$.

    $x = \dfrac{-b \pm \sqrt{b^2 - 4ac}}{2a} = \dfrac{-2 \pm \sqrt{4 - 4(1)(-24)}}{2(1)}$

    $x = \dfrac{-2 \pm \sqrt{100}}{2} = \dfrac{-2 \pm 10}{2}$.

    $x = \dfrac{-2 + 10}{2} = 4$ or $x = \dfrac{-2 - 10}{2} = -6$

33. $4x^2 - 12x + 9 = 0$. $a = 4$, $b = -12$, $c = 9$.

$$x = \frac{-b \pm \sqrt{b^2 - 4ac}}{2a}$$

$$= \frac{-(-12) \pm \sqrt{144 - 4(4)(9)}}{2(4)} = \frac{12 \pm \sqrt{0}}{8} = \frac{12 \pm 0}{8} = \frac{3}{2}$$

35. $p^2 - 5p + 3 = 0$. $a = 1$, $b = -5$, $c = 3$.

$$p = \frac{-b \pm \sqrt{b^2 - 4ac}}{2a} = \frac{-(-5) \pm \sqrt{25 - 4(1)(3)}}{2(1)}$$

$$= \frac{5 \pm \sqrt{13}}{2}$$

37. $4 - 2n + n^2 = 0$, $n^2 - 2n + 4 = 0$. $a = 1$, $b = -2$, $c = 4$.

$$n = \frac{-b \pm \sqrt{b^2 - 4ac}}{2a}$$

$$= \frac{-(-2) \pm \sqrt{4 - 4(1)(4)}}{2(1)} = \frac{2 \pm \sqrt{-12}}{2},$$

so there are no real roots.

39. $6x^2 + 7x - 5 = 0$. $a = 6$, $b = 7$, $c = -5$.

$$x = \frac{-b \pm \sqrt{b^2 - 4ac}}{2a} = \frac{-7 \pm \sqrt{49 - 4(6)(-5)}}{2(6)} = \frac{-7 \pm \sqrt{169}}{12}$$

$$= \frac{-7 \pm 13}{12}. \text{ Thus } x = \frac{-7 + 13}{12} = \frac{1}{2} \text{ or } x = \frac{-7 - 13}{12} = -\frac{5}{3}.$$

41. $2x^2 - 3x = 20$, $2x^2 - 3x - 20 = 0$. $a = 2$, $b = -3$, $c = -20$.

$$x = \frac{-b \pm \sqrt{b^2 - 4ac}}{2a}$$

$$= \frac{-(-3) \pm \sqrt{9 - 4(2)(-20)}}{2(2)} = \frac{3 \pm \sqrt{169}}{4} = \frac{3 \pm 13}{4}.$$

$$x = \frac{3 + 13}{4} = \frac{16}{4} = 4 \text{ or } x = \frac{3 - 13}{4} = \frac{-10}{4} = -\frac{5}{2}$$

43. $2x^2 + 4x - 5 = 0$. $a = 2$, $b = 4$, $c = -5$.

$$x = \frac{-b \pm \sqrt{b^2 - 4ac}}{2a} = \frac{-4 \pm \sqrt{16 - 4(2)(-5)}}{2(2)}$$

$$= \frac{-4 \pm \sqrt{56}}{4} = \frac{-4 \pm 2\sqrt{14}}{4}$$

$$= \frac{-2 \pm \sqrt{14}}{2}$$

45. $(x^2)^2 - 5(x^2) + 6 = 0$. Let $w = x^2$. Then $w^2 - 5w + 6 = 0$, $(w - 3)(w - 2) = 0$. Thus $w = 3, 2$. Hence $x^2 = 3$ or $x^2 = 2$, so $x = \pm\sqrt{3}, \pm\sqrt{2}$, which check.

47. $2\left(\frac{1}{x}\right)^2 + 3\left(\frac{1}{x}\right) - 2 = 0$. Let $w = \frac{1}{x}$. Then $2w^2 + 3w - 2 = 0$, $(2w - 1)(w + 2) = 0$, so $w = \frac{1}{2}, -2$. Thus $x = 2, -\frac{1}{2}$.

49. $(x^{-2})^2 - 9(x^{-2}) + 14 = 0$. Set $w = x^{-2} \Rightarrow w^2 - 9w + 14 = 0$, $(w - 7)(w - 2) = 0$, so $w = 7, 2$. Thus $\frac{1}{x^2} = 7$ or $\frac{1}{x^2} = 2$, so $x^2 = \frac{1}{7}$ or $x^2 = \frac{1}{2}$. This gives $x = \pm\frac{\sqrt{7}}{7}, \pm\frac{\sqrt{2}}{2}$.

51. $(x - 3)^2 + 9(x - 3) + 14 = 0$. Let $w = x - 3$. Then $w^2 + 9w + 14 = 0$, $(w + 7)(w + 2) = 0$, so $w = -7, -2$. If $x - 3 = -7$, then $x = -4$, and if $x - 3 = -2$, then $x = 1$. Thus $x = -4, 1$.

53. If $w = \frac{1}{x - 2}$, then $w^2 - 12w + 35 = 0$, $(w - 7)(w - 5) = 0$, so $w = 7, 5$. Thus $\frac{1}{x - 2} = 7$ or $\frac{1}{x - 2} = 5$, from which $x - 2 = \frac{1}{7}$ or $x - 2 = \frac{1}{5}$. This gives $x = \frac{15}{7}, \frac{11}{5}$.

55. $x^2 = \frac{x + 3}{2}$, $2x^2 = x + 3$, $2x^2 - x - 3 = 0$, $(2x - 3)(x + 1) = 0$, from which $x = \frac{3}{2}, -1$.

57. $\frac{3}{x - 4} + \frac{x - 3}{x} = 2$.
Multiplying both sides by the L.C.D., $x(x - 4)$, gives
$$3x + (x - 3)(x - 4) = 2x(x - 4),$$
$$3x + x^2 - 7x + 12 = 2x^2 - 8x,$$
$$x^2 - 4x + 12 = 2x^2 - 8x,$$
$$0 = x^2 - 4x - 12,$$
$$0 = (x - 6)(x + 2).$$
$$x = 6, -2.$$

59. $\frac{6x + 7}{2x + 1} - \frac{6x + 1}{2x} = 1$.

   Multiplying both sides by the L.C.D., $2x(2x + 1)$, gives
   $2x(6x + 7) - (2x + 1)(6x + 1) = 1(2x)(2x + 1)$,
   $12x^2 + 14x - (12x^2 + 8x + 1) = 4x^2 + 2x$,
   $6x - 1 = 4x^2 + 2x$, $\quad 0 = 4x^2 - 4x + 1$, $\quad 0 = (2x - 1)^2$,
   $2x - 1 = 0$, $\quad 2x = 1$, $\quad x = \frac{1}{2}$

61. $\frac{2}{r - 2} - \frac{r + 1}{r + 4} = 0$. Mult. both sides by $(r - 2)(r + 4)$:

   $2(r + 4) - (r - 2)(r + 1) = 0$, $\quad 2r + 8 - (r^2 - r - 2) = 0$,
   $-r^2 + 3r + 10 = 0$, $\quad r^2 - 3r - 10 = 0$, $\quad (r - 5)(r + 2) = 0$.
   Thus $r = 5, -2$.

63. $\frac{y + 1}{y + 3} + \frac{y + 5}{y - 2} = \frac{14y + 7}{(y + 3)(y - 2)}$. Multiplying both sides by

   the L.C.D., $(y + 3)(y - 2)$, gives
   $(y + 1)(y - 2) + (y + 5)(y + 3) = 14y + 7$,
   $y^2 - y - 2 + y^2 + 8y + 15 = 14y + 7$,
   $2y^2 + 7y + 13 = 14y + 7$, $\quad 2y^2 - 7y + 6 = 0$,
   $(2y - 3)(y - 2) = 0$, $\quad y = \frac{3}{2}$ or $y = 2$. But $y = 2$ does not

   check. The solution is $y = \frac{3}{2}$.

65. $\frac{2}{x^2 - 1} - \frac{1}{x(x - 1)} = \frac{2}{x^2}$. Multiplying both sides by the

   L.C.D., $x^2(x + 1)(x - 1)$, gives
   $2x^2 - x(x + 1) = 2(x + 1)(x - 1)$,
   $2x^2 - x^2 - x = 2x^2 - 2$, $\quad x^2 - x = 2x^2 - 2$,
   $0 = x^2 + x - 2 = (x + 2)(x - 1)$, $\quad x = -2$ or $x = 1$. But
   $x = 1$ does not check. The solution is $x = -2$.

67. $(\sqrt{x + 2})^2 = (x - 4)^2$, $\quad x + 2 = x^2 - 8x + 16$,
   $0 = x^2 - 9x + 14$, $0 = (x - 7)(x - 2)$, $\quad x = 7$ or $x = 2$. Only
   $x = 7$ checks.

69. $(q + 2)^2 = (2\sqrt{4q - 7})^2$, $q^2 + 4q + 4 = 16q - 28$,
$q^2 - 12q + 32 = 0$, $(q - 4)(q - 8) = 0$. Thus $q = 4, 8$ which both check.

71. $\sqrt{x + 7} = 1 + \sqrt{2x}$, $(\sqrt{x + 7})^2 = (1 + \sqrt{2x})^2$,
$x + 7 = 1 + 2\sqrt{2x} + 2x$, $6 - x = 2\sqrt{2x}$. Squaring both sides again gives $(6 - x)^2 = 4(2x)$, $36 - 12x + x^2 = 8x$,
$x^2 - 20x + 36 = 0$, $(x - 18)(x - 2) = 0$, $x = 18$ or $x = 2$.
Only $x = 2$ checks.

73. $\sqrt{x} + 1 = \sqrt{2x+1}$, $(\sqrt{x} + 1)^2 = (\sqrt{2x+1})^2$,
$x + 2\sqrt{x} + 1 = 2x + 1$, $2\sqrt{x} = x$, $(2\sqrt{x})^2 = x^2$, $4x = x^2$,
$0 = x^2 - 4x$, $0 = x(x - 4)$. Thus $x = 0, 4$ which check.

75. $\sqrt{x + 5} + 1 = 2\sqrt{x}$, $(\sqrt{x + 5} + 1)^2 = (2\sqrt{x})^2$,
$x + 5 + 2\sqrt{x + 5} + 1 = 4x$, $2\sqrt{x + 5} = 3x - 6$,
$(2\sqrt{x + 5})^2 = (3x - 6)^2$, $4(x + 5) = 9x^2 - 36x + 36$,
$0 = 9x^2 - 40x + 16$, $0 = (9x - 4)(x - 4)$, $x = \frac{4}{9}$ or $x = 4$.
Only $x = 4$ checks.

77. $x = \dfrac{-(-4.7) \pm \sqrt{(-4.7)^2 - 4(0.02)(8.6)}}{2(0.02)} = 233.16$ or 1.84

79. $\overline{M} = \dfrac{Q(Q + 10)}{44}$, $44\overline{M} = Q^2 + 10Q$, $0 = Q^2 + 10Q - 44\overline{M}$.
From the quadratic formula with $a = 1$, $b = 10$, $c = -44\overline{M}$,
$$Q = \frac{-10 \pm \sqrt{100 - 4(1)(-44\overline{M})}}{2(1)} = \frac{-10 \pm 2\sqrt{25 + 44\overline{M}}}{2}$$
$= -5 \pm \sqrt{25 + 44\overline{M}}$. Thus $-5 + \sqrt{25 + 44\overline{M}}$ is a root.

81. Given $c = \dfrac{A}{A + 12}d$ and $c = \dfrac{A + 1}{24}d$, we set $\dfrac{A}{A + 12}d = \dfrac{A + 1}{24}d$.
Dividing both sides by d and then multiplying both sides by $24(A + 12)$ give
$$24A = (A + 12)(A + 1)$$
$$24A = A^2 + 13A + 12$$

$$0 = A^2 - 11A + 12$$

From the quadratic formula,

$$A = \frac{11 \pm \sqrt{121 - 48}}{2} = \frac{11 \pm \sqrt{73}}{2} = \frac{11 \pm 8.54}{2}..$$

Thus $A = \frac{11 + 8.54}{2} = \frac{19.54}{2} = 9.77$ or

$$A = \frac{11 - 8.54}{2} = \frac{2.46}{2} = 1.23,$$

which are not extraneous. Rounding these answers gives 1 year and 10 years.

83. $\frac{1}{p} + \frac{1}{120 - p} = \frac{1}{24}$. Multiplying both sides by $24p(120 - p)$

gives

$$24(120 - p) + 24p = p(120 - p),$$
$$2880 - 24p + 24p = 120p - p^2,$$
$$p^2 - 120p + 2880 = 0.$$

By the quadratic formula,

$$p = \frac{-(-120) \pm \sqrt{(-120)^2 - 4(1)(2880)}}{2} = \frac{120 \pm \sqrt{2880}}{2}$$

$$= 86.8 \text{ cm or } 33.2 \text{ cm}.$$

85. (a) When the object strikes the ground, t must be 0, so
$0 = 44.1t - 4.9t^2 = t(44.1 - 4.9t)$, $t = 0$ or $t = 9$.
Choose $t = 9$ s.

(b) Setting h = 88.2 gives
$88.2 = 44.1t - 4.9t^2$, $4.9t^2 - 44.1t + 88.2 = 0$,

$$t = \frac{44.1 \pm \sqrt{(-44.1)^2 - 4(4.9)(88.2)}}{2(4.9)}$$

$$= \frac{44.1 \pm 14.7}{9.8}.$$

$t = 3$ s or $t = 6$ s.

CHAPTER 1 - REVIEW PROBLEMS

1. $4 - 3x = 2 + 5x$, $2 = 8x$, $x = \frac{2}{8} = \frac{1}{4}$

3. $3[2 - 4(1 + x)] = 5 - 3(3 - x)$, $3[2 - 4 - 4x] = 5 - 9 + 3x$,
$-6 - 12x = -4 + 3x$, $-2 = 15x$, $x = -\frac{2}{15}$

5. $2 - w = 3 + w$, $-2w = 1$, $w = -\frac{1}{2}$

7. $x = 2x - (7 + x)$, $x = x - 7$, $0 = -7 \Rightarrow$ No solution.

9. $2\left(4 - \frac{3}{5}p\right) = 5$, $8 - \frac{6}{5}p = 5$, $-\frac{6}{5}p = -3$, $-6p = -15$, $p = \frac{5}{2}$

11. $\frac{3x - 1}{x + 4} = 0$, $3x - 1 = 0$, $3x = 1$, $x = \frac{1}{3}$

13. $\frac{2x}{x - 3} - \frac{x + 1}{x + 2} = 1$. Multiplying both sides by the L.C.D.,

   $(x - 3)(x + 2)$, gives

   $(x - 3)(x + 2)\left[\frac{2x}{x - 3} - \frac{x + 1}{x + 2}\right] = (x - 3)(x + 2)[1]$,

   $2x(x + 2) - (x - 3)(x + 1) = (x - 3)(x + 2)$,

   $2x^2 + 4x - (x^2 - 2x - 3) = x^2 - x - 6$,

   $2x^2 + 4x - x^2 + 2x + 3 = x^2 - x - 6$,

   $x^2 + 6x + 3 = x^2 - x - 6$, $7x = -9$, $x = -\frac{9}{7}$

15. $(3x + 5)(x - 1) = 0$.

   $3x + 5 = 0$ or $x - 1 = 0$

   $x = -\frac{5}{3}$ or $x = 1$

17. $5q^2 = 7q$, $5q^2 - 7q = 0$, $q(5q - 7) = 0$. Thus $q = 0$ or

   $5q - 7 = 0$, from which $q = 0, \frac{7}{5}$.

19. $(x - 5)^2 = 0$, $x - 5 = 0$, $x = 5$

21. $3x^2 - 5 = 0$, $3x^2 = 5$, $x^2 = \frac{5}{3}$, $x = \pm\sqrt{\frac{5}{3}} = \frac{\pm\sqrt{15}}{3}$

23. $(8t - 5)(2t + 6) = 0$, $8t - 5 = 0$ or $2t + 6 = 0$,
   $t = 5/8$ or $t = -3$

25. $-3x^2 + 5x - 1 = 0$, $3x^2 - 5x + 1 = 0$. Using the quadratic
   formula with $a = 3$, $b = -5$ and $c = 1$ gives

   $$x = \frac{-b \pm \sqrt{b^2 - 4ac}}{2a} = \frac{-(-5) \pm \sqrt{25 - 4(3)(1)}}{2(3)} = \frac{5 \pm \sqrt{13}}{6}$$

27. $x(x^2 - 9) - 4(x^2 - 9) = 0$, $(x^2 - 9)(x - 4) = 0$,
$(x + 3)(x - 3)(x - 4) = 0$. Setting each factor equal to
0 gives $x = \pm 3$, 4

29. $\frac{6w + 7}{2w + 1} - \frac{6w + 1}{2w} = 1$. Multiplying both sides by the
L.C.D., $(2w + 1)(2w)$, gives
$2w(6w + 7) - (2w + 1)(6w + 1) = (2w + 1)(2w)$,
$12w^2 + 14w - (12w^2 + 8w + 1) = 4w^2 + 2w$,
$6w - 1 = 4w^2 + 2w$, $0 = 4w^2 - 4w + 1$, $0 = (2w - 1)^2$,
$2w - 1 = 0$, $2w = 1$, $w = \frac{1}{2}$

31. Multiplying both sides by the L.C.D., $(x + 3)(x - 3)$, gives
$2 - 3x(x - 3) = 1(x + 3)$, $2 - 3x^2 + 9x = x + 3$,
$0 = 3x^2 - 8x + 1$,
$$x = \frac{-b \pm \sqrt{b^2 - 4ac}}{2a} = \frac{-(-8) \pm \sqrt{64 - 4(3)(1)}}{2(3)}$$
$$= \frac{8 \pm \sqrt{52}}{6} = \frac{8 \pm 2\sqrt{13}}{6} = \frac{4 \pm \sqrt{13}}{3}$$

33. $\sqrt{2x + 5} = 5$, $(\sqrt{2x + 5})^2 = 5^2$, $2x + 5 = 25$, $2x = 20$,
$x = 10$, which checks.

35. $(\sqrt[3]{11x + 9})^3 = 4^3$, $11x + 9 = 64$, $11x = 55$, $x = 5$

37. $\sqrt{y} + 6 = 5$, $\sqrt{y} = -1$, which has no solution because
the square root of a real number cannot be negative.

39. $(\sqrt{x-1})^2 = (7 - \sqrt{x+6})^2$, $x - 1 = 49 - 14\sqrt{x + 6} + x + 6$,
$14\sqrt{x + 6} = 56$, $\sqrt{x + 6} = 4$, $x + 6 = 16$, $x = 10$

41. $x + 2 = 2\sqrt{4x - 7}$, $(x + 2)^2 = (2\sqrt{4x - 7})^2$,
$x^2 + 4x + 4 = 4(4x - 7)$, $x^2 + 4x + 4 = 16x - 28$,
$x^2 - 12x + 32 = 0$, $(x - 4)(x - 8) = 0$, $x = 4$ or $x = 8$

43. Set $w = y^{1/3}$. Then $w^2 + w - 2 = 0$, $(w + 2)(w - 1) = 0$,
so $w = -2$, 1. Thus $y^{1/3} = -2$ or $y^{1/3} = 1$, from which
$y = -8$, 1.

45. $E = \frac{4\pi kQ}{A}$, $EA = (4\pi k)Q$, $Q = \frac{EA}{4\pi k}$

47. $n - 1 = C + \frac{C'}{\lambda^2}$, $n - 1 - C = \frac{C'}{\lambda^2}$, $C' = \lambda^2(n - 1 - C)$

49. $T^2 = 4\pi^2\left(\frac{L}{g}\right)$, $T = \pm\sqrt{4\pi^2\left(\frac{L}{g}\right)} = \pm 2\pi\sqrt{\frac{L}{g}}$

51. $mgh = \frac{1}{2}mv^2 + \frac{1}{2}I\omega^2$, $2mgh = mv^2 + I\omega^2$, $2mgh - mv^2 = I\omega^2$,

$\omega^2 = \frac{2mgh - mv^2}{I}$, $\omega = \pm\sqrt{\frac{2mgh - mv^2}{I}}$

53. $S^2 + \frac{R}{L}S + \frac{1}{LC} = 0$. By the quadratic formula,

$$S = \frac{-(R/L) \pm \sqrt{(R/L)^2 - 4(1)[1/(LC)]}}{2(1)}$$

$$= -\frac{R}{2L} \pm \frac{\sqrt{\left(\frac{R}{L}\right)^2 - \frac{4}{LC}}}{2}.$$

Introducing the denominator 2 into the radical gives

$$S = -\frac{R}{2L} \pm \sqrt{\frac{\left(\frac{R}{L}\right)^2 - \frac{4}{LC}}{4}}$$

$$= -\frac{R}{2L} \pm \sqrt{\left(\frac{R}{2L}\right)^2 - \frac{1}{LC}}.$$

MATHEMATICAL SNAPSHOT - CHAPTER 1

1. (a) $\$100 + \$100(0.114) = \$111.40$
   (b) $\$1 + \$1(0.135) = \$1.135$
   (c) $\$100 \cdot \frac{1 \text{ lb}}{\$1} = 100 \text{ lb}$
   (d) $\$111.40 \cdot \frac{1 \text{ lb}}{\$1.135} \approx 98.15 \text{ lb}$
   (e) $g = \frac{98.15 - 100}{100} = -0.0185 = -1.85\%$ (a loss of 1.85%)
   (f) $g = \frac{y - i}{1 + i} = \frac{0.114 - 0.135}{1 + 0.135} \approx -0.0185 = -1.85\%$

2

Applications of Equations and Inequalities

1. Let w be the width and 2w be the length of the plot. Then

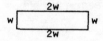

   area = 800 $\Rightarrow$ (2w)w = 800, $2w^2$ = 800, w^2 = 400, w = 20 ft. Thus the length is 40 ft, so the amount of fencing needed is 2(40) + 2(20) = 120 ft.

3. Let n = number of ounces in each part. Then we have 3n + 5n = 128, 8n = 128, n = 16. Thus there should be 3(16) = 48 ounces of A and 5(16) = 80 ounces of B.

5. Let n = number of ounces in each part. Then we have 2n + (1)n = 16, 3n = 16, n = 16/3. Thus the turpentine needed is (1)n = 16/3 = $5\frac{1}{3}$ ounces.

7. Let w = width (in meters) of pavement. The remaining

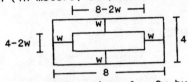

plot for flowers has dimensions 8 - 2w by 4 - 2w. Thus

$$(8 - 2w)(4 - 2w) = 12$$
$$32 - 24w + 4w^2 = 12$$
$$4w^2 - 24w + 20 = 0$$
$$w^2 - 6w + 5 = 0$$
$$(w - 1)(w - 5) = 0.$$

Hence w = 1, 5. But w = 5 is impossible since one dimension of the original plot is 4 m. Thus the width of the pavement should be 1 m.

9. Let q = number of units for $50,000 profit.

$$\text{Profit} = \text{Total Revenue} - \text{Total Cost}$$
$$50,000 = 3q - (2.20q + 95,000)$$
$$50,000 = 0.80q - 95,000$$
$$145,000 = 0.8q$$
$$\frac{145,000}{0.8} = q$$
$$q = 181,250$$

11. Let x = amount at 6% and 20,000 - x = amount at $7\frac{1}{2}$%.

$$x(0.06) + (20,000 - x)(0.075) = 1440$$
$$-0.015x + 1500 = 1440$$
$$-0.015x = -60$$
$$x = 4000,$$

so 20,000 - x = 16,000. Thus the investment consisted of $4000 at 6% and $16,000 at $7\frac{1}{2}$%.

13. Let p = selling price. Then profit = 0.2p.

$$\text{selling price} = \text{cost} + \text{profit}$$
$$p = 3.40 + 0.2p$$
$$0.8p = 3.40$$
$$p = \frac{3.40}{0.8} = \$4.25$$

15. Following the procedure in Example 4 we obtain

$$2,000,000(1 + r)^2 = 2,163,200$$
$$(1 + r)^2 = \frac{676}{625}$$
$$(1 + r) = \pm\frac{26}{25}$$
$$r = -1 \pm \frac{26}{25}$$

We choose $r = 1/25 = 0.04 = 4\%$.

17. Let n = number of room applications sent out.
$$0.95n = 76, \quad n = \frac{76}{0.95} = 80$$

19. Let s = monthly salary of deputy sheriff.
$$0.30s = 200, \quad s = \frac{200}{0.30}$$
Yearly salary $= 12s = 12\left(\frac{200}{0.30}\right) = \$8000$

21. Let q = number of cartridges sold to break even.
$$\text{total revenue} = \text{total cost}$$
$$19.95q = 14.95q + 8000$$
$$5q = 8000$$
$$q = 1600$$

23. Let v = total annual vision-care expenses (in dollars) covered by program. Then $10 + 0.80(v - 10) = 60$, $0.80v + 2 = 60$, $0.80v = 58$, $v = \$72.50$

25. Revenue = (number of units sold)(price per unit). Thus $400 = q\left[\frac{80 - q}{4}\right]$, $1600 = 80q - q^2$, $q^2 - 80q + 1600 = 0$. $(q - 40)^2 = 0$, $q = 40$.

27. Let q = required number of units. We equate incomes under both proposals.
$$2000 + 0.50q = 25,000, \quad 0.50q = 23,000, \quad q = 46,000 \text{ units.}$$

29. Let n = number of \$20 increases. Then at the rental charge of $400 + 20n$ dollars per suite, the number of suites that can be rented is $50 - 2n$. The total of all monthly rents is $(400 + 20n)(50 - 2n)$, which must equal 20,240. Thus

$$20,240 = (400 + 20n)(50 - 2n)$$
$$20,240 = 20,000 + 200n - 40n^2$$
$$40n^2 - 200n + 240 = 0$$
$$n^2 - 5n + 6 = 0$$
$$(n - 2)(n - 3) = 0$$
$$n = 2, 3$$

Thus the rent should be either $400 + 2($20) = $440 or
$400 + 3($20) = $460.

31. $10,000 = 800p - 7p^2$, $\quad 7p^2 - 800p + 10,000 = 0$

$$p = \frac{800 \pm \sqrt{640,000 - 280,000}}{14}$$

$$= \frac{800 \pm \sqrt{360,000}}{14} = \frac{800 \pm 600}{14}.$$

For $p > 50$ we choose $p = (800 + 600)/14 = 100

33. To have supply = demand, then $2p - 8 = 300 - 2p$, $4p = 308$,
$p = 77$.

35. Let w = width (in ft) of enclosed area. Then length of

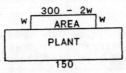

enclosed area is $300 - w - w = 300 - 2w$. Thus
$$w(300 - 2w) = 11,200$$
$$2w(150 - w) = 11,200$$
$$w(150 - w) = 5600$$
$$0 = w^2 - 150w + 5600$$
$$0 = (w - 80)(w - 70).$$

Hence $w = 80, 70$. If $w = 70$, then length is $300 - 2w = 300 - 2(70) = 160$. Since building has length of only 150 ft, we reject $w = 70$. If $w = 80$, then length is $300 - 2w = 300 - 2(80) = 140$. Thus the dimensions are 80 ft by 140 ft.

37. Original volume = $(10)(5)(2) = 100$ cm^3. Volume cut from
    bar = $0.28(100) = 28$ cm^3. Volume of new bar = $100 - 28 = 72$ cm^3. Let x = number of centimeters that the length
    and width are each reduced. Then

$$(10 - x)(5 - x)2 = 72$$
$$(10 - x)(5 - x) = 36$$
$$x^2 - 15x + 50 = 36$$
$$x^2 - 15x + 14 = 0$$
$$(x - 1)(x - 14) = 0, \quad \text{so } x = 1 \text{ or } 14.$$

Because of the length and width of the original bar, we
reject $x = 14$ and choose $x = 1$. The new bar has length
$10 - x = 10 - 1 = 9$ cm and width is $5 - x = 5 - 1 = 4$ cm.

39. Let x = amount of loan. Then the funds actually received
    are $x - 0.15x$. Hence $x - 0.15x = 95,000$, $0.85x = 95,000$,
    $x = \$112,000$ (to nearest thousand)

41. Let n = number of acres sold. Then $n + 20$ acres were
    originally purchased at a cost of $7200/(n + 20)$ each.
    The price of each acre sold was $30 + [7200/(n + 20)]$.
    Since the revenue from selling n acres is $7200 (the
    original cost of the parcel), we have

$$n\left[30 + \frac{7200}{n + 20}\right] = 7200$$
$$n\left[\frac{30n + 600 + 7200}{n + 20}\right] = 7200$$
$$n(30n + 600 + 7200) = 7200(n + 20)$$
$$30n^2 + 7800n = 7200n + 144,000$$
$$n^2 + 20n - 4800 = 0$$
$$(n + 80)(n - 60) = 0, \quad n = -80 \text{ or } n = 60.$$

We choose $n = 60$ (since $n > 0$).

43. Let q = number of units of B and $q + 25$ = number of units
    of A. Each unit of B costs $\frac{1000}{q}$, and each unit of A
    costs $\frac{1500}{q + 25}$. Therefore,

$$\frac{1500}{q + 25} = \frac{1000}{q} + 2$$
$$1500q = 1000(q + 25) + 2(q)(q + 25)$$

$$O = 2q^2 - 450q + 25,000$$
$$O = q^2 - 225q + 12,500 = (q - 100)(q - 125).$$
$$q = 100 \text{ or } q = 125.$$

If q = 100, then q + 25 = 125; if q = 125, q + 25 = 150.
Thus the company should produce either 125 units of A and
100 units of B, or 150 units of A and 125 units of B.

EXERCISE 2.2

1. $3x > 12$, $\quad x > \frac{12}{3}$, $\quad x > 4$.

$(4, \infty)$:

3. $4x - 13 \leq 7$, $\quad 4x \leq 20$, $\quad x \leq 5$.

$(-\infty, 5]$:

5. $-4x \geq 2$, $\quad x \leq \frac{2}{-4}$, $\quad x \leq -\frac{1}{2}$.

$\left(-\infty, -\frac{1}{2}\right]$:

7. $3 - 5s > 5$, $\quad -5s > 2$, $\quad s < -\frac{2}{5}$.

$\left(-\infty, -\frac{2}{5}\right)$:

9. $3 < 2y + 3$, $\quad 0 < 2y$, $\quad 0 < y$, $\quad y > 0$.

$(0, \infty)$:

11. $2x - 3 \leq 4 + 7x$, $\quad -5x \leq 7$, $\quad x \geq -\frac{7}{5}$.

$\left[-\frac{7}{5}, \infty\right)$:

13. $3(2 - 3x) > 4(1 - 4x)$, $6 - 9x > 4 - 16x$, $7x > -2$, $x > -\frac{2}{7}$.

$\left(-\frac{2}{7}, \infty\right)$:

15. $2(3x - 2) > 3(2x - 1)$, $6x - 4 > 6x - 3$, $0 > 1$, which is false for all x. Thus the solution set is $\emptyset$.

17. $x + 2 < \sqrt{3} - x$, $2x < \sqrt{3} - 2$, $x < \frac{\sqrt{3} - 2}{2}$.

$$\left(-\infty, \frac{\sqrt{3} - 2}{2}\right):$$
$$\frac{\sqrt{3} - 2}{2}$$

19. $\frac{5}{3}x < 10$, $5x < 30$, $x < 6$.

$$(-\infty, 6):$$
6

21. $\frac{9y + 1}{4} \leq 2y - 1$, $9y + 1 \leq 8y - 4$, $y \leq -5$.

$$(-\infty, -5]:$$
-5

23. $4x - 1 \geq 4(x - 2) + 7$, $4x - 1 \geq 4x - 1$, $0 \geq 0$, which is true for all x. The solution is $-\infty < x < \infty$.

$$(-\infty, \infty):$$

25. $\frac{1 - t}{2} < \frac{3t - 7}{3}$, $3 - 3t < 6t - 14$, $-9t < -17$, $t > \frac{17}{9}$

$$\left(\frac{17}{9}, \infty\right):$$
$17/9$

27. $2x + 3 \geq \frac{1}{2}x - 4$, $4x + 6 \geq x - 8$, $3x \geq -14$, $x \geq -\frac{14}{3}$

$$\left[-\frac{14}{3}, \infty\right):$$
$-14/3$

29. $\frac{2}{3}r < \frac{5}{6}r$, $4r < 5r$, $0 < r$, $r > 0$

$$(0, \infty):$$
0

31. $\frac{y}{2} + \frac{y}{3} > y + \frac{y}{5}$, $15y + 10y > 30y + 6y$, $25y > 36y$, $0 > 11y$, $0 > y$, $y < 0$

$$(-\infty, 0):$$
0

33. $0.1(0.03x + 4) \geq 0.02x + 0.434,$

     $0.003x + 0.4 \geq 0.02x + 0.434,$ $-0.017x \geq 0.034,$ $x \leq -2$

         $(-\infty, -2]$:

35. $444,000 < S < 636,000$

EXERCISE 2.3

1. Let q = number of units sold. For Profit > 0, we have

         Tot. Rev. - Total Cost > 0

         $20q - (15q + 600,000) > 0$

            $5q - 600,000 > 0$

                $5q > 600,000$, so $q > 120,000$.

Thus at least 120,001 units must be sold.

3. Let x = number of miles driven per year. If auto is rented, the annual cost is $12(135) + 0.05x$. If auto is purchased, the annual cost is $1000 + 0.10x$. We want

         Rental cost $\leq$ Purchase cost

     $12(135) + 0.05x \leq 1000 + 0.10x$

       $1620 + 0.05x \leq 1000 + 0.10x$

             $620 \leq 0.05x$

         $12,400 \leq x$

The least number of miles driven per year must be 12,400.

5. Let q be the required number of magazines. Then cost of publication is $q(0.65)$. Revenue from dealers is $0.60q$ and from advertising it is $0.10(q - 10,000)(0.60)$. Thus

                 Profit ≥ 0

         Total Revenue - Total Cost ≥ 0

$0.60q + 0.10(q - 10,000)(0.60) - 0.65q \geq 0$

               $0.01q - 600 \geq 0$

                   $0.01q \geq 600$

                     $q \geq 60,000$

Thus at least 60,000 magazines are required.

7. Let x = amount at $6\frac{3}{4}$% and 30,000 - x = amount at 5%. Then

interest at $6\frac{3}{4}$% + interest at 5% $\geqq$ interest at $6\frac{1}{2}$%

$$x(0.0675) + (30,000 - x)(0.05) \geqq (0.065)(30,000)$$
$$0.0175x + 1500 \geqq 1950$$
$$0.0175x \geqq 450$$
$$x \geqq 25,714.29.$$

Thus at least $25,714.29 must be invested at $6\frac{3}{4}$%.

9. Let q be the number of units sold this month at $4.00 each. Then 2500 - q will be sold at $4.50 each. Then

$$\text{Total revenue} \geqq 10,750$$
$$4q + 4.5(2500 - q) \geqq 10,750$$
$$-0.5q + 11,250 \geqq 10,750$$
$$500 \geqq 0.5q$$
$$1000 \geqq q$$

The maximum number of units sold this month is 1000.

11. For t < 40, we want

income on hourly basis > income on per-job basis
$$8.50t > 300 + 3(40 - t)$$
$$8.50t > 420 - 3t$$
$$11.50t > 420$$
$$t > 36.5 \text{ hr}$$

13. Let c be the cost (in dollars) of a ticket.

$$1000 + 0.40(800c) \leqq 2440$$
$$320c \leqq 1440$$
$$c \leqq 4.50$$

Thus the dean could charge at most $4.50 per ticket. With this charge, the amount left for expenses is
$$800(4.50) - [1000 + 0.40(800)(4.50)] = \$1160.$$

EXERCISE 2.4

1. $|-13| = 13$

3. $|8 - 2| = |6| = 6$

5. $\left|3\left(-\frac{5}{3}\right)\right| = |-5| = 5$

7. $|x| < 3, \quad -3 < x < 3$

9. Because $2 - \sqrt{5} < 0$, $|2 - \sqrt{5}| = -(2 - \sqrt{5}) = \sqrt{5} - 2$

11. (a) $|x - 7| < 3$ (b) $|x - 2| < 3$ (c) $|x - 7| \leq 5$
    (d) $|x - 7| = 4$ (e) $|x + 4| < 2$ (f) $|x| < 3$
    (g) $|x| > 6$ (h) $|x - 6| > 4$ (i) $|x - 105| < 3$
    (j) $|x - 850| < 100$

13. $|p_1 - p_2| \leq 2$

15. $|x| = 7 \Rightarrow x = \pm 7$

17. $\left|\frac{x}{3}\right| = 2 \Rightarrow \frac{x}{3} = \pm 2 \Rightarrow x = \pm 6$

19. $|x - 5| = 8, \quad x - 5 = \pm 8, \quad x = 5 \pm 8, \quad x = 13 \text{ or } x = -3$

21. $|5x - 2| = 0 \Rightarrow 5x - 2 = 0 \Rightarrow x = \frac{2}{5}$

23. $|7 - 4x| = 5, \quad 7 - 4x = \pm 5, \quad -4x = -7 \pm 5, \quad -4x = -2 \text{ or } -12,$
    $x = \frac{1}{2} \text{ or } x = 3$

25. $|x| < 4 \Rightarrow -4 < x < 4 \Rightarrow (-4, 4)$

27. $\left|\frac{x}{4}\right| > 2, \quad \frac{x}{4} < -2 \text{ or } \frac{x}{4} > 2, \quad x < -8 \text{ or } x > 8, \text{ so the}$
    solution is $(-\infty, -8) \cup (8, \infty)$

29. $|x + 7| < 2, \quad -2 < x + 7 < 2, \quad -9 < x < -5 \Rightarrow (-9, -5)$

31. $\left|x - \frac{1}{2}\right| > \frac{1}{2}, \quad x - \frac{1}{2} < -\frac{1}{2} \text{ or } x - \frac{1}{2} > \frac{1}{2}, \quad x < 0 \text{ or } x > 1 \Rightarrow$
    $(-\infty, 0) \cup (1, \infty)$

33. $|5 - 2x| \leq 1, \quad -1 \leq 5 - 2x \leq 1, \quad -6 \leq -2x \leq -4, \quad 3 \geq x \geq 2,$
    which may be rewritten as $2 \leq x \leq 3 \Rightarrow [2, 3]$.

35. $\left|\frac{3x - 8}{2}\right| \geq 4$, $\frac{3x - 8}{2} \leq -4$ or $\frac{3x - 8}{2} \geq 4$

$\qquad\qquad\qquad 3x - 8 \leq -8$ or $3x - 8 \geq 8$

$\qquad\qquad\qquad\quad 3x \leq 0$ or $3x \geq 16$

$\qquad\qquad\qquad\quad\ \ x \leq 0$ or $x \geq \frac{16}{3}$

The solution is $(-\infty, 0] \cup \left[\frac{16}{3}, \infty\right)$

37. $|x - \mu| > h\sigma$. Either $x - \mu < -h\sigma$, or $x - \mu > h\sigma$. Thus either $x < \mu - h\sigma$ or $x > \mu + h\sigma$, so the solution is $(-\infty, \mu - h\sigma) \cup (\mu + h\sigma, \infty)$.

CHAPTER 2 - REVIEW PROBLEMS

1. $3x - 8 \geq 4(x - 2)$, $3x - 8 \geq 4x - 8$, $-x \geq 0$, $x \leq 0 \Rightarrow (-\infty, 0]$

3. $-(5x + 2) < -(2x + 4)$, $-5x - 2 < -2x - 4$, $-3x < -2$, $x > \frac{2}{3} \Rightarrow \left(\frac{2}{3}, \infty\right)$

5. $3p(1 - p) > 3(2 + p) - 3p^2$, $3p - 3p^2 > 6 + 3p - 3p^2$, $0 > 6$, which is false for all x. The solution set is $\emptyset$.

7. Multiplying both sides by 6 gives $2(x + 1) - 3(1) \leq 6(2)$, $2x + 2 - 3 \leq 12$, $2x \leq 13$, $x \leq \frac{13}{2} \Rightarrow \left(-\infty, \frac{13}{2}\right]$

9. Multiplying both sides by 8 gives $2s - 24 \leq 3 + 2s$, $0 \leq 27$, which is true for all s. Thus $-\infty < s < \infty \Rightarrow (-\infty, \infty)$

11. $|3 - 2x| = 7$, $3 - 2x = 7$ or $3 - 2x = -7$

$\qquad\qquad\qquad\qquad -2x = 4$ or $-2x = -10$

$\qquad\qquad\qquad\qquad\quad\ x = -2$ or $x = 5$

13. $|4t - 1| < 1$, $-1 < 4t - 1 < 1$, $0 < 4t < 2$, $0 < t < \frac{1}{2} \Rightarrow$ $\left(0, \frac{1}{2}\right)$

15. $|3 - 2x| \geq 4$, $\quad 3 - 2x \geq 4 \qquad$ or $\qquad 3 - 2x \leq -4$

$$-2x \geq 1 \qquad \text{or} \qquad -2x \leq -7$$

$$x \leq -\tfrac{1}{2} \qquad \text{or} \qquad x \geq \tfrac{7}{2}$$

The solution is $\left(-\infty, -\tfrac{1}{2}\right] \cup \left[\tfrac{7}{2}, \infty\right)$.

17. Let x be the number of issues with decline, and x + 48 be the number of issues with increase. Then

$$x + (x + 48) = 1132, \quad 2x = 1084, \quad x = 542$$

19. Let q units be produced at plant A, and 10,000 - q units be produced at B.

$$\text{Cost at A + Cost at B} \leq 117,000$$

$$[5q + 30,000] + [5.50(10,000 - q) + 35,000] \leq 117,000$$

$$-0.5q + 120,000 \leq 117,000$$

$$-0.5q \leq -3000$$

$$q \geq 6000$$

Thus at least 6000 units must be produced at plant A.

MATHEMATICAL SNAPSHOT - CHAPTER 2

1. $t = 2l - 4 = 2\left(2\tfrac{1}{2}\right) - 4 = 1$ hour

3. $l = 2\tfrac{2}{3}$ hours. Thus $t = \tfrac{3}{2}l - 3 = \tfrac{3}{2}\left(2\tfrac{2}{3}\right) - 3 = 1$ hour.

5. An appropriate formula for this situation is

$$\frac{t}{4} + \frac{l - \frac{lc}{60} - t}{2} = 1.$$

Letting $l = 3$ and $c = 8$ gives

$$\frac{t}{4} + \frac{3 - \frac{3(8)}{60} - t}{2} = 1$$

$$t + 2\left(3 - \tfrac{2}{5} - t\right) = 4$$

$$-t + \tfrac{26}{5} = 4$$

$$t = \tfrac{6}{5} \text{ hr} = 1 \text{ hr and 12 min.}$$

3

Functions and Graphs

1. The denominator is zero when $x = 0$. Any other real number can be used for x.   <u>Ans.</u> all real numbers except 0

3. For $\sqrt{x - 3}$ to be real, $x - 3 \geqq 0$, so $x \geqq 3$.
   <u>Ans.</u> all real numbers $\geqq 3$

5. Any real number can be used for t.   <u>Ans.</u> all real numbers

7. We exclude values of x where $2x + 5 = 0$, $2x = -5$, $x = -\dfrac{5}{2}$.
   <u>Ans.</u> all real numbers except $-\dfrac{5}{2}$

9. We exclude values of y for which $y^2 - y = 0$,
   $y(y - 1) = 0$, $y = 0$ or 1.
   <u>Ans.</u> all real numbers except 0 and 1

11. We exclude all values of s for which $2s^2 - 7s - 4 = 0$,
    $(s - 4)(2s + 1) = 0$, $s = 4, -\frac{1}{2}$.

    <u>Ans.</u> all real numbers except 4 and $-\frac{1}{2}$

13. $f(x) = 2x + 1$
    $f(0) = 2(0) + 1 = 1$
    $f(3) = 2(3) + 1 = 7$
    $f(-4) = 2(-4) + 1 = -7$

15. $G(x) = 2 - x^2$
    $G(-8) = 2 - (-8)^2 = 2 - 64 = -62$
    $G(u) = 2 - u^2$
    $G(u^2) = 2 - (u^2)^2 = 2 - u^4$

17. $g(u) = u^2 + u$
    $g(-2) = (-2)^2 + (-2) = 4 - 2 = 2$
    $g(2v) = (2v)^2 + (2v) = 4v^2 + 2v$
    $g(-x^2) = (-x^2)^2 + (-x^2) = x^4 - x^2$

19. $f(x) = x^2 + 2x + 1$
    $f(1) = 1^2 + 2(1) + 1 = 1 + 2 + 1 = 4$
    $f(-1) = (-1)^2 + 2(-1) + 1 = 1 - 2 + 1 = 0$
    $f(x + h) = (x + h)^2 + 2(x + h) + 1$
    $\qquad = x^2 + 2xh + h^2 + 2x + 2h + 1$

21. $g(x) = \dfrac{x - 5}{x^2 + 4}$
    $g(5) = \dfrac{5 - 5}{5^2 + 4} = 0$
    $g(3x) = \dfrac{3x - 5}{(3x)^2 + 4} = \dfrac{3x - 5}{9x^2 + 4}$
    $g(x + h) = \dfrac{(x + h) - 5}{(x + h)^2 + 4} = \dfrac{x + h - 5}{x^2 + 2xh + h^2 + 4}$

23. $f(x) = x^{4/3}$

    $f(0) = 0^{4/3} = 0$

    $f(64) = 64^{4/3} = (\sqrt[3]{64})^4 = (4)^4 = 256$

    $f(\frac{1}{8}) = (\frac{1}{8})^{4/3} = (\sqrt[3]{1/8})^4 = (\frac{1}{2})^4 = \frac{1}{16}$

25. $f(x) = 4x - 5$

    (a) $f(x + h) = 4(x + h) - 5 = 4x + 4h - 5$

    (b) $\dfrac{f(x + h) - f(x)}{h} = \dfrac{(4x + 4h - 5) - (4x - 5)}{h} = \dfrac{4h}{h} = 4$

27. $f(x) = x^2 + 2x$

    (a) $f(x + h) = (x + h)^2 + 2(x + h)$

                 $= x^2 + 2xh + h^2 + 2x + 2h$

    (b) $\dfrac{f(x+h) - f(x)}{h} = \dfrac{(x^2+2xh+h^2+2x+2h) - (x^2+2x)}{h}$

                       $= \dfrac{2xh+h^2+2h}{h} = 2x+h+2$

29. $f(x) = 2 - 4x - 3x^2$

    (a) $f(x + h) = 2 - 4(x + h) - 3(x + h)^2$

                  $= 2 - 4x - 4h - 3(x^2 + 2xh + h^2)$

                  $= 2 - 4x - 4h - 3x^2 - 6xh - 3h^2$

    (b) $\dfrac{f(x + h) - f(x)}{h}$

       $= \dfrac{2 - 4x - 4h - 3x^2 - 6xh - 3h^2 - (2 - 4x - 3x^2)}{h}$

       $= \dfrac{-4h - 6xh - 3h^2}{h} = \dfrac{h(-4 - 6x - 3h)}{h} = -4 - 6x - 3h$

31. $f(x) = \dfrac{1}{x}$

    (a) $f(x + h) = \dfrac{1}{x + h}$

    (b) $\dfrac{f(x + h) - f(x)}{h} = \dfrac{\frac{1}{x + h} - \frac{1}{x}}{h} = \dfrac{\frac{x - (x + h)}{x(x + h)}}{h}$

                       $= \dfrac{-h}{x(x + h)h} = -\dfrac{1}{x(x + h)}$

33. $f(x) = 9x + 7$

$$\frac{f(2 + h) - f(2)}{h} = \frac{9(2 + h) + 7 - [9(2) + 7]}{h}$$

$$= \frac{25 + 9h - [25]}{h} = \frac{9h}{h} = 9$$

35. $y - 3x - 4 = 0$. The equivalent form $y = 3x + 4$ shows that for each input x there is exactly one output, $3x - 4$. Thus y is a function of x. Solving for x gives $x = \frac{y - 4}{3}$. This shows that for each input y there is exactly one output, $\frac{y - 4}{3}$. Thus x is a function of y.

37. $y = 7x^2$. For each input x, there is exactly one output, $7x^2$. Thus y is a function of x. Solving for x gives $x = \pm\sqrt{\frac{y}{7}}$. If, for example, $y = 7$, then $x = \pm1$, so x is not a function of y.

39. Yes, because corresponding to each input r there is exactly one output, πr^2.

41. Weekly excess of income over expenses is 2000-1600 = 400. After t weeks the excess accumulates to 400t. Thus the value V of the business at the end of t weeks is given by $V = f(t) = 10,000 + 400t$.

43. Yes. For each input q there corresponds exactly one output, 1.25q, so P is a function of q. The dependent variable is P and the independent variable is q.

45. (a) $f(1000) = \frac{(\sqrt[3]{1000})^4}{2500} = \frac{10^4}{2500} = \frac{10,000}{2500} = 4$

   (b) $f(2000) = \frac{[\sqrt[3]{1000(2)}]^4}{2500} = \frac{(10\sqrt[3]{2})^4}{2500} = \frac{10,000\sqrt[3]{2^4}}{2500}$

   $$= 4\sqrt[3]{2^3 \cdot 2} = 8\sqrt[3]{2}$$

   (c) $f(2I_0) = \frac{(2I_0)^{4/3}}{2500} = \frac{2^{4/3}I_0^{4/3}}{2500} = 2\sqrt[3]{2}\left[\frac{I_0^{4/3}}{2500}\right] = 2\sqrt[3]{2}f(I_0)$.

   Thus $f(2I_0) = 2\sqrt[3]{2}f(I_0)$, which means that doubling the intensity increases the response by a factor of $2\sqrt[3]{2}$.

47. (a) Domain: 3000, 2900, 2300, 2000.
    $f(2900) = 12$, $f(3000) = 10$
    (b) Domain: 10, 12, 17, 20.
    $g(10) = 3000$, $g(17) = 2300$

49. (a) 11.33; (b) 50.62

EXERCISE 3.2

1. all real numbers

3. all real numbers

5. (a) 3;   (b) 7

7. (a) 4;   (b) -3

9. $f(x) = 8$
   $f(2) = 8$, $f(t + 8) = 8$, $f(-\sqrt{17}) = 8$

11. $F(10) = 1$, $F(-\sqrt{3}) = -1$, $F(0) = 0$, $F(-18/5) = -1$

13. $G(8) = 8$, $G(3) = 3$, $G(-1) = 2-(-1) = 3$, $G(1) = 2-1 = 1$

15. $6! = 6 \cdot 5 \cdot 4 \cdot 3 \cdot 2 \cdot 1 = 720$

17. $(4 - 2)! = 2! = 2 \cdot 1 = 2$

19. $\frac{5!}{4!} = \frac{5 \cdot 4 \cdot 3 \cdot 2 \cdot 1}{4 \cdot 3 \cdot 2 \cdot 1} = 5$
    $8 \cdot 7 = 56$

21. (a) $C = 850 + 3q$
    (b) $1600 = 850 + 3q$, $750 = 3q$, $q = 250$

23. $P(2) = \frac{3!\left(\frac{1}{4}\right)^2\left(\frac{3}{4}\right)^1}{2!(1!)} = \frac{6\left(\frac{1}{16}\right)\left(\frac{3}{4}\right)}{2(1)} = \frac{9}{64}$

25. (a) all T such that $30 \leq T \leq 39$

(b) $f(30) = \frac{1}{24}(30) + \frac{11}{4} = \frac{5}{4} + \frac{11}{4} = \frac{16}{4} = 4$

$f(36) = \frac{1}{24}(36) + \frac{11}{4} = \frac{6}{4} + \frac{11}{4} = \frac{17}{4}$

$f(39) = \frac{4}{3}(39) - \frac{175}{4} = 52 - \frac{175}{4} = \frac{33}{4}$

27. (a) 214.61; (b) -52.89

EXERCISE 3.3

1. $f(x) = x + 3$, $g(x) = x + 5$

(a) $(f + g)(x) = f(x) + g(x) = (x + 3) + (x + 5) = 2x + 8$

(b) $(f + g)(0) = 2(0) + 8 = 8$

(c) $(f - g)(x) = f(x) - g(x) = (x + 3) - (x + 5) = -2$

(d) $(fg)(x) = f(x)g(x) = (x + 3)(x + 5) = x^2 + 8x + 15$

(e) $(fg)(-2) = (-2)^2 + 8(-2) + 15 = 3$

(f) $\frac{f}{g}(x) = \frac{f(x)}{g(x)} = \frac{x + 3}{x + 5}$

(g) $(f \circ g)(x) = f(g(x)) = f(x + 5) = (x + 5) + 3 = x + 8$

(h) $(f \circ g)(3) = 3 + 8 = 11$

(i) $(g \circ f)(x) = g(f(x)) = g(x + 3) = (x + 3) + 5 = x + 8$

3. $f(x) = x^2$, $g(x) = x^2 + x$

(a) $(f + g)(x) = f(x) + g(x) = x^2 + (x^2 + x) = 2x^2 + x$

(b) $(f - g)(x) = f(x) - g(x) = x^2 - (x^2 + x) = -x$

(c) $(f - g)\left(- \frac{1}{2}\right) = -\left(- \frac{1}{2}\right) = \frac{1}{2}$

(d) $(fg)(x) = f(x)g(x) = x^2(x^2 + x) = x^4 + x^3$

(e) $\frac{f}{g}(x) = \frac{f(x)}{g(x)} = \frac{x^2}{x^2 + x} = \frac{x}{x + 1}$ [for $x \neq 0$]

(f) $\frac{f}{g}\left(- \frac{1}{2}\right) = \frac{-1/2}{(-1/2) + 1} = \frac{-1/2}{1/2} = -1$

(g) $(f \circ g)(x) = f(g(x)) = f(x^2+x) = (x^2+x)^2 = x^4+2x^3+x^2$

(h) $(g \circ f)(x) = g(f(x)) = g(x^2) = (x^2)^2 + x^2 = x^4 + x^2$

(i) $(g \circ f)(-3) = (-3)^4 + (-3)^2 = 81 + 9 = 90$

5. $f(g(2)) = f(4 - 4) = f(0) = 0 + 6 = 6$
   $g(f(2)) = g(12 + 6) = g(18) = 4 - 36 = -32$

7. $(F \circ G)(t) = F(G(t)) = F\left(\dfrac{2}{t - 1}\right) = \left(\dfrac{2}{t - 1}\right)^2 + 3\left(\dfrac{2}{t - 1}\right) + 1$

   $\qquad\qquad = \dfrac{4}{(t - 1)^2} + \dfrac{6}{t - 1} + 1$

   $(G \circ F)(t) = G(F(t)) = G(t^2 + 3t + 1) = \dfrac{2}{(t^2 + 3t + 1) - 1}$

   $\qquad\qquad = \dfrac{2}{t^2 + 3t}$

9. $(f \circ g)(v) = f(g(v)) = f(\sqrt{\sqrt{v + 2}}) = \dfrac{2}{(\sqrt{\sqrt{v + 2}})^2 + 1}$

   $\qquad\qquad = \dfrac{1}{v + 2 + 1} = \dfrac{1}{v + 3}$

   $(g \circ f)(w) = g(f(w)) = g\left(\dfrac{1}{w^2 + 1}\right) = \sqrt{\dfrac{1}{w^2 + 1} + 2}$

   $\qquad\qquad = \sqrt{\dfrac{1 + 2(w^2 + 1)}{w^2 + 1}} = \sqrt{\dfrac{2w^2 + 3}{w^2 + 1}}$

11. Let $g(x) = 4x - 3$ and $f(x) = x^5$. Then

    $\qquad h(x) = (4x - 3)^5 = [g(x)]^5 = f(g(x))$

13. Let $g(x) = x^2 - 2$ and $f(x) = \dfrac{1}{x}$. Then

    $\qquad h(x) = \dfrac{1}{x^2 - 2} = \dfrac{1}{g(x)} = f(g(x))$

15. Let $g(x) = \dfrac{x + 1}{3}$ and $f(x) = \sqrt[5]{x}$. Then

    $\qquad h(x) = \sqrt[5]{g(x)} = f(g(x))$

17. $(g \circ f)(m) = g(f(m)) = g\left(\dfrac{40m - m^2}{4}\right) = 40\left(\dfrac{40m - m^2}{4}\right)$

    $\qquad = 10(40m - m^2) = 400m - 10m^2.$

    This represents the total revenue received when the total output of m employees is sold.

EXERCISE 3.4

1.

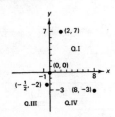

3. (a) $f(0) = 1$, $f(2) = 2$, $f(4) = 3$, $f(-2) = 0$
   (b) Domain: all real numbers
   (c) Range: all real numbers
   (d) $f(x) = 0$ for $x = -2$. So real zero is -2.

5. (a) $f(0) = 0$, $f(1) = -1$, $f(-1) = -1$
   (b) Domain: all real numbers
   (c) Range: all nonpositive reals
   (d) $f(x) = 0$ for $x = 0$. So real zero is 0.

7. $y = x$. If $y = 0$, then $x = 0$. If $x = 0$, then $y = 0$.
   Int.: $(0,0)$. y is a function of x. Domain: all real
   numbers. Range: all real numbers. See graph below.

9. $y = 3x-5$. If $y = 0$, then $0 = 3x-5$, $x = \frac{5}{3}$. If $x = 0$,
   then $y = -5$. Int.: $\left(\frac{5}{3},0\right)$, $(0,-5)$. y is a function of x.
   Domain: all real numbers. Range: all real numbers. See
   graph below.

7.

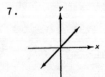

9.

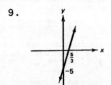

11. $y = x^2$. If $y = 0$, then $0 = x^2$, $x = 0$. If $x = 0$, then
    $y = 0$. Int.: $(0,0)$. y is a function of x. Domain: all
    real numbers. Range: all reals ≥ 0. See graph on next
    page.

13. x = 0. If y = 0, then x = 0. If x = 0, then y can be
    any real number. Int.: every point on y-axis.
    y is not a function of x. See graph below.

11.   13.

15. $y = x^3$. If y = 0, then $0 = x^3$, x = 0. If x = 0, then
    y = 0. Int.: (0,0). y is a function of x. Domain: all
    real numbers. Range: all real numbers. See graph below.

17. $x = -3y^2$. If y = 0, then x = 0. If x = 0, then
    $0 = -3y^2$, y = 0. Int.: (0,0).
    y is not a function of x. See graph below.

15.   17.

19. 2x+y-2 = 0. If y = 0, then 2x-2 = 0, x = 1. If x = 0,
    then y-2 = 0, y = 2. Int.: (1,0), (0,2). Note that
    y = 2-2x. y is a function of x. Domain: all real
    numbers. Range: all real numbers. See graph below.

21. $s = f(t) = 4-t^2$. If s = 0, then $0 = 4-t^2$,
    0 = (2+t)(2-t), t = ±2. If t = 0, then s = 4.
    Int.: (2,0), (-2,0), (0,4). Domain: all real numbers.
    Range: all reals ≦ 4. See graph below.

19.   21.

23. $y = g(x) = 2$. Because y cannot be 0, there is no
x-intercept. If $x = 0$, then $y = 2$. Int.: $(0,2)$.
Domain: all real numbers. Range: 2. See graph below.

25. $y = h(x) = x^2 - 4x + 1$. If $y = 0$, then $0 = x^2 - 4x + 1$, and by
the quadratic formula, $x = \frac{4 \pm \sqrt{12}}{2} = 2 \pm \sqrt{3}$. If $x = 0$,
then, $y = 1$. Int.: $(2 \pm \sqrt{3}, 0)$, $(0,1)$. Domain: all real
numbers. Range: all reals ≥ -3. See graph below.

23.

25.

27. $f(t) = -t^3$. If $f(t) = 0$, then $0 = -t^3$, $t = 0$. If $t = 0$,
then $f(t) = 0$. Int.: $(0,0)$. Domain: all real numbers.
Range: all real numbers. See graph below.

29. $s = F(r) = \sqrt{r-5}$. Note that for $\sqrt{r-5}$ to be a real
number, $r-5 \geq 0$, or $r \geq 5$. If $s = 0$, then $0 = \sqrt{r-5}$,
$0 = r-5$, or $r = 5$. Because $r \geq 5$, then $r \neq 0$, so no
s-intercept exists. Int.: $(5,0)$. Domain: all reals ≥ 5.
Range: all reals ≥ 0. See graph below.

27.

29.

31. $f(x) = |2x-1|$. If $f(x) = 0$, then $0 = |2x-1|$, $2x-1 = 0$,
$x = \frac{1}{2}$. If $x = 0$, then $f(x) = |-1| = 1$.
Int.: $\left(\frac{1}{2}, 0\right)$, $(0,1)$. Domain: all real numbers.
Range: all reals ≥ 0. See graph on next page.

33. $F(t) = \frac{16}{t^2}$. If $F(t) = 0$, then $0 = \frac{16}{t^2}$, which has no

solution. Because $t \neq 0$, there is no vertical-axis
intercept. Int.: none. Domain: all nonzero reals.
Range: all positive reals. See graph below.

31.

33.
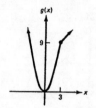

35. Domain: all reals ≥ 0
    Range: all real c such
    that $0 \leq c \leq 2$

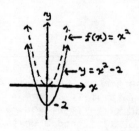

37. Domain: all real numbers
    Range: all reals ≥ 0

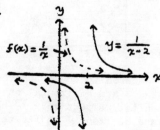

39. From the vertical-line test, the graphs that represent
functions of x are (a), (b), and (d)

41.
![f(x)=x², y=x²-2 graph]

43.
![f(x)=1/x, y=1/(x-2) graph]

45.

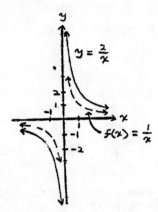

47.

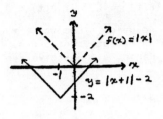

49.

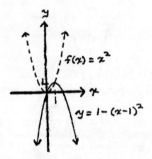

51.

As price decreases,
quantity increases;
p is a function of q.

53.

55. Compared to the graph for k = 0, the graphs for k = 1, 2,
and 3 are horizontal shifts to the left of 1, 2, and 3
units, respectively. The graphs for k = -1, -2, and -3
are horizontal shifts to the right of 1, 2, and 3 units,
respectively.

57. (a) Range: $(-\infty, \infty)$;   (b) Int.: (-1.73, 0), (0, 4.00)

59. (a) Range: [1.93, ∞);   (b) Min. value of f(x): 1.93;
    (c) Int.: (0, 2.48);   (d) No real zero

EXERCISE 3.5

In Problems 1-23, parts a, b, and c, refer to the testing for symmetry about the x-axis, y-axis, and origin, respectively.

1. $y = 5x$. Intercepts: If $y = 0$, then $5x = 0$, or $x = 0$;
if $x = 0$, then $y = 5 \cdot 0 = 0$. Testing for symmetry about
(a) x-axis, (b) y-axis, and (c) origin gives:
  a. $-y = 5x$, $y = -5x$;   b. $y = 5(-x) = -5x$;
  c. $-y = 5(-x)$, $y = 5x$.
Since equivalent equation is obtained in case (c) only,
the graph only has symmetry about the origin.
<u>Ans.</u> (0,0); sym. about origin

3. $2x^2 + y^2 x^4 = 8 - y$. Intercepts: If $y = 0$, then $2x^2 = 8$,
$x^2 = 4$, $x = \pm 2$; if $x = 0$, then $0 = 8 - y$, $y = 8$. Testing
for symmetry gives:
  a. $2x^2 + (-y)^2 x^4 = 8 - (-y)$, $2x^2 + y^2 x^4 = 8 + y$;
  b. $2(-x)^2 + y^2 (-x)^4 = 8 - y$, $2x^2 + y^2 x^4 = 8 - y$;
  c. $2(-x)^2 + (-y)^2 (-x)^4 = 8 - (-y)$, $2x^2 + y^2 x^4 = 8 + y$.
<u>Ans.</u> (± 2,0), (0,8); sym. about y-axis

5. $4x^2 - 9y^2 = 36$. Intercepts: If $y = 0$, then $4x^2 = 36$,
$x^2 = 9$, $x = \pm 3$; if $x = 0$, then $-9y^2 = 36$, $y^2 = -4$, which
has no real root. Testing for symmetry gives:
  a. $4x^2 - 9(-y)^2 = 36$, $4x^2 - 9y^2 = 36$;
  b. $4(-x)^2 - 9y^2 = 36$, $4x^2 - 9y^2 = 36$.
Since equivalent equations are obtained in (a) and (b),
the graph has symmetry about x- and y-axes. Thus there
is also symmetry about origin.
<u>Ans.</u> (± 3,0); sym. about x-axis, y-axis, origin

7. $x = -2$. Intercepts: If $y = 0$, then $x = -2$; because
$x \neq 0$, there is no y-intercept. Symmetry: a. $x = -2$;

Exercise 3.5

—56—

b. $-x = -2$, $x = 2$;   c. $-x = -2$, $x = 2$.
Ans. $(-2,0)$; sym. about x-axis

9. $x = -y^{-4}$. Intercepts: Because $y \neq 0$, there is no x-intercept; if $x = 0$, then $0 = -1/y^4$, which has no solution. Symmetry: a. $x = -(-y)^{-4}$, $x = -y^{-4}$; b. $-x = -y^{-4}$, $x = y^{-4}$; c. $-x = -(-y)^{-4}$, $x = y^{-4}$.
Ans. no intercepts; sym. about x-axis

11. $x-4y-y^2+21 = 0$. Intercepts: If $y = 0$, then $x+21 = 0$, $x = -21$; if $x = 0$, then $-4y-y^2+21 = 0$, $y^2+4y-21 = 0$, $(y+7)((y-3) = 0$, $y = -7$ or $y = 3$. Symmetry:
a. $x-4(-y)-(-y)^2+21 = 0$, $x+4y-y^2+21 = 0$;
b. $(-x)-4y-y^2+21 = 0$, $-x-4y-y^2+21 = 0$;
c. $(-x)-4(-y)-(-y)^2+21 = 0$, $-x+4y-y^2+21 = 0$.
Ans. $(-21,0)$, $(0,-7)$, $(0,3)$; no symmetry

13. $y = f(x) = \dfrac{x^3}{x^2+5}$. Intercepts: If $y = 0$, then $\dfrac{x^3}{x^2+5} = 0$, $x = 0$; if $x = 0$, then $y = 0$. Symmetry: a. because f is not the zero function, there is no x-axis symmetry;
b. $-y = \dfrac{x^3}{x^2+5}$, $y = -\dfrac{x^3}{x^2+5}$; c. $-y = \dfrac{(-x)^3}{(-x)^2+5}$, $y = \dfrac{x^3}{x^2+5}$
Ans. $(0,0)$; sym. about origin

15. $y = \dfrac{1}{x^3+1}$. Inercepts: If $y = 0$, then $\dfrac{1}{x^3+1} = 0$, which has no solution; if $x = 0$, then $y = 1$.
Symmetry: a. $-y = \dfrac{1}{x^3+1}$, $y = -\dfrac{1}{x^3+1}$; b. $y = \dfrac{1}{(-x)^3+1}$, $y = \dfrac{1}{-x^3+1}$; c. $-y = \dfrac{1}{(-x)^3+1}$, $-y = \dfrac{1}{-x^3+1}$, $y = \dfrac{1}{x^3-1}$.
Ans. $(0,1)$; no symmetry

17. $2x+y^2 = 4$. Intercepts: If $y = 0$, then $2x = 4$, $x = 2$; if $x = 0$, then $y^2 = 4$, $y = \pm2$. Symmetry: a. $2x+(-y)^2 =$

4, $2x+y^2 = 4$; b. $2(-x)+y^2 = 4$, $-2x+y^2 = 4$;

c. $2(-x)+(-y)^2 = 4$, $-2x+y^2 = 4$.   <u>Ans.</u> (2,0), (0,±2);

sym. about x-axis; see graph below

19. $y = f(x) = x^3-4x$. Intercepts: If $y = 0$, then $x^3-4x = 0$,
    $x(x+2)(x-2) = 0$, $x = 0$ or $x = ±2$; if $x = 0$, then $y = 0$.
    Symmetry: a. because f is not the zero function, there
    is no x-axis symmetry; b. $y = (-x)^3-4(-x)$, $y = -x^3+4x$;
    c. $-y = (-x)^3-4(-x)$, $y = x^3-4x$.
    <u>Ans.</u> (0,0), (±2,0); sym. about origin; see graph below

17.

19.

21. $|x|-|y| = 0$. Intercepts: If $y = 0$, then $|x| = 0$, $x = 0$,
    if $x = 0$, then $-|y| = 0$, $y = 0$. Symmetry: a. $|x|-|-y| =$
    0, $|x|-|y| = 0$; b. $|-x|-|y| = 0$, $|x|-|y| = 0$; c. from (a)
    and (b) there is symmetry about x- and y-axes, so sym-
    metry about origin exists.
    <u>Ans.</u> (0,0); sym. about x-axis, y-axis, origin; see graph
        on next page

23. $4x^2+y^2 = 16$. Intercepts: If $y = 0$, then $4x^2 = 16$,
    $x^2 = 4$, $x = ±2$; if $x = 0$, then $y^2 = 16$, $y = ±4$.
    Symmetry: a. $4x^2+(-y)^2 = 16$, $4x^2+y^2 = 16$; b. $4(-x)^2+y^2 =$
    16, $4x^2+y^2 = 16$; c. from (a) and (b) there is symmetry
    about x- and y-axes, so symmetry about origin exists.
    <u>Ans.</u> (±2,0), (0,±4); sym. about x-axis, y-axis, origin;
        see graph on next page

21. 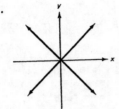   23.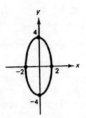

25. $y = f(x) = 2 - 0.03x^2 - x^4$. Replacing x by -x gives
$y = 2 - 0.03(-x)^2 - (-x)^4$ or $y = 2 - 0.03x^2 - x^4$, which
is equivalent to original equation. Thus graph is
symmetric about y-axis.
(a) Int.: (±1.18, 0), (0, 2.00)
(b) Range: $(-\infty, 2]$
(c) Max. value of f(x): 2.00

CHAPTER 3 - REVIEW PROBLEMS

1. Denominator is 0 when $x^2 - 3x + 2 = 0$, $(x - 1)(x - 2) = 0$,
x = 1, 2.   Domain: all real numbers except 1 and 2.

3. all real numbers

5. For $\sqrt{x}$ to be real, x must be nonnegative. For the
denominator x - 1 to be different from 0, x cannot be 1.
Both conditions are satisfied by all nonnegative numbers
except 1.   Domain: all nonnegative reals except 1.

7. $f(x) = 3x^2 - 4x + 7$
$f(0) = 3(0)^2 - 4(0) + 7 = 7$
$f(-3) = 3(-3)^2 - 4(-3) + 7 = 27 + 12 + 7 = 46$
$f(5) = 3(5)^2 - 4(5) + 7 = 75 - 20 + 7 = 62$
$f(t) = 3t^2 - 4t + 7$

9. $G(x) = \sqrt{x - 1}$

   $G(1) = \sqrt{1 - 1} = \sqrt{0} = 0$

   $G(10) = \sqrt{10 - 1} = \sqrt{9} = 3$

   $G(t + 1) = \sqrt{(t + 1) - 1} = \sqrt{t}$

   $G(x^2) = \sqrt{x^2 - 1}$

11. $h(u) = \dfrac{\sqrt{u + 4}}{u}$

    $h(5) = \dfrac{\sqrt{5 + 4}}{5} = \dfrac{\sqrt{9}}{5} = \dfrac{3}{5}$

    $h(-4) = \dfrac{\sqrt{-4 + 4}}{-4} = \dfrac{0}{-4} = 0$

    $h(x) = \dfrac{\sqrt{x + 4}}{x}$

    $h(u - 4) = \dfrac{\sqrt{(u - 4) + 4}}{u - 4} = \dfrac{\sqrt{u}}{u - 4}$

13. $f(4) = 8 - 4^2 = 8 - 16 = -8$

    $f(-2) = 4$

    $f(0) = 4$

    $f(10) = 8 - 10^2 = 8 - 100 = -92$

15. (a) $f(x + h) = 3 - 7(x + h) = 3 - 7x - 7h$

    (b) $\dfrac{f(x + h) - f(x)}{h} = \dfrac{(3 - 7x - 7h) - (3 - 7x)}{h} = \dfrac{-7h}{h} = -7$

17. (a) $f(x + h) = 4(x + h)^2 + 2(x + h) - 5$

    $= 4x^2 + 8xh + 4h^2 + 2x + 2h - 5$

    (b) $\dfrac{f(x + h) - f(x)}{h} = \dfrac{(4x^2 + 8xh + 4h^2 + 2x + 2h - 5) - (4x^2 + 2x - 5)}{h}$

    $= \dfrac{8xh + 4h^2 + 2h}{h} = \dfrac{h(8x + 4h + 2)}{h} = 8x + 4h + 2$

19. $f(x) = 3x - 1$, $g(x) = 2x + 3$

    (a) $(f + g)(x) = f(x) + g(x) = (3x-1)+(2x+3) = 5x + 2$

    (b) $(f + g)(4) = 5(4) + 2 = 22$

    (c) $(f - g)(x) = f(x) - g(x) = (3x-1)-(2x+3) = x - 4$

    (d) $(fg)(x) = f(x)g(x) = (3x-1)(2x+3) = 6x^2 + 7x - 3$

(e) $(fg)(1) = 6(1)^2 + 7(1) - 3 = 10$

(f) $\frac{f}{g}(x) = \frac{f(x)}{g(x)} = \frac{3x - 1}{2x + 3}$

(g) $(f \circ g)(x) = f(g(x)) = f(2x+3) = 3(2x+3)-1 = 6x + 8$

(h) $(f \circ g)(5) = 6(5) + 8 = 38$

(i) $(g \circ f)(x) = g(f(x)) = g(3x-1) = 2(3x-1)+3 = 6x + 1$

21. $f(x) = \frac{1}{x}$, $g(x) = x - 1$

$(f \circ g)(x) = f(g(x)) = f(x - 1) = \frac{1}{x - 1}$

$(g \circ f)(x) = g(f(x)) = g(\frac{1}{x}) = \frac{1}{x} - 1 = \frac{1 - x}{x}$

23. $f(x) = x + 2$, $g(x) = x^3$

$(f \circ g)(x) = f(g(x)) = f(x^3) = x^3 + 2$

$(g \circ f)(x) = g(f(x)) = g(x + 2) = (x + 2)^3$

25. $y = 2x-3x^3$. Intercepts: If $y = 0$, then $0 = 2x-3x^3 = 3x\left(\frac{2}{3} - x^2\right) = 3x(\sqrt{2/3} + x)(\sqrt{2/3} - x)$, or $x = 0$, $\pm\sqrt{2/3}$; if $x = 0$, then $y = 0$. Testing for symmetry about (a) x-axis, (b) y-axis, and (c) origin gives:

a. $-y = 2x-3x^3$, $y = -2x+3x^3$, which is not orig. eq.;

b. $y = 2(-x)-3(-x)^3$, $y = -2x+3x^3$, which is not orig. eq.;

c. $-y = 2(-x)-3(-x)^3$, $y = 2x-3x^3$, which is orig. eq.

Ans. $(0,0)$, $(\pm\sqrt{2/3},0)$; sym. about origin

27. $y = 9-x^2$. Intercepts: If $y = 0$, then $0 = 9-x^2 = (3+x)(3-x)$, or $x = \pm3$; if $x = 0$, then $y = 9$. Testing for symmetry about (a) x-axis, (b) y-axis, and (c) origin gives:

a. $-y = 9-x^2$, $y = -9+x^2$, which is not orig. eq.;

b. $y = 9-(-x)^2$, $y = 9-x^2$, which is orig. eq.;

c. from (b) there is y-axis sym. and from (a) there is no x-axis sym. So there can be no sym. about origin.

Ans. $(0,9)$, $(\pm3,0)$; sym. about y-axis; see graph on
next page

29. $G(u) = \sqrt{u+4}$. Intercepts: If $G(u) = 0$, the $0 = \sqrt{u+4}$, $0 = u+4$, or $u = -4$; if $u = 0$, then $G(u) = \sqrt{4} = 2$. Domain: all reals u such that $u \geq -4$. Range: all reals ≥ 0.
Ans. $(0,2)$, $(-4,0)$; all reals u such that $u \geq -4$; all reals ≥ 0; see graph below

27.

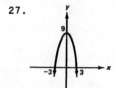

29.

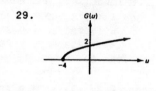

31. $y = g(t) = \frac{2}{t-4}$. Intercepts: If $y = 0$, then $0 = \frac{2}{t-4}$, which has no solution; if $t = 0$, then $y = 2/(-4) = -1/2$. Domain: all real numbers t such that $t \neq 4$. Range: all reals $\neq 0$.
Ans. $(0, -\frac{1}{2})$; all $t \neq 4$; all nonzero real numbers;

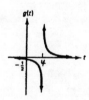

33. Domain: all real numbers.   35.
Range: all reals ≥ 1.

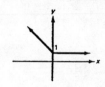

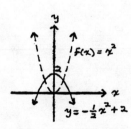

37. From the vertical-line test, the graphs that represent functions of x are (a) and (c).

39. (a) $(-\infty, \infty)$;   (b) (1.92, 0), (0, 7.00)

41. (a) none;   (b) 1, 3

CHAPTER 3 — MATHEMATICAL SNAPSHOT

1. $f(100,000) = 17,733.50 + 0.33(100,000 - 78,400)$
    $= 17,733.50 + 0.33(21,600)$
    $= 17,733.50 + 7128 = 24,861.50.$
    Ans. $24,861.50

3. $f(240,000) = 47,871.60 + 0.28(240,000 - 162,770)$
    $= 47,871.60 + 0.28(77,230)$
    $= 47,871.60 + 21,624.40 = 69,496.$
    Ans. $69,496

4

Lines, Parabolas and Systems

1. $m = \dfrac{10 - 1}{7 - 4} = \dfrac{9}{3} = 3$

3. $m = \dfrac{3 - (-2)}{-6 - 4} = \dfrac{5}{-10} = -\dfrac{1}{2}$

5. The difference in the x-coordinates is $5 - 5 = 0$, so the slope is undefined.

7. $m = \dfrac{-2 - (-2)}{4 - 5} = \dfrac{0}{-1} = 0$

9. $y - 8 = 6(x - 2)$
   $y - 8 = 6x - 12$
   $6x - y - 4 = 0$

11. $y - 5 = -\dfrac{1}{4}(x + 2)$

   $4(y - 5) = -(x + 2)$

   $4y - 20 = -x - 2$

   $x + 4y - 18 = 0$

13. $m = \frac{4-1}{1-(-6)} = \frac{3}{7}$.

    $y - 4 = \frac{3}{7}(x - 1)$

    $7(y - 4) = 3(x - 1)$

    $7y - 28 = 3x - 3$

    $3x - 7y + 25 = 0$

15. $m = \frac{-9-(-1)}{-2-3} = \frac{-8}{-5}$.

    $y - (-1) = \frac{8}{5}(x - 3)$

    $5(y + 1) = 8(x - 3)$

    $5y + 5 = 8x - 24$

    $8x - 5y - 29 = 0$

17. $y = 2x + 4$

    $2x - y + 4 = 0$

19. $y = -\frac{1}{2}x - 3$

    $2y = 2\left(-\frac{1}{2}x - 3\right)$

    $2y = -x - 6$

    $x + 2y + 6 = 0$

21. A horizontal line has the form $y = b$. Thus $y = -2$, or $y + 2 = 0$.

23. x = 2, or x - 2 = 0.

25. y = 2x - 1 has the form
    y = mx + b, where m = 2
    and b = -1.

27. x + 2y - 3 = 0
    2y = -x + 3
    y = $-\frac{1}{2}$x + $\frac{3}{2}$.
    m = $-\frac{1}{2}$, b = $\frac{3}{2}$

29. x = -5 is a vertical
    line. Thus the slope
    is undefined. There
    is no y-intercept.

31. y = 3x,
    y = 3x + 0.
    m = 3, b = 0

33. $y = 1$,
    $y = 0x + 1$.
    $m = 0$, $b = 1$

35. $2x = 5 - 3y$, or $2x + 3y - 5 = 0$ (general form).
    $3y = -2x + 5$, or $y = -\frac{2}{3}x + \frac{5}{3}$ (slope-intercept form).

37. $4x + 9y - 5 = 0$ is a general form.
    $9y = -4x + 5$, or $y = -\frac{4}{9}x + \frac{5}{9}$ (slope-intercept form).

39. $\frac{x}{2} - \frac{y}{3} = -4$, $6\left(\frac{x}{2} - \frac{y}{3}\right) = 6(-4)$, $3x - 2y = -24$,
    $3x - 2y + 24 = 0$ (general form).
    $-2y = -3x-24$, or $y = \frac{3}{2}x + 12$ (slope-intercept form).

41. The lines $y = 7x + 2$ and $y = 7x - 3$ have the same slope, 7. Thus they are parallel.

43. The lines $y = 5x + 2$ and $-5x + y - 3 = 0$ (or $y = 5x + 3$) have the same slope, 5. Thus they are parallel.

45. The line $x + 2y + 1 = 0$ $\left(\text{or } y = -\frac{1}{2}x - \frac{1}{2}\right)$ has slope $m_1 = -\frac{1}{2}$ and the line $y = -2x$ has slope $m_2 = -2$. Since $m_1 \neq m_2$ and $m_1 \neq -\frac{1}{m_2}$, the lines are neither parallel nor perpendicular.

47. The line $y = 3$ is horizontal and the line $x = -\frac{1}{3}$ is vertical, so the lines are perpendicular.

49. The line $3x + y = 4$ (or $y = -3x + 4$) has slope $m = -3$, and the line $x - 3y + 1 = 0$ $\left(\text{or } y = \frac{1}{3}x + \frac{1}{3}\right)$ has slope $m_2 = \frac{1}{3}$. Since $m_2 = -\frac{1}{m_2}$, the lines are perpendicular.

51. The slope of $y = 4x - 5$ is 4, so the slope of a line parallel to it must also be 4. An equation of the desired line is $y - 3 = 4[x - (-1)]$, or $y = 4x + 7$.

53. $y = 2$ is a horizontal line. A line parallel to it has the form $y = b$. Since the line must pass through (2, 1) its equation is $y = 1$.

55. $y = 3x - 5$ has slope 3 $\Rightarrow$ perpendicular line has slope $-\frac{1}{3}$ and its equation is $y - 4 = -\frac{1}{3}(x - 3)$, or $y = -\frac{1}{3}x + 5$.

57. $y = -4$ is a horizontal line, so the perpendicular line must be vertical with equation of the form $x = a$. Since that line passes through (7, 4), its equation is $x = 7$.

59. $2x + 3y + 6 = 0$ $\left(\text{or } y = -\frac{2}{3}x - 2\right)$ has slope $-\frac{2}{3}$. Parallel line has $m = -\frac{2}{3}$. Equation is $y - (-5) = -\frac{2}{3}[x - (-7)]$, or $y = -\frac{2}{3}x - \frac{29}{3}$.

61. (1, 2), (-3, 8). $m = (8 - 2)/(-3 - 1) = 6/(-4) = -3/2$. Point-slope form: $y - 2 = -\frac{3}{2}(x - 1)$. If first coordinate is 5, then $x = 5$ and $y - 2 = -\frac{3}{2}(5 - 1)$, $y - 2 = -\frac{3}{2}(4)$, $y - 2 = -6$, or $y = -4$. Thus the point is (5, -4).

63. (a) Using the points (0.5, 0) and (3.5, -1) gives a slope of $m = \frac{-1 - 0}{3.5 - 0.5} = -\frac{1}{3}$. An equation is
$$y - 0 = -\frac{1}{3}(x - 0.5), \quad \text{or} \quad y = -\frac{1}{3}x + \frac{1}{6}.$$
(b) Using the points (0.5, 0) and (-1, -4.5) gives a slope of $\frac{-4.5 - 0}{-1 - 0.5} = 3$, so an equation is
$$y - 0 = 3(x - 0.5), \quad \text{or} \quad y = 3x - \frac{3}{2}.$$
The paths are perpendicular to each other because the slope of the line in Part (a), -1/3, is the negative reciprocal of the slope in Part (b), 3.

65. The lines are parallel, which is expected because they have the same slope, 1.5.

EXERCISE 4.2

1. $y = f(x) = -4x = -4x + 0$
   has the form $f(x) = ax + b$
   with $a = -4$ and $b = 0$. Thus
   the slope is -4 and the
   vertical-axis intercept is 0.

3. $g(t) = 2t - 4$ has the form
   $g(t) = at + b$ with $a = 2$ (the
   slope) and $b = -4$ (the
   vertical-axis intercept).

5. $h(q) = \frac{7-q}{2} = \frac{7}{2} - \frac{1}{2}q$ has the
   form $h(q) = aq + b$ where $a = -\frac{1}{2}$
   (the slope) and $b = \frac{7}{2}$ (the
   vertical-axis intercept).

7. $f(x) = ax + b = 4x + b$. Since $f(2) = 8$, $8 = 4(2) + b$,
   $8 = 8 + b$, $b = 0 \Rightarrow f(x) = 4x$.

9. Let $y = f(x)$. The points $(1, 2)$ and $(-2, 8)$ lie on the
   graph of f. $m = (8-2)/(-2-1) = -2$. Thus $y-2 = -2(x-1)$,
   so $y = -2x + 4 \Rightarrow f(x) = -2x + 4$.

11. $f(x) = ax + b = -\frac{1}{2}x + b$. Since $f\left(-\frac{1}{2}\right) = 4$, we have
    $4 = -\frac{1}{2}\left(-\frac{1}{2}\right) + b$, $b = 4 - \frac{1}{4} = \frac{15}{4}$, so $f(x) = -\frac{1}{2}x + \frac{15}{4}$.

13. Let $y = f(x)$. The points $(-2, -1)$ and $(-4, -3)$ lie on the
    graph of f. $m = (-3+1)/(-4+2) = 1$. Thus $y + 1 = 1(x + 2)$,
    so $y = x + 1 \Rightarrow f(x) = x + 1$.

15. The points (40, 12) and (25, 18) lie on the graph of the
    equation, which is a line.   m = (18-12)/(25-40) = -2/5.
    Hence an equation of the line is p - 12 = (-2/5)(q - 40),
    which can be written p = (-2/5)q + 28. When q = 30, then
    p =(-2/5)(30) + 28 = -12 + 28 = $16.

17. The line passing through (10, 40) and (20, 70) has slope
    (70-40)/(20-10) = 3, so an equation for the line is
    c - 40 = 3(q - 10), or c = 3q + 10. If q = 35, then c =
    3(35) + 10 = 105 + 10 = $115.

19. Each year the value decreases by 0.10(8000). After t
    years the total decrease is 0.10(8000)t. Thus
    v = 8000 - 0.10(8000)t, or v = -800t + 8000. The slope
    is -800.

21. At $200/ton, x tons cost 200x, and at $2000/acre, y acres
    cost 2000y. Hence the required equation is 200x + 2000y =
    20,000, which can be written as x + 10y = 100.

23. (a) m = (100 - 80)/(100 - 56) = 20/44 = 5/11.

    $y - 100 = \frac{5}{11}(x - 100)$, $y = \frac{5}{11}x - \frac{500}{11} + 100$, or

    $y = \frac{5}{11}x + \frac{600}{11}$.

    (b) $60 = \frac{5}{11}x + \frac{600}{11}$, $\frac{5}{11}x = 60 - \frac{600}{11} = \frac{660}{11} - \frac{600}{11} = \frac{60}{11}$, or
    x = 12, which is the lowest passing score on original
    scale.

25. p = f(t) = at + b, f(5) = 0.32, a = slope = 0.059.
    (a) p = f(t) = 0.059t + b. Since f(5) = 0.32,
        0.32 = 0.059(5) + b, 0.32 = 0.295 + b, so b = 0.025.
        Thus p = 0.059t + 0.025.
    (b) When t = 9, then p = 0.059(9) + 0.025 = 0.556.

27. (a) $m = \dfrac{t_2 - t_1}{c_2 - c_1} = \dfrac{80 - 68}{172 - 124} = \dfrac{12}{48} = \dfrac{1}{4}$.

    $t - 68 = \frac{1}{4}(c - 124)$, $t - 68 = \frac{1}{4}c - 31$, or $t = \frac{1}{4}c + 37$.

  (b) Since c is the number of chirps per minute, then $\frac{1}{4}c$
     is the number of chirps in 1/4 minute or 15 seconds.
     Thus from part (a), to estimate temperature add 37 to
     the number of chirps in 15 seconds.

29. $m = \dfrac{P_2 - P_1}{T_2 - T_1} = \dfrac{100 - 90}{80 - 40} = \dfrac{1}{4}$, so an equation is given by

    $P - 100 = \frac{1}{4}(T - 80)$, $P - 100 = \frac{1}{4}T - 20$, $P = \frac{T}{4} + 80$.

31. (a) Yes: $Q = (1.8704b^2)x + (3.340b^3)$, where b is a
     constant, is the equation of a straight line in
     slope-intercept form. The slope is $1.8704b^2$ and
     the Q-intercept is $3.340b^3$.

  (b) If b = 1, the slope is $1.8704(1^2) = 1.8704$.

EXERCISE 4.3

1. $f(x) = 5x^2$ has the form $f(x) = ax^2 + bx + c$ where $a = 5$,
   $b = 0$, and $c = 0$ $\Rightarrow$ quadratic

3. $g(x) = 7 - 6x$ cannot be put the form $g(x) = ax^2 + bx + c$
   where $a \neq 0$ $\Rightarrow$ not quadratic

5. $h(q) = (q+4)^2 = q^2 + 8q + 16$ has form $h(q) = aq^2 + bq + c$
   where $a = 1$, $b = 8$, and $c = 16$ $\Rightarrow$ quadratic

7. $f(s) = \dfrac{s^2 - 4}{2} = \frac{1}{2}s^2 - 2$ has the form $f(s) = as^2 + bs + c$
   where $a = 1/2$, $b = 0$, and $c = -2$ $\Rightarrow$ quadratic

9. $y = f(x) = -4x^2 + 8x + 7$. $a = -4$, $b = 8$, $c = 7$.
  (a) Vertex occurs when $x = -b/(2a) = -8/[2(-4)] = 1$.
     When $x = 1$, then $y = f(1) = -4(1)^2 + 8(1) + 7 = 11$.
     Vertex: (1, 11).
  (b) $a = -4 < 0$, so vertex corresponds to highest point.

11. $y = f(x) = x^2 + 2x - 8$. $a = 1$, $b = 2$, $c = -8$.

(a) $c = -8$. Thus the y-intercept is -8.

(b) $x^2 + 2x - 8 = (x + 4)(x - 2) = 0$, so $x = -4$, 2.
x-intercepts: -4, 2

(c) $\frac{-b}{2a} = \frac{-2}{2 \cdot 1} = -1$. $f(-1) = (-1)^2 + 2(-1) - 8 = -9$.
Vertex: $(-1, -9)$.

13. $y = f(x) = x^2 - 6x + 5$. $a = 1$, $b = -6$, $c = 5$.

Vertex: $-b/(2a) = -(-6)/(2 \cdot 1) = 3$.

$\qquad f(3) = 3^2 - 6(3) + 5 = -4$.
$\qquad$ Vertex $= (3, -4)$.

y-intercept: $c = 5$.

x-intercepts: $x^2 - 6x + 5$
$\qquad\qquad = (x-1)(x-5) = 0$,
$\qquad\qquad$ so $x = 1$, 5.

Range: all $y \geq -4$.

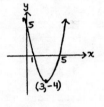

15. $y = g(x) = -2x^2 - 6x$. $a = -2$, $b = -6$, $c = 0$.

Vertex: $-\frac{b}{2a} = \frac{-(-6)}{2(-2)} = \frac{6}{-4} = -\frac{3}{2}$

$\qquad f(-3/2) = -2(-3/2)^2 - 6(-3/2)$
$\qquad\qquad = (-9/2) + 9 = 9/2$.
$\qquad$ Vertex $= (-3/2, 9/2)$.

y-intercept: $c = 0$.

x-intercepts: $-2x^2 - 6x = -2x(x + 3) = 0$, so $x = 0$, -3.

Range: all $y \leq 9/2$.

17. $s = h(t) = t^2 + 2t + 1$. $a = 1$, $b = 2$, $c = 1$.

Vertex: $-b/(2a) = -2/(2 \cdot 1) = -1$.

$\qquad h(-1) = (-1)^2 + 2(-1) + 1 = 0$.
$\qquad$ Vertex $= (-1,0)$.

s-intercept: $c = 1$.

t-intercepts: $t^2 + 2t + 1 = (t+1)^2 = 0$,
$\qquad\qquad$ so $t = -1$.

Range: all $s \geq 0$.

19. $y = f(x) = -9 + 8x - 2x^2$. $a - -2$, $b = 8$, $c = -9$.
    Vertex: $-b/(2a) = -8/[2(-2)] = 2$.

$\qquad$ $f(2) = -9 + 8(2) - 2(2)^2 = -1$.
$\qquad$ Vertex = $(2, -1)$.
    y-intercept: $c = -9$.
    x-intercepts: Because the parabola opens
    $\qquad\qquad$ downward ($a < 0$) and vertex
    $\qquad\qquad$ is below x-axis, there is no
    $\qquad\qquad$ x-intercept.
    Range: $y \leq -1$.

21. $t = f(s) = s^2 - 8s + 13$. $a = 1$, $b = -8$, $c = 13$.
    Vertex: $-b/(2a) = -(-8)/(2 \cdot 1) = 4$.

$\qquad$ $f(4) = 4^2 - 8(4) + 13 = -3$.
$\qquad$ Vertex = $(4, -3)$.
    t-intercept: $c = 13$.

    s-intercepts: Solving $s^2 - 8s + 13 = 0$
    $\qquad\qquad$ by the quadratic formula:

$$s = \frac{-(-8) \pm \sqrt{(-8)^2 - 4(1)(13)}}{2(1)}$$

$$= \frac{8 \pm \sqrt{12}}{2} = \frac{8 \pm 2\sqrt{3}}{2} = 4 \pm \sqrt{3}.$$

    Range: all $t \geq -3$.

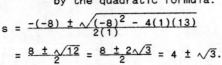

23. $f(x) = 100x^2 - 20x + 25$.   Since $a = 100 > 0$, the parabola
    opens upward and $f(x)$ has a minimum value that occurs at
    the vertex where $x = -b/(2a) = -(-20)/(2 \cdot 100) = 1/10$.   The
    minimum value is

$\qquad$ $f(1/10) = 100(1/10)^2 - 20(1/10) + 25 = 24$.

25. $f(x) = 4x - 50 - 0.1x^2$.   Since $a = -0.1 < 0$, the parabola
    opens downward and $f(x)$ has a maximum value that occurs
    when $\dot{x} = -b/(2a) = -4/[2(-0.1)] = 20$.   The maximum value
    is

$\qquad$ $f(20) = 4(20) - 50 - 0.1(20)^2 = -10$.

27. $r = pq = (1200 - 3q)q = 1200q - 3q^2$.   Because $a = -3 < 0$,
    r has a maximum value that occurs at the vertex where
    $q = -b/(2a) = = -1200/[2(-3)] = 200$.   When $q = 200$, then
    $r = 1200(200) - 3(200)^2 = 240,000 - 120,000 = \$120,000$.

29. $f(P) = (-1/50)P^2 + 2P + 20$, where $0 \leq P \leq 100$. Because
$a = -1/50 < 0$, $f(P)$ has a maximum value that occurs at the
vertex. $-\frac{b}{2a} = -\frac{2}{2(-1/50)} = 50$. The maximum value of
$f(P)$ is $f(50) = (-1/50)(50)^2 + 2(50) + 20 = 70$ grams.

31. $s = 3.2t^2 - 16t + 28.7$ (quadratic with $a = 3.2 > 0$).
    (a) s is minimum when $t = -\frac{b}{2a} = -\frac{-16}{2(3.2)} = 2.50$ s
    (b) $t = 2.50 \Rightarrow s = 3.2(2.50)^2 - 16(2.50) + 28.7 = 8.70$ m

33. $M = \frac{w\ell x}{2} - \frac{wx^2}{2}$, which is quadratic (in x) with $a = -\frac{w}{2} < 0$.
    (a) M is maximum when $x = -\frac{b}{2a} = -\frac{w\ell/2}{2(-w/2)} = \frac{\ell}{2}$
    (b) If $x = \frac{\ell}{2}$, then $M = \frac{w\ell}{2}\left(\frac{\ell}{2}\right) - \frac{w}{2}\left(\frac{\ell}{2}\right)^2 = \frac{w\ell^2}{4} - \frac{w\ell^2}{8} = \frac{w\ell^2}{8}$
    (c) $0 = \frac{w\ell x}{2} - \frac{wx^2}{2} = \frac{wx}{2}(\ell - x) \Rightarrow x = 0$ or $x = \ell$.

35. Since the total length of fencing is 200, the side opposite

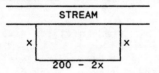

    the stream has length $200 - 2x$. The area A is given by
$$A = x(200 - 2x) = 200x - 2x^2,$$
    which is quadratic with $a = -2 < 0$. Thus A is maximum
    when $x = -\frac{200}{2(-2)} = 50$. Then the side opposite the stream
    is $200 - 2x = 200 - 2(50) = 100$. Thus the dimensions are
    50 ft by 100 ft.

37. (1.11, 2.88)

39. (a) none; (b) one; (c) two

41. 6.89

EXERCISE 4.4

1. $\begin{cases} x + 4y = 3, & (1) \\ 3x - 2y = -5. & (2) \end{cases}$

   From Eq.(1), $x = 3 - 4y$. Substituting in Eq.(2) gives
   $3(3 - 4y) - 2y = -5$, $9 - 12y - 2y = -5$, $-14y = -14$, or
   $y = 1 \Rightarrow x = 3 - 4y = 3 - 4(1) = -1$. Thus $x = -1$, $y = 1$

3. $\begin{cases} 3x - 4y = 13, & (1) \\ 2x + 3y = 3. & (2) \end{cases}$ Multiplying Eq.(1) by 3 and Eq.(2)

   by 4 gives $\begin{cases} 9x - 12y = 39, \\ 8x + 12y = 12. \end{cases}$ Adding gives $17x = 51$, or

   $x = 3$. From Eq. (2) we have $2(3) + 3y = 3$, $3y = -3$, or
   $y = -1$. Thus $x = 3$, $y = -1$

5. $\begin{cases} 5v + 2w = 36, & (1) \\ 8v - 3w = -54. & (2) \end{cases}$ Multiplying Eq.(1) by 3 and Eq.(2)

   by 2 gives $\begin{cases} 15v + 6w = 108, \\ 16v - 6w = -108. \end{cases}$ Adding gives $31v = 0 \Rightarrow v = 0$.

   From Eq. (1), $5(0) + 2w = 36$, so $w = 18$. Thus $v = 0$, $w = 18$

7. $\begin{cases} x - 2y = 8, & (1) \\ 5x + 3y = 1. & (2) \end{cases}$ From Eq.(1), $x = 2y + 8$. Substituting

   in Eq.(2) gives $5(2y + 8) + 3y = 1$, $13y = -39$, or $y = -3 \Rightarrow$
   $x = 2y + 8 = 2(-3) + 8 = 2$. Thus $x = 2$, $y = -3$

9. $\begin{cases} 4x - 3y - 2 = 3x - 7y, \\ x + 5y - 2 = y + 4. \end{cases}$ Simplifying, we have $\begin{cases} x + 4y = 2, \\ x + 4y = 6. \end{cases}$

   Subtracting the second equation from the first gives
   $0 = -4$, which is never true. Thus there is no solution

11. $\begin{cases} \frac{2}{3}x + \frac{1}{2}y = 2, \\ \frac{3}{8}x + \frac{5}{6}y = -\frac{11}{2}. \end{cases}$ Clearing fractions gives the system

   $\begin{cases} 4x + 3y = 12, \\ 9x + 20y = -132. \end{cases}$ Multiplying the first equation by 9 and

   the second equation by -4 gives $\begin{cases} 36x+27y = 108, \\ -36x-80y = 528. \end{cases}$ Adding

   gives $-53y = 636$, or $y = -12$. From $4x + 3y = 12$, we have
   $4x + 3(-12) = 12$, $4x = 48 \Rightarrow x = 12$. Thus $x = 12$, $y = -12$

13. $\begin{cases} 4p + 12q = 6, & (1) \\ 2p + 6q = 3. & (2) \end{cases}$ Multiplying Eq.(2) by -2 gives

$\begin{cases} 4p + 12q = 6, \\ -4p - 12q = -6. \end{cases}$ Adding gives 0 = 0, which implies that

both equations represent the same line, $4p + 12q = 6$ or $p = \frac{3}{2} - 3q$. There are infinitely many solutions, given parametrically by $p = \frac{3}{2} - 3r$, $q = r$, where r is any real number.

15. $\begin{cases} 2x + y + 6z = 3, & (1) \\ x - y + 4z = 1, & (2) \\ 3x + 2y - 2z = 2. & (3) \end{cases}$ Adding Eqs.(1) and (2), and

adding 2 times Eq.(2) to Eq.(3) give $\begin{cases} 3x + 10z = 4, \\ 5x + 6z = 4. \end{cases}$

Multiplying the first equation by 5 and the second equation by -3 give $\begin{cases} 15x + 50z = 20, \\ -15x - 18z = -12. \end{cases}$ Adding $\Rightarrow$ 32z = 8,

or z = 1/4. From $3x + 10z = 4$, we have $3x + 10(1/4) = 4$, $3x = 3/2$, or x = 1/2. From $2x + y + 6z = 3$, we have $2 \cdot \frac{1}{2} + y + 6 \cdot \frac{1}{4} = 3$, or $y = \frac{1}{2}$. Thus $x = \frac{1}{2}$, $y = \frac{1}{2}$, $z = \frac{1}{4}$.

17. $\begin{cases} 5x - 7y + 4z = 2, & (1) \\ 3x + 2y - 2z = 3, & (2) \\ 2x - y + 3z = 4. & (3) \end{cases}$ From Eq.(3), $y = 2x + 3z - 4$.

Substituting in Eqs.(1) and (2) gives

$\begin{cases} 5x - 7(2x + 3z - 4) + 4z = 2, \\ 3x + 2(2x + 3z - 4) - 2z = 3, \end{cases}$ or $\begin{cases} -9x - 17z = -26, \\ 7x + 4z = 11. \end{cases}$

Multiplying the first equation by 7 and the second by 9 give $\begin{cases} -63x - 119z = -182, \\ 63x + 36z = 99. \end{cases}$ By adding we have -83z = -83, or

z = 1. From $7x + 4z = 11$, we have $7x + 4(1) = 11$, $7x = 7$, or x = 1. From $2x - y + 3z = 4$ we get $2(1) - y + 3(1) = 4$, $-y = -1$, or y = 1. Therefore x = 1, y = 1, z = 1

19. $\begin{cases} x - 2z = 1 & (1) \\ y + z = 3. & (2) \end{cases}$ From Eq.(1), $x = 1 + 2z$; from Eq.(2),

$y = 3 - z$. Setting $z = r$ gives the parametric solution $x = 1 + 2r$, $y = 3 - r$, $z = r$, where r is any real number.

21. $\begin{cases} x - y + 2z = 0, & (1) \\ 2x + y - z = 0 & (2) \\ x + 2y - 3z = 0 & (3) \end{cases}$ Adding Eq.(1) to Eq. (3) gives

$\begin{cases} x - y + 2z = 0, \\ 2x + y - z = 0 \\ 2x + y - z = 0 \end{cases}$ We can ignore the third equation because

the second equation can be used to reduce it to $0 = 0$. We

have $\begin{cases} x - y + 2z = 0, \\ 2x + y - z = 0. \end{cases}$ Adding the first equation to the

second gives $3x + z = 0$, or $x = -\frac{1}{3}z$. Substituting in

the first equation we have $-\frac{1}{3}z - y + 2z = 0$, or $y = \frac{5}{3}z$.

Letting $z = r$ gives the parametric solution $x = -\frac{1}{3}r$,

$y = \frac{5}{3}r$, $z = r$, where r is any real number.

23. $\begin{cases} 2x + 2y - z = 3, & (1) \\ 4x + 4y - 2z = 6. & (2) \end{cases}$ Multiplying Eq.(2) by $-1/2$ gives

$\begin{cases} 2x + 2y - z = 3, \\ -2x -24y + z = -3. \end{cases}$ Adding the first equation to the

second equation gives $\begin{cases} 2x + 2y - z = 3, \\ 0 = 0. \end{cases}$ Solving the first

equation for x, we have $x = \frac{3}{2} - y + \frac{1}{2}z$. Letting $y = r$ and

$z = 3$ gives a parametric solution $x = \frac{3}{2} - r + \frac{1}{2}s$, $y = r$,

$z = s$, where r and s are any real numbers.

25. Substituting the data and aligning terms, we get the system

$$\begin{cases} 5i_1 \qquad + 5i_3 = -15, & (1) \\ \qquad 3i_2 + 5i_3 = -2, & (2) \\ i_1 + \quad i_2 - \quad i_3 = 0. & (3) \end{cases}$$

Equation (1) may be written $i_1 + i_3 = -3$. Thus we have $i_1 = -i_3 - 3$. Substituting into Eq.(3), we have $-i_3 - 3 + i_2 - i_3 = 0$, or $i_2 - 2i_3 = 3$. This equation and Eq.(2) form the system $\begin{cases} i_2 - 2i_3 = 3, \\ 3i_2 + 5i_3 = -2. \end{cases}$ Multiplying the first equation by -3 and adding the result to the second give $11i_3 = -11$, $i_3 = -1$. Thus $i_2 - 2(-1) = 3$, or $i_2 = 1$. From $i_1 + i_3 = -3$, we get $i_1 + (-1) = -3$, or $i_1 = -2$. Thus $i_1 = -2$, $i_2 = 1$, $i_3 = -1$ (all in amperes).

27. Let x = number of gallons of 20% solution and y = number of gallons of 30% solution. Then

$$\begin{cases} x + \quad y = 700, & (1) \\ 0.20x + 0.30y = 0.24(700). & (2) \end{cases}$$

From Eq.(1), $y = 700 - x$. Substituting in Eq.(2) gives $0.20x + 0.30(700 - x) = 0.24(700)$, $-0.10x + 210 = 168$, $-0.10x = -42$, or $x = 420$. $y = 700 - x = 700 - 420 = 280$. Thus 420 gal of 20% solution and 280 gal of 30% solution must be mixed.

29. Let a = speed of airplane in still air and w = speed of wind. Then rate of airplane with tail wind is $a + w$, and rate against wind is $a - w$. Since (rate)(time) = distance, and 3 h 36 min = $3\frac{36}{60}$ h = $\frac{18}{5}$ h, we have

$$\begin{cases} (a + w)(3) = 900, \\ (a - w)\left(\frac{18}{5}\right) = 900, \end{cases} \quad \text{or, more simply,} \quad \begin{cases} a + w = 300, \\ a - w = 250. \end{cases}$$

Adding the equations gives $2a = 550$, so $a = 275$. Since $a + w = 300$, we have $w = 25$. Thus the speed of the airplane in still air is 275 mi/h and the speed of the wind is 25 mi/h.

31. Let x = number of Early American units and y = number of
    Contemporary units. The fact that 20% more of Early
    American styles are sold than Contemporary styles means
    that x = y + 0.20y, or x = 1.20y. An analysis of profit
    gives 250x + 350y = 130,000. Thus we have the system

    $$\begin{cases} x = 1.20y, & (1) \\ 250x + 350y = 130,000. & (2) \end{cases}$$ Substituting 1.20y

    for x in Eq.(2) gives 250(1.20y) + 350y = 130,000,
    300y + 350y = 130,000, 650y = 130,000, or y = 200.
    Thus x = 1.20y = 1.20(200) = 240. Therefore 240 units of
    Early American and 200 units of Contemporary must be
    sold.

33. Let x = number of calculators produced at Exton, and
    y = number of calculators produced at Whyton. The total
    cost at Exton is 7.50x + 7000, and the total cost at
    Whyton is 6.00y + 8800. Thus 7.50x + 7000 = 6.00y + 8800.
    Also, x + y = 1500. This gives the system

    $$\begin{cases} x + y = 1500, & (1) \\ 7.50x + 7000 = 6.00y + 8800. & (2) \end{cases}$$

    From Eq.(1), y = 1500 - x. Substituting in Eq.(2) gives
              7.50x + 7000 = 6.00(1500 - x) + 8800,
              7.50x + 7000 = 9000 - 6x + 8800,
                    13.5x = 10,800, or x = 800.
    Thus y = 1500 - x = 1500 - 800 = 700. Therefore 800
    calculators must be made at the Exton plant and 700
    calculators at the Whyton plant.

35. Let x = rate on first $100,000 and y = rate on sales over
    over $100,000. Then

    $$\begin{cases} 100,000x + 75,000y = 8500, & (1) \\ 100,000x + 180,000y = 14,800. & (2) \end{cases}$$

    Subtracting Eq.(1) from Eq.(2) gives 105,000y = 6300 or
    y = 0.06. Substituting in Eq.(1) gives
    100,000x + 75,000(0.06) = 8500, 100,000x + 4500 = 8500,
    100,000x = 4000, or x = 0.04. Thus the rates are 4% on
    the first $100,000 and 6% on the remainder.

37. Let x = number of units of Argon I and y = number of
units of Argon II that the company can make. These
require 6x + 10y doodles and 3x + 8y skeeters. Thus

$$\begin{cases} 6x + 10y = 760, \quad (1) \\ 3x + 8y = 500. \quad (2) \end{cases}$$ Multiplying Eq.(2) by -2 gives

$$\begin{cases} 6x + 10y = 760, \\ -6x - 16y = -1000. \end{cases}$$ By adding we obtain -6y = -240, or

y = 40. From Eq.(1), 6x + 10(40) = 760, 6x = 360, or
x = 60. Thus 60 units of Argon I and 40 units of Argon
II can be made.

39. Let c = number of chairs company makes, r = number of
rockers, and L = number of chaise lounges.

Wood used: (1)c + (1)r + (1)L = 400
Plastic used: (1)c + (1)r + (2)L = 600
Aluminum used: (2)c + (3)r + (5)L = 1500

Thus we have the system $\begin{cases} c + r + L = 400, \quad (1) \\ c + r + 2L = 600, \quad (2) \\ 2c + 3r + 5L = 1500. \quad (3) \end{cases}$

Subtracting Eq.(1) from Eq.(2) gives L = 200. Adding -2
times Eq.(1) to Eq.(3) gives r + 3L = 700, from which
r + 3(200) = 700, or r = 100. From Eq.(1) we have
c + 100 + 200 = 400, or c = 100. Thus 100 chairs, 100
rockers and 200 chaise lounges should be made.

41. Let x = number of skilled workers employed,
      y = number of semiskilled workers employed,
      z = number of shipping clerks employed.
Then we have the system

number of workers: x + y + z = 70,   (1)
wages: 8x + 4y + 5z = 370,   (2)
semiskilled: y = 2x.   (3)

From Eq.(3), y = 2x. Substituting in Eqs.(1) and (2)

gives $\begin{cases} x + (2x) + z = 70, \\ 8x + 4(2x) + 5z = 370, \end{cases}$ or $\begin{cases} 3x + z = 70, \\ 16x + 5z = 370. \end{cases}$

Adding -5 times the first equation to the second gives
x = 20. Thus y = 2x = 2(20) = 40. From x + y + z = 70
we get 20 + 40 + z = 70, or z = 10. The company should
hire 40 semiskilled workers, 20 skilled workers, and 10
shipping clerks.

EXERCISE 4.5

*In the following solutions, any reference to Eq.(1) or Eq.(2)
refers to the first or second equation, respectively, in the
given system.*

1. From Eq.(2), $y = -3x$. Substituting in Eq.(1) gives
   $-3x = 4 - x^2$, $x^2 - 3x - 4 = 0$, $(x - 4)(x + 1) = 0$. Thus
   $x = 4, -1$. From $y = -3x$, if $x = 4$, then $y = -3(4) = -12$;
   if $x = -1$, then $y = -3(-1) = 3$. Thus there are two
   solutions: $x = 4$, $y = -12$; $x = -1$, $y = 3$.

3. From Eq.(2), $q = p - 2$. Substituting in Eq.(1) gives
   $p^2 = 4 - (p - 2)$, $p^2 + p - 6 = 0$, $(p + 3)(p - 2) = 0$.
   Thus $p = -3, 2$. From $q = p - 2$, if $p = -3$, we have
   $q = -3 - 2 = -5$; if $p = 2$, then $q = 2 - 2 = 0$. There are
   two solutions: $p = -3$, $q = -5$; $p = 2$, $q = 0$.

5. Substituting $y = x^2$ into $x = y^2$ gives $x = x^4$, $x^4 - x = 0$,
   $x(x^3 - 1) = 0$. Thus $x = 0, 1$. From $y = x^2$, if $x = 0$,
   then $y = 0^2 = 0$; if $x = 1$, then $y = 1^2 = 1$. There are two
   solutions; $x = 0$, $y = 0$; $x = 1$, $y = 1$.

7. Substituting $y = x^2-2x$ in Eq.(1) $\Rightarrow x^2 - 2x = 4x - x^2 + 8$,
   $2x^2 - 6x - 8 = 0$, $x^2 - 3x - 4 = 0$, $(x - 4)(x + 1) = 0$.
   Thus $x = 4, -1$. From $y = x^2 - 2x$, if $x = 4$, then we have
   $y = 4^2 - 2(4) = 8$; if $x = -1$, then $y = (-1)^2 - 2(-1) = 3$.
   There are two solutions: $x = 4$, $y = 8$; $x = -1$, $y = 3$.

9. Substituting $p = \sqrt{q}$ in Eq.(2) gives $\sqrt{q} = q^2$. Squaring
   both sides, we have $q = q^4$, $q^4 - q = 0$, $q(q^3 - 1) = 0$.
   Thus $q = 0, 1$. From $p = \sqrt{q}$, if $q = 0$, then $p = \sqrt{0} = 0$;
   if $q = 1$, then $p = \sqrt{1} = 1$. There are two solutions:
   $p = 0$, $q = 0$; $p = 1$, $q = 1$.

11. Replacing x^2 by y^2+14 in Eq.(2) gives $y = (y^2 + 14) - 16$,
    $y^2 - y - 2 = 0$, $(y - 2)(y + 1) = 0$. Thus $y = 2, -1$. If
    $y = 2$, then $x^2 = y^2 + 14 = 2^2 + 14 = 18$, so $x = \pm\sqrt{18} =$
    $\pm3\sqrt{2}$. If $y = -1$, then $x^2 = y^2 + 14 = (-1)^2 + 14 = 15$,

so $x = \pm\sqrt{15}$. Therefore, the system has four solutions:
$x = 3\sqrt{2}$, $y = 2$; $x = -3\sqrt{2}$, $y = 2$; $x = \sqrt{15}$, $y = -1$;
$x = -\sqrt{15}$, $y = -1$.

13. Substituting $x = y + 6$ in Eq.(2) gives
$$y = 3\sqrt{(y + 6) + 4} = 3\sqrt{y + 10}.$$
Squaring, we have $y^2 = 9(y + 10)$, $y^2 - 9y - 90 = 0$,
$(y - 15)(y + 6) = 0$. Thus $y = 15, -6$. But $y = -6$ does
not satify Eq.(2) [a negative number does not equal a nonnegative number]. If $y = 15$, then $x = y + 6 = 15 + 6 = 21$.
These values of x and y satisfy the system. Thus the
only solution is $x = 21$, $y = 15$.

15. Three

17. $x = -1.3$, $y = 5.1$

EXERCISE 4.6

1. Equating p-values gives $\frac{3}{100}q + 2 = -\frac{7}{100}q + 12$, from which
$\frac{1}{10}q = 10$, or $q = 100$. If $q = 100$, then $p = \frac{3}{100}q + 2 = $
$\frac{3}{100}(100) + 2 = 5$. The equilibrium point is $(100, 5)$.

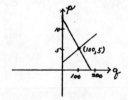

3. $\begin{cases} 35q - 2p + 250 = 0, \quad (1) \\ 65q + p - 537.5 = 0. \quad (2) \end{cases}$ Multiplying Eq.(2) by 2
and adding equations give $165q - 825 = 0$, or $q = 5$. From
Eq.(2), $65(5) + p - 537.5 = 0$, or $p = 212.50$. Thus the
equilibrium point is $(5, 212.50)$.

5. Equating p-values: $2q + 20 = 200 - 2q^2$, $2q^2 + 2q - 180 = 0$,
   $q^2 + q - 90 = 0$, $(q + 10)(q - 9) = 0$. Thus $q = -10, 9$.
   Since $q \geq 0$, choose $q = 9$. Then $p = 2q + 20 = 2(9) + 20 =$
   38. The equilibrium point is $(9, 38)$.

7. Equating p-values gives $20 - q = \sqrt{q + 10}$. Squaring both
   sides gives $400 - 40q + q^2 = q + 10$, $q^2 - 41q + 390 = 0$,
   $(q - 26)(q - 15) = 0$. Thus $q = 26, 15$. If $q = 26$, then
   $p = 20 - q = 20 - 26 = -6$. But p cannot be negative.
   If $q = 15$, then $p = 20 - q = 20 - 15 = 5$. The
   equilibrium point is $(15, 5)$.

9. Letting $y_{TR} = y_{TC}$ gives $3q = 2q + 4500$, or $q = 4500$ units.

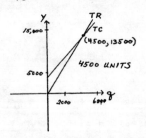

11. Letting $y_{TR} = y_{TC}$ gives $0.05q = 0.85q + 600$, $-0.80q = 600$,
    or $q = -750$, which is negative. Thus cannot break-even at
    any level of production.

13. Letting $y_{TR} = y_{TC}$ gives $100 - \frac{1000}{q+10} = q + 40$. Multiplying
    both sides by $q + 10$ gives
    $$100(q + 10) - 1000 = (q + 10)(q + 40),$$
    $q^2 - 50q + 400 = 0$, $(q - 10)(q - 40) = 0$. Thus $q = 10$ or
    40 units.

15. $\begin{cases} 3q - 200p + 1800 = 0, & (1) \\ 3q + 100p - 1800 = 0. & (2) \end{cases}$

(a) Subtracting Eq.(2) from Eq.(1) gives $-300p + 3600 = 0$, or $p = \$12$.

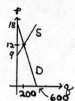

(b) Before the tax the supply equation is

$3q - 200p + 1800 = 0$, $-200p = -3q - 1800$,

or $p = \frac{3}{200}q + 9$. After the tax the supply equation

is $p = \frac{3}{200}q + 9 + 0.27$, or $p = \frac{3}{200}q + 9.27$. This

equation can be written $-3q + 200p - 1854 = 0$, and the new system to solve is

$\begin{cases} -3q + 200p - 1854 = 0, \\ 3q + 100p - 1800 = 0. \end{cases}$

Adding gives $300p - 3654 = 0 \Rightarrow p = 3654/300 = \$12.18$.

17. Since profit = total revenue - total cost, then
$4600 = 8.35q - (2116 + 7.20q)$. Solving gives
$4600 = 1.15q - 2116$, $1.15q = 6716$, $q = \frac{6716}{1.15} = \underline{5840 \text{ units}}$.
For a loss (negative profit) of $1150, we solve
$-1150 = 8.35q - (2116 + 7.20q)$.
Thus $-1150 = 1.15q - 2116$, $1.15q = 966$, or $q = \underline{840 \text{ units}}$.
To break even, we have $y_{TR} = y_{TC}$, or $8.35q = 2116 + 7.20q$,
$1.15q = 2167$, or $q = \underline{1840 \text{ units}}$.

19. Let q = break-even quantity. Since total revenue is $5q$,
we have $5q = 200,000$, which yields $q = 40,000$. Let c be
the variable cost per unit. Then
    Tot. Rev. = Tot. Cost = Variable Cost + Fixed Cost.
Thus $200,000 = 40,000c + 40,000$, $160,000 = 40,000c$, or
$c = \$4$.

21. $y_{TC} = 2q + 1050$; $y_{TR} = 50\sqrt{q}$. Letting $y_{TR} = y_{TC}$ gives
$50\sqrt{q} = 2q + 1050$, or $25\sqrt{q} = q + 525$. Squaring gives
$625q = q^2 + 1050q + (525)^2$, or $q^2 + 425q + (525)^2 = 0$.
By the quadratic formula

$$q = \frac{-425 \pm \sqrt{(425)^2 - 4(1)(525)^2}}{2},$$

which is not real. Thus total cost always exceeds total
revenue $\Rightarrow$ no break-even point.

23. After the subsidy the supply equation is
$$p = \left[\frac{8}{100}q + 50\right] - 1.50, \text{ or } p = \frac{8}{100}q + 48.50.$$

The system to consider is $\begin{cases} p = \frac{8}{100}q + 48.50, \\ p = -\frac{7}{100}q + 65. \end{cases}$ Equating

p-values gives $\frac{8}{100}q + 48.50 = -\frac{7}{100}q + 65$, $\frac{15}{100}q = 16.5$,
or $q = 110$. When $q = 110$, then $p = \frac{8}{100}q + 48.50 =$
$\frac{8}{100}(110) + 48.50 = 8.8 + 48.50 = 57.30$. Thus the original
equilibrium price decreases by \$0.70.

25. Equating q_A-values gives $8 - p_A + p_B = -2 + 5p_A - p_B$,
$10 = 6p_A - 2p_B$, or $5 = 3p_A - p_B$. Equating q_B-values
gives $26 + p_A - p_B = -4 - p_A + 3p_B$, $30 = -2p_A + 4p_B$, or
$15 = -p_A + 2p_B$.
Now we solve $\begin{cases} 3p_A - p_B = 5, \\ -p_A + 2p_B = 15. \end{cases}$ Adding 2 times the first
equation to the second gives $5p_A = 25$, or $p_A = 5$. From
$3p_A - p_B = 5$, $3(5) - p_B = 5$, or $p_B = 10$. Thus $p_A = 5$ and
$p_B = 10$.

27. 2.4 and 11.3

CHAPTER 4 - REVIEW PROBLEMS

1. Solving $\frac{k-5}{3-2} = 4$ gives $k - 5 = 4$, $k = 9$.

3. $(3, -2)$ and $(0, 1)$ lie on the line, so $m = \frac{1 - (-2)}{0 - 3} = -1$.
   Slope-intercept form: $y = mx + b \Rightarrow y = -x + 1$.
   A general form: $x + y - 1 = 0$.

5. $y - 4 = \frac{1}{2}(x - 10)$, $y - 4 = \frac{1}{2}x - 5$, $y = \frac{1}{2}x - 1$, which is
   slope-intercept form. Clearing fractions, we have
   $2y = 2\left(\frac{1}{2}x - 1\right)$, $2y = x - 2$, $x - 2y - 2 = 0$, which is a
   general form.

7. Slope of a horizontal line is 0. Thus $y - 4 = 0[x - (-2)]$,
   $y - 4 = 0$, so slope-intercept form is $y = 4$. A general
   form is $y - 4 = 0$.

9. $y + 3x = 2$ (or $y = -3x + 2$) has slope -3, so the line
   perpendicular to it has slope $\frac{1}{3}$. Since the y-intercept
   is 2, the equation is $y = \frac{1}{3}x + 2$. A general form is
   $x - 3y + 6 = 0$.

*In Problems 11-15, m_1 = slope of first line, and m_2 = slope of
second line.*

11. $x + 4y + 2 = 0$ $\left(\text{or } y = -\frac{1}{4}x - \frac{1}{2}\right)$ has slope $m_1 = -\frac{1}{4}$
    and $8x - 2y - 2 = 0$ (or $y = 4x - 1$) has slope $m_2 = 4$.
    Since $m_1 = -\frac{1}{m_2}$, the lines are perpendicular to each
    other.

13. $x - 3 = 2(y + 4)$ $\left(\text{or } y = \frac{1}{2}x - \frac{11}{2}\right)$ has slope $m_1 = \frac{1}{2}$, and
    $y = 4x + 2$ has slope $m_2 = 4$. Since $m_1 \neq m_2$ and $m_1 \neq -\frac{1}{m_2}$,
    the lines are neither parallel nor perpendicular to each
    other.

15. $y = \frac{1}{2}x + 5$ has slope $\frac{1}{2}$, and $2x = 4y - 3$ $\left(\text{or } y = \frac{1}{2}x + \frac{3}{4}\right)$ has slope $\frac{1}{2}$. Since $m_1 = m_2$, the lines are parallel.

17. $3x - 2y = 4$,
$-2y = -3x + 4$,
$y = \frac{3}{2}x - 2$.
$m = \frac{3}{2}$.

19. $4 - 3y = 0$,
$-3y = -4$,
$y = \frac{4}{3}$.
$m = 0$.

21. $y = f(x) = 4 - 2x$ has the
linear form $f(x) = ax + b$,
where $a = -2$ and $b = 4$.
Slope $= -2$; y-intercept $(0, 4)$.

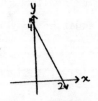

23. $y = f(x) = 9 - x^2$ has the
quadratic form $f(x) = ax^2 + bx + c$,
where $a = -1$, $b = 0$ and $c = 9$.
Vertex: $-b/(2a) = -0/[2(-1)] = 0$.

$\quad\quad f(0) = 9 - 0^2 = 9$.
$\quad\quad \Rightarrow$ Vertex $= (0,9)$.
y-intercept: $c = 9$.
x-intercepts: $9 - x^2 = (3 - x)(3 + x) = 0$, so $x = 3, -3$.

25. $y = h(t) = t^2 - 4t - 5$ has the
    quadratic form $h(t) = at^2 + bt + c$,
    where $a = 1$, $b = -4$, and $c = -5$.
    Vertex: $-b/(2a) = -(-4)/(2 \cdot 1) = 2$.

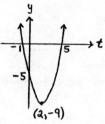

$$h(2) = 2^2 - 4(2) - 5 = -9.$$
    $\Rightarrow$ Vertex = $(2, -9)$.
    y-intercept: $c = -5$.
    t-intercepts: t^2-4t-5
    $\qquad\qquad = (t-5)(t+1) = 0 \Rightarrow t = 5, -1$.

27. $p = g(t) = 3t$ has the linear
    form $g(t) = at + b$, where
    $a = 3$ and $b = 0$.
    Slope = 3; p-intercept $(0,0)$.

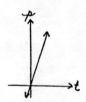

29. $y = F(x) = -(x^2 + 2x + 3)$
    $\qquad\qquad = -x^2 - 2x - 3$
    has the quadratic form

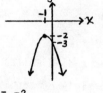

    $F(x) = ax^2 + bx + c$, where
    $a = -1$, $b = -2$, and $c = -3$.
    Vertex: $-b/(2a) = -(-2)/[2(-1)] = -1$.
    $\qquad\qquad F(-1) = -[(-1)^2 + 2(-1) + 3] = -2$.
    $\qquad\qquad \Rightarrow$ Vertex = $(-1, -2)$.
    y-intercept: $c = -3$.
    x-intercepts: Because the parabola opens downward $(a < 0)$
    $\qquad\qquad\qquad$ and the vertex is below the x-axis, there is
    $\qquad\qquad\qquad$ no x-intercept.

31. $\begin{cases} 2x - y = 6, & (1) \\ 3x + 2y = 5. & (2) \end{cases}$ From Eq.(1), $y = 2x-6$. Substituting in
    Eq.(2) gives $3x + 2(2x - 6) = 5$, $7x - 12 = 5$, $7x = 17$, or
    $x = \frac{17}{7} \Rightarrow y = 2x - 6 = 2 \cdot \frac{17}{7} - 6 = -\frac{8}{7}$. Thus $x = \frac{17}{7}$, $y = -\frac{8}{7}$.

33. $\begin{cases} 4x + 5y = 3, & (1) \\ 3x + 4y = 2. & (2) \end{cases}$ Multiplying Eq.(1) by 3 and Eq.(2) by

-4 gives $\begin{cases} 12x + 15y = 9, \\ -12x - 16y = -8. \end{cases}$ Adding gives $-y = 1$, or $y = -1$.

From Eq.(1), $4x + 5(-1) = 3$, $4x = 8$, or $x = 2$. Thus $x = 2$, $y = -1$.

35. $\begin{cases} \frac{1}{4}x - \frac{3}{2}y = -4, & (1) \\ \frac{3}{4}x + \frac{1}{2}y = 8. & (2) \end{cases}$ Multiplying Eq.(2) by 3 gives

$\begin{cases} \frac{1}{4}x - \frac{3}{2}y = -4, \\ \frac{9}{4}x + \frac{3}{2}y = 24. \end{cases}$ Adding the first equation to the second

gives $\frac{5}{2}x = 20$, or $x = 8$. From Eq.(1), $\frac{1}{4}(8) - \frac{3}{2}y = -4$,

$-\frac{3}{2}y = -6$, or $y = 4$. Thus $x = 8$, $y = 4$.

37. $\begin{cases} 3x - 2y + z = -2, & (1) \\ 2x + y + z = 1, & (2) \\ x + 3y - z = 3. & (3) \end{cases}$ Subtracting Eq.(2) from Eq.(1),

and adding Eq.(2) to Eq.(3) give $\begin{cases} x - 3y = -3, \\ 3x + 4y = 4. \end{cases}$

Multiplying the first equation by -3 gives

$\begin{cases} -3x + 9y = 9, \\ 3x + 4y = 4. \end{cases}$ Adding the first equation to the second

gives $13y = 13$, or $y = 1$. From the equation $x - 3y = -3$

we get $x - 3(1) = -3$, or $x = 0$. From $3x - 2y + z = -2$,

$3(0) - 2(1) + z = -2$, or $z = 0$. Thus $x = 0$, $y = 1$, $z = 0$.

39. $\begin{cases} x^2 - y + 2x = 7, & (1) \\ x^2 + y = 5. & (2) \end{cases}$ From Eq.(2), $y = 5-x^2$. Substituting

in Eq.(1) gives $x^2 - (5 - x^2) + 2x = 7$, $2x^2 + 2x - 12 = 0$,

$x^2 + x - 6 = 0$, $(x + 3)(x - 2) = 0$. Thus $x = -3, 2$.

Since $y = 5 - x^2$, if $x = -3$, then $y = 5 - (-3)^2 = -4$; if

$x = 2$, then $y = 5 - 2^2 = 1$. Thus the two solutions are

$x = -3$, $y = -4$, and $x = 2$, $y = 1$.

41. $\begin{cases} x \quad\quad + 2z = -2, & (1) \\ x + y + \; z = 5. & (2) \end{cases}$ From Eq.(1) we have $x = -2 - 2z$.
Substituting in Eq.(2) gives $-2 - 2z + y + z = 5$, so
$y = 7 + z$. Letting $z = r$ gives the parametric solution
$x = -2 - 2r$, $y = 7 + r$, $z = r$, where r is any real number.

43. $\begin{cases} x - \; y - \; z = 0, & (1) \\ 2x - 2y + 3z = 0. & (2) \end{cases}$ Multiplying Eq.(1) by -2 gives

$\begin{cases} -2x + 2y + 2z = 0, \\ \;\; 2x - 2y + 3z = 0. \end{cases}$ Adding Eq.(1) to Eq.(2) gives

$\begin{cases} -2x + 2y + 2z = 0, \\ \quad\quad\quad\quad\;\; 5z = 0. \end{cases}$ From the second equation, $z = 0$.

Substituting in Eq.(1) gives $x - y - 0 = 0$, so $x = y$.
Letting $y = r$ gives the parametric solution $x = r$, $y = r$,
$z = 0$, where r is any real number.

45. $a = 1$ when $b = 2$; $a = 2$ when $b = 1$, so
$$m = \frac{a_2 - a_1}{b_2 - b_1} = \frac{2 - 1}{1 - 2} = \frac{1}{-1} = -1.$$
Thus an equation relating a and b is $a - 1 = -1(b - 2)$,
$a - 1 = -b + 2$, or $a + b - 3 = 0$.
When $b = 3$, then $a = -b + 3 = -(3) + 3 = 0$.

47. Slope is $\frac{-4}{3} \Rightarrow f(x) = ax + b = \frac{-4}{3}x + b$. Since $f(1) = 5$,
$5 = -\frac{4}{3}(1) + b$, $b = \frac{19}{3}$. Thus $f(x) = -\frac{4}{3}x + \frac{19}{3}$.

49. $r = pq = (200 - 2q)q = 200q - 2q^2$, which is a quadratic
function with $a = -2$, $b = 200$, $c = 0$. Since $a < 0$, r has
a maximum value when $q = -b/(2a) = -200/(-4) = 50$ units.
If $q = 50$, then $r = [200 - 2(50)](50) = \$5000$.

51. $\begin{cases} 125p - q - \;\; 250 = 0, \\ 100p + q - 1100 = 0. \end{cases}$ Adding gives $225p - 1350 = 0$, or
$p = \frac{1350}{225} = 6$.

53. (a) R = aL + b. If L = 0, then R = 1310. Thus we have
1310 = 0·L + b, or b = 1310. So R = aL + 1310. Since
R = 1460 when L = 2, 1460 = a(2) + 1310, 150 = 2a, or
a = 75. Thus R = 75L + 1310.
(b) If L = 1, then R = 75(1) + 1310 = 1385 milliseconds.
(c) Since R = 75L + 1310, the slope is 75.
The slope gives the change in R for each 1-unit
increase in L. Thus the time necessary to travel from
one level to the next level is 75 milliseconds.

55. Considering cost we get 5x + 3(2y) = 3000 $\Rightarrow$ y = $\frac{3000 - 5x}{6}$.

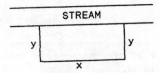

Area = A = xy, so A = $x\left(\frac{3000 - 5x}{6}\right)$, A = 500x - $\frac{5}{6}x^2$, which
is quadratic with a < 0. Thus A has a maximum value when
x = $-\frac{500}{2(-5/6)}$ = 300, which gives y = $\frac{3000 - 5(300)}{6}$ = 250.
Thus the dimensions are 300 ft by 250 ft.

CHAPTER 4 — MATHEMATICAL SNAPSHOT

1. m = 16, h = 1, t = 45, c = 4.
   (a) z = $\left[\frac{(m - c)(60h + t)}{30c} + 1.5\right]$

   = $\left[\frac{(16 - 4)(60·1 + 45)}{30·4} + 1.5\right]$

   = $\left[\frac{12(105)}{120} + 1.5\right]$ = [10.5 + 1.5] = [12] = 12

(b) $F(x) = \begin{cases} \left[\frac{12}{11}c + 0.5\right], & \text{if } x = 2, \\[2mm] \left[\frac{16}{11}c + 0.5\right], & \text{if } x = 3, \\[2mm] \left[\frac{m-c}{z-1}(x-1) + c + 0.5\right], & \text{otherwise} \end{cases}$

$= \begin{cases} \left[\frac{12}{11}(4) + 0.5\right], & \text{if } x = 2, \\[2mm] \left[\frac{16}{11}(4) + 0.5\right], & \text{if } x = 3, \\[2mm] \left[\frac{16-4}{12-1}(x-1) + 4 + 0.5\right], & \text{otherwise} \end{cases}$

$= \begin{cases} [4.864], & \text{if } x = 2, \\[1mm] [6.318], & \text{if } x = 3, \\[2mm] \left[\frac{12}{11}(x-1) + 4.5\right], & \text{otherwise} \end{cases}$

$= \begin{cases} 4, & \text{if } x = 2, \\[1mm] 6, & \text{if } x = 3, \\[2mm] \left[\frac{12}{11}(x-1) + 4.5\right], & \text{otherwise} \end{cases}$

(c) $F(1) = [0 + 4.5] = [4.5] = 4,$
$F(2) = 4,$
$F(3) = 6,$
$F(4) = \left[\frac{12}{11}(3) + 4.5\right] = [7.773] = 7,$
$F(5) = \left[\frac{12}{11}(4) + 4.5\right] = [8.864] = 8,$
$F(6) = \left[\frac{12}{11}(5) + 4.5\right] = [9.955] = 9,$
$F(7) = \left[\frac{12}{11}(6) + 4.5\right] = [11.045] = 11,$
$F(8) = \left[\frac{12}{11}(7) + 4.5\right] = [12.136] = 12,$
$F(9) = \left[\frac{12}{11}(8) + 4.5\right] = [13.227] = 13,$
$F(10) = \left[\frac{12}{11}(9) + 4.5\right] = [14.318] = 14,$
$F(11) = \left[\frac{12}{11}(10) + 4.5\right] = [15.409] = 15,$
$F(12) = \left[\frac{12}{11}(11) + 4.5\right] = [16.5] = 16.$

5

Exponential and Logarithmic Functions

1.

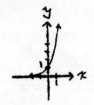

3.

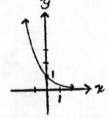

5.

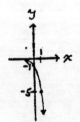

7.

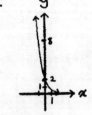

9.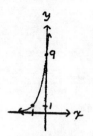

11.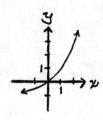

13.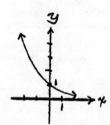

15.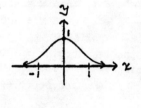

17. For 2010 we have t = 20, so
$$P = 125,000(1.12)^{20/20} = 125,000(1.12)^1 = 140,000$$

19. $P = 1 - \frac{1}{2}\left(1 - c\right)^0 = 1 - \frac{1}{2}(1) = \frac{1}{2}$

21. (a) $4000(1.06)^7 \approx 4000(1.503630) = \$6014.52$
    (b) $6014.52 - 4000 = \$2014.52$

23. (a) $700(1.035)^{30} \approx 700(2.806794) \approx \$1964.76$
    (b) $1964.76 - 700 = \$1264.76$

25. (a) $10,000(1.02)^{34} \approx 10,000(1.960676) = \$19,606.76$
    (b) $19,606.76 - 10,000 = \$9606.76$

27. (a) $5000(1.0075)^{30} \approx 5000(1.251272) = \$6256.36$
    (b) $6256.36 - 5000 = \$1256.36$

29. $4000\left(1 + \frac{0.085}{4}\right)^{60} \approx \$14,124.86$

31. $8000\left(1 + \frac{0.0625}{365}\right)^{3(365)} \approx \$9649.69$

33. $6000(1.02)^{28} \approx 6000(1.741024) \approx \$10,446.14$

35. (a) $N = 400(1.05)^{t}$

     (b) When $t = 1$, then $N = 400(1.05)^{1} = 420.$

     (c) When $t = 4$, then $N = 400(1.05)^{4} \approx 486.$

37. $P = 100,000(1 - 0.01)^{t} = 100,000(0.99)^{t}$, where P is the population after t years. When $t = 3$,
$$P = 100,000(0.99)^{3} \approx 97,030.$$

39. 4.4817

41. 0.67032   43.

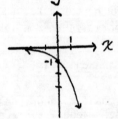

45. For $x = 3$, $P = \frac{e^{-3}3^{3}}{3!} \approx \frac{(0.04979)(27)}{6} \approx 0.2241$

47. $e^{kt} = (e^{k})^{t} = b^{t}$, where $b = e^{k}$

49. (a) When $t = 0$, $N = 10e^{-0.028(0)} = 10 \cdot 1 = 10$

     (b) When $t = 10$,
$$N = 10e^{-0.028(10)} = 10e^{-0.28} \approx 10(0.75578) \approx 7.6$$

     (c) When $t = 50$,

$N = 10e^{-0.028(50)} = 10e^{-1.4} \approx 10(0.24660) \approx 2.5$
(d) After 50 hours, approximately 1/4 of the initial
amount remains. Because $1/4 = (1/2)(1/2)$, 50 hours
corresponds to 2 half-lives. Thus the half-life is
approximately 25 hours.   <u>Ans.</u> 25 hours

51. After one half-life, $\frac{1}{2}$ gram remains. After two half-

lives, $\frac{1}{2} \cdot \frac{1}{2} = \left(\frac{1}{2}\right)^2 = \frac{1}{4}$ gram remains. Continuing in this

manner, after n half-lives, $\left(\frac{1}{2}\right)^n$ gram remains. Because

$\frac{1}{16} = \left(\frac{1}{2}\right)^4$, after 4 half-lives, $\frac{1}{16}$ gram remains. This

corresponds to $4 \cdot 8 = 32$ years.   <u>Ans.</u> 32 years

53. $f(x) = \frac{e^{-4}4^x}{x!}$. $f(2) = \frac{e^{-4}4^2}{2!} \approx \frac{0.01832(16)}{2} \approx 0.1466$.

55. 1.58

57. The first integer t for which the graph of
    $P = 1000(1.05)^t$ lies on or above the horizontal line
    $P = 2000$ is 15.   <u>Ans.</u> 15

<u>EXERCISE 5.2</u>

1. $\log 10,000 = 4$ 3. $2^6 = 64$

5. $\ln 7.3891 = 2$ 7. $e^{1.09861} = 3$

9.   11.

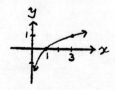   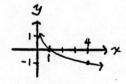

13.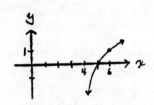

15.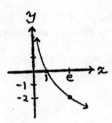

17. Because $6^2 = 36$, $\log_6 36 = 2$

19. Because $3^3 = 27$, $\log_3 27 = 3$

21. Because $7^1 = 7$, $\log_7 7 = 1$

23. Because $10^{-2} = 0.01$, $\log 0.01 = -2$

25. Because $5^0 = 1$, $\log_5 1 = 0$

27. Because $2^{-3} = \frac{1}{8}$, $\log_2 \frac{1}{8} = -3$

29. $3^2 = x$, $x = 9$ 31. $5^3 = x$, $x = 125$

33. $10^{-1} = x$, $x = \frac{1}{10}$ 35. $e^2 = x$

37. $x^3 = 8$, $x = 2$ 39. $x^{-1} = \frac{1}{6}$, $x = 6$

41. $3^{-4} = x$, $x = \frac{1}{81}$

43. $x^2 = 6 - x$, $x^2 + x - 6 = 0$, $(x + 3)(x - 2) = 0$. The roots of this equation are -3 and 2. But since $x > 0$, we choose $x = 2$.

45. $2 + \log_2 4 = 3x - 1$, $2 + 2 = 3x - 1$, $5 = 3x$, $x = \frac{5}{3}$

47. $x^2 = 2x + 8$, $x^2 - 2x - 8 = 0$, $(x - 4)(x + 2) = 0$.
The roots of this equation are 4 and -2. But since
$x > 0$, we choose $x = 4$.

49. $e^{3x} = 2$. In logarithmic form, $3x = \ln 2$, $x = \frac{\ln 2}{3}$

51. $e^{2x-5} + 1 = 4$, $e^{2x-5} = 3$, $2x - 5 = \ln 3$, $x = \frac{5 + \ln 3}{2}$

53. 1.60944

55. 2.00013

57. $c = (12 \ln 6) + 20 \approx 12(1.79176) + 20 \approx 41.50$

59. $1.5M = \log\left(\frac{E}{2.5 \times 10^{11}}\right)$, $10^{1.5M} = \frac{E}{2.5 \times 10^{11}}$,
$E = (2.5 \times 10^{11})(10^{1.5M})$, $E = 2.5 \times 10^{11+1.5M}$

61. (a) $p = 760e^{-0.125(7.3)} \approx 305.2$ mm of mercury
    (b) $400 = 760e^{-0.125h}$, $\frac{400}{760} = e^{-0.125h}$,
    $\ln \frac{400}{760} = -0.125h$, $h = \frac{\ln \frac{400}{760}}{-0.125} \approx 5.13$ km

63. $u_0 = A \ln(x_1) + \frac{x_2^2}{2}$, $u_0 - \frac{x_2^2}{2} = A \ln(x_1)$,
    $\ln(x_1) = \frac{u_0 - \frac{x_2^2}{2}}{A}$, $x_1 = e^{[u_0-(x_2^2/2)]/A}$

65. $T = \frac{\ln 2}{0.03194} \approx \frac{0.69315}{0.03194} \approx 21.7$ years

67. $x + 2e^{3y} - 10 = 0$, $2e^{3y} = 10 - x$, $e^{3y} = \frac{10 - x}{2}$,
    $3y = \ln \frac{10 - x}{2}$, $y = \frac{1}{3} \ln \frac{10 - x}{2}$

69. 1.41, 3.06

EXERCISE 5.3

1. $\log 15 = \log(5 \cdot 3) = \log 5 + \log 3 \approx 0.6990 + 0.4771$
$$= 1.1761$$

3. $\log \frac{8}{3} = \log 8 - \log 3 = 0.9031 - 0.4771 = 0.4260$

5. $\log 36 = \log 6^2 = 2 \log 6 \approx 2(0.7782) = 1.5564$

7. $\log 2000 = \log(2 \cdot 10^3) = \log 2 + \log 10^3 \approx 0.3010 + 3$
$$= 3.3010$$

9. $\log_7 7^{48} = 48$

11. $\ln e^{5.01} = \log_e e^{5.01} = 5.01$

13. $\log_2 3 = \frac{\log 3}{\log 2} \approx \frac{0.4771}{0.3010} = 1.5850$

15. $\ln \frac{1}{e} = -\ln e = -\log_e e = -1$

17. $\log 10 + \ln e^3 = \log_{10} 10 + \log_e e^3 = 1 + 3 = 4$

19. $\ln[x(x + 1)^2] = \ln x + \ln(x + 1)^2 = \ln x + 2 \ln(x + 1)$

21. $\ln \frac{x^2}{(x + 1)^3} = \ln x^2 - \ln(x + 1)^3 = 2 \ln x - 3 \ln(x + 1)$

23. $\ln\left(\frac{x}{x + 1}\right)^3 = 3 \ln \frac{x}{x + 1} = 3[\ln x - \ln(x + 1)]$

25. $\ln \frac{x}{(x + 1)(x + 2)} = \ln x - \ln[(x + 1)(x + 2)]$
$$= \ln x - [\ln(x + 1) + \ln (x + 2)]$$
$$= \ln x - \ln(x + 1) - \ln(x + 2)$$

27. $\ln \dfrac{\sqrt{x}}{(x+1)^2(x+2)^3} = \ln x^{1/2} - \ln[(x+1)^2(x+2)^3]$

$= \tfrac{1}{2} \ln x - [\ln(x+1)^2 + \ln(x+2)^3]$

$= \tfrac{1}{2} \ln x - [2 \ln(x+1) + 3 \ln(x+2)]$

$= \tfrac{1}{2} \ln x - 2 \ln(x+1) - 3 \ln(x+2)$

29. $\ln\left[\dfrac{1}{x+2} \sqrt[5]{\dfrac{x^2}{x+1}}\right] = \ln\left[\dfrac{1}{x+2}\left(\dfrac{x^2}{x+1}\right)^{1/5}\right]$

$= \ln \dfrac{x^{2/5}}{(x+2)(x+1)^{1/5}}$

$= \ln x^{2/5} - \ln[(x+2)(x+1)^{1/5}]$

$= \tfrac{2}{5} \ln x - [\ln(x+2) + \ln(x+1)^{1/5}]$

$= \tfrac{2}{5} \ln x - \ln(x+2) - \tfrac{1}{5} \ln(x+1)$

31. $\log(7 \cdot 4) = \log 28$

33. $\log_2 \dfrac{2x}{x+1}$

35. $\log 7^9 + \log 23^5 = \log[7^9(23)^5]$

37. $2 + 10 \log(1.05) = \log 100 + \log(1.05)^{10}$

$= \log[100(1.05)^{10}]$

39. $e^{4\ln 3 - 3\ln 4} = e^{\ln 3^4 - \ln 4^3} = e^{\ln(3^4/4^3)} = \dfrac{3^4}{4^3} = \dfrac{81}{64}$

41. $\log_6 54 - \log_6 9 = \log_6 \dfrac{54}{9} = \log_6 6 = 1$

43. $e^{\ln(2x)} = 5, \ 2x = 5, \ x = \dfrac{5}{2}$

45. $10^{\log x^2} = 4, \ x^2 = 4, \ x = \pm 2$

47. From the change of base formula with $b = 10$, $m = x + 8$,

and a = e, we have $\log(x + 8) = \dfrac{\log_e(x + 8)}{\log_e 2} = \dfrac{\ln(x + 8)}{\ln 10}$.

49. From the change of base formula with b = 3, m = $x^2 + 1$, and a = e, we have

$$\log_3(x^2 + 1) = \dfrac{\log_e(x^2 + 1)}{\log_e 3} = \dfrac{\ln(x^2 + 1)}{\ln 3}.$$

51. $e^{\ln z} = 7e^y$, $z = 7e^y$, $\dfrac{z}{7} = e^y$, $y = \ln \dfrac{z}{7}$

53. $C = B + E$, $C = B\left(1 + \dfrac{E}{B}\right)$, $\ln C = \ln\left[B\left(1 + \dfrac{E}{B}\right)\right]$,

$\ln C = \ln B + \ln\left(1 + \dfrac{E}{B}\right)$

55. $M = \log(A) + 3$
   (a) $M = \log(1) + 3 = 0 + 3 = 3$
   (b) Given $M_1 = \log(A_1) + 3$, let

$$M = \log(100A_1) + 3$$
$$M = \log 100 + \log(A_1) + 3$$
$$M = 2 + [\log(A_1) + 3]$$
$$M = 2 + M_1$$

57. $pH = -\log(3 \times 10^{-4}) = -[\log 3 + \log 10^{-4}]$
   $= -[\log(3) - 4] = 4 - \log 3 \approx 4 - 0.4771 = 3.5229$

59. $[H^+] \cdot (3 \times 10^{-2}) = 10^{-14}$, $[H^+] = \dfrac{1}{3} \times 10^{-12}$.

$pH = -\log\left(\dfrac{1}{3} \times 10^{-12}\right) = -\left[\log \dfrac{1}{3} + \log 10^{-12}\right]$
   $= -[-\log(3) - 12] = 12 + \log 3 \approx 12 + 0.4771 = 12.4771$

EXERCISE 5.4

1. $\log(2x + 1) = \log(x + 6)$, $2x + 1 = x + 6$, $x = 5.000$

3. $\log x - \log(x - 1) = \log 4$, $\log \dfrac{x}{x - 1} = \log 4$, $\dfrac{x}{x - 1} = 4$,
   $x = 4x - 4$, $4 = 3x$, $x = \dfrac{4}{3} \approx 1.333$

5. $\ln(-x) = \ln(x^2 - 6)$, $-x = x^2 - 6$, $x^2 + x - 6 = 0$,
$(x + 3)(x - 2) = 0$, $x = -3$ or $x = 2$.   However, $x = -3$ is the only value that satisfies the original equation.
<u>Ans.</u>   -3.000

7. $e^{2x}e^{5x} = e^{14}$, $e^{7x} = e^{14}$, $7x = 14$, $x = 2.000$

9. $(16)^{3x} = 2$, $(2^4)^{3x} = 2$, $2^{12x} = 2^1$, $12x = 1$,
$x = \frac{1}{12} \approx 0.083$

11. $e^{2x} = 5$, $2x = \ln 5$, $x = \frac{\ln 5}{2} \approx \frac{1.60944}{2} \approx 0.805$

13. $3e^{3x+1} = 15$, $e^{3x+1} = 5$, $3x + 1 = \ln 5$,
$x = \frac{\ln(5) - 1}{3} \approx \frac{1.60944 - 1}{3} \approx 0.203$

15. $10^{4/x} = 6$, $\frac{4}{x} = \log 6$, $x = \frac{4}{\log 6} \approx \frac{4}{0.7782} \approx 5.140$

17. $\frac{5}{10^{2x}} = 7$, $10^{2x} = \frac{5}{7}$, $2x = \log \frac{5}{7} = \log 5 - \log 7$,
$x = \frac{\log 5 - \log 7}{2} = \frac{0.6990 - 0.8451}{2} \approx -0.073$

19. $2^x = 5$, $\ln 2^x = \ln 5$, $x \ln 2 = \ln 5$,
$x = \frac{\ln 5}{\ln 2} \approx \frac{1.60944}{0.69315} \approx 2.322$

21. $5^{2x-5} = 9$, $\ln 5^{2x-5} = \ln 9$, $(2x-5)\ln 5 = \ln 9$,
$2x-5 = \frac{\ln 9}{\ln 5}$, $2x = \frac{\ln 9}{\ln 5} + 5$,
$x = \frac{\frac{\ln 9}{\ln 5} + 5}{2} \approx \frac{\frac{2.19722}{1.60944} + 5}{2} \approx 3.183$

23. $2^{-2x/3} = \frac{4}{5}$, $\ln 2^{-2x/3} = \ln \frac{4}{5}$, $-\frac{2x}{3} \ln 2 = \ln \frac{4}{5}$,
$-\frac{2x}{3} = \frac{\ln \frac{4}{5}}{\ln 2} = \frac{\ln 4 - \ln 5}{\ln 2}$,
$x = -\frac{3}{2}\left(\frac{\ln 4 - \ln 5}{\ln 2}\right) \approx -\frac{3}{2}\left(\frac{1.38629 - 1.60944}{0.69315}\right) \approx 0.483$

25. $(4)5^{3-x} - 7 = 2$, $5^{3-x} = \frac{9}{4}$, $\ln 5^{3-x} = \ln \frac{9}{4}$,

$(3-x) \ln 5 = \ln 9 - \ln 4$, $3-x = \frac{\ln 9 - \ln 4}{\ln 5}$,

$x = 3 - \frac{\ln 9 - \ln 4}{\ln 5} \approx 3 - \frac{2.19722 - 1.38629}{1.60944} \approx 2.496$

27. $\log(x-3) = 3$, $10^3 = x - 3$, $x = 10^3 + 3 = 1003.000$

29. $\log_4(3x-4) = 2$, $4^2 = 3x - 4$, $3x = 4^2 + 4$,

$x = \frac{4^2 + 4}{3} = \frac{20}{3} = 6.667$

31. $\log(3x-1) - \log(x-3) = 2$, $\log \frac{3x-1}{x-3} = 2$, $10^2 = \frac{3x-1}{x-3}$,

$100(x - 3) = 3x - 1$, $97x = 301$, $x = \frac{299}{97} \approx 3.082$

33. $\log_3(2x+3) = 4 - \log_3(x+6)$, $\log_3(2x+3) + \log_3(x+6) = 4$,

$\log_3[(2x+3)(x+6)] = 4$, $3^4 = (2x + 3)(x + 6)$,

$2x^2 + 15x + 18 = 64$, $2x^2 + 15x - 63 = 0$,

$(2x + 21)(x - 3) = 0$, $x = -\frac{21}{2}$ or $x = 3$. However, $x = 3$

is the only value that satisfies the original equation.
Ans. 3.000

35. $\log_2\left(\frac{2}{x}\right) = 3 + \log_2 x$, $\log_2\left(\frac{2}{x}\right) - \log_2 x = 3$, $\log_2 \frac{2/x}{x} = 3$,

$\log_2 \frac{2}{x^2} = 3$, $2^3 = \frac{2}{x^2}$, $x^2 = \frac{1}{4}$, $x = \pm\frac{1}{2}$. However, $x = \frac{1}{2}$ is

the only value that satisfies the original equation.
Ans. 0.500

37. $\log S = \log 12.4 + 0.26 \log A$,
$\log S = \log 12.4 + \log A^{0.26}$
$\log S = \log[12.4A^{0.26}]$
$S = 12.4A^{0.26}$

39. (a) When $t = 0$, $Q = 100e^{-0.035(0)} = 100e^0 = 100 \cdot 1 = 100$
(b) If $Q = 20$, then $20 = 100e^{-0.035t}$. Solving for t gives

$$\frac{20}{100} = e^{-0.035t}, \qquad \frac{1}{5} = e^{-0.035t}, \qquad \ln \frac{1}{5} = -0.035t,$$

$$-\ln 5 = -0.035t, \qquad t = \frac{\ln 5}{0.035} \approx \frac{1.60944}{0.035} \approx 46.$$

41. $q = 500(1 - e^{-0.2t})$

    (a) If $t = 1$, then $q = 500(1 - e^{-0.2}) \approx 91$

    (b) If $t = 10$, then $q = 500(1 - e^{-2}) \approx 432$

    (c) We solve the equation

$$400 = 500(1 - e^{-0.2t})$$

$$\frac{4}{5} = 1 - e^{-0.2t}$$

$$e^{-0.2t} = \frac{1}{5} = 0.2$$

$$-0.2t = \ln \frac{1}{5} = -\ln 5$$

$$t = \frac{\ln 5}{0.2} = \frac{1.60944}{0.2} \approx 8$$

43. If $P = 1,500,000$, then $1,500,000 = 1,000,000(1.02)^t$.
Solving for t gives

$$\frac{1,500,000}{1,000,000} = (1.02)^t$$

$$1.5 = (1.02)^t$$

$$\ln 1.5 = \ln(1.02)^t$$

$$\ln 1.5 = t \ln 1.02$$

$$t = \frac{\ln 1.5}{\ln 1.02} \approx \frac{0.40547}{0.01980} \approx 20.48$$

45. $q = 80 - 2^p$, $2^p = 80 - q$, $\log 2^p = \log(80 - q)$,

    $p \log 2 = \log(80 - q)$, $p = \frac{\log(80 - q)}{\log 2}$. When $q = 60$, then

$$p = \frac{\log 20}{\log 2} = \frac{\log(2 \cdot 10)}{\log 2} = \frac{\log 2 + \log 10}{\log 2}$$

$$\approx \frac{0.3010 + 1}{0.3010} \approx 4.32$$

47. Let I be the original intensity. The intensity of light
passing through 1 sheet is 0.9I. The intensity of light
passing through 2 sheets is $0.9[(0.9)I[= (0.9)^2 I$. Continuing in this manner, the light passing through n sheets
is $(0.9)^n I$. The equation to solve is $(0.9)^n I = 0.5I$ or,

more simply, $(0.9)^n = 0.5$. Solving gives
$$\ln(0.9)^n = \ln 0.5$$
$$n \ln 0.9 = \ln 0.5$$
$$n = \frac{\ln 0.5}{\ln 0.9} \approx \frac{\ln(1/2)}{\ln(9/10)} = \frac{-\ln 2}{\ln 9 - \ln 10}$$
$$n \approx \frac{-0.69315}{2.19722 - 2.30259} \approx 7$$

49. 0.69

CHAPTER 5 - REVIEW PROBLEMS

1. $\log_3 243 = 5$ 3. $16^{1/4} = 2$ 5. $\ln 54.598 = 4$

7. Because $5^3 = 125$, $\log_5 125 = 3$

9. Because $2^{-4} = \frac{1}{16}$, $\log_2 \frac{1}{16} = -4$

11. Because $\left(\frac{1}{3}\right)^{-2} = 3^2 = 9$, $\log_{1/3} 9 = -2$

13. $5^x = 125$, $x = 3$

15. $10^{-2} = x$, $x = \frac{1}{10^2} = \frac{1}{100}$

17. $\log_x(2x + 3) = 2$, $x^2 = 2x + 3$, $x^2 - 2x - 3 = 0$,
    $(x - 3)(x + 1) = 0$, $x = 3$ or $x = -1$. However, $x = 3$ is
    the only value that satisfies original equation. <u>Ans.</u> 3

19. $\log 2500 = \log(25 \cdot 100) = \log(5^2 10^2) = \log 5^2 + \log 10^2$
    $\qquad = 2 \log(5) + 2 \approx 2(0.6990) + 2 = 3.3980$

21. $2 \log 5 - 3 \log 3 = \log 5^2 - \log 3^3 = \log \frac{5^2}{3^3} = \log \frac{25}{27}$

23. $2 \ln x + \ln y - 3 \ln z = \ln x^2 + \ln y - \ln z^3$

$$= \ln x^2 y - \ln z^3 = \ln \frac{x^2 y}{z^3}$$

25. $\frac{1}{2} \log_2 x + 2 \log_2 x^2 - 3 \log_2 (x+1) - 4 \log_2 (x+2) =$

$\log_2 x^{1/2} + \log_2 (x^2)^2 - [\log_2 (x+1)^3 + \log_2 (x+2)^4] =$

$\log_2 (x^{1/2} x^4) - \log_2 [(x+1)^3 (x+2)^4] = \log_2 \dfrac{x^{9/2}}{(x+1)^3 (x+2)^4}$

27. $\ln \dfrac{x^2 y}{z^3} = \ln x^2 y - \ln z^3 = \ln x^2 + \ln y - \ln z^3$

$$= 2 \ln x + \ln y - 3 \ln z$$

29. $\ln \sqrt[3]{xyz} = \ln (xyz)^{1/3} = \frac{1}{3} \ln (xyz) = \frac{1}{3}(\ln x + \ln y + \ln z)$

31. $\ln\left[\frac{1}{x}\sqrt{\frac{y}{z}}\right] = \ln \dfrac{\left(\frac{y}{z}\right)^{1/2}}{x} = \ln\left(\frac{y}{z}\right)^{1/2} - \ln x$

$$= \frac{1}{2} \ln \frac{y}{z} - \ln x = \frac{1}{2}(\ln y - \ln z) - \ln x$$

33. $\log_3 (x + 5) = \dfrac{\log_e (x + 5)}{\log_e 3} = \dfrac{\ln (x + 5)}{\ln 3}$

35. $\log_5 19 = \dfrac{\log_2 19}{\log_2 5} = \dfrac{4.2479}{2.3219} = 1.8295$

37. $\log(16\sqrt{3}) = \log 4^2 + \log\sqrt{3} = 2 \log 4 + \frac{1}{2} \log 3 = 2y + \frac{1}{2}x$

39. $e^{\ln x} + \ln e^x + \ln 1 = x + x + 0 = 2x$

41. In exponential form,   43.

$y = e^{x^2 + 2}$.

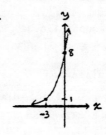

45. $4x + 1 = x + 2$, $3x = 1$, $x = \frac{1}{3}$

47. $3^{4x} \, 9^{x+1}$, $3^{4x} = (3^2)^{x+1}$, $3^{4x} = 3^{2(x+1)}$, $4x = 2(x + 1)$,
    $4x = 2x + 2$, $2x = 2$, $x = 1$

49. $\log x + \log(10x) = 3$, $\log x + \log 10 + \log x = 3$,
    $2 \log(x) + 1 = 3$, $2 \log(x) = 2$, $\log x = 1$, $x = 10^1 = 10$

51. $\ln(\log_x 2) = -1$, $\log_x 2 = e^{-1}$, $x^{e^{-1}} = 2$, $\left(x^{e^{-1}}\right)^e = 2^e$,
    $x^{e^{-1}e} = 2^e$, $x^1 = 2^e$, $x = 2^e$

53. $e^{3x} = 2$, $3x = \ln 2$, $x = \frac{\ln 2}{3} \approx \frac{0.69315}{3} \approx 0.231$

55. $3(10^{x+4} - 3) = 9$, $10^{x+4} - 3 = 3$, $10^{x+4} = 6$,
    $x + 4 = \log 6$, $x = \log(6) - 4 \approx 0.7782 - 4 \approx -3.222$

57. $4^{x+3} = 7$, $\ln 4^{x+3} = \ln 7$, $(x + 3)\ln 4 = \ln 7$,
    $x + 3 = \frac{\ln 7}{\ln 4}$, $x = \frac{\ln 7}{\ln 4} - 3$, $x \approx \frac{1.94591}{1.38629} - 3 \approx -1.596$

59. Quarterly rate = $0.06/4 = 0.015$. $6\frac{1}{2}$ yrs = 26 quarters.

   (a) $2600(1.015)^{26} \approx 2600(1.472710) \approx \$3829.05$
   (b) $3829.05 - 2600 = \$1229.05$

61. Monthly rate = $0.11/12$.   5 yrs = 60 mos.
    $4000\left(1 + \frac{0.11}{12}\right)^{60} \approx \$6915.66$

63. (a) $P = 8000(1 + 0.02)^t$ or $P = 8000(1.02)^t$
    (b) When $t = 2$, then $P = 8000(1.02)^2 \approx 8323$

65. $N = 10e^{-0.41t}$.
    (a) When $t = 0$, then $N = 10e^0 = 10 \cdot 1 = 10$ mg
    (b) When $t = 2$, then $N = 10e^{-0.82} \approx 10(0.44043) \approx 4.4$ mg
    (c) When $t = 10$, then $N = 10e^{-4.1} \approx 10(0.01657) \approx 0.2$ mg

(d) $\frac{\ln 2}{0.41} \approx \frac{0.69315}{0.41} \approx 1.7$

(e) If $N = 1$, then $1 = 10e^{-0.41t}$. Solving for t gives

$\frac{1}{10} = e^{-0.41t}$, $-0.41t = \ln \frac{1}{10} = -\ln 10$,

$t = \frac{\ln 10}{0.41} \approx \frac{2.30259}{0.41} \approx 5.6$

67. $R = 10e^{-t/40}$

(a) If $t = 20$,

$R = 10e^{-20/40} = 10e^{-1/2} \approx 10(0.60653) \approx 6$

(b) $5 = 10e^{-t/40}$, $\frac{1}{2} = e^{-t/40}$. Thus $-\frac{t}{40} = \ln \frac{1}{2} = -\ln 2$,

so $t = 40 \ln 2 \approx 40(0.69315) \approx 28$

69. $T_t - T_e = (T_t - T_e)_o e^{-at}$

$e^{-at} = \frac{T_t - T_e}{(T_t - T_e)_o}$, $\qquad -at = \ln \frac{T_t - T_e}{(T_t - T_e)_o}$

$a = -\frac{1}{t} \ln \frac{T_t - T_e}{(T_t - T_e)_o}$, $\qquad a = \frac{1}{t} \ln \frac{(T_t - T_e)_o}{T_t - T_e}$

71. $(-1.69, -1.16)$, $(2.34, 1.47)$

73. 2.93

CHAPTER 5 — MATHEMATICAL SNAPSHOT

1. $T = \frac{P(1 - e^{-dkI})}{e^{kI} - 1}$.

(a) $T(e^{kI} - 1) = P(1 - e^{-dkI})$, $\frac{T(e^{kI} - 1)}{1 - e^{-dkI}} = P$ or

$P = \frac{T(e^{kI} - 1)}{1 - e^{-dkI}}$

(b) $T(e^{kI} - 1) = P - Pe^{-dkI}$, $Pe^{-dkI} = P - T(e^{kI} - 1)$,

$e^{-dkI} = \frac{P - T(e^{kI} - 1)}{P}$, $-dkI = \ln\left[\frac{P - T(e^{kI} - 1)}{P}\right]$,

$d = -\frac{1}{kI} \ln\left[\frac{P - T(e^{kI} - 1)}{P}\right]$, $d = \frac{1}{kI} \ln\left[\frac{P}{P - T(e^{kI} - 1)}\right]$

3. $P = 100$, $I = 4$, $d = 3$, $H = 8$, $k = \dfrac{\ln 2}{H} = \dfrac{\ln 2}{8}$.

(a) $T = \dfrac{P(1 - e^{-dkI})}{e^{kI} - 1} = \dfrac{100(1 - e^{-3 \cdot \frac{\ln 2}{8} \cdot 4})}{e^{\frac{\ln 2}{8} \cdot 4} - 1}$

$\quad = \dfrac{100(1 - [e^{\ln 2}]^{-3/2})}{[e^{\ln 2}]^{1/2} - 1} = \dfrac{100(1 - 2^{-3/2})}{2^{1/2} - 1} \approx 156$

(b) $R = P(1 - e^{-dkI})$. From part (a), $P(1 - e^{-dkI}) = 100(1 - 2^{-3/2})$. Thus $R = 100(1 - 2^{-3/2}) \approx 65$

6

Matrix Algebra

1. (a) The order is the number of rows by the number of
   columns. Thus A is 2 × 3, B is 3 × 3, C is 3 × 2,
   D is 2 × 2, E is 4 × 4, F is 1 × 2, G is 3 × 1, H is
   3 × 3, J is 1 × 1.

   (b) A square matrix has the same number of rows as number
   of columns. Thus the square matrices are B, D, E, H,
   and J.

   (c) An upper triangular matrix is a square matrix where
   all entries *below* the main diagonal are zeros. Thus H
   and J are upper triangular. A lower triangular matrix
   is a square matrix where all entries *above* the main
   diagonal are zeros. Thus D and J are lower triangular.

   (d) A row vector (or row matrix) has only one row. Thus F
   and J are row vectors.

   (e) A column vector (or column matrix has only one column.
   Thus G and J are column vectors.

3. a_{43} is the entry in the 4th row and 3rd column, namely 2.

5. a_{32} is the entry in the 3rd row and 2nd column, namely 4.

7. a_{14} is the entry in the 1st row and 4th column, namely 6.

9. The main diagonal entries are the entries on the diagonal extending from the upper left corner to the lower right corner. Thus the main diagonal entries are 7, 2, 1, 0.

11. $\begin{bmatrix} 2\cdot1+3\cdot1 & 2\cdot1+3\cdot2 & 2\cdot1+3\cdot3 & 2\cdot1+3\cdot4 \\ 2\cdot2+3\cdot1 & 2\cdot2+3\cdot2 & 2\cdot2+3\cdot3 & 2\cdot2+3\cdot4 \\ 2\cdot3+3\cdot1 & 2\cdot3+3\cdot2 & 2\cdot3+3\cdot3 & 2\cdot3+3\cdot4 \end{bmatrix} = \begin{bmatrix} 5 & 8 & 11 & 14 \\ 7 & 10 & 13 & 16 \\ 9 & 12 & 15 & 18 \end{bmatrix}$

13. $12\cdot10 = 120$, so A has 120 entries. For a_{33}, $i = 3 = j$, so $a_{33} = 1$. Since $5 \neq 2$, $a_{52} = 0$. For $a_{10,10}$, $i = 10 = j$, so $a_{10,10} = 1$. Since $12 \neq 10$, $a_{12,10} = 0$.

15. A zero matrix is a matrix in which all entries are zeros.

(a) $\begin{bmatrix} 0 & 0 & 0 & 0 \\ 0 & 0 & 0 & 0 \\ 0 & 0 & 0 & 0 \\ 0 & 0 & 0 & 0 \end{bmatrix}$ (b) $\begin{bmatrix} 0 & 0 & 0 & 0 & 0 & 0 \\ 0 & 0 & 0 & 0 & 0 & 0 \\ 0 & 0 & 0 & 0 & 0 & 0 \\ 0 & 0 & 0 & 0 & 0 & 0 \\ 0 & 0 & 0 & 0 & 0 & 0 \\ 0 & 0 & 0 & 0 & 0 & 0 \end{bmatrix}$

17. $A^T = \begin{bmatrix} 6 & -3 \\ 2 & 4 \end{bmatrix}^T = \begin{bmatrix} 6 & 2 \\ -3 & 4 \end{bmatrix}$

19. $A^T = \begin{bmatrix} 1 & 3 & 2 & 3 \\ 3 & 2 & -2 & 0 \\ -4 & 2 & 0 & 1 \end{bmatrix}^T = \begin{bmatrix} 1 & 3 & -4 \\ 3 & 2 & 2 \\ 2 & -2 & 0 \\ 3 & 0 & 1 \end{bmatrix}$

21. (a) A and C are diagonal matrices.
    (b) All are them are triangular matrices.

23. $A^T = \begin{bmatrix} 1 & 2 & 3 \\ 4 & 5 & 6 \\ 7 & 8 & 9 \end{bmatrix}^T = \begin{bmatrix} 1 & 4 & 7 \\ 2 & 5 & 8 \\ 3 & 6 & 9 \end{bmatrix}$.

    $(A^T)^T = \begin{bmatrix} 1 & 4 & 7 \\ 2 & 5 & 8 \\ 3 & 6 & 9 \end{bmatrix}^T = \begin{bmatrix} 1 & 2 & 3 \\ 4 & 5 & 6 \\ 7 & 8 & 9 \end{bmatrix} = A$.

25. Equating corresponding entries gives 6 = 6, 2 = 2, x = 6, 7 = 7, 3y = 2, and 2z = 7. Thus x = 6, $y = \frac{2}{3}$, $z = \frac{7}{2}$.

27. Equating corresponding entries gives 2x = y, 7 = 7, 7 = 7, and 2y = y. Now, $2y = y \Rightarrow y = 0$. Thus from 2x = y we get 2x = 0, so x = 0. The solution is x = 0, y = 0.

29. (a) From J, the entry in row 3 (super-duper) and column 2 (white) is 7. Thus in January, 7 white super-duper models were sold.

    (b) From F, the entry in row 2 (deluxe) and column 3 (blue) is 3. Thus in February, 3 blue deluxe models were sold.

    (c) The entries in row 1 (regular) and column 4 (purple) give the number of purple regular models sold. For J the entry is 2 and for F the entry is 4. Thus more purple regular models were sold in February.

    (d) In both January and February, the deluxe blue models (row 2, column 3) sold the same number of units (3).

    (e) In January a total of 0 + 1 + 3 + 5 = 9 deluxe models were sold. In February a total of 2 + 3 + 3 + 2 = 10 deluxe models were sold. Thus more deluxe models were sold in February.

    (f) In January a total of 2 + 0 + 2 = 4 red widgets were sold, while in February a total of 0 + 2 + 4 = 6 red widgets were sold. Thus more red widgets were sold in February.

    (g) Adding all entries in matrix J yields that a total of 35 widgets were sold in January.

31. By equating entries we find that x must satisfy

$$x^2 + 1993x = 1994 \quad \text{and} \quad \sqrt{x^2} = -x.$$

The second equation implies that $x < 0$. From the first equation, $x^2 + 1993x - 1994 = 0$, $(x + 1994)(x - 1) = 0$, so $x = -1994$.

EXERCISE 6.2

1. $\begin{bmatrix} 2 & 0 & -3 \\ -1 & 4 & 0 \\ 1 & -6 & 5 \end{bmatrix} + \begin{bmatrix} 2 & -3 & 4 \\ -1 & 6 & 5 \\ 9 & 11 & -2 \end{bmatrix} = \begin{bmatrix} 2+2 & 0+(-3) & -3+4 \\ -1+(-1) & 4+6 & 0+5 \\ 1+9 & -6+(11) & 5+(-2) \end{bmatrix}$

$$= \begin{bmatrix} 4 & -3 & 1 \\ -2 & 10 & 5 \\ 10 & 5 & 3 \end{bmatrix}$$

3. $\begin{bmatrix} 1 & 4 \\ -2 & 7 \\ 6 & 9 \end{bmatrix} - \begin{bmatrix} 6 & -1 \\ 7 & 2 \\ 1 & 0 \end{bmatrix} = \begin{bmatrix} 1-6 & 4-(-1) \\ -2-7 & 7-2 \\ 6-1 & 9-0 \end{bmatrix} = \begin{bmatrix} -5 & 5 \\ -9 & 5 \\ 5 & 9 \end{bmatrix}$

5. $3[1 \quad -3 \quad 1] + 2[-6 \quad 1 \quad 4] - 0[-2 \quad 7 \quad 4]$
   $= [3 \quad -9 \quad 3] + [-12 \quad 2 \quad 8] - [0 \quad 0 \quad 0]$
   $= [3-12-0 \quad -9+2-0 \quad 3+8-0] = [-9 \quad -7 \quad 11]$

7. $\begin{bmatrix} 1 & 2 \\ 3 & 4 \end{bmatrix}$ has order 2×2, and $\begin{bmatrix} 5 \\ 6 \end{bmatrix}$ has order 2×1. Thus the sum is not defined.

9. $-6\begin{bmatrix} 2 & -6 & 7 & 1 \\ 7 & 1 & 6 & -2 \end{bmatrix} = \begin{bmatrix} -6\cdot 2 & -6(-6) & -6\cdot 7 & -6\cdot 1 \\ -6\cdot 7 & -6\cdot 1 & -6\cdot 6 & -6(-2) \end{bmatrix}$

$$= \begin{bmatrix} -12 & 36 & -42 & -6 \\ -42 & -6 & -36 & 12 \end{bmatrix}$$

11. $\begin{bmatrix} 2 & -4 & 0 \\ 0 & 6 & -2 \\ -4 & 0 & 10 \end{bmatrix} + \frac{1}{3}\begin{bmatrix} 9 & 0 & 3 \\ 0 & 3 & 0 \\ 3 & 9 & 9 \end{bmatrix} = \begin{bmatrix} 2 & -4 & 0 \\ 0 & 6 & -2 \\ -4 & 0 & 10 \end{bmatrix} + \begin{bmatrix} 3 & 0 & 1 \\ 0 & 1 & 0 \\ 1 & 3 & 3 \end{bmatrix}$

$$= \begin{bmatrix} 5 & -4 & 1 \\ 0 & 7 & -2 \\ -3 & 3 & 13 \end{bmatrix}$$

13. $-B = -\begin{bmatrix} -6 & -5 \\ 2 & -3 \end{bmatrix} = (-1)\begin{bmatrix} -6 & -5 \\ 2 & -3 \end{bmatrix} = \begin{bmatrix} -1(-6) & -1(-5) \\ -1(2) & -1(-3) \end{bmatrix} = \begin{bmatrix} 6 & 5 \\ -2 & 3 \end{bmatrix}$

15. $2O = 2\begin{bmatrix} 0 & 0 \\ 0 & 0 \end{bmatrix} = \begin{bmatrix} 2\cdot0 & 2\cdot0 \\ 2\cdot0 & 2\cdot0 \end{bmatrix} = \begin{bmatrix} 0 & 0 \\ 0 & 0 \end{bmatrix} = O$

17. $2(A - 2B) = 2\left\{\begin{bmatrix} 2 & 1 \\ 3 & -3 \end{bmatrix} - 2\begin{bmatrix} -6 & -5 \\ 2 & -3 \end{bmatrix}\right\}$

$= 2\left\{\begin{bmatrix} 2 & 1 \\ 3 & -3 \end{bmatrix} - \begin{bmatrix} -12 & -10 \\ 4 & -6 \end{bmatrix}\right\}$

$= 2\begin{bmatrix} 14 & 11 \\ -1 & 3 \end{bmatrix} = \begin{bmatrix} 28 & 22 \\ -2 & 6 \end{bmatrix}$

19. $3(A - C)$ is a 2×2 matrix and 6 is a number. Therefore $3(A - C) + 6$ is not defined.

21. $2B - 3A + 2C = 2\begin{bmatrix} -6 & -5 \\ 2 & -3 \end{bmatrix} - 3\begin{bmatrix} 2 & 1 \\ 3 & -3 \end{bmatrix} + 2\begin{bmatrix} -2 & -1 \\ -3 & 3 \end{bmatrix}$

$= \begin{bmatrix} -12 & -10 \\ 4 & -6 \end{bmatrix} - \begin{bmatrix} 6 & 3 \\ 9 & -9 \end{bmatrix} + \begin{bmatrix} -4 & -2 \\ -6 & 6 \end{bmatrix}$

$= \begin{bmatrix} -18 & -13 \\ -5 & 3 \end{bmatrix} + \begin{bmatrix} -4 & -2 \\ -6 & 6 \end{bmatrix} = \begin{bmatrix} -22 & -15 \\ -11 & 9 \end{bmatrix}$

23. $\frac{1}{2}A - 2(B + 2C) = \frac{1}{2}\begin{bmatrix} 2 & 1 \\ 3 & -3 \end{bmatrix} - 2\left\{\begin{bmatrix} -6 & -5 \\ 2 & -3 \end{bmatrix} + 2\begin{bmatrix} -2 & -1 \\ -3 & 3 \end{bmatrix}\right\}$

$= \begin{bmatrix} 1 & \frac{1}{2} \\ \frac{3}{2} & -\frac{3}{2} \end{bmatrix} - 2\left\{\begin{bmatrix} -6 & -5 \\ 2 & -3 \end{bmatrix} + \begin{bmatrix} -4 & -2 \\ -6 & 6 \end{bmatrix}\right\}$

$= \begin{bmatrix} 1 & \frac{1}{2} \\ \frac{3}{2} & -\frac{3}{2} \end{bmatrix} - 2\begin{bmatrix} -10 & -7 \\ -4 & 3 \end{bmatrix} = \begin{bmatrix} 1 & \frac{1}{2} \\ \frac{3}{2} & -\frac{3}{2} \end{bmatrix} - \begin{bmatrix} -20 & -14 \\ -8 & 6 \end{bmatrix}$

$= \begin{bmatrix} 21 & \frac{29}{2} \\ \frac{19}{2} & \frac{-15}{2} \end{bmatrix}$

25. $3(A + B) = 3\begin{bmatrix} -4 & -4 \\ 5 & -6 \end{bmatrix} = \begin{bmatrix} -12 & -12 \\ 15 & -18 \end{bmatrix}$.

$3A + 3B = \begin{bmatrix} 6 & 3 \\ 9 & -9 \end{bmatrix} + \begin{bmatrix} -18 & -15 \\ 6 & -9 \end{bmatrix} = \begin{bmatrix} -12 & -12 \\ 15 & -18 \end{bmatrix}$.

Thus $3(A + B) = 3A + 3B$.

27. $k_1(k_2A) = k_1\begin{bmatrix} 2k_2 & k_2 \\ 3k_2 & -3k_2 \end{bmatrix} = \begin{bmatrix} 2k_1k_2 & k_1k_2 \\ 3k_1k_2 & -3k_1k_2 \end{bmatrix}$.

$(k_1k_2)A = \begin{bmatrix} 2k_1k_2 & k_1k_2 \\ 3k_1k_2 & -3k_1k_2 \end{bmatrix}$. Thus $k_1(k_2A) = (k_1k_2)A$.

29. $3A^T + D = 3\begin{bmatrix} 1 & 0 & 2 \\ 2 & -1 & 0 \end{bmatrix} + \begin{bmatrix} 1 & 2 & -1 \\ 1 & 0 & 2 \end{bmatrix}$

$= \begin{bmatrix} 3 & 0 & 6 \\ 6 & -3 & 0 \end{bmatrix} + \begin{bmatrix} 1 & 2 & -1 \\ 1 & 0 & 2 \end{bmatrix} = \begin{bmatrix} 4 & 2 & 5 \\ 7 & -3 & 2 \end{bmatrix}$

31. $2B^T - 3C^T = 2\begin{bmatrix} 1 & 4 \\ 3 & -1 \end{bmatrix} - 3\begin{bmatrix} 1 & 1 \\ 0 & 2 \end{bmatrix}$

$= \begin{bmatrix} 2 & 8 \\ 6 & -2 \end{bmatrix} - \begin{bmatrix} 3 & 3 \\ 0 & 6 \end{bmatrix} = \begin{bmatrix} -1 & 5 \\ 6 & -8 \end{bmatrix}$

33. $C^T - D = \begin{bmatrix} 1 & 0 \\ 1 & 2 \end{bmatrix}^T - \begin{bmatrix} 1 & 2 & -1 \\ 1 & 0 & 2 \end{bmatrix}$ is impossible because C^T and D are not of the same size.

35. $x\begin{bmatrix} 2 \\ 1 \end{bmatrix} - y\begin{bmatrix} -3 \\ 5 \end{bmatrix} = \begin{bmatrix} 2x \\ x \end{bmatrix} - \begin{bmatrix} -3y \\ 5y \end{bmatrix} = \begin{bmatrix} 2x + 3y \\ x - 5y \end{bmatrix} = \begin{bmatrix} 16 \\ 22 \end{bmatrix}$.

Equating corresponding entries gives $\begin{cases} 2x + 3y = 16, \\ x - 5y = 22. \end{cases}$

From the second equation, $x = 22 + 5y$. Substituting in the first equation gives $2(22 + 5y) + 3y = 16$, $13y = -28$, or $y = \frac{-28}{13}$. Therefore $x = 22 + 5y = 22 + 5\left(-\frac{28}{13}\right) = \frac{146}{13}$. The solution is $x = 146/13$, $y = -28/13$.

37. $3\begin{bmatrix} x \\ y \end{bmatrix} - 3\begin{bmatrix} -2 \\ 4 \end{bmatrix} = 4\begin{bmatrix} 6 \\ -2 \end{bmatrix}$, $\begin{bmatrix} 3x + 6 \\ 3y - 12 \end{bmatrix} = \begin{bmatrix} 24 \\ -8 \end{bmatrix}$.

   $3x + 6 = 24$, $3x = 18$, or $x = 6$. $3y - 12 = -8$, $3y = 4$, or
   $y = 4/3$.   Thus $x = 6$, $y = 4/3$.

39. $\begin{bmatrix} 2 \\ 4 \\ 6 \end{bmatrix} + 2\begin{bmatrix} x \\ y \\ 4z \end{bmatrix} = \begin{bmatrix} -10 \\ -24 \\ 14 \end{bmatrix}$, $\begin{bmatrix} 2 + 2x \\ 4 + 2y \\ 6 + 8z \end{bmatrix} = \begin{bmatrix} -10 \\ -24 \\ 14 \end{bmatrix}$.

   $2 + 2x = -10$, $2x = -12$, or $x = -6$. $4 + 2y = -24$, $2y = -28$,
   or $y = -14$. $6 + 8z = 14$, $8z = 8$, or $z = 1$.   Thus $x = -6$,
   $y = -14$, $z = 1$.

41. $[1.1p_1 \quad 1.1p_2 \quad 1.1p_3] = 1.1[p_1 \quad p_2 \quad p_3] = 1.1P$.   Thus P
   must be multiplied by 1.1.

EXERCISE 6.3

 1. $c_{11} = 1(0) + 3(-2) + (-2)(3) = -12$

 3. $c_{32} = 0(-2) + 4(4) + 3(1) = 19$

 5. $c_{22} = -2(-2) + 1(4) + (-1)(1) = 7$

 7. A is 2 × 3 and E is 3 × 2, so AE is 2 × 2; 2·2 = 4 entries

 9. E is 3 × 2 and C is 2 × 5, so EC is 3 × 5; 3·5 = 15 entries

11. F is 2 × 3 and B is 3 × 1, so FB is 2 × 1; 2·1 = 2 entries

13. E is 3 × 2 and A is 2 × 3, so EA is 3 × 3; 3·3 = 9 entries

15. E is 3 × 2. F is 2 × 3 and B is 3 × 1, so FB is 2 × 1.
    Thus E(FB) is 3 × 1; 3·1 = 3 entries.

17. An identity matrix is a square
    matrix (in this case 4 × 4) with $\implies$ $\begin{bmatrix} 1 & 0 & 0 & 0 \\ 0 & 1 & 0 & 0 \\ 0 & 0 & 1 & 0 \\ 0 & 0 & 0 & 1 \end{bmatrix}$
    1's on the main diagonal and all
    other entries 0's.

19. $\begin{bmatrix} 2 & -4 \\ 3 & 2 \end{bmatrix} \begin{bmatrix} 3 & 0 \\ -1 & 4 \end{bmatrix} = \begin{bmatrix} 2(3)+(-4)(-1) & 2(0)+(-4)(4) \\ 3(3)+2(-1) & 3(0)+2(4) \end{bmatrix}$

$= \begin{bmatrix} 10 & -16 \\ 7 & 8 \end{bmatrix}$

21. $\begin{bmatrix} 2 & 0 & 3 \\ -1 & 4 & 5 \end{bmatrix} \begin{bmatrix} 1 \\ 4 \\ 7 \end{bmatrix} = \begin{bmatrix} 2(1)+0(4)+3(7) \\ -1(1)+4(4)+5(7) \end{bmatrix} = \begin{bmatrix} 23 \\ 50 \end{bmatrix}$

23. $\begin{bmatrix} 1 & 4 & -1 \\ 0 & 0 & 2 \\ -2 & 1 & 1 \end{bmatrix} \begin{bmatrix} -2 & 1 & 0 \\ 0 & 1 & 1 \\ 1 & 1 & 2 \end{bmatrix}$

$= \begin{bmatrix} 1(-2)+4(0)+(-1)1 & 1(1)+4(1)+(-1)(1) & 1(0)+4(1)+(-1)(2) \\ 0(-2)+0(0)+2(1) & 0(1)+0(1)+2(1) & 0(0)+0(1)+2(2) \\ -2(-2)+1(0)+1(1) & -2(1)+1(1)+1(1) & . \quad -2(0)+1(1)+1(2) \end{bmatrix}$

$= \begin{bmatrix} -3 & 4 & 2 \\ 2 & 2 & 4 \\ 5 & 0 & 3 \end{bmatrix}$

25. $[-1 \quad 2 \quad 3] \begin{bmatrix} 3 & 1 & -1 & 2 \\ 0 & 4 & 3 & 1 \\ -1 & 3 & 1 & -2 \end{bmatrix}$

$= [-3+0-3 \quad -1+8+9 \quad 1+6+3 \quad -2+2-6] = [-6 \quad 16 \quad 10 \quad -6]$

27. $\begin{bmatrix} 2 \\ 3 \\ -4 \\ 1 \end{bmatrix} [2 \quad 3 \quad -2 \quad 3] = \begin{bmatrix} 2(2) & 2(3) & 2(-2) & 2(3) \\ 3(2) & 3(3) & 3(-2) & 3(3) \\ -4(2) & -4(3) & -4(-2) & -4(3) \\ 1(2) & 1(3) & 1(-2) & 1(3) \end{bmatrix}$

$= \begin{bmatrix} 4 & 6 & -4 & 6 \\ 6 & 9 & -6 & 9 \\ -8 & -12 & 8 & -12 \\ 2 & 3 & -2 & 3 \end{bmatrix}$

29. $3\left\{ \begin{bmatrix} -2 & 0 & 2 \\ 3 & -1 & 1 \end{bmatrix} + 2\begin{bmatrix} -1 & 0 & 2 \\ 1 & 1 & -2 \end{bmatrix} \right\} \begin{bmatrix} 1 & 2 \\ 3 & 4 \\ 5 & 6 \end{bmatrix}$

$= 3\left\{ \begin{bmatrix} -2 & 0 & 2 \\ 3 & -1 & 1 \end{bmatrix} + \begin{bmatrix} -2 & 0 & 4 \\ 2 & 2 & -4 \end{bmatrix} \right\} \begin{bmatrix} 1 & 2 \\ 3 & 4 \\ 5 & 6 \end{bmatrix}$

$$= 3\left\{ \begin{bmatrix} -4 & 0 & 6 \\ 5 & 1 & -3 \end{bmatrix} \right\} \begin{bmatrix} 1 & 2 \\ 3 & 4 \\ 5 & 6 \end{bmatrix} = \begin{bmatrix} -12 & 0 & 18 \\ 15 & 3 & -9 \end{bmatrix} \begin{bmatrix} 1 & 2 \\ 3 & 4 \\ 5 & 6 \end{bmatrix}$$

$$= \begin{bmatrix} -12(1)+0(3)+18(5) & -12(2)+0(4)+18(6) \\ 15(1)+3(3)+(-9)(5) & 15(2)+3(4)+(-9)(6) \end{bmatrix} = \begin{bmatrix} 78 & 84 \\ -21 & -12 \end{bmatrix}$$

31. $$\begin{bmatrix} 1 & 2 \\ 3 & 4 \end{bmatrix} \left\{ \begin{bmatrix} 2 & 0 & 1 \\ 1 & 0 & -2 \end{bmatrix} \begin{bmatrix} 1 & -2 \\ 2 & 1 \\ 3 & 0 \end{bmatrix} \right\} = \begin{bmatrix} 1 & 2 \\ 3 & 4 \end{bmatrix} \left\{ \begin{bmatrix} 2+0+3 & -4+0+0 \\ 1+0-6 & -2+0+0 \end{bmatrix} \right\}$$

$$= \begin{bmatrix} 1 & 2 \\ 3 & 4 \end{bmatrix} \begin{bmatrix} 5 & -4 \\ -5 & -2 \end{bmatrix} = \begin{bmatrix} 5-10 & -4-4 \\ 15-20 & -12-8 \end{bmatrix} = \begin{bmatrix} -5 & -8 \\ -5 & -20 \end{bmatrix}$$

33. $$\begin{bmatrix} 1 & 0 & 0 \\ 0 & 1 & 0 \\ 0 & 0 & 1 \end{bmatrix} \begin{bmatrix} x \\ y \\ z \end{bmatrix} = I \begin{bmatrix} x \\ y \\ z \end{bmatrix} = \begin{bmatrix} x \\ y \\ z \end{bmatrix}$$

35. $$\begin{bmatrix} 2 & 1 & 3 \\ 4 & 9 & 7 \end{bmatrix} \begin{bmatrix} x_1 \\ x_2 \\ x_3 \end{bmatrix} = \begin{bmatrix} 2x_1+x_2+3x_3 \\ 4x_1+9x_2+7x_3 \end{bmatrix}$$

37. $DI - \frac{1}{3}E = D - \frac{1}{3}E = \begin{bmatrix} 1 & 0 & 0 \\ 0 & 1 & 1 \\ 1 & 2 & 1 \end{bmatrix} - \frac{1}{3}\begin{bmatrix} 3 & 0 & 0 \\ 0 & 6 & 0 \\ 0 & 0 & 3 \end{bmatrix}$

$$= \begin{bmatrix} 1 & 0 & 0 \\ 0 & 1 & 1 \\ 1 & 2 & 1 \end{bmatrix} - \begin{bmatrix} 1 & 0 & 0 \\ 0 & 2 & 0 \\ 0 & 0 & 1 \end{bmatrix} = \begin{bmatrix} 0 & 0 & 0 \\ 0 & -1 & 1 \\ 1 & 2 & 0 \end{bmatrix}$$

39. $3A - 2BC = 3\begin{bmatrix} 1 & -2 \\ 0 & 3 \end{bmatrix} - 2\begin{bmatrix} -2 & 3 & 0 \\ 1 & -4 & 2 \end{bmatrix} \begin{bmatrix} -1 & 1 \\ 0 & 3 \\ 2 & 4 \end{bmatrix}$

$$= \begin{bmatrix} 3 & -6 \\ 0 & 9 \end{bmatrix} - 2\begin{bmatrix} 2+0+0 & -2+9+0 \\ -1+0+2 & 1-12+4 \end{bmatrix}$$

$$= \begin{bmatrix} 3 & -6 \\ 0 & 9 \end{bmatrix} - \begin{bmatrix} 4 & 14 \\ 2 & -14 \end{bmatrix} = \begin{bmatrix} -1 & -20 \\ -2 & 23 \end{bmatrix}$$

41. $2I - \frac{1}{2}EF = 2I - \frac{1}{2}\begin{bmatrix} 3 & 0 & 0 \\ 0 & 6 & 0 \\ 0 & 0 & 3 \end{bmatrix}\begin{bmatrix} \frac{1}{3} & 0 & 0 \\ 0 & \frac{1}{6} & 0 \\ 0 & 0 & \frac{1}{3} \end{bmatrix}$

$= 2I - \begin{bmatrix} \frac{3}{2} & 0 & 0 \\ 0 & 3 & 0 \\ 0 & 0 & \frac{3}{2} \end{bmatrix}\begin{bmatrix} \frac{1}{3} & 0 & 0 \\ 0 & \frac{1}{6} & 0 \\ 0 & 0 & \frac{1}{3} \end{bmatrix}$

$= 2I - \begin{bmatrix} \frac{1}{2}+0+0 & 0+0+0 & 0+0+0 \\ 0+0+0 & 0+\frac{1}{2}+0 & 0+0+0 \\ 0+0+0 & 0+0+0 & 0+0+\frac{1}{2} \end{bmatrix}$

$= \begin{bmatrix} 2 & 0 & 0 \\ 0 & 2 & 0 \\ 0 & 0 & 2 \end{bmatrix} - \begin{bmatrix} \frac{1}{2} & 0 & 0 \\ 0 & \frac{1}{2} & 0 \\ 0 & 0 & \frac{1}{2} \end{bmatrix} = \begin{bmatrix} \frac{3}{2} & 0 & 0 \\ 0 & \frac{3}{2} & 0 \\ 0 & 0 & \frac{3}{2} \end{bmatrix}$

43. $(DC)A = \left\{ \begin{bmatrix} 1 & 0 & 0 \\ 0 & 1 & 1 \\ 1 & 2 & 1 \end{bmatrix}\begin{bmatrix} -1 & 1 \\ 0 & 3 \\ 2 & 4 \end{bmatrix} \right\}A$

$= \begin{bmatrix} -1+0+0 & 1+0+0 \\ 0+0+2 & 0+3+4 \\ -1+0+2 & 1+6+4 \end{bmatrix}A = \begin{bmatrix} -1 & 1 \\ 2 & 7 \\ 1 & 11 \end{bmatrix}\begin{bmatrix} 1 & -2 \\ 0 & 3 \end{bmatrix}$

$= \begin{bmatrix} -1+0 & 2+3 \\ 2+0 & -4+21 \\ 1+0 & -2+33 \end{bmatrix} = \begin{bmatrix} -1 & 5 \\ 2 & 17 \\ 1 & 31 \end{bmatrix}$

45. Impossible: A is not a square matrix, so A^2 is not defined.

47. $B^3 = (B^2)B = \begin{bmatrix} 0 & 0 & -1 \\ 2 & -1 & 0 \\ 0 & 0 & 2 \end{bmatrix}^2 B = \begin{bmatrix} 0 & 0 & -1 \\ 2 & -1 & 0 \\ 0 & 0 & 2 \end{bmatrix}\begin{bmatrix} 0 & 0 & -1 \\ 2 & -1 & 0 \\ 0 & 0 & 2 \end{bmatrix}B$

$= \begin{bmatrix} 0 & 0 & -2 \\ -2 & 1 & -2 \\ 0 & 0 & 4 \end{bmatrix}\begin{bmatrix} 0 & 0 & -1 \\ 2 & -1 & 0 \\ 0 & 0 & 2 \end{bmatrix} = \begin{bmatrix} 0 & 0 & -4 \\ 2 & -1 & -2 \\ 0 & 0 & 8 \end{bmatrix}$

49. $(AC)^2 = \left\{ \begin{bmatrix} 1 & -1 & 0 \\ 0 & 1 & 1 \end{bmatrix} \begin{bmatrix} 1 & 0 \\ 2 & -1 \\ 0 & 1 \end{bmatrix} \right\}^2 = \begin{bmatrix} -1 & 1 \\ 2 & 0 \end{bmatrix}^2$

$= \begin{bmatrix} -1 & 1 \\ 2 & 0 \end{bmatrix} \begin{bmatrix} -1 & 1 \\ 2 & 0 \end{bmatrix} = \begin{bmatrix} 3 & -1 \\ -2 & 2 \end{bmatrix}$

51. $(BA^T)^T = \left\{ \begin{bmatrix} 0 & 0 & -1 \\ 2 & -1 & 0 \\ 0 & 0 & 2 \end{bmatrix} \begin{bmatrix} 1 & 0 \\ -1 & 1 \\ 0 & 1 \end{bmatrix} \right\}^T = \begin{bmatrix} 0 & -1 \\ 3 & -1 \\ 0 & 2 \end{bmatrix}^T = \begin{bmatrix} 0 & 3 & 0 \\ -1 & -1 & 2 \end{bmatrix}$

53. $(2I)^2 - 2I^2 = (2I)^2 - 2I = \begin{bmatrix} 2 & 0 & 0 \\ 0 & 2 & 0 \\ 0 & 0 & 2 \end{bmatrix}^2 - \begin{bmatrix} 2 & 0 & 0 \\ 0 & 2 & 0 \\ 0 & 0 & 2 \end{bmatrix}$

$= \begin{bmatrix} 2 & 0 & 0 \\ 0 & 2 & 0 \\ 0 & 0 & 2 \end{bmatrix} \begin{bmatrix} 2 & 0 & 0 \\ 0 & 2 & 0 \\ 0 & 0 & 2 \end{bmatrix} - \begin{bmatrix} 2 & 0 & 0 \\ 0 & 2 & 0 \\ 0 & 0 & 2 \end{bmatrix}$

$= \begin{bmatrix} 4 & 0 & 0 \\ 0 & 4 & 0 \\ 0 & 0 & 4 \end{bmatrix} - \begin{bmatrix} 2 & 0 & 0 \\ 0 & 2 & 0 \\ 0 & 0 & 2 \end{bmatrix} = \begin{bmatrix} 2 & 0 & 0 \\ 0 & 2 & 0 \\ 0 & 0 & 2 \end{bmatrix}$

55. $A(I - \emptyset) = A(I) = AI$. Since I is 3×3 and A has three columns, $AI = A$. Thus

$$A(I - \emptyset) = A = \begin{bmatrix} 1 & -1 & 0 \\ 0 & 1 & 1 \end{bmatrix}$$

57. $(AB)(AB)^T = \begin{bmatrix} 1 & -1 & 0 \\ 0 & 1 & 1 \end{bmatrix} \begin{bmatrix} 0 & 0 & -1 \\ 2 & -1 & 0 \\ 0 & 0 & 2 \end{bmatrix} (AB)^T = \begin{bmatrix} -2 & 1 & -1 \\ 2 & -1 & 2 \end{bmatrix} (AB)^T$

$= \begin{bmatrix} -2 & 1 & -1 \\ 2 & -1 & 2 \end{bmatrix} \begin{bmatrix} -2 & 2 \\ 1 & -1 \\ -1 & 2 \end{bmatrix} = \begin{bmatrix} 6 & -7 \\ -7 & 9 \end{bmatrix}$

59. $AX = C$. $A = \begin{bmatrix} 3 & 1 \\ 7 & -2 \end{bmatrix}$, $X = \begin{bmatrix} x \\ y \end{bmatrix}$, $C = \begin{bmatrix} 6 \\ 5 \end{bmatrix} \Rightarrow \begin{bmatrix} 3 & 1 \\ 7 & -2 \end{bmatrix} \begin{bmatrix} x \\ y \end{bmatrix} = \begin{bmatrix} 6 \\ 5 \end{bmatrix}$

61. AX = C. $A = \begin{bmatrix} 4 & -1 & 3 \\ 3 & 0 & -1 \\ 0 & 3 & 2 \end{bmatrix}$, $X = \begin{bmatrix} r \\ s \\ t \end{bmatrix}$, $C = \begin{bmatrix} 9 \\ 7 \\ 15 \end{bmatrix}$. Thus the

system is represented by $\begin{bmatrix} 4 & -1 & 3 \\ 3 & 0 & -1 \\ 0 & 3 & 2 \end{bmatrix} \begin{bmatrix} r \\ s \\ t \end{bmatrix} = \begin{bmatrix} 9 \\ 7 \\ 15 \end{bmatrix}$

63. $Q = [7 \quad 3 \quad 5]$, $R = \begin{bmatrix} 5 & 20 & 16 & 7 & 17 \\ 7 & 18 & 12 & 9 & 21 \\ 6 & 25 & 8 & 5 & 13 \end{bmatrix}$, $C = \begin{bmatrix} 1500 \\ 800 \\ 500 \\ 100 \\ 1000 \end{bmatrix}$.

$QRC = Q(RC) = Q\begin{bmatrix} 5\cdot1500 + 20\cdot800 + 16\cdot500 + 7\cdot100 + 17\cdot1000 \\ 7\cdot1500 + 18\cdot800 + 12\cdot500 + 9\cdot100 + 21\cdot1000 \\ 6\cdot1500 + 25\cdot800 + 8\cdot500 + 5\cdot100 + 13\cdot1000 \end{bmatrix}$

$= [7 \quad 3 \quad 5] \begin{bmatrix} 49,200 \\ 52,800 \\ 46,500 \end{bmatrix} = [7\cdot49,200 + 3\cdot52,800 + 5\cdot46,500]$

$= [735,300] \implies \$735,300$

65. (a) Amount spent for goods:

coal industry: $D_C P = [0 \quad 1 \quad 4] \begin{bmatrix} 10,000 \\ 20,000 \\ 40,000 \end{bmatrix} = [180,000]$

$\implies \$180,000$

elec. industry: $D_E P = [20 \quad 0 \quad 8] \begin{bmatrix} 10,000 \\ 20,000 \\ 40,000 \end{bmatrix} = [520,000]$

$\implies \$520,000$

steel industry: $D_S P = [30 \quad 5 \quad 0] \begin{bmatrix} 10,000 \\ 20,000 \\ 40,000 \end{bmatrix} = [400,000]$

$\implies \$400,000$

consumer 1: $D_1 P = [3 \quad 2 \quad 5] \begin{bmatrix} 10,000 \\ 20,000 \\ 40,000 \end{bmatrix} = [270,000]$

$\implies \$270,000$

consumer 2: $D_2 P = [0 \quad 17 \quad 1] \begin{bmatrix} 10,000 \\ 20,000 \\ 40,000 \end{bmatrix} = [380,000]$

$\implies \$380,000$

$$\text{consumer 3: } D_3P = [4 \quad 6 \quad 12] \begin{bmatrix} 10,000 \\ 20,000 \\ 40,000 \end{bmatrix} = [640,000]$$

$$\Rightarrow \$640,000$$

(b) From Example 3 of Sec. 6.2, the number of units sold of coal, electricity, and steel are 57, 31, and 30, respectively. Thus

profit for coal is $10,000(57) - 180,000 = \$390,000$
profit for elec. is $20,000(31) - 520,000 = \$100,000$
profit for steel is $40,000(30) - 400,000 = \$800,000$

(c) From (a), the total amount of money that is paid by all the industries and consumers is

$180,000 + 520,000 + 400,000 + 270,000 + 380,000$
$+ 640,000$

$= \$2,390,000$.

(d) The proportion of the total amount in (c) paid out by the industries is

$$\frac{180,000 + 520,000 + 400,000}{2,390,000} = \frac{110}{239}$$

Proportion of total amount in (c) paid by consumers is

$$\frac{270,000 + 380,000 + 640,000}{2,390,000} = \frac{129}{239}$$

67. $\begin{bmatrix} 1 & 2 \\ 1 & 2 \end{bmatrix} \begin{bmatrix} 2 & -3 \\ -1 & 3/2 \end{bmatrix} = \begin{bmatrix} 1(2)+(2)(-1) & 1(-3)+2(3/2) \\ 1(2)+2(-1) & 1(-3)+2(3/2) \end{bmatrix} = \begin{bmatrix} 0 & 0 \\ 0 & 0 \end{bmatrix}$

EXERCISE 6.4

1. The first nonzero entry in row 2 is not to the right of the first nonzero entry in row 1 $\Rightarrow$ not reduced.

3. Reduced.

5. The first row consists entirely of zeros and is not below each row containing a nonzero entry $\Rightarrow$ not reduced.

7. $\begin{bmatrix} 1 & 3 \\ 4 & 0 \end{bmatrix} \xrightarrow{-4R_1+R_2} \begin{bmatrix} 1 & 3 \\ 0 & -12 \end{bmatrix} \xrightarrow{-\frac{1}{12}R_2} \begin{bmatrix} 1 & 3 \\ 0 & 1 \end{bmatrix} \xrightarrow{-3R_2+R_1} \begin{bmatrix} 1 & 0 \\ 0 & 1 \end{bmatrix}$

9. $\begin{bmatrix} 2 & 4 & 6 \\ 1 & 2 & 3 \\ 1 & 2 & 3 \end{bmatrix} \xrightarrow{R_1 \leftrightarrow R_3} \begin{bmatrix} 1 & 2 & 3 \\ 1 & 2 & 3 \\ 2 & 4 & 6 \end{bmatrix} \xrightarrow{(-1)R_1 + R_2} \begin{bmatrix} 1 & 2 & 3 \\ 0 & 0 & 0 \\ 2 & 4 & 6 \end{bmatrix}$

$\xrightarrow{-2R_1 + R_3} \begin{bmatrix} 1 & 2 & 3 \\ 0 & 0 & 0 \\ 0 & 0 & 0 \end{bmatrix}$

11. $\begin{bmatrix} 2 & 0 & 3 & 1 \\ 1 & 4 & 2 & 2 \\ -1 & 3 & 1 & 4 \\ 0 & 2 & 1 & 0 \end{bmatrix} \xrightarrow{R_1 \leftrightarrow R_2} \begin{bmatrix} 1 & 4 & 2 & 2 \\ 2 & 0 & 3 & 1 \\ -1 & 3 & 1 & 4 \\ 0 & 2 & 1 & 0 \end{bmatrix}$

$\xrightarrow[R_1 + R_3]{-2R_1 + R_2} \begin{bmatrix} 1 & 4 & 2 & 2 \\ 0 & -8 & -1 & -3 \\ 0 & 7 & 3 & 6 \\ 0 & 2 & 1 & 0 \end{bmatrix} \xrightarrow{-\frac{1}{8}R_2} \begin{bmatrix} 1 & 4 & 2 & 2 \\ 0 & 1 & 1/8 & 3/8 \\ 0 & 7 & 3 & 6 \\ 0 & 2 & 1 & 0 \end{bmatrix}$

$\xrightarrow[-2R_2 + R_4]{-7R_2 + R_3} \begin{bmatrix} 1 & 0 & 3/2 & 1/2 \\ 0 & 1 & 1/8 & 3/8 \\ 0 & 0 & 17/8 & 27/8 \\ 0 & 0 & 3/4 & -3/4 \end{bmatrix} \xrightarrow{\frac{8}{17}R_2} \begin{bmatrix} 1 & 0 & 3/2 & 1/2 \\ 0 & 1 & 1/8 & 3/8 \\ 0 & 0 & 1 & 27/17 \\ 0 & 0 & 3/4 & -3/4 \end{bmatrix}$

$\xrightarrow[\substack{(-3/2)R_3 + R_1 \\ (-1/8)R_3 + R_2 \\ (-3/4)R_3 + R_4}]{} \begin{bmatrix} 1 & 0 & 0 & -32/17 \\ 0 & 1 & 0 & 3/17 \\ 0 & 0 & 1 & 27/17 \\ 0 & 0 & 0 & -33/17 \end{bmatrix}$

$\xrightarrow{(-17/33)R_4} \begin{bmatrix} 1 & 0 & 0 & -32/17 \\ 0 & 1 & 0 & 3/17 \\ 0 & 0 & 1 & 27/17 \\ 0 & 0 & 0 & 1 \end{bmatrix} \xrightarrow[\substack{(32/17)R_4 + R_1 \\ (-3/17)R_4 + R_2 \\ (-27/17)R_4 + R_3}]{} \begin{bmatrix} 1 & 0 & 0 & 0 \\ 0 & 1 & 0 & 0 \\ 0 & 0 & 1 & 0 \\ 0 & 0 & 0 & 1 \end{bmatrix}$

13. $\begin{bmatrix} 2 & 3 & | & 5 \\ 1 & -2 & | & -1 \end{bmatrix} \rightarrow \begin{bmatrix} 1 & -2 & | & -1 \\ 2 & 3 & | & 5 \end{bmatrix} \rightarrow \begin{bmatrix} 1 & -2 & | & -1 \\ 0 & 7 & | & 7 \end{bmatrix}$

$\rightarrow \begin{bmatrix} 1 & -2 & | & -1 \\ 0 & 1 & | & 1 \end{bmatrix} \rightarrow \begin{bmatrix} 1 & 0 & | & 1 \\ 0 & 1 & | & 1 \end{bmatrix}$. Thus $x = 1$, $y = 1$.

15. $\begin{bmatrix} 3 & 1 & | & 4 \\ 12 & 4 & | & 2 \end{bmatrix} \rightarrow \begin{bmatrix} 3 & 1 & | & 4 \\ 0 & 0 & | & -14 \end{bmatrix} \rightarrow \begin{bmatrix} 1 & 1/3 & | & 4/3 \\ 0 & 0 & | & -14 \end{bmatrix}$

$\rightarrow \begin{bmatrix} 1 & 1/3 & | & 4/3 \\ 0 & 0 & | & 1 \end{bmatrix} \rightarrow \begin{bmatrix} 1 & 1/3 & | & 0 \\ 0 & 0 & | & 1 \end{bmatrix}$.

Last row indicates 0 = 1, which is never true, so there is no solution.

17. $\begin{bmatrix} 1 & 2 & 1 & | & 4 \\ 3 & 0 & 2 & | & 5 \end{bmatrix} \longrightarrow \begin{bmatrix} 1 & 2 & 1 & | & 4 \\ 0 & -6 & -1 & | & -7 \end{bmatrix} \longrightarrow \begin{bmatrix} 1 & 2 & 1 & | & 4 \\ 0 & 1 & 1/6 & | & 7/6 \end{bmatrix}$

$\longrightarrow \begin{bmatrix} 1 & 0 & 2/3 & | & 5/3 \\ 0 & 1 & 1/6 & | & 7/6 \end{bmatrix}$, which gives $\begin{cases} x + \frac{2}{3}z = \frac{5}{3}, \\ y + \frac{1}{6}z = \frac{7}{6}. \end{cases}$

Thus $x = -\frac{2}{3}r + \frac{5}{3}$, $y = -\frac{1}{6}r + \frac{7}{6}$, $z = r$, where r is any real number.

19. $\begin{bmatrix} 1 & -3 & | & 0 \\ 2 & 2 & | & 3 \\ 5 & -1 & | & 1 \end{bmatrix} \longrightarrow \begin{bmatrix} 1 & -3 & | & 0 \\ 0 & 8 & | & 3 \\ 0 & 14 & | & 1 \end{bmatrix} \longrightarrow \begin{bmatrix} 1 & -3 & | & 0 \\ 0 & 1 & | & 3/8 \\ 0 & 14 & | & 1 \end{bmatrix}$

$\longrightarrow \begin{bmatrix} 1 & 0 & | & 9/8 \\ 0 & 1 & | & 3/8 \\ 0 & 0 & | & -17/4 \end{bmatrix}$. From third row, $0 = -17/4$, which is never true, so there is no solution.

21. $\begin{bmatrix} 1 & -1 & -3 & | & -4 \\ 2 & -1 & -4 & | & -7 \\ 1 & 1 & -1 & | & -2 \end{bmatrix} \longrightarrow \begin{bmatrix} 1 & -1 & -3 & | & -4 \\ 0 & 1 & 2 & | & 1 \\ 0 & 2 & 2 & | & 2 \end{bmatrix} \longrightarrow \begin{bmatrix} 1 & 0 & -1 & | & -3 \\ 0 & 1 & 2 & | & 1 \\ 0 & 0 & -2 & | & 0 \end{bmatrix}$

$\longrightarrow \begin{bmatrix} 1 & 0 & -1 & | & -3 \\ 0 & 1 & 2 & | & 1 \\ 0 & 0 & 1 & | & 0 \end{bmatrix} \longrightarrow \begin{bmatrix} 1 & 0 & 0 & | & -3 \\ 0 & 1 & 0 & | & 1 \\ 0 & 0 & 1 & | & 0 \end{bmatrix}$.

Thus $x = -3$, $y = 1$, $z = 0$.

23. $\begin{bmatrix} 2 & 0 & -4 & | & 8 \\ 1 & -2 & -2 & | & 14 \\ 1 & 1 & -2 & | & -1 \\ 3 & 1 & 1 & | & 0 \end{bmatrix} \longrightarrow \begin{bmatrix} 1 & 0 & -2 & | & 4 \\ 1 & -2 & -2 & | & 14 \\ 1 & 1 & -2 & | & -1 \\ 3 & 1 & 1 & | & 0 \end{bmatrix} \longrightarrow \begin{bmatrix} 1 & 0 & -2 & | & 4 \\ 0 & -2 & 0 & | & 10 \\ 0 & 1 & 0 & | & -5 \\ 0 & 1 & 7 & | & -12 \end{bmatrix}$

$\longrightarrow \begin{bmatrix} 1 & 0 & -2 & | & 4 \\ 0 & 1 & 0 & | & -5 \\ 0 & -2 & 0 & | & 10 \\ 0 & 1 & 7 & | & -12 \end{bmatrix} \longrightarrow \begin{bmatrix} 1 & 0 & -2 & | & 4 \\ 0 & 1 & 0 & | & -5 \\ 0 & 0 & 0 & | & 0 \\ 0 & 0 & 7 & | & -7 \end{bmatrix}$

$\longrightarrow \begin{bmatrix} 1 & 0 & -2 & | & 4 \\ 0 & 1 & 0 & | & -5 \\ 0 & 0 & 1 & | & -1 \\ 0 & 0 & 0 & | & 0 \end{bmatrix} \longrightarrow \begin{bmatrix} 1 & 0 & 0 & | & 2 \\ 0 & 1 & 0 & | & -5 \\ 0 & 0 & 1 & | & -1 \\ 0 & 0 & 0 & | & 0 \end{bmatrix}$.

Thus $x = 2$, $y = -5$, $z = -1$.

25.
$$\begin{bmatrix} 1 & 1 & -1 & 1 & 1 & | & 0 \\ 1 & 1 & 1 & -1 & 1 & | & 0 \\ 1 & -1 & -1 & 1 & -1 & | & 0 \\ 1 & 1 & -1 & -1 & -1 & | & 0 \end{bmatrix} \longrightarrow \begin{bmatrix} 1 & 1 & -1 & 1 & 1 & | & 0 \\ 0 & 0 & 2 & -2 & 0 & | & 0 \\ 0 & -2 & 0 & 0 & -2 & | & 0 \\ 0 & 0 & 0 & -2 & -2 & | & 0 \end{bmatrix}$$

$$\longrightarrow \begin{bmatrix} 1 & 1 & -1 & 1 & 1 & | & 0 \\ 0 & -2 & 0 & 0 & -2 & | & 0 \\ 0 & 0 & 2 & -2 & 0 & | & 0 \\ 0 & 0 & 0 & -2 & -2 & | & 0 \end{bmatrix} \longrightarrow \begin{bmatrix} 1 & 1 & -1 & 1 & 1 & | & 0 \\ 0 & 1 & 0 & 0 & 1 & | & 0 \\ 0 & 0 & 1 & -1 & 0 & | & 0 \\ 0 & 0 & 0 & 1 & 1 & | & 0 \end{bmatrix}$$

$$\longrightarrow \begin{bmatrix} 1 & 0 & -1 & 1 & 0 & | & 0 \\ 0 & 1 & 0 & 0 & 1 & | & 0 \\ 0 & 0 & 1 & -1 & 0 & | & 0 \\ 0 & 0 & 0 & 1 & 1 & | & 0 \end{bmatrix} \longrightarrow \begin{bmatrix} 1 & 0 & 0 & 0 & 0 & | & 0 \\ 0 & 1 & 0 & 0 & 1 & | & 0 \\ 0 & 0 & 1 & -1 & 0 & | & 0 \\ 0 & 0 & 0 & 1 & 1 & | & 0 \end{bmatrix}$$

$$\longrightarrow \begin{bmatrix} 1 & 0 & 0 & 0 & 0 & | & 0 \\ 0 & 1 & 0 & 0 & 1 & | & 0 \\ 0 & 0 & 1 & 0 & 1 & | & 0 \\ 0 & 0 & 0 & 1 & 1 & | & 0 \end{bmatrix}.$$

Thus $x_1 = 0$, $x_2 = -r$, $x_3 = -r$, $x_4 = -r$, $x_5 = r$, where r is any real number.

27. Let x = federal tax and y = state tax. Then
$x = 0.25(312{,}000 - y)$ and $y = 0.10(312{,}000 - x)$.
Equivalently,

$$\begin{cases} x + 0.25y = 78{,}000, \\ 0.10x + y = 31{,}200. \end{cases}$$

$$\begin{bmatrix} 1 & 0.25 & | & 78{,}000 \\ 0.10 & 1 & | & 31{,}200 \end{bmatrix} \longrightarrow \begin{bmatrix} 1 & 0.25 & | & 78{,}000 \\ 0 & 0.975 & | & 23{,}400 \end{bmatrix}$$

$$\longrightarrow \begin{bmatrix} 1 & 0.25 & | & 78{,}000 \\ 0 & 1 & | & 24{,}000 \end{bmatrix} \longrightarrow \begin{bmatrix} 1 & 0 & | & 72{,}000 \\ 0 & 1 & | & 24{,}000 \end{bmatrix}.$$

Thus $x = 72{,}000$ and $y = 24{,}000$, so the federal tax is
$\$72{,}000$ and the state tax is $\$24{,}000$.

29. Let x = number of units of A produced, y = number of units
of B produced, and z = number of units of C produced. Then

no. of units: $x + y + z = 11{,}000$,
total cost: $4x + 5y + 7z + 17{,}000 = 80{,}000$,
total profit: $x + 2y + 3z = 25{,}000$.

Equivalently,

$$\begin{cases} x + y + z = 11{,}000, \\ 4x + 5y + 7z = 63{,}000, \\ x + 2y + 3z = 25{,}000. \end{cases}$$

$$\begin{bmatrix} 1 & 1 & 1 & | & 11{,}000 \\ 4 & 5 & 7 & | & 63{,}000 \\ 1 & 2 & 3 & | & 25{,}000 \end{bmatrix} \longrightarrow \begin{bmatrix} 1 & 1 & 1 & | & 11{,}000 \\ 0 & 1 & 3 & | & 19{,}000 \\ 0 & 1 & 2 & | & 14{,}000 \end{bmatrix}$$

$$\longrightarrow \begin{bmatrix} 1 & 0 & -2 & | & -8{,}000 \\ 0 & 1 & 3 & | & 19{,}000 \\ 0 & 0 & -1 & | & -5{,}000 \end{bmatrix} \longrightarrow \begin{bmatrix} 1 & 0 & -2 & | & -8{,}000 \\ 0 & 1 & 3 & | & 19{,}000 \\ 0 & 0 & 1 & | & 5{,}000 \end{bmatrix}$$

$$\longrightarrow \begin{bmatrix} 1 & 0 & 0 & | & 2000 \\ 0 & 1 & 0 & | & 4000 \\ 0 & 0 & 1 & | & 5000 \end{bmatrix}.$$

Thus x = 2000, y = 4000, and z = 5000, so 2000 units of A, 4000 units of B and 5000 units of C should be produced.

31. Let x = number of brand X pills, y = number of brand Y pills, and z = number of brand Z pills. Considering the unit requirements gives the system

$$\begin{cases} 2x + 1y + 1z = 10 & \text{(vitamin A)}, \\ 3x + 3y + 0z = 9 & \text{(vitamin D)}, \\ 5x + 4y + 1z = 19 & \text{(vitamin E)}. \end{cases}$$

$$\begin{bmatrix} 2 & 1 & 1 & | & 10 \\ 3 & 3 & 0 & | & 9 \\ 5 & 4 & 1 & | & 19 \end{bmatrix} \longrightarrow \begin{bmatrix} 1 & 1/2 & 1/2 & | & 5 \\ 3 & 3 & 0 & | & 9 \\ 5 & 4 & 1 & | & 19 \end{bmatrix} \longrightarrow \begin{bmatrix} 1 & 1/2 & 1/2 & | & 5 \\ 0 & 3/2 & -3/2 & | & -6 \\ 0 & 3/2 & -3/2 & | & -6 \end{bmatrix}$$

$$\longrightarrow \begin{bmatrix} 1 & 1/2 & 1/2 & | & 5 \\ 0 & 1 & -1 & | & -4 \\ 0 & 0 & 0 & | & 0 \end{bmatrix} \longrightarrow \begin{bmatrix} 1 & 0 & 1 & | & 7 \\ 0 & 1 & -1 & | & -4 \\ 0 & 0 & 0 & | & 0 \end{bmatrix}.$$

Thus $\begin{cases} x = 7 - r, \\ y = r - 4, \\ z = r, \end{cases}$ where r = 4, 5, 6, 7.

The only solutions for the problem are z = 4, x = 3, and y = 0; z = 5, x = 2, and y = 1; z = 6, x = 1, and y = 2; z = 7, x = 0, and y = 3. Their respective costs (in cents) are 15, 23, 31, and 39.

(a) The possible combinations are
    3 of X, 4 of Z;   2 of X, 1 of Y,   5 of Z; 1 of X,
    2 of Y, 6 of Z;   3 of Y, 7 of Z.
(b) The combination 3 of X, 4 of Z costs 15 cents a day.
(c) The least expensive combination is 3 of X, 4 of Z; the most expensive is 3 of Y, 7 of Z.

33. (a) Let x, y, and z represent the number of units of S, D and G, respectively. Then we obtain the system

$$\begin{cases} 12x + 20y + 32z = 200 & \text{(stock A)}, \\ 16x + 12y + 28z = 176 & \text{(stock B)}, \\ 8x + 28y + 36z = 264 & \text{(stock C)}. \end{cases}$$

$$\begin{bmatrix} 12 & 20 & 32 & | & 220 \\ 16 & 12 & 28 & | & 176 \\ 8 & 28 & 36 & | & 264 \end{bmatrix} \xrightarrow[\substack{(1/4)R_1 \\ (1/4)R_2 \\ (1/8)R_3}]{} \begin{bmatrix} 3 & 5 & 8 & | & 55 \\ 4 & 3 & 7 & | & 44 \\ 1 & 7/2 & 9/2 & | & 33 \end{bmatrix}$$

$$\xrightarrow{R_1 \leftrightarrow R_3} \begin{bmatrix} 1 & 7/2 & 9/2 & | & 33 \\ 4 & 3 & 7 & | & 44 \\ 3 & 5 & 8 & | & 55 \end{bmatrix}$$

$$\xrightarrow[\substack{-4R_1+R_2 \\ -3R_1+R_3}]{} \begin{bmatrix} 1 & 7/2 & 9/2 & | & 33 \\ 0 & -11 & -11 & | & -88 \\ 0 & -11/2 & -11/2 & | & -44 \end{bmatrix}$$

$$\xrightarrow{(-1/11)R_2} \begin{bmatrix} 1 & 7/2 & 9/2 & | & 33 \\ 0 & 1 & 1 & | & 8 \\ 0 & -11/2 & -11/2 & | & -44 \end{bmatrix}$$

$$\xrightarrow[\substack{(-7/2)R_2+R_1 \\ (11/2)R_2+R_3}]{} \begin{bmatrix} 1 & 0 & 1 & | & 5 \\ 0 & 1 & 1 & | & 8 \\ 0 & 0 & 0 & | & 0 \end{bmatrix}.$$

Thus x = 5 - r, y = 8 - r, and z = r, where r = 0, 1, 2, 3, 4, 5. The six possible combinations are given by the following table:

	COMBINATION					
	1	2	3	4	5	6
x	5	4	3	2	1	0
y	8	7	6	5	4	3
z	0	1	2	3	4	5

(b) Computing the cost of each combination, we find that they are 4700, 4600, 4500, 4400, 4300 and 4200 dollars, respectively. Thus combination 6, namely x = 0, y = 3, z = 5, minimizes cost.

EXERCISE 6.5

1. $\begin{bmatrix} 1 & -1 & -1 & 4 & | & 5 \\ 2 & -3 & -4 & 9 & | & 13 \\ 2 & 1 & 4 & 5 & | & 1 \end{bmatrix} \rightarrow \begin{bmatrix} 1 & -1 & -1 & 4 & | & 5 \\ 0 & -1 & -2 & 1 & | & 3 \\ 0 & 3 & 6 & -3 & | & -9 \end{bmatrix}$

$\rightarrow \begin{bmatrix} 1 & -1 & -1 & 4 & | & 5 \\ 0 & 1 & 2 & -1 & | & -3 \\ 0 & 3 & 6 & -3 & | & -9 \end{bmatrix} \rightarrow \begin{bmatrix} 1 & 0 & 1 & 3 & | & 2 \\ 0 & 1 & 2 & -1 & | & -3 \\ 0 & 0 & 0 & 0 & | & 0 \end{bmatrix}.$

Thus $w = -r - 3s + 2$, $x = -2r + s - 3$, $y = r$, $z = s$
(where r and s are any real numbers).

3. $\begin{bmatrix} 3 & -1 & -3 & -1 & | & -2 \\ 2 & -2 & -6 & -6 & | & -4 \\ 2 & -1 & -3 & -2 & | & -2 \\ 3 & 1 & 3 & 7 & | & 2 \end{bmatrix} \rightarrow \begin{bmatrix} 1 & -1/3 & -1 & -1/3 & | & -2/3 \\ 2 & -2 & -6 & -6 & | & -4 \\ 2 & -1 & -3 & -2 & | & -2 \\ 3 & 1 & 3 & 7 & | & 2 \end{bmatrix}$

$\rightarrow \begin{bmatrix} 1 & -1/3 & -1 & -1/3 & | & -2/3 \\ 0 & -4/3 & -4 & -16/3 & | & -8/3 \\ 0 & -1/3 & -1 & -4/3 & | & -2/3 \\ 0 & 2 & 6 & 8 & | & 4 \end{bmatrix} \rightarrow \begin{bmatrix} 1 & -1/3 & -1 & -1/3 & | & -2/3 \\ 0 & 1 & 3 & 4 & | & 2 \\ 0 & -1/3 & -1 & -4/3 & | & -2/3 \\ 0 & 2 & 6 & 8 & | & 4 \end{bmatrix}$

$\rightarrow \begin{bmatrix} 1 & 0 & 0 & 1 & | & 0 \\ 0 & 1 & 3 & 4 & | & 2 \\ 0 & 0 & 0 & 0 & | & 0 \\ 0 & 0 & 0 & 0 & | & 0 \end{bmatrix}.$

Thus $w = -s$, $x = -3r - 4s + 2$, $y = r$, $z = s$.

5. $\begin{bmatrix} 1 & 1 & 3 & -1 & | & 2 \\ 2 & 1 & 5 & -2 & | & 0 \\ 2 & -1 & 3 & -2 & | & -8 \\ 3 & 2 & 8 & -3 & | & 2 \\ 1 & 0 & 2 & -1 & | & -2 \end{bmatrix} \rightarrow \begin{bmatrix} 1 & 1 & 3 & -1 & | & 2 \\ 0 & -1 & -1 & 0 & | & -4 \\ 0 & -3 & -3 & 0 & | & -12 \\ 0 & -1 & -1 & 0 & | & -4 \\ 0 & -1 & -1 & 0 & | & -4 \end{bmatrix}$

$\rightarrow \begin{bmatrix} 1 & 1 & 3 & -1 & | & 2 \\ 0 & 1 & 1 & 0 & | & 4 \\ 0 & -3 & -3 & 0 & | & -12 \\ 0 & -1 & -1 & 0 & | & -4 \\ 0 & -1 & -1 & 0 & | & -4 \end{bmatrix} \rightarrow \begin{bmatrix} 1 & 0 & 2 & -1 & | & -2 \\ 0 & 1 & 1 & 0 & | & 4 \\ 0 & 0 & 0 & 0 & | & 0 \\ 0 & 0 & 0 & 0 & | & 0 \\ 0 & 0 & 0 & 0 & | & 0 \end{bmatrix}.$

Thus $w = -2r + s - 2$, $x = -r + 4$, $y = r$, $z = s$.

7. $\begin{bmatrix} 4 & -3 & 5 & -10 & 11 & | & -8 \\ 2 & 1 & 5 & 0 & 3 & | & 6 \end{bmatrix} \rightarrow \begin{bmatrix} 0 & -5 & -5 & -10 & 5 & | & -20 \\ 2 & 1 & 5 & 0 & 3 & | & 6 \end{bmatrix}$

$\rightarrow \begin{bmatrix} 2 & 1 & 5 & 0 & 3 & | & 6 \\ 0 & -5 & -5 & -10 & 5 & | & -20 \end{bmatrix} \rightarrow \begin{bmatrix} 2 & 1 & 5 & 0 & 3 & | & 6 \\ 0 & 1 & 1 & 2 & -1 & | & 4 \end{bmatrix}$

$\rightarrow \begin{bmatrix} 2 & 0 & 4 & -2 & 4 & | & 2 \\ 0 & 1 & 1 & 2 & -1 & | & 4 \end{bmatrix} \rightarrow \begin{bmatrix} 1 & 0 & 2 & -1 & 2 & | & 1 \\ 0 & 1 & 1 & 2 & -1 & | & 4 \end{bmatrix}$.

Thus $x_1 = -2r + s - 2t + 1$, $x_2 = -r - 2s + t + 4$, $x_3 = r$,

$x_4 = s$, $x_5 = t$.

9. The system is homogeneous with fewer equations than unknowns (2 < 3), so there are infinitely many solutions.

11. $\begin{bmatrix} 3 & -4 \\ 1 & 5 \\ 4 & -1 \end{bmatrix} \rightarrow \begin{bmatrix} 1 & 5 \\ 3 & -4 \\ 4 & -1 \end{bmatrix} \rightarrow \begin{bmatrix} 1 & 5 \\ 0 & -19 \\ 0 & -21 \end{bmatrix} \rightarrow \begin{bmatrix} 1 & 5 \\ 0 & 1 \\ 0 & -21 \end{bmatrix} \rightarrow \begin{bmatrix} 1 & 0 \\ 0 & 1 \\ 0 & 0 \end{bmatrix} = A.$

A has k = 2 nonzero rows. Number of unknowns is n = 2. Thus k = n, so the system has the trivial solution only.

13. $\begin{bmatrix} 1 & 1 & 1 \\ 1 & 0 & -1 \\ 1 & -2 & -5 \end{bmatrix} \rightarrow \begin{bmatrix} 1 & 1 & 1 \\ 0 & -1 & -2 \\ 0 & -3 & -6 \end{bmatrix} \rightarrow \begin{bmatrix} 1 & 1 & 1 \\ 0 & 1 & 2 \\ 0 & -3 & -6 \end{bmatrix} \rightarrow \begin{bmatrix} 1 & 0 & -1 \\ 0 & 1 & 2 \\ 0 & 0 & 0 \end{bmatrix} = A.$

A has k = 2 nonzero rows. Number of unknowns is n = 3. Thus k < n $\Rightarrow$ infinitely many solutions.

15. $\begin{bmatrix} 1 & 1 \\ 3 & -4 \end{bmatrix} \rightarrow \begin{bmatrix} 1 & 1 \\ 0 & -7 \end{bmatrix} \rightarrow \begin{bmatrix} 1 & 1 \\ 0 & 1 \end{bmatrix} \rightarrow \begin{bmatrix} 1 & 0 \\ 0 & 1 \end{bmatrix}$. Thus x = 0, y = 0.

17. $\begin{bmatrix} 1 & 6 & -2 \\ 2 & -3 & 4 \end{bmatrix} \rightarrow \begin{bmatrix} 1 & 6 & -2 \\ 0 & -15 & 8 \end{bmatrix} \rightarrow \begin{bmatrix} 1 & 6 & -2 \\ 0 & 1 & -8/15 \end{bmatrix}$

$\rightarrow \begin{bmatrix} 1 & 0 & 6/5 \\ 0 & 1 & -8/15 \end{bmatrix}$. Thus $x = -\frac{6}{5}r$, $y = \frac{8}{15}r$, $z = r$.

19. $\begin{bmatrix} 1 & 1 \\ 3 & -4 \\ 5 & -8 \end{bmatrix} \rightarrow \begin{bmatrix} 1 & 1 \\ 0 & -7 \\ 0 & -13 \end{bmatrix} \rightarrow \begin{bmatrix} 1 & 1 \\ 0 & 1 \\ 0 & -13 \end{bmatrix} \rightarrow \begin{bmatrix} 1 & 0 \\ 0 & 1 \\ 0 & 0 \end{bmatrix} \Rightarrow x = 0, y = 0.$

21. $\begin{bmatrix} 1 & 1 & 1 \\ 5 & -2 & -9 \\ 3 & 1 & -1 \\ 3 & -2 & -7 \end{bmatrix} \rightarrow \begin{bmatrix} 1 & 1 & 1 \\ 0 & -7 & -14 \\ 0 & -2 & -4 \\ 0 & -5 & -10 \end{bmatrix} \rightarrow \begin{bmatrix} 1 & 1 & 1 \\ 0 & 1 & 2 \\ 0 & -2 & -4 \\ 0 & -5 & -10 \end{bmatrix} \rightarrow \begin{bmatrix} 1 & 0 & -1 \\ 0 & 1 & 2 \\ 0 & 0 & 0 \\ 0 & 0 & 0 \end{bmatrix}.$

Thus $x = r$, $y = -2r$, $z = r$.

23. $\begin{bmatrix} 1 & 1 & 1 & 4 \\ 1 & 1 & 0 & 5 \\ 2 & 1 & 3 & 4 \\ 1 & -3 & 2 & -9 \end{bmatrix} \rightarrow \begin{bmatrix} 1 & 1 & 1 & 4 \\ 0 & 0 & -1 & 1 \\ 0 & -1 & 1 & -4 \\ 0 & -4 & 1 & -13 \end{bmatrix} \rightarrow \begin{bmatrix} 1 & 1 & 1 & 4 \\ 0 & -1 & 1 & -4 \\ 0 & 0 & -1 & 1 \\ 0 & -4 & 1 & -13 \end{bmatrix}$

$\rightarrow \begin{bmatrix} 1 & 1 & 1 & 4 \\ 0 & 1 & -1 & 4 \\ 0 & 0 & -1 & 1 \\ 0 & -4 & 1 & -13 \end{bmatrix} \rightarrow \begin{bmatrix} 1 & 0 & 2 & 0 \\ 0 & 1 & -1 & 4 \\ 0 & 0 & -1 & 1 \\ 0 & 0 & -3 & 3 \end{bmatrix} \rightarrow \begin{bmatrix} 1 & 0 & 2 & 0 \\ 0 & 1 & -1 & 4 \\ 0 & 0 & 1 & -1 \\ 0 & 0 & -3 & 3 \end{bmatrix}$

$\rightarrow \begin{bmatrix} 1 & 0 & 0 & 2 \\ 0 & 1 & 0 & 3 \\ 0 & 0 & 1 & -1 \\ 0 & 0 & 0 & 0 \end{bmatrix}.$ Thus $w = -2r$, $x = -3r$, $y = r$, $z = r$.

EXERCISE 6.6

1. $\begin{bmatrix} 6 & 1 & | & 1 & 0 \\ 5 & 1 & | & 0 & 1 \end{bmatrix} \rightarrow \begin{bmatrix} 1 & 1/6 & | & 1/6 & 0 \\ 5 & 1 & | & 0 & 1 \end{bmatrix} \rightarrow \begin{bmatrix} 1 & 1/6 & | & 1/6 & 0 \\ 0 & 1/6 & | & -5/6 & 1 \end{bmatrix}$

$\rightarrow \begin{bmatrix} 1 & 0 & | & 1 & -1 \\ 0 & 1/6 & | & -5/6 & 1 \end{bmatrix} \rightarrow \begin{bmatrix} 1 & 0 & | & 1 & -1 \\ 0 & 1 & | & -5 & 6 \end{bmatrix}.$ Thus the

inverse is $\begin{bmatrix} 1 & -1 \\ -5 & 6 \end{bmatrix}.$

3. $\begin{bmatrix} 1 & 1 & | & 1 & 0 \\ 1 & 1 & | & 0 & 1 \end{bmatrix} \rightarrow \begin{bmatrix} 1 & 1 & | & 1 & 0 \\ 0 & 0 & | & -1 & 1 \end{bmatrix} \implies$ not invertible.

5. $\begin{bmatrix} 1 & 0 & 0 & | & 1 & 0 & 0 \\ 0 & -3 & 0 & | & 0 & 1 & 0 \\ 0 & 0 & 4 & | & 0 & 0 & 1 \end{bmatrix} \rightarrow \begin{bmatrix} 1 & 0 & 0 & | & 1 & 0 & 0 \\ 0 & 1 & 0 & | & 0 & -1/3 & 0 \\ 0 & 0 & 1 & | & 0 & 0 & 1/4 \end{bmatrix}.$

The inverse is $\begin{bmatrix} 1 & 0 & 0 \\ 0 & -1/3 & 0 \\ 0 & 0 & 1/4 \end{bmatrix}.$

7. $\begin{bmatrix} 1 & 2 & 3 & | & 1 & 0 & 0 \\ 0 & 0 & 4 & | & 0 & 1 & 0 \\ 0 & 0 & 5 & | & 0 & 0 & 1 \end{bmatrix} \rightarrow \begin{bmatrix} 1 & 2 & 3 & | & 1 & 0 & 0 \\ 0 & 0 & 1 & | & 0 & 1/4 & 0 \\ 0 & 0 & 5 & | & 0 & 0 & 1 \end{bmatrix}$

$\rightarrow \begin{bmatrix} 1 & 2 & 0 & | & 1 & -3/4 & 0 \\ 0 & 0 & 1 & | & 0 & 1/4 & 0 \\ 0 & 0 & 0 & | & 0 & -5/4 & 1 \end{bmatrix} \implies$ not invertible.

9. The matrix is not square, so it is not invertible.

11. $\begin{bmatrix} 1 & 1 & 1 & | & 1 & 0 & 0 \\ 0 & 1 & 1 & | & 0 & 1 & 0 \\ 0 & 0 & 1 & | & 0 & 0 & 1 \end{bmatrix} \rightarrow \begin{bmatrix} 1 & 0 & 0 & | & 1 & -1 & 0 \\ 0 & 1 & 1 & | & 0 & 1 & 0 \\ 0 & 0 & 1 & | & 0 & 0 & 1 \end{bmatrix}$

$\rightarrow \begin{bmatrix} 1 & 0 & 0 & | & 1 & -1 & 0 \\ 0 & 1 & 0 & | & 0 & 1 & -1 \\ 0 & 0 & 1 & | & 0 & 0 & 1 \end{bmatrix}$. The inverse is $\begin{bmatrix} 1 & -1 & 0 \\ 0 & 1 & -1 \\ 0 & 0 & 1 \end{bmatrix}$.

13. $\begin{bmatrix} 7 & 0 & -2 & | & 1 & 0 & 0 \\ 0 & 1 & 0 & | & 0 & 1 & 0 \\ -3 & 0 & 1 & | & 0 & 0 & 1 \end{bmatrix} \rightarrow \begin{bmatrix} 1 & 0 & -2/7 & | & 1/7 & 0 & 0 \\ 0 & 1 & 0 & | & 0 & 1 & 0 \\ -3 & 0 & 1 & | & 0 & 0 & 1 \end{bmatrix}$

$\rightarrow \begin{bmatrix} 1 & 0 & -2/7 & | & 1/7 & 0 & 0 \\ 0 & 1 & 0 & | & 0 & 1 & 0 \\ 0 & 0 & 1/7 & | & 3/7 & 0 & 1 \end{bmatrix} \rightarrow \begin{bmatrix} 1 & 0 & 0 & | & 1 & 0 & 2 \\ 0 & 1 & 0 & | & 0 & 1 & 0 \\ 0 & 0 & 1/7 & | & 3/7 & 0 & 1 \end{bmatrix}$

$\rightarrow \begin{bmatrix} 1 & 0 & 0 & | & 1 & 0 & 2 \\ 0 & 1 & 0 & | & 0 & 1 & 0 \\ 0 & 0 & 1 & | & 3 & 0 & 7 \end{bmatrix}$. The inverse is $\begin{bmatrix} 1 & 0 & 2 \\ 0 & 1 & 0 \\ 3 & 0 & 7 \end{bmatrix}$.

15. $\begin{bmatrix} 2 & 1 & 0 & | & 1 & 0 & 0 \\ 4 & -1 & 5 & | & 0 & 1 & 0 \\ 1 & -1 & 2 & | & 0 & 0 & 1 \end{bmatrix} \rightarrow \begin{bmatrix} 1 & -1 & 2 & | & 0 & 0 & 1 \\ 4 & -1 & 5 & | & 0 & 1 & 0 \\ 2 & 1 & 0 & | & 1 & 0 & 0 \end{bmatrix}$

$\rightarrow \begin{bmatrix} 1 & -1 & 2 & | & 0 & 0 & 1 \\ 0 & 3 & -3 & | & 0 & 1 & -4 \\ 0 & 3 & -4 & | & 1 & 0 & -2 \end{bmatrix} \rightarrow \begin{bmatrix} 1 & -1 & 2 & | & 0 & 0 & 1 \\ 0 & 1 & -1 & | & 0 & 1/3 & -4/3 \\ 0 & 3 & -4 & | & 1 & 0 & -2 \end{bmatrix}$

$\rightarrow \begin{bmatrix} 1 & 0 & 1 & | & 0 & 1/3 & -1/3 \\ 0 & 1 & -1 & | & 0 & 1/3 & -4/3 \\ 0 & 0 & -1 & | & 1 & -1 & 2 \end{bmatrix} \rightarrow \begin{bmatrix} 1 & 0 & 1 & | & 0 & 1/3 & -1/3 \\ 0 & 1 & -1 & | & 0 & 1/3 & -4/3 \\ 0 & 0 & 1 & | & -1 & 1 & -2 \end{bmatrix}$

$\rightarrow \begin{bmatrix} 1 & 0 & 0 & | & 1 & -2/3 & 5/3 \\ 0 & 1 & 0 & | & -1 & 4/3 & -10/3 \\ 0 & 0 & 1 & | & -1 & 1 & -2 \end{bmatrix}$. Inverse is $\begin{bmatrix} 1 & -2/3 & 5/3 \\ -1 & 4/3 & -10/3 \\ -1 & 1 & -2 \end{bmatrix}$.

17. $\begin{bmatrix} 1 & 2 & 3 & | & 1 & 0 & 0 \\ 1 & 3 & 5 & | & 0 & 1 & 0 \\ 1 & 5 & 12 & | & 0 & 0 & 1 \end{bmatrix} \rightarrow \begin{bmatrix} 1 & 2 & 3 & | & 1 & 0 & 0 \\ 0 & 1 & 2 & | & -1 & 1 & 0 \\ 0 & 3 & 9 & | & -1 & 0 & 1 \end{bmatrix}$

$\rightarrow \begin{bmatrix} 1 & 0 & -1 & | & 3 & -2 & 0 \\ 0 & 1 & 2 & | & -1 & 1 & 0 \\ 0 & 0 & 3 & | & 2 & -3 & 1 \end{bmatrix} \rightarrow \begin{bmatrix} 1 & 0 & -1 & | & 3 & -2 & 0 \\ 0 & 1 & 2 & | & -1 & 1 & 0 \\ 0 & 0 & 1 & | & 2/3 & -1 & 1/3 \end{bmatrix}$

$\rightarrow \begin{bmatrix} 1 & 0 & 0 & | & 11/3 & -3 & 1/3 \\ 0 & 1 & 0 & | & -7/3 & 3 & -2/3 \\ 0 & 0 & 1 & | & 2/3 & -1 & 1/3 \end{bmatrix}.$

The inverse is $\begin{bmatrix} 11/3 & -3 & 1/3 \\ -7/3 & 3 & -2/3 \\ 2/3 & -1 & 1/3 \end{bmatrix}.$

19. $X = \begin{bmatrix} x_1 \\ x_2 \end{bmatrix} = A^{-1}B = \begin{bmatrix} 1 & 2 \\ 1 & 1 \end{bmatrix}\begin{bmatrix} 2 \\ 4 \end{bmatrix} = \begin{bmatrix} 10 \\ 6 \end{bmatrix} \Longrightarrow x_1 = 10, \ x_2 = 6$

21. $\begin{bmatrix} 6 & 5 & | & 1 & 0 \\ 1 & 1 & | & 0 & 1 \end{bmatrix} \rightarrow \begin{bmatrix} 1 & 1 & | & 0 & 1 \\ 6 & 5 & | & 1 & 0 \end{bmatrix} \rightarrow \begin{bmatrix} 1 & 1 & | & 0 & 1 \\ 0 & -1 & | & 1 & -6 \end{bmatrix}$

$\rightarrow \begin{bmatrix} 1 & 1 & | & 0 & 1 \\ 0 & 1 & | & -1 & 6 \end{bmatrix} \rightarrow \begin{bmatrix} 1 & 0 & | & 1 & -5 \\ 0 & 1 & | & -1 & 6 \end{bmatrix}.$

$\begin{bmatrix} x \\ y \end{bmatrix} = A^{-1}B = \begin{bmatrix} 1 & -5 \\ -1 & 6 \end{bmatrix}\begin{bmatrix} 2 \\ -3 \end{bmatrix} = \begin{bmatrix} 17 \\ -20 \end{bmatrix} \Longrightarrow x = 17, \ y = -20$

23. $\begin{bmatrix} 2 & 1 & | & 1 & 0 \\ 3 & -1 & | & 0 & 1 \end{bmatrix} \rightarrow \begin{bmatrix} 1 & 1/2 & | & 1/2 & 0 \\ 3 & -1 & | & 0 & 1 \end{bmatrix} \rightarrow \begin{bmatrix} 1 & 1/2 & | & 1/2 & 0 \\ 0 & -5/2 & | & -3/2 & 1 \end{bmatrix}$

$\rightarrow \begin{bmatrix} 1 & 1/2 & | & 1/2 & 0 \\ 0 & 1 & | & 3/5 & -2/5 \end{bmatrix} \rightarrow \begin{bmatrix} 1 & 0 & | & 1/5 & 1/5 \\ 0 & 1 & | & 3/5 & -2/5 \end{bmatrix}.$

$\begin{bmatrix} x \\ y \end{bmatrix} = A^{-1}B = \begin{bmatrix} 1/5 & 1/5 \\ 3/5 & -2/5 \end{bmatrix}\begin{bmatrix} 5 \\ 0 \end{bmatrix} = \begin{bmatrix} 1 \\ 3 \end{bmatrix} \Longrightarrow x = 1, \ y = 3$

25. The coefficient matrix is not invertible. The method of reduction yields

$$\begin{bmatrix} 2 & 6 & | & 2 \\ 3 & 9 & | & 3 \end{bmatrix} \rightarrow \begin{bmatrix} 1 & 3 & | & 1 \\ 3 & 9 & | & 3 \end{bmatrix} \rightarrow \begin{bmatrix} 1 & 3 & | & 1 \\ 0 & 0 & | & 0 \end{bmatrix}.$$

Thus $x = -3r + 1, \ y = r.$

27.
$$\begin{bmatrix} 1 & 2 & 1 & | & 1 & 0 & 0 \\ 3 & 0 & 1 & | & 0 & 1 & 0 \\ 1 & -1 & 1 & | & 0 & 0 & 1 \end{bmatrix} \rightarrow \begin{bmatrix} 1 & 2 & 1 & | & 1 & 0 & 0 \\ 0 & -6 & -2 & | & -3 & 1 & 0 \\ 0 & -3 & 0 & | & -1 & 0 & 1 \end{bmatrix}$$

$$\rightarrow \begin{bmatrix} 1 & 2 & 1 & | & 1 & 0 & 0 \\ 0 & 1 & 1/3 & | & 1/2 & -1/6 & 0 \\ 0 & -3 & 0 & | & -1 & 0 & 1 \end{bmatrix} \rightarrow \begin{bmatrix} 1 & 0 & 1/3 & | & 0 & 1/3 & 0 \\ 0 & 1 & 1/3 & | & 1/2 & -1/6 & 0 \\ 0 & 0 & 1 & | & 1/2 & -1/2 & 1 \end{bmatrix}$$

$$\rightarrow \begin{bmatrix} 1 & 0 & 0 & | & -1/6 & 1/2 & -1/3 \\ 0 & 1 & 0 & | & 1/3 & 0 & -1/3 \\ 0 & 0 & 1 & | & 1/2 & -1/2 & 1 \end{bmatrix}.$$

$$\begin{bmatrix} x \\ y \\ z \end{bmatrix} = A^{-1}B = \begin{bmatrix} -1/6 & 1/2 & -1/3 \\ 1/3 & 0 & -1/3 \\ 1/2 & -1/2 & 1 \end{bmatrix} \begin{bmatrix} 4 \\ 2 \\ 1 \end{bmatrix} = \begin{bmatrix} 0 \\ 1 \\ 2 \end{bmatrix}.$$

Thus $x = 0$, $y = 1$, $z = 2$.

29.
$$\begin{bmatrix} 1 & 1 & 1 & | & 1 & 0 & 0 \\ 1 & -1 & 1 & | & 0 & 1 & 0 \\ 1 & -1 & -1 & | & 0 & 0 & 1 \end{bmatrix} \rightarrow \begin{bmatrix} 1 & 1 & 1 & | & 1 & 0 & 0 \\ 0 & -2 & 0 & | & -1 & 1 & 0 \\ 0 & -2 & -2 & | & -1 & 0 & 1 \end{bmatrix}$$

$$\rightarrow \begin{bmatrix} 1 & 1 & 1 & | & 1 & 0 & 0 \\ 0 & 1 & 0 & | & 1/2 & -1/2 & 0 \\ 0 & -2 & -2 & | & -1 & 0 & 1 \end{bmatrix} \rightarrow \begin{bmatrix} 1 & 0 & 1 & | & 1/2 & 1/2 & 0 \\ 0 & 1 & 0 & | & 1/2 & -1/2 & 0 \\ 0 & 0 & -2 & | & 0 & -1 & 1 \end{bmatrix}$$

$$\rightarrow \begin{bmatrix} 1 & 0 & 1 & | & 1/2 & 1/2 & 0 \\ 0 & 1 & 0 & | & 1/2 & -1/2 & 0 \\ 0 & 0 & 1 & | & 0 & 1/2 & -1/2 \end{bmatrix} \rightarrow \begin{bmatrix} 1 & 0 & 0 & | & 1/2 & 0 & 1/2 \\ 0 & 1 & 0 & | & 1/2 & -1/2 & 0 \\ 0 & 0 & 1 & | & 0 & 1/2 & -1/2 \end{bmatrix}.$$

$$\begin{bmatrix} x \\ y \\ z \end{bmatrix} = A^{-1}B = \begin{bmatrix} 1/2 & 0 & 1/2 \\ 1/2 & -1/2 & 0 \\ 0 & 1/2 & -1/2 \end{bmatrix} \begin{bmatrix} 2 \\ 1 \\ 0 \end{bmatrix} = \begin{bmatrix} 1 \\ 1/2 \\ 1/2 \end{bmatrix}.$$

Thus $x = 1$, $y = 1/2$, $z = 1/2$.

31. The coefficient matrix is not invertible. The method of reduction yields

$$\begin{bmatrix} 1 & 3 & 3 & | & 7 \\ 2 & 1 & 1 & | & 4 \\ 1 & 1 & 1 & | & 4 \end{bmatrix} \rightarrow \begin{bmatrix} 1 & 3 & 3 & | & 7 \\ 0 & -5 & -5 & | & -10 \\ 0 & -2 & -2 & | & -3 \end{bmatrix} \rightarrow \begin{bmatrix} 1 & 3 & 3 & | & 7 \\ 0 & 1 & 1 & | & 2 \\ 0 & -2 & -2 & | & -3 \end{bmatrix}$$

$$\rightarrow \begin{bmatrix} 1 & 0 & 0 & | & 1 \\ 0 & 1 & 1 & | & 2 \\ 0 & 0 & 0 & | & 1 \end{bmatrix}.$$
The third row indicates that $0 = 1$, which is never true, so there is no solution.

33.
$$\left[\begin{array}{cccc|cccc} 1 & 0 & 2 & 1 & 1 & 0 & 0 & 0 \\ 1 & -1 & 0 & 2 & 0 & 1 & 0 & 0 \\ 2 & 1 & 0 & 1 & 0 & 0 & 1 & 0 \\ 1 & 2 & 1 & 1 & 0 & 0 & 0 & 1 \end{array}\right] \rightarrow \left[\begin{array}{cccc|cccc} 1 & 0 & 2 & 1 & 1 & 0 & 0 & 0 \\ 0 & -1 & -2 & 1 & -1 & 1 & 0 & 0 \\ 0 & 1 & -4 & -1 & -2 & 0 & 1 & 0 \\ 0 & 2 & -1 & 0 & -1 & 0 & 0 & 1 \end{array}\right]$$

$$\rightarrow \left[\begin{array}{cccc|cccc} 1 & 0 & 2 & 1 & 1 & 0 & 0 & 0 \\ 0 & 1 & 2 & -1 & 1 & -1 & 0 & 0 \\ 0 & 0 & -6 & 0 & -3 & 1 & 1 & 0 \\ 0 & 0 & -5 & 2 & -3 & 2 & 0 & 1 \end{array}\right]$$

$$\rightarrow \left[\begin{array}{cccc|cccc} 1 & 0 & 2 & 1 & 1 & 0 & 0 & 0 \\ 0 & 1 & 2 & -1 & 1 & -1 & 0 & 0 \\ 0 & 0 & 1 & 0 & 1/2 & -1/6 & -1/6 & 0 \\ 0 & 0 & -5 & 2 & -3 & 2 & 0 & 1 \end{array}\right]$$

$$\rightarrow \left[\begin{array}{cccc|cccc} 1 & 0 & 0 & 1 & 0 & 1/3 & 1/3 & 0 \\ 0 & 1 & 0 & -1 & 0 & -2/3 & 1/3 & 0 \\ 0 & 0 & 1 & 0 & 1/2 & -1/6 & -1/6 & 0 \\ 0 & 0 & 0 & 2 & -1/2 & 7/6 & -5/6 & 1 \end{array}\right]$$

$$\rightarrow \left[\begin{array}{cccc|cccc} 1 & 0 & 0 & 1 & 0 & 1/3 & 1/3 & 0 \\ 0 & 1 & 0 & -1 & 0 & -2/3 & 1/3 & 0 \\ 0 & 0 & 1 & 0 & 1/2 & -1/6 & -1/6 & 0 \\ 0 & 0 & 0 & 1 & -1/4 & 7/12 & -5/12 & 1/2 \end{array}\right]$$

$$\rightarrow \left[\begin{array}{cccc|cccc} 1 & 0 & 0 & 0 & 1/4 & -1/4 & 3/4 & -1/2 \\ 0 & 1 & 0 & 0 & -1/4 & -1/12 & -1/12 & 1/2 \\ 0 & 0 & 1 & 0 & 1/2 & -1/6 & -1/6 & 0 \\ 0 & 0 & 0 & 1 & -1/4 & 7/12 & -5/12 & 1/2 \end{array}\right].$$

$$\begin{bmatrix} w \\ x \\ y \\ z \end{bmatrix} = A^{-1}B = \begin{bmatrix} 1/4 & -1/4 & 3/4 & -1/2 \\ -1/4 & -1/12 & -1/12 & 1/2 \\ 1/2 & -1/6 & -1/6 & 0 \\ -1/4 & 7/12 & -5/12 & 1/2 \end{bmatrix} \begin{bmatrix} 4 \\ 12 \\ 12 \\ 12 \end{bmatrix} = \begin{bmatrix} 1 \\ 3 \\ -2 \\ 7 \end{bmatrix}.$$

Thus $w = 1$, $x = 3$, $y = -2$, $z = 7$.

35. $I - A = \begin{bmatrix} 1 & 0 \\ 0 & 1 \end{bmatrix} - \begin{bmatrix} 2 & -1 \\ 1 & 3 \end{bmatrix} = \begin{bmatrix} -1 & 1 \\ -1 & -2 \end{bmatrix}.$

$$\left[\begin{array}{cc|cc} -1 & 1 & 1 & 0 \\ -1 & -2 & 0 & 1 \end{array}\right] \rightarrow \left[\begin{array}{cc|cc} 1 & -1 & -1 & 0 \\ -1 & -2 & 0 & 1 \end{array}\right] \rightarrow \left[\begin{array}{cc|cc} 1 & -1 & -1 & 0 \\ 0 & -3 & -1 & 1 \end{array}\right]$$

$$\rightarrow \left[\begin{array}{cc|cc} 1 & -1 & -1 & 0 \\ 0 & 1 & 1/3 & -1/3 \end{array}\right] \rightarrow \left[\begin{array}{cc|cc} 1 & 0 & -2/3 & -1/3 \\ 0 & 1 & 1/3 & -1/3 \end{array}\right].$$

Thus $(I - A)^{-1} = \begin{bmatrix} -2/3 & -1/3 \\ 1/3 & -1/3 \end{bmatrix}.$

37. Let x = number of first model and y = number of second model.

(a) The system is
$$\begin{cases} x + y = 100 & \text{(painting)}, \\ \frac{1}{2}x + y = 80 & \text{(polishing)}. \end{cases}$$

Let $A = \begin{bmatrix} 1 & 1 \\ 1/2 & 1 \end{bmatrix}$.

$$\begin{bmatrix} 1 & 1 & | & 1 & 0 \\ 1/2 & 1 & | & 0 & 1 \end{bmatrix} \rightarrow \begin{bmatrix} 1 & 1 & | & 1 & 0 \\ 0 & 1/2 & | & -1/2 & 1 \end{bmatrix}$$

$$\rightarrow \begin{bmatrix} 1 & 1 & | & 1 & 0 \\ 0 & 1 & | & -1 & 2 \end{bmatrix} \rightarrow \begin{bmatrix} 1 & 0 & | & 2 & -2 \\ 0 & 1 & | & -1 & 2 \end{bmatrix}.$$

$$\begin{bmatrix} x \\ y \end{bmatrix} = A^{-1} \begin{bmatrix} 100 \\ 80 \end{bmatrix} = \begin{bmatrix} 2 & -2 \\ -1 & 2 \end{bmatrix} \begin{bmatrix} 100 \\ 80 \end{bmatrix} = \begin{bmatrix} 40 \\ 60 \end{bmatrix}.$$

Thus 40 of first model and 60 of second model.

(b) The system is
$$\begin{cases} 10x + 7y = 800 & \text{(widgets)}, \\ 14x + 10y = 1130 & \text{(shims)}. \end{cases}$$

Let $A = \begin{bmatrix} 10 & 7 \\ 14 & 10 \end{bmatrix}$.

$$\begin{bmatrix} 10 & 7 & | & 1 & 0 \\ 14 & 10 & | & 0 & 1 \end{bmatrix} \rightarrow \begin{bmatrix} 1 & 7/10 & | & 1/10 & 0 \\ 14 & 10 & | & 0 & 1 \end{bmatrix}$$

$$\rightarrow \begin{bmatrix} 1 & 7/10 & | & 1/10 & 0 \\ 0 & 1/5 & | & -7/5 & 1 \end{bmatrix} \rightarrow \begin{bmatrix} 1 & 7/10 & | & 1/10 & 0 \\ 0 & 1 & | & -7 & 5 \end{bmatrix}$$

$$\rightarrow \begin{bmatrix} 1 & 0 & | & 5 & -7/2 \\ 0 & 1 & | & -7 & 5 \end{bmatrix}.$$

$$\begin{bmatrix} x \\ y \end{bmatrix} = A^{-1} \begin{bmatrix} 800 \\ 1130 \end{bmatrix} = \begin{bmatrix} 5 & -7/2 \\ -7 & 5 \end{bmatrix} \begin{bmatrix} 800 \\ 1130 \end{bmatrix} = \begin{bmatrix} 45 \\ 50 \end{bmatrix}.$$

Thus 45 of first model and 50 of second model.

39. (a) $(B^{-1}A^{-1})(AB) = B^{-1}(A^{-1}A)B = B^{-1}IB = B^{-1}B = I.$ Since an invertible matrix has exactly one inverse, $B^{-1}A^{-1}$ is the inverse of AB.

(b) From part (a),
$$(AB)^{-1} = B^{-1}A^{-1} = \begin{bmatrix} 1 & 1 \\ 1 & 2 \end{bmatrix} \begin{bmatrix} 1 & 2 \\ 3 & 4 \end{bmatrix} = \begin{bmatrix} 4 & 6 \\ 7 & 10 \end{bmatrix}.$$

41. $P^T P = \begin{bmatrix} 3/5 & 4/5 \\ -4/5 & 3/5 \end{bmatrix} \begin{bmatrix} 3/5 & -4/5 \\ 4/5 & 3/5 \end{bmatrix} = \begin{bmatrix} 1 & 0 \\ 0 & 1 \end{bmatrix} = I$, so $P^T = P^{-1}$.

Yes, P is orthogonal.

EXERCISE 6.7

1. $2(2) - 1(3) = 4 - 3 = 1$

3. $(-2)(-6) - (-3)(-4) = 12 - 12 = 0$

5. $1(y) - x(0) = y$

7. $\frac{1(4) - 2(3)}{2(6) - 1(5)} = \frac{-2}{7} = -\frac{2}{7}$

9. $2(k) - 3(4) = 2k - 12 = 12$, $2k = 24$, $k = 12$.

11. $\begin{vmatrix} 1 & 3 \\ 7 & 9 \end{vmatrix} = 1(9) - 3(7) = -12$

13. $(-1)^{3+2} \begin{vmatrix} 1 & 3 \\ 4 & 6 \end{vmatrix} = -[1(6) - 3(4)] = -[-6] = 6$

15. $\begin{vmatrix} a_{11} & a_{13} & a_{14} \\ a_{21} & a_{23} & a_{24} \\ a_{41} & a_{43} & a_{44} \end{vmatrix}$

17. $(-1)^{1+3} \begin{vmatrix} a_{21} & a_{22} & a_{24} \\ a_{31} & a_{32} & a_{34} \\ a_{41} & a_{42} & a_{44} \end{vmatrix} = (-1)^4 \cdot \begin{vmatrix} a_{21} & a_{22} & a_{24} \\ a_{31} & a_{32} & a_{34} \\ a_{41} & a_{42} & a_{44} \end{vmatrix}$

$= \begin{vmatrix} a_{21} & a_{22} & a_{24} \\ a_{31} & a_{32} & a_{34} \\ a_{41} & a_{42} & a_{44} \end{vmatrix}$

19. Expanding along column 2 gives

$1(-1)^3 \begin{vmatrix} 2 & 1 \\ -4 & 6 \end{vmatrix} + 0 + 0 = 1(-1)[12 - (-4)] = -16$.

21. Expanding along row 1 gives

$$1(-1)^2 \begin{vmatrix} 5 & 4 \\ -2 & 1 \end{vmatrix} + 2(-1)^3 \begin{vmatrix} 4 & 4 \\ 3 & 1 \end{vmatrix} - 3(-1)^4 \begin{vmatrix} 4 & 5 \\ 3 & -2 \end{vmatrix}$$

$$= 1(1)[5 - (-8)] + 2(-1)[4 - 12] - 3(1)[-8 - 15]$$

$$= 13 + 16 + 69 = 98.$$

23. Expanding along row 3 gives

$$0 + 6(-1)^5 \begin{vmatrix} 2 & 5 \\ -3 & -1 \end{vmatrix} - 1(-1)^6 \begin{vmatrix} 2 & 1 \\ -3 & 4 \end{vmatrix}$$

$$= 0 + 6(-1)[-2 - (-15)] - 1(1)[8 - (-3)]$$

$$= 0 - 78 - 11 = -89.$$

25. Expanding along row 1 gives

$$2(-1)^2 \begin{vmatrix} 1 & -1 \\ 2 & -3 \end{vmatrix} - 1(-1)^3 \begin{vmatrix} 1 & -1 \\ 1 & -3 \end{vmatrix} + 3(-1)^4 \begin{vmatrix} 1 & 1 \\ 1 & 2 \end{vmatrix}$$

$$= 2(1)[-3 - (-2)] - 1(-1)[-3 - (-1)] + 3(1)[2 - 1]$$

$$= -2 - 2 + 3 = -1.$$

27. Expanding along row 1 gives

$$\tfrac{1}{2}(-1)^2 \begin{vmatrix} 1/3 & 2/3 \\ -4 & 1 \end{vmatrix} + \tfrac{2}{3}(-1)^3 \begin{vmatrix} -1 & 2/3 \\ 3 & 1 \end{vmatrix} - \tfrac{1}{2}(-1)^4 \begin{vmatrix} -1 & 1/3 \\ 3 & -4 \end{vmatrix}$$

$$= \tfrac{1}{2}(1)\left[\tfrac{1}{3} - \left(-\tfrac{8}{3}\right)\right] + \tfrac{2}{3}(-1)[-1 - 2] - \tfrac{1}{2}(1)[4 - 1]$$

$$= \tfrac{3}{2} + 2 - \tfrac{3}{2} = 2.$$

29. Expanding along column 3 gives

$$3(-1)^4 \begin{vmatrix} 4 & -1 & 1 \\ 2 & 1 & 3 \\ -1 & 2 & -1 \end{vmatrix} + 0 + 0 + 3(-1)^7 \begin{vmatrix} 1 & 0 & 2 \\ 4 & -1 & 1 \\ 2 & 1 & 3 \end{vmatrix}$$

$$= 3\left\{ 4(-1)^2 \begin{vmatrix} 1 & 3 \\ 2 & -1 \end{vmatrix} - 1(-1)^3 \begin{vmatrix} 2 & 3 \\ -1 & -1 \end{vmatrix} + 1(-1)^4 \begin{vmatrix} 2 & 1 \\ -1 & 2 \end{vmatrix} \right\} -$$

$$3\left\{ 1(-1)^2 \begin{vmatrix} -1 & 1 \\ 1 & 3 \end{vmatrix} + 0 + 2(-1)^4 \begin{vmatrix} 4 & -1 \\ 2 & 1 \end{vmatrix} \right\}$$

$$= 3\{4(1)[-1-6] - 1(-1)[-2-(-3)] + 1(1)[4-(-1)]\} -$$
$$3\{1(1)[-3-1] + 2(1)[4-(-2)]\}$$

$$= 3\{-28 + 1 + 5\} - 3\{-4 + 12\} = -66 - 24 = -90.$$

31. Since all entries below the main diagonal are zeros, the determinant is $(1)(1)(1)(1) = 1$.

33. Since all entries below (as well as above) main diagonal are zeros, the determinant is $(1)(-2)(4)(-3) = 24$.

35. $\begin{vmatrix} 1 & -1 & 2 & -1 \\ 2 & 3 & -1 & 2 \\ 1 & 4 & -3 & 3 \\ 4 & 1 & 3 & 0 \end{vmatrix} \overset{R_3+R_1}{=} \begin{vmatrix} 2 & 3 & -1 & 1 \\ 2 & 3 & -1 & 2 \\ 1 & 4 & -3 & 3 \\ 4 & 1 & 3 & 0 \end{vmatrix} = 0$ since row 1 and row two are identical

37. Multiplying column 2 by 2 and adding the result to column 5 gives all zeros. Thus the determinant is 0.

39. $\begin{vmatrix} x & -2 \\ 7 & 7-x \end{vmatrix} = 26$, $x(7 - x) - (-2)(7) = 26$, $7x - x^2 + 14 = 26$,

$x^2 - 7x + 12 = 26$, $(x-3)(x-4) = 0 \implies x = 3$ or $x = 4$

41. Multiplying each entry of A by 2 gives 2A. Since A has order 4, property 7 gives $|2A| = 2^4|A| = 16(12) = 192$.

43. (a) Since $A^{-1}A = I$, we have $|A^{-1}A| = |I| \implies |A^{-1}||A| = 1$.
    Thus $|A^{-1}| = \frac{1}{|A|}$.

    (b) From part (a), if $|A| = 3$, then $|A^{-1}| = \frac{1}{|A|} = \frac{1}{3}$.

45. The given system is equivalent to the system
$$\begin{cases} x + 3y + 2z = 0, \\ x + cy + 4z = 0, \\ \quad\quad 2y + cz = 0. \end{cases}$$

For the system to have infinitely many solutions, the coefficient matrix must not be invertible; that is, its determinant must be zero.

$\begin{vmatrix} 1 & 3 & 2 \\ 1 & c & 4 \\ 0 & 2 & c \end{vmatrix} = 0 \implies (c^2 + 4 + 0) - (0 + 8 + 3c) = 0$
$\implies c^2 - 3c - 4 = 0$, $(c + 1)(c - 4) = 0$
$\implies c = -1$ or $c = 4$

EXERCISE 6.8

1. $\Delta = \begin{vmatrix} 2 & -1 \\ 3 & 1 \end{vmatrix} = 5$, $\Delta_x = \begin{vmatrix} 4 & -1 \\ 5 & 1 \end{vmatrix} = 9$, $\Delta_y = \begin{vmatrix} 2 & 4 \\ 3 & 5 \end{vmatrix} = -2$.

Thus $x = \frac{\Delta_x}{\Delta} = \frac{9}{5}$, $y = \frac{\Delta_y}{\Delta} = -\frac{2}{5}$.

3. $\begin{cases} -2x = 4 - 3y \\ y = 6x - 1 \end{cases}$ is equivalent to $\begin{cases} -2x + 3y = 4 \\ -6x + y = -1. \end{cases}$

$\Delta = \begin{vmatrix} -2 & 3 \\ -6 & 1 \end{vmatrix} = 16$, $\Delta_x = \begin{vmatrix} 4 & 3 \\ -1 & 1 \end{vmatrix} = 7$, $\Delta_y = \begin{vmatrix} -2 & 4 \\ -6 & -1 \end{vmatrix} = 26$.

Thus $x = \frac{\Delta_x}{\Delta} = \frac{7}{16}$, $y = \frac{\Delta_y}{\Delta} = \frac{26}{16} = \frac{13}{8}$.

5. $\begin{cases} 3(x + 2) = 5 \\ 6(x + y) = -8 \end{cases}$ is equivalent to $\begin{cases} 3x = -1 \\ 6x + 6y = -8. \end{cases}$

$\Delta = \begin{vmatrix} 3 & 0 \\ 6 & 6 \end{vmatrix} = 18$, $\Delta_x = \begin{vmatrix} -1 & 0 \\ -8 & 6 \end{vmatrix} = -6$, $\Delta_y = \begin{vmatrix} 3 & -1 \\ 6 & -8 \end{vmatrix} = -18$.

Thus $x = \Delta_x/\Delta = -6/18 = -1/3$, $y = \Delta_y/\Delta = -18/18 = -1$.

7. $\Delta = \begin{vmatrix} 3/2 & -1/4 \\ 1/3 & 1/2 \end{vmatrix} = \frac{5}{6}$,

$\Delta_x = \begin{vmatrix} 1 & -1/4 \\ 2 & 1/2 \end{vmatrix} = 1$, $\Delta_z = \begin{vmatrix} 3/2 & 1 \\ 1/3 & 2 \end{vmatrix} = \frac{8}{3}$.

Thus $x = \Delta_x/\Delta = 1/(5/6) = 6/5$,

$z = \Delta_z/\Delta = (8/3)/(5/6) = 16/5$.

9. $\Delta = \begin{vmatrix} 1 & 1 & 1 \\ 1 & -1 & 1 \\ 2 & -1 & 3 \end{vmatrix} = -2$, $\Delta_x = \begin{vmatrix} 6 & 1 & 1 \\ 2 & -1 & 1 \\ 6 & -1 & 3 \end{vmatrix} = -8$,

$\Delta_y = \begin{vmatrix} 1 & 6 & 1 \\ 1 & 2 & 1 \\ 2 & 6 & 3 \end{vmatrix} = -4$, $\Delta_z = \begin{vmatrix} 1 & 1 & 6 \\ 1 & -1 & 2 \\ 2 & -1 & 6 \end{vmatrix} = 0$.

Thus $x = \Delta_x/\Delta = -8/(-2) = 4$,

$y = \Delta_y/\Delta = -4/(-2) = 2$,

$z = \Delta_z/\Delta = 0/(-2) = 0$.

11. $\Delta = \begin{vmatrix} 2 & -3 & 4 \\ 1 & 1 & -3 \\ 3 & 2 & -1 \end{vmatrix} = 30$, $\Delta_x = \begin{vmatrix} 0 & -3 & 4 \\ 4 & 1 & -3 \\ 0 & 2 & -1 \end{vmatrix} = 20$,

$\Delta_y = \begin{vmatrix} 2 & 0 & 4 \\ 1 & 4 & -3 \\ 3 & 0 & -1 \end{vmatrix} = -56$, $\Delta_z = \begin{vmatrix} 2 & -3 & 0 \\ 1 & 1 & 4 \\ 3 & 2 & 0 \end{vmatrix} = -52$.

Thus $x = \Delta_x/\Delta = 20/30 = 2/3$,

$y = \Delta_y/\Delta = -56/30 = -28/15$,

$z = \Delta_z/\Delta = -52/30 = -26/15$.

13. $\Delta = \begin{vmatrix} 1 & -2 & 1 \\ 2 & 1 & 2 \\ 1 & 8 & 1 \end{vmatrix} = 0$, so Cramer's rule does not apply. We

solve the system by matrix reduction.

$\begin{bmatrix} 1 & -2 & 1 & | & 3 \\ 2 & 1 & 2 & | & 6 \\ 1 & 8 & 1 & | & 3 \end{bmatrix} \rightarrow \begin{bmatrix} 1 & -2 & 1 & | & 3 \\ 0 & 5 & 0 & | & 0 \\ 0 & 10 & 0 & | & 0 \end{bmatrix}$

$\rightarrow \begin{bmatrix} 1 & -2 & 1 & | & 3 \\ 0 & 1 & 0 & | & 0 \\ 0 & 0 & 0 & | & 0 \end{bmatrix}$

$\rightarrow \begin{bmatrix} 1 & 0 & 1 & | & 3 \\ 0 & 1 & 0 & | & 0 \\ 0 & 0 & 0 & | & 0 \end{bmatrix}$.

Thus $x = 3 - r$, $y = 0$, $z = r$.

15. $\Delta = \begin{vmatrix} 2 & -3 & 1 \\ 1 & -6 & 3 \\ 3 & 3 & -2 \end{vmatrix} = -6$, $\Delta_x = \begin{vmatrix} -2 & -3 & 1 \\ -2 & -6 & 3 \\ 2 & 3 & -2 \end{vmatrix} = -6$,

$\Delta_y = \begin{vmatrix} 2 & -2 & 1 \\ 1 & -2 & 3 \\ 3 & 2 & -2 \end{vmatrix} = -18$, $\Delta_z = \begin{vmatrix} 2 & -3 & -2 \\ 1 & -6 & -2 \\ 3 & 3 & 2 \end{vmatrix} = -30$.

Thus $x = \Delta_x/\Delta = -6/(-6) = 1$,

$y = \Delta_y/\Delta = -18/(-6) = 3$,

$z = \Delta_z/\Delta = -30/(-6) = 5$.

17. $\Delta = \begin{vmatrix} 1 & -1 & 3 & 1 \\ 1 & 2 & 0 & -3 \\ 2 & 3 & 6 & 1 \\ 1 & 1 & 1 & 1 \end{vmatrix}$

$= 1(-1)^3 \begin{vmatrix} -1 & 3 & 1 \\ 3 & 6 & 1 \\ 1 & 1 & 1 \end{vmatrix} + 2(-1)^4 \begin{vmatrix} 1 & 3 & 1 \\ 2 & 6 & 1 \\ 1 & 1 & 1 \end{vmatrix} + 0 - 3(-1)^6 \begin{vmatrix} 1 & -1 & 3 \\ 2 & 3 & 6 \\ 1 & 1 & 1 \end{vmatrix}$

$= 1(-1)(-14) + 2(1)(-2) - 3(1)(-10) = 14 - 4 + 30 = 40.$

$\Delta_y = \begin{vmatrix} 1 & -14 & 3 & 1 \\ 1 & 12 & 0 & -3 \\ 2 & 1 & 6 & 1 \\ 1 & 6 & 1 & 1 \end{vmatrix}$

$= 3(-1)^4 \begin{vmatrix} 1 & 12 & -3 \\ 2 & 1 & 1 \\ 1 & 6 & 1 \end{vmatrix} + 0 + 6(-1)^6 \begin{vmatrix} 1 & -14 & 1 \\ 1 & 12 & -3 \\ 1 & 6 & 1 \end{vmatrix} + 1(-1)^7 \begin{vmatrix} 1 & -14 & 1 \\ 1 & 12 & -3 \\ 2 & 1 & 1 \end{vmatrix}$

$= 3(1)(-50) + 6(1)(80) + 1(-1)(90) = -150 + 480 - 90 = 240.$

$\Delta_w = \begin{vmatrix} 1 & -1 & 3 & -14 \\ 1 & 2 & 0 & 12 \\ 2 & 3 & 6 & 1 \\ 1 & 1 & 1 & 6 \end{vmatrix}$

$= 1(-1)^3 \begin{vmatrix} -1 & 3 & -14 \\ 3 & 6 & 1 \\ 1 & 1 & 6 \end{vmatrix} + 2(-1)^4 \begin{vmatrix} 1 & 3 & -14 \\ 2 & 6 & 1 \\ 1 & 1 & 6 \end{vmatrix} + 0 + 12(-1)^6 \begin{vmatrix} 1 & -1 & 3 \\ 2 & 3 & 6 \\ 1 & 1 & 1 \end{vmatrix}$

$= 1(-1)(-44) + 2(1)(58) + 12(1)(-10) = 44 + 116 - 120 = 40.$

Thus $y = \Delta_y/\Delta = 240/40 = 6, \quad w = \Delta_w/\Delta = 40/40 = 1.$

19. $\begin{cases} 2 - y = x \\ 3 + x = -y \end{cases}$ is equivalent to $\begin{cases} x + y = 2 \\ x + y = -3. \end{cases}$

Since $\Delta = \begin{vmatrix} 1 & 1 \\ 1 & 1 \end{vmatrix} = 0$, Cramer's rule does not apply.

However, the equations represent distinct parallel lines, so no solution exists.

21. Let c, h, t denote the number of concerts, hockey games, and theater productions she should attend, respectively. Then we are given that $c + h + t = 24$, $c = 2h$, and $c = \dfrac{h + t}{2}$. Simplifying gives the following system:

Exercise 6.8

$$\begin{cases} c + h + t = 24, \\ c - 2h = 0, \\ 2c - h - t = 0. \end{cases}$$

$$\Delta = \begin{vmatrix} 1 & 1 & 1 \\ 1 & -2 & 0 \\ 2 & -1 & -1 \end{vmatrix} = 6, \quad \Delta_h = \begin{vmatrix} 1 & 24 & 1 \\ 1 & 0 & 0 \\ 2 & 0 & -1 \end{vmatrix} = 24.$$

$h = \Delta_h/\Delta = 24/6 = 4$ hockey games.

EXERCISE 6.9

1. $[c_{ij}] = \begin{bmatrix} 2 & -1 \\ 2 & 3 \end{bmatrix}$. adj $A = [c_{ij}]^T = \begin{bmatrix} 2 & 2 \\ -1 & 3 \end{bmatrix}$. $|A| = 8$.

$A^{-1} = \frac{1}{|A|}$ adj $A = \frac{1}{8}\begin{bmatrix} 2 & 2 \\ -1 & 3 \end{bmatrix} = \begin{bmatrix} 1/4 & 1/4 \\ -1/8 & 3/8 \end{bmatrix}$.

3. $[c_{ij}] = \begin{bmatrix} 1/6 & 0 \\ -3/8 & 1/4 \end{bmatrix}$. adj $A = [c_{ij}]^T = \begin{bmatrix} 1/6 & -3/8 \\ 0 & 1/4 \end{bmatrix}$.

$|A| = \frac{1}{24}$. $A^{-1} = \frac{1}{|A|}$ adj $A = 24\begin{bmatrix} 1/6 & -3/8 \\ 0 & 1/4 \end{bmatrix} = \begin{bmatrix} 4 & -9 \\ 0 & 6 \end{bmatrix}$.

5. $[c_{ij}] = \begin{bmatrix} 7 & -4 & 1 \\ -8 & 5 & -1 \\ 5 & -3 & 1 \end{bmatrix}$. adj $A = [c_{ij}]^T = \begin{bmatrix} 7 & -8 & 5 \\ -4 & 5 & -3 \\ 1 & -1 & 1 \end{bmatrix}$.

$|A| = 1$. $A^{-1} = \frac{1}{|A|}$ adj $A = 1\begin{bmatrix} 7 & -8 & 5 \\ -4 & 5 & -3 \\ 1 & -1 & 1 \end{bmatrix} = \begin{bmatrix} 7 & -8 & 5 \\ -4 & 5 & -3 \\ 1 & -1 & 1 \end{bmatrix}$.

7. $[c_{ij}] = \begin{bmatrix} 2/3 & 4/3 & 1/3 \\ 1/3 & -1/3 & -1/3 \\ 0 & 5/3 & 2/3 \end{bmatrix}$.

adj $A = [c_{ij}]^T = \begin{bmatrix} 2/3 & 1/3 & 0 \\ 4/3 & -1/3 & 5/3 \\ 1/3 & -1/3 & 2/3 \end{bmatrix}$. $|A| = \frac{1}{3}$.

$A^{-1} = \frac{1}{|A|}$ adj $A = 3\begin{bmatrix} 2/3 & 1/3 & 0 \\ 4/3 & -1/3 & 5/3 \\ 1/3 & -1/3 & 2/3 \end{bmatrix} = \begin{bmatrix} 2 & 1 & 0 \\ 4 & -1 & 5 \\ 1 & -1 & 2 \end{bmatrix}$.

9. $[c_{ij}] = \begin{bmatrix} -1/4 & 0 & -1/4 \\ 1/8 & -1/4 & -1/8 \\ -3/8 & 0 & -1/8 \end{bmatrix}$.

adj $A = [c_{ij}]^T = \begin{bmatrix} -1/4 & 1/8 & -3/8 \\ 0 & -1/4 & 0 \\ -1/4 & -1/8 & -1/8 \end{bmatrix}$. $|A| = -\frac{1}{8}$.

$A^{-1} = \frac{1}{|A|}$ adj $A = -8 \begin{bmatrix} -1/4 & 1/8 & -3/8 \\ 0 & -1/4 & 0 \\ -1/4 & -1/8 & -1/8 \end{bmatrix} = \begin{bmatrix} 2 & -1 & 3 \\ 0 & 2 & 0 \\ 2 & 1 & 1 \end{bmatrix}$.

11. $[c_{ij}] = \begin{bmatrix} 1/15 & 0 & 1/15 \\ 2/15 & 1/15 & -1/15 \\ -1/15 & 4/15 & 2/15 \end{bmatrix}$.

adj $A = [c_{ij}]^T = \begin{bmatrix} 1/15 & 2/15 & -1/15 \\ 0 & 1/15 & 4/15 \\ 1/15 & -1/15 & 2/15 \end{bmatrix}$. $|A| = \frac{1}{15}$.

$A^{-1} = \frac{1}{|A|}$ adj $A = 15 \begin{bmatrix} 1/15 & 2/15 & -1/15 \\ 0 & 1/15 & 4/15 \\ 1/15 & -1/15 & 2/15 \end{bmatrix} = \begin{bmatrix} 1 & 2 & -1 \\ 0 & 1 & 4 \\ 1 & -1 & 2 \end{bmatrix}$.

13. $|A| = 3$.

$[c_{ij}] = \begin{bmatrix} 1 & -2 & -3 \\ 0 & 3 & 3 \\ 1 & -8 & -6 \end{bmatrix}$.   adj $A = [c_{ij}]^T = \begin{bmatrix} 1 & 0 & 1 \\ -2 & 3 & -8 \\ -3 & 3 & -6 \end{bmatrix}$.

$A^{-1} = \frac{1}{|A|}$ adj $A = \frac{1}{3} \begin{bmatrix} 1 & 0 & 1 \\ -2 & 3 & -8 \\ -3 & 3 & -6 \end{bmatrix} = \begin{bmatrix} 1/3 & 0 & 1/3 \\ -2/3 & 1 & -8/3 \\ -1 & 1 & -2 \end{bmatrix}$.

$X = A^{-1}B = \begin{bmatrix} 1/3 & 0 & 1/3 \\ -2/3 & 1 & -8/3 \\ -1 & 1 & -2 \end{bmatrix} \begin{bmatrix} 2 \\ 1 \\ 1 \end{bmatrix} = \begin{bmatrix} 1 \\ -3 \\ -3 \end{bmatrix}$.

Thus $x = 1$, $y = -3$, $z = -3$

15. (a) $|A| = \begin{vmatrix} 1 & 2 & 1 & 3 \\ 1 & 3 & -1 & 2 \\ -2 & 4 & 1 & 2 \\ -1 & 0 & 4 & 2 \end{vmatrix} = \begin{vmatrix} 1 & 2 & 1 & 3 \\ 0 & 1 & -2 & -1 \\ 0 & 8 & 3 & 8 \\ 0 & 2 & 5 & 5 \end{vmatrix}$

$= 1 \cdot (-1)^2 \begin{vmatrix} 1 & -2 & -1 \\ 8 & 3 & 8 \\ 2 & 5 & 5 \end{vmatrix} = -11$

(b) If $A^{-1} = [d_{ij}]_{4 \times 4}$, then the entries in the second row of A^{-1} are

$$d_{21} = \frac{1}{|A|}c_{12} = \frac{1}{-11}(-1)^3 \begin{vmatrix} 1 & -1 & 2 \\ -2 & 1 & 2 \\ -1 & 4 & 2 \end{vmatrix} = \frac{1}{11}(-22) = -2.$$

$$d_{22} = \frac{1}{|A|}c_{22} = \frac{1}{-11}(-1)^4 \begin{vmatrix} 1 & 1 & 3 \\ -2 & 1 & 2 \\ -1 & 4 & 2 \end{vmatrix} = -\frac{1}{11}(-25) = \frac{25}{11}.$$

$$d_{23} = \frac{1}{|A|}c_{32} = \frac{1}{-11}(-1)^5 \begin{vmatrix} 1 & 1 & 3 \\ 1 & -1 & 2 \\ -1 & 4 & 2 \end{vmatrix} = \frac{1}{11}(-5) = -\frac{5}{11}.$$

$$d_{24} = \frac{1}{|A|}c_{42} = \frac{1}{-11}(-1)^6 \begin{vmatrix} 1 & 1 & 3 \\ 1 & -1 & 2 \\ -2 & 1 & 2 \end{vmatrix} = -\frac{1}{11}(-13) = \frac{13}{11}.$$

17. (a) Let x, y, and z represent the number of ladders of length 4, 5, and 6 ft, respectively. Then

$$\begin{cases} 0.5x + y + 1.5z = 15, & \text{(sanding)} \\ x + y + z = 15, & \text{(staining)} \\ x + y + 2z = 20. & \text{(varnishing)} \end{cases}$$

(b) $|A| = -0.5.$

$$[c_{ij}] = \begin{bmatrix} 1 & -1 & 1 \\ -0.5 & -0.5 & 0.5 \\ -0.5 & 1 & -0.5 \end{bmatrix}.$$

$$\text{adj } A = [c_{ij}]^T = \begin{bmatrix} 1 & -0.5 & -0.5 \\ -1 & -0.5 & 1 \\ 0 & 0.5 & -0.5 \end{bmatrix}.$$

$$A^{-1} = \frac{1}{|A|} \text{ adj } A = \frac{1}{-0.5}\begin{bmatrix} 1 & -0.5 & -0.5 \\ -1 & -0.5 & 1 \\ 0 & 0.5 & -0.5 \end{bmatrix} = \begin{bmatrix} -2 & 1 & 1 \\ 2 & 1 & -2 \\ 0 & -1 & 1 \end{bmatrix}.$$

(c) $X = A^{-1}B = \begin{bmatrix} -2 & 1 & 1 \\ 2 & 1 & -2 \\ 0 & -1 & 1 \end{bmatrix}\begin{bmatrix} 15 \\ 15 \\ 20 \end{bmatrix} = \begin{bmatrix} 5 \\ 5 \\ 5 \end{bmatrix}.$ Thus $x = y = z = 5.$

<u>EXERCISE 6.10</u>

1. $A = \begin{bmatrix} \frac{200}{1200} & \frac{500}{1500} \\ \frac{400}{1200} & \frac{200}{1500} \end{bmatrix} = \begin{bmatrix} \frac{1}{6} & \frac{1}{3} \\ \frac{1}{3} & \frac{2}{15} \end{bmatrix}.$

$I - A = \begin{bmatrix} 1 & 0 \\ 0 & 1 \end{bmatrix} - \begin{bmatrix} 1/6 & 1/3 \\ 1/3 & 2/15 \end{bmatrix} = \begin{bmatrix} 5/6 & -1/3 \\ -1/3 & 13/15 \end{bmatrix}.$

$(I - A)^{-1} = \begin{bmatrix} 78/55 & 6/11 \\ 6/11 & 15/11 \end{bmatrix}.$

Thus $X = (I - A)^{-1}C = \begin{bmatrix} 78/55 & 6/11 \\ 6/11 & 15/11 \end{bmatrix}\begin{bmatrix} 600 \\ 805 \end{bmatrix} = \begin{bmatrix} 1290 \\ 1425 \end{bmatrix}.$

The total value of other production costs is

$$P_A + P_B = \frac{600}{1200}(1290) + \frac{800}{1500}(1425) = 645 + 760 = 1405.$$

3. $A = \begin{bmatrix} \frac{18}{108} & \frac{30}{120} & \frac{45}{180} \\ \frac{27}{108} & \frac{30}{120} & \frac{60}{180} \\ \frac{54}{108} & \frac{40}{120} & \frac{60}{180} \end{bmatrix} = \begin{bmatrix} 1/6 & 1/4 & 1/4 \\ 1/4 & 1/4 & 1/3 \\ 1/2 & 1/3 & 1/3 \end{bmatrix}.$

$I - A = \begin{bmatrix} 5/6 & -1/4 & -1/4 \\ -1/4 & 3/4 & -1/3 \\ -1/2 & -1/3 & 2/3 \end{bmatrix}.$

$(I - A)^{-1} = \dfrac{1}{|I - A|}\text{adj}(I - A)$

$= \dfrac{1}{\frac{109}{864}}\begin{bmatrix} 7/18 & 1/4 & 13/48 \\ 1/3 & 31/72 & 49/144 \\ 11/24 & 29/72 & 9/16 \end{bmatrix} = \begin{bmatrix} \frac{336}{109} & \frac{216}{109} & \frac{234}{109} \\ \frac{288}{109} & \frac{372}{109} & \frac{294}{109} \\ \frac{396}{109} & \frac{348}{109} & \frac{486}{109} \end{bmatrix}.$

(a) $X = (I - A)^{-1}C = (I - A)^{-1}\begin{bmatrix} 50 \\ 40 \\ 30 \end{bmatrix} = \begin{bmatrix} 297.80 \\ 349.54 \\ 443.12 \end{bmatrix}$

(b) $X = (I - A)^{-1}C = (I - A)^{-1}\begin{bmatrix} 10 \\ 10 \\ 24 \end{bmatrix} = \begin{bmatrix} 102.17 \\ 125.28 \\ 175.27 \end{bmatrix}$

CHAPTER 6 - REVIEW PROBLEMS

1. $3\begin{bmatrix} 3 & 4 \\ -5 & 1 \end{bmatrix} - 2\begin{bmatrix} 1 & 0 \\ 2 & 4 \end{bmatrix} = \begin{bmatrix} 9 & 12 \\ -15 & 3 \end{bmatrix} - \begin{bmatrix} 2 & 0 \\ 4 & 8 \end{bmatrix} = \begin{bmatrix} 7 & 12 \\ -19 & -5 \end{bmatrix}$

3. $\begin{bmatrix} 1 & 7 \\ 2 & -3 \\ 1 & 0 \end{bmatrix}\begin{bmatrix} 1 & 0 & -2 \\ 0 & 5 & 1 \end{bmatrix} = \begin{bmatrix} 1+0 & 0+35 & -2+7 \\ 2+0 & 0-15 & -4-3 \\ 1+0 & 0+0 & -2+0 \end{bmatrix} = \begin{bmatrix} 1 & 35 & 5 \\ 2 & -15 & -7 \\ 1 & 0 & -2 \end{bmatrix}$

5. $\begin{bmatrix} 1 & 0 \\ -1 & 4 \end{bmatrix}\left\{\begin{bmatrix} 1 & 4 \\ 6 & 5 \end{bmatrix} - \begin{bmatrix} 2 & 6 \\ 5 & 0 \end{bmatrix}\right\} = \begin{bmatrix} 1 & 0 \\ -1 & 4 \end{bmatrix}\begin{bmatrix} -1 & -2 \\ 1 & 5 \end{bmatrix}$

$$= \begin{bmatrix} -1+0 & -2+0 \\ 1+4 & 2+20 \end{bmatrix} = \begin{bmatrix} -1 & -2 \\ 5 & 22 \end{bmatrix}$$

7. $2\begin{bmatrix} 1 & -2 \\ 3 & 1 \end{bmatrix}^2 [1 \quad -2]^T = 2\begin{bmatrix} -5 & -4 \\ 6 & -5 \end{bmatrix}\begin{bmatrix} 1 \\ -2 \end{bmatrix} = 2\begin{bmatrix} 3 \\ 16 \end{bmatrix} = \begin{bmatrix} 6 \\ 32 \end{bmatrix}$

9. $(2A)^T - 3I^2 = 2A^T - 3I = 2\begin{bmatrix} 1 & -1 \\ 1 & 2 \end{bmatrix} - \begin{bmatrix} 3 & 0 \\ 0 & 3 \end{bmatrix}$

$$= \begin{bmatrix} 2 & -2 \\ 2 & 4 \end{bmatrix} - \begin{bmatrix} 3 & 0 \\ 0 & 3 \end{bmatrix} = \begin{bmatrix} -1 & -2 \\ 2 & 1 \end{bmatrix}$$

11. $B^4 + I^4 = \begin{bmatrix} 1 & 0 \\ 0 & 2 \end{bmatrix}^4 + \begin{bmatrix} 1 & 0 \\ 0 & 1 \end{bmatrix}^4 = \begin{bmatrix} 1 & 0 \\ 0 & 16 \end{bmatrix} + \begin{bmatrix} 1 & 0 \\ 0 & 1 \end{bmatrix} = \begin{bmatrix} 2 & 0 \\ 0 & 17 \end{bmatrix}$

13. $\begin{bmatrix} 5x \\ 2x \end{bmatrix} = \begin{bmatrix} 15 \\ y \end{bmatrix}$. $5x = 15$, or $x = 3$.
$2x = y$, $2 \cdot 3 = y$, or $y = 6$

15. $\begin{bmatrix} 1 & 4 \\ 5 & 8 \end{bmatrix} \rightarrow \begin{bmatrix} 1 & 4 \\ 0 & -12 \end{bmatrix} \rightarrow \begin{bmatrix} 1 & 4 \\ 0 & 1 \end{bmatrix} \rightarrow \begin{bmatrix} 1 & 0 \\ 0 & 1 \end{bmatrix}$

17. $\begin{bmatrix} 2 & 4 & 3 \\ 1 & 2 & 3 \\ 4 & 8 & 6 \end{bmatrix} \rightarrow \begin{bmatrix} 1 & 2 & 3 \\ 2 & 4 & 3 \\ 4 & 8 & 6 \end{bmatrix} \rightarrow \begin{bmatrix} 1 & 2 & 3 \\ 0 & 0 & -3 \\ 0 & 0 & -6 \end{bmatrix} \rightarrow \begin{bmatrix} 1 & 2 & 3 \\ 0 & 0 & 1 \\ 0 & 0 & -6 \end{bmatrix}$

$$\rightarrow \begin{bmatrix} 1 & 2 & 0 \\ 0 & 0 & 1 \\ 0 & 0 & 0 \end{bmatrix}$$

19. $\begin{bmatrix} 2 & -5 \\ 4 & 3 \end{bmatrix} \rightarrow \begin{bmatrix} 2 & -5 \\ 0 & 13 \end{bmatrix} \rightarrow \begin{bmatrix} 1 & -1/5 \\ 0 & 1 \end{bmatrix} \rightarrow \begin{bmatrix} 1 & 0 \\ 0 & 1 \end{bmatrix}.$

Thus x = 0, y = 0

21. $\begin{bmatrix} 1 & 1 & 2 & | & 1 \\ 3 & -2 & -4 & | & -7 \\ 2 & -1 & -2 & | & 2 \end{bmatrix} \rightarrow \begin{bmatrix} 1 & 1 & 2 & | & 1 \\ 0 & -5 & -10 & | & -10 \\ 0 & -3 & -6 & | & 0 \end{bmatrix} \rightarrow \begin{bmatrix} 1 & 1 & 2 & | & 1 \\ 0 & 1 & 2 & | & 2 \\ 0 & ^-3 & -6 & | & 0 \end{bmatrix}$

$\rightarrow \begin{bmatrix} 1 & 0 & 0 & | & -1 \\ 0 & 1 & 2 & | & 2 \\ 0 & 0 & 0 & | & 6 \end{bmatrix}.$ Row three indicates that 0 = 6, which is never true, so there is no solution.

23. $\begin{bmatrix} 1 & 5 & | & 1 & 0 \\ 3 & 9 & | & 0 & 1 \end{bmatrix} \rightarrow \begin{bmatrix} 1 & 5 & | & 1 & 0 \\ 0 & -6 & | & -3 & 1 \end{bmatrix} \rightarrow \begin{bmatrix} 1 & 5 & | & 1 & 0 \\ 0 & 1 & | & 1/2 & -1/6 \end{bmatrix}$

$\rightarrow \begin{bmatrix} 1 & 0 & | & -3/2 & 5/6 \\ 0 & 1 & | & 1/2 & -1/6 \end{bmatrix} \implies A^{-1} = \begin{bmatrix} -3/2 & 5/6 \\ 1/2 & -1/6 \end{bmatrix}$

25. $\begin{bmatrix} 1 & 3 & -2 & | & 1 & 0 & 0 \\ 4 & 1 & 0 & | & 0 & 1 & 0 \\ 3 & -2 & 2 & | & 0 & 0 & 1 \end{bmatrix} \rightarrow \begin{bmatrix} 1 & 3 & -2 & | & 1 & 0 & 0 \\ 0 & -11 & 8 & | & -4 & 1 & 0 \\ 0 & -11 & 8 & | & -3 & 0 & 0 \end{bmatrix}$

$\begin{bmatrix} 1 & 3 & -2 & | & 1 & 0 & 0 \\ 0 & -11 & 8 & | & -4 & 1 & 0 \\ 0 & 0 & 0 & | & 1 & 0 & 0 \end{bmatrix} \rightarrow \begin{bmatrix} 1 & 3 & -2 & | & 1 & 0 & 0 \\ 0 & 1 & -8/11 & | & 4/11 & -1/11 & 0 \\ 0 & 0 & 0 & | & 1 & 0 & 0 \end{bmatrix}$

$\rightarrow \begin{bmatrix} 1 & 0 & 2/11 & | & -1/11 & 3/11 & 0 \\ 0 & 1 & -8/11 & | & 4/11 & -1/11 & 0 \\ 0 & 0 & 0 & | & 1 & 0 & 0 \end{bmatrix} \implies$ no inverse exists

27. $\begin{bmatrix} 3 & 1 & 4 & | & 1 & 0 & 0 \\ 1 & 0 & 1 & | & 0 & 1 & 0 \\ 0 & 2 & 1 & | & 0 & 0 & 1 \end{bmatrix} \rightarrow \begin{bmatrix} 1 & 0 & 1 & | & 0 & 1 & 0 \\ 3 & 1 & 4 & | & 1 & 0 & 0 \\ 0 & 2 & 1 & | & 0 & 0 & 1 \end{bmatrix}$

$\rightarrow \begin{bmatrix} 1 & 0 & 1 & | & 0 & 1 & 0 \\ 0 & 1 & 1 & | & 1 & -3 & 0 \\ 0 & 2 & 1 & | & 0 & 0 & 1 \end{bmatrix} \rightarrow \begin{bmatrix} 1 & 0 & 1 & | & 0 & 1 & 0 \\ 0 & 1 & 1 & | & 1 & -3 & 0 \\ 0 & 0 & -1 & | & -2 & 6 & 1 \end{bmatrix}$

$\rightarrow \begin{bmatrix} 1 & 0 & 1 & | & 0 & 1 & 0 \\ 0 & 1 & 1 & | & 1 & -3 & 0 \\ 0 & 0 & 1 & | & 2 & -6 & -1 \end{bmatrix} \rightarrow \begin{bmatrix} 1 & 0 & 0 & | & -2 & 7 & 1 \\ 0 & 1 & 0 & | & -1 & 3 & 1 \\ 0 & 0 & 1 & | & 2 & -6 & -1 \end{bmatrix}.$

$\begin{bmatrix} x \\ y \\ z \end{bmatrix} = A^{-1}B = \begin{bmatrix} -2 & 7 & 1 \\ -1 & 3 & 1 \\ 2 & -6 & -1 \end{bmatrix} \begin{bmatrix} 1 \\ 0 \\ 2 \end{bmatrix} = \begin{bmatrix} 0 \\ 1 \\ 0 \end{bmatrix}.$

Thus x = 0, y = 1, z = 0.

29. $2(7) - (-1)4 = 18.$

31. Expanding along row 2 gives

$$0 + 1(-1)^4 \begin{vmatrix} 1 & -1 \\ 1 & 2 \end{vmatrix} + 4(-1)^5 \begin{vmatrix} 1 & 2 \\ 1 & 2 \end{vmatrix}$$

$$= 1(1)[2 - (-1)] + 4(-1)[2 - 2] = 3 + 0 = 3.$$

33. Since all entries below the main diagonal are zeros, the the determinant is $(r)(i)(c)(h)$ = rich

35. $\Delta = \begin{vmatrix} 3 & -1 \\ 2 & 3 \end{vmatrix} = 11,$ $\Delta_x = \begin{vmatrix} 1 & -1 \\ 8 & 3 \end{vmatrix} = 11,$ $\Delta_y = \begin{vmatrix} 3 & 1 \\ 2 & 8 \end{vmatrix} = 22.$

$$x = \Delta_x/\Delta = 11/11 = 1,$$

$$y = \Delta_y/\Delta = 22/11 = 2.$$

37. $[c_{ij}] = \begin{bmatrix} -1 & 1 & -2 \\ 2 & 0 & -2 \\ -1 & -1 & 2 \end{bmatrix}.$ adj $A = [c_{ij}]^T = \begin{bmatrix} -1 & 2 & -1 \\ 1 & 0 & -1 \\ -2 & -2 & 2 \end{bmatrix}.$

$|A| = -2.$

$$A^{-1} = \frac{1}{|A|} \text{ adj } A = -\frac{1}{2} \begin{bmatrix} -1 & 2 & -1 \\ 1 & 0 & -1 \\ -2 & -2 & 2 \end{bmatrix} = \begin{bmatrix} 1/2 & -1 & 1/2 \\ -1/2 & 0 & 1/2 \\ 1 & 1 & -1 \end{bmatrix}.$$

39. $A = \begin{bmatrix} 0/3 & 2/4 \\ 1/3 & 0/4 \end{bmatrix} = \begin{bmatrix} 0 & 1/2 \\ 1/3 & 0 \end{bmatrix}.$

$I - A = \begin{bmatrix} 1 & 0 \\ 0 & 1 \end{bmatrix} - \begin{bmatrix} 0 & 1/2 \\ 1/3 & 0 \end{bmatrix} = \begin{bmatrix} 1 & -1/2 \\ -1/3 & 1 \end{bmatrix}.$

$(I - A)^{-1} = \begin{bmatrix} 6/5 & 3/5 \\ 2/5 & 6/5 \end{bmatrix}.$

$X = (I - A)^{-1} C = \begin{bmatrix} 6/5 & 3/5 \\ 2/5 & 6/5 \end{bmatrix} \begin{bmatrix} 2 \\ 2 \end{bmatrix} = \begin{bmatrix} 3.6 \\ 3.2 \end{bmatrix}.$

41. $|A^{-1}B^T| = |A^{-1}||B^T| = \frac{1}{|A|} \cdot |B| = \frac{1}{-2} \cdot 4 = -2$

43. $A^2 = \begin{bmatrix} 0 & 0 & 1 \\ 0 & 1 & 0 \\ 1 & 0 & 0 \end{bmatrix} \begin{bmatrix} 0 & 0 & 1 \\ 0 & 1 & 0 \\ 1 & 0 & 0 \end{bmatrix} = \begin{bmatrix} 1 & 0 & 0 \\ 0 & 1 & 0 \\ 0 & 0 & 1 \end{bmatrix} = I_3.$ Since $AA = I$,

it follows that $A^{-1} = A.$ $A^{1994} = (A^2)^{997} = I^{997} = I_3.$

45. $\Delta = \begin{vmatrix} a & 0 & c \\ b & b & 0 \\ a & a & c \end{vmatrix} = (abc + abc + 0) - (abc + 0 + 0) = abc.$

$\Delta_x = \begin{vmatrix} a & 0 & c \\ b & b & 0 \\ c & a & c \end{vmatrix} = (abc + abc + 0) - (bc^2 + 0 + 0)$
$$= 2abc - bc^2.$$

$\Delta_y = \begin{vmatrix} a & a & c \\ b & b & 0 \\ a & c & c \end{vmatrix} = (abc + bc^2 + 0) - (abc + 0 + abc)$
$$= bc^2 - abc.$$

$\Delta_z = \begin{vmatrix} a & 0 & a \\ b & b & b \\ a & a & c \end{vmatrix} = (abc + a^2b + 0) - (a^2b + a^2b + 0)$
$$= abc - a^2b.$$

$x = \dfrac{\Delta_x}{\Delta} = \dfrac{2abc - bc^2}{abc} = 2 - \dfrac{c}{a},$ $y = \dfrac{\Delta_y}{\Delta} = \dfrac{bc^2 - abc}{abc} = \dfrac{c}{a} - 1,$

$z = \dfrac{\Delta_y}{\Delta} = \dfrac{abc - a^2b}{abc} = 1 - \dfrac{a}{c}$

47. (a) Let x, y, and z represent the weekly doses of capsules of brand I, II, and III, respectively. Then

$$\begin{cases} x + y + 4z = 13 & \text{(vitamin A),} \\ x + 2y + 7z = 22 & \text{(vitamin B),} \\ x + 3y + 10z = 31 & \text{(vitamin C).} \end{cases}$$

$\begin{bmatrix} 1 & 1 & 4 & | & 13 \\ 1 & 2 & 7 & | & 22 \\ 1 & 3 & 10 & | & 31 \end{bmatrix} \xrightarrow[-1R_1+R_3]{-1R_1+R_2} \begin{bmatrix} 1 & 1 & 4 & | & 13 \\ 0 & 1 & 3 & | & 9 \\ 0 & 2 & 6 & | & 18 \end{bmatrix}$

$\xrightarrow[(-2)R_2+R_3]{(-1)R_2+R_1} \begin{bmatrix} 1 & 0 & 1 & | & 4 \\ 0 & 1 & 3 & | & 9 \\ 0 & 0 & 0 & | & 0 \end{bmatrix}.$ Thus x = 4 - r, y = 9 - 3r, and z = r, where r = 0, 1, 2, 3

The four possible combinations are given by

Combination	x	y	z
1	4	9	0
2	3	6	1
3	2	3	2
4	1	0	3

(b) Computing the cost of each combination, we find that they are 83, 77, 71, and 65 cents, respectively. Thus combination 4, namely x = 1, y = 0, z = 3, minimizes

CHAPTER 6 — MATHEMATICAL SNAPSHOT

1. $A = \begin{bmatrix} 20 & 40 & 30 & 10 \\ 30 & 0 & 10 & 10 \\ 10 & 0 & 30 & 50 \end{bmatrix}$, $T = \begin{bmatrix} 7 \\ 10 \\ 7 \\ 5 \end{bmatrix}$, $C = \begin{bmatrix} 9 \\ 8 \\ 10 \end{bmatrix}$.

$$C^T(AT) = C^T \left\{ \begin{bmatrix} 20 & 40 & 30 & 10 \\ 30 & 0 & 10 & 10 \\ 10 & 0 & 30 & 50 \end{bmatrix} \begin{bmatrix} 7 \\ 10 \\ 7 \\ 5 \end{bmatrix} \right\} = C^T \begin{bmatrix} 800 \\ 330 \\ 530 \end{bmatrix}$$

$$= \begin{bmatrix} 9 & 8 & 10 \end{bmatrix} \begin{bmatrix} 800 \\ 330 \\ 530 \end{bmatrix} = [15,140]. \quad \underline{Ans.} \quad \$151.40$$

7

Linear Programming

1.

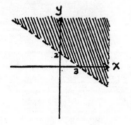

3.

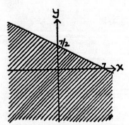

5.

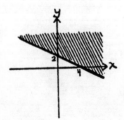

7.

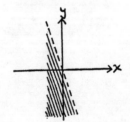

9.

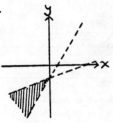

11.

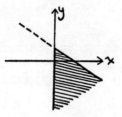

13.

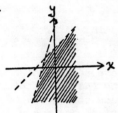

15.

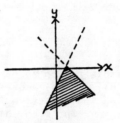

17.

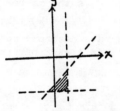

19.

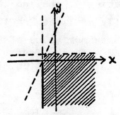

21.

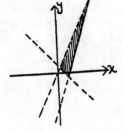

23.

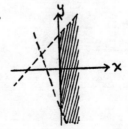

Exercise 7.1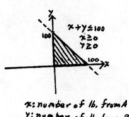

25.   27.

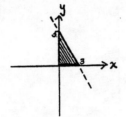

x: number of lb. from A
y: number of lb. from B

EXERCISE 7.2

1. The feasible region appears below. The corner points
   are (0, 0), (60, 0) and (40, 20). Using the method of
   evaluating the objective function at each corner point, we
   have P(0, 0) = 0, P(60, 0) = 600, and P(40, 20) = 640.
   Thus P has a maximum value of 640 at (40, 20). That is, P
   has a maximum value of 640 when x = 40 and y = 20.
   [Alternatively, the indicated dashed line is the member of
   the family of lines P = 10x + 12y $\left(\text{or } y = -\frac{5}{6}x + \frac{P}{6}\right)$ that
   gives a maximum value of P and that has at least one point
   in common with the feasible region. It is clear that the
   line contains corner point (40, 20).]

1.   3.

3. The feasible region appears above. The corner points are (2, 3), (0, 5), (0, 7) and (10/3, 7). Evaluating Z at each point, we find that Z has a maximum value of -10 when x = 2 and y = 3.

5. The feasible region (see below) is empty, so there is no optimum solution.

5.

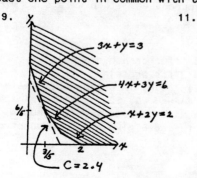

7.

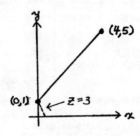

7. The feasible region (see above) is a line segment. The corner points are (0, 1) and (4, 5). Z has a minimum value of 3 when x = 0 and y = 1.

9. The feasible region (see below) is unbounded with 4 corner points. The member (see dashed line) of the family of lines C = 2x + y which gives a minimum value of C, subject to the constraints, intersects the feasible region at corner point (3/5, 6/5) where C = 2.4. Thus C has a minimum value of 2.4 when x = 3/5 and y = 6/5. [Note: Here we chose the member of the family y = -2x + C whose y-intercept was *closest* to the origin and which had at least one point in common with the feasible region.]

9.

11.

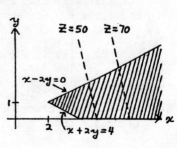

11. The feasible region (see above) is unbounded with 2
    corner points. The family of lines given by Z = 10x + 2y
    has members (see dashed lines for two sample members) that
    have arbitrarily large values of Z and that also intersect
    the feasible region. Thus no optimum solution exists.

13. Let x and y be the numbers of widgets and wadgits made per
    week, respectively. We maximize P = 4x + 6y subject to
    $$\begin{cases} x \geq 0, \\ y \geq 0, \\ 2x + y \leq 70 \quad \text{(for machine A),} \\ x + y \leq 40 \quad \text{(for machine B),} \\ x + 3y \leq 90 \quad \text{(for finishing).} \end{cases}$$
    The feasible region appears below. The corner points are
    (0, 0), (0, 30), (15, 25), (30, 10) and (35, 0). By
    evaluating P at each corner point, we find that P is
    maximized at corner point (15, 25), where its value is
    210. Thus 15 widgets and 25 wadgits should be made each
    week to give a maximum profit of $210.

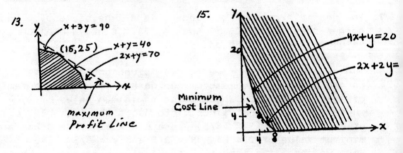

15. Let x and y be the numbers of units of Food A and Food B,
    respectively, that are purchased. Then we are to minimize
    C = 1.20x + 0.80y where
    $$\begin{cases} x \geq 0, \\ y \geq 0, \\ 2x + 2y \geq 16 \quad \text{(for carbohydrates),} \\ 4x + y \geq 20 \quad \text{(for protein).} \end{cases}$$
    The feasible region (see above) is unbounded. The corner
    points are (8, 0), (4, 4) and (0, 20). C is minimized at
    (4, 4) where C = 8 (see minimum cost line). Thus 4 units
    of Food A and 4 units of Food B give a minimum cost of $8.

17. Let x and y be the numbers of tons of ores I and II, respectively, that are processed. Then we are to minimize C = 50x + 60y where

$$\begin{cases} x \geq 0, \\ y \geq 0, \\ 100x + 200y \geq 3000 \quad \text{(for mineral A)}, \\ 200x + 50y \geq 2500 \quad \text{(for mineral B)}, \end{cases}$$

The feasible region (see below) is unbounded with 3 corner points. C is minimized at the corner point (10, 10) where C = 1100 (see the minimum cost line). Thus 10 tons of ore I and 10 tons of ore II give a minimum cost of $1100.

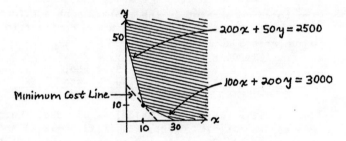

19. Let x (y) be the number of chambers of type A (B). Then we are to minimize C = 600,000x + 300,000y, where

$$\begin{cases} x \geq 4, \\ y \geq 4, \\ 10x + 4y \geq 100 \quad \text{(for polymer } P_1\text{)}, \\ 20x + 30y \geq 420 \quad \text{(for polymer } P_2\text{)}. \end{cases}$$

The feasible region (see below) is unbounded with 3 corner points. Evaluating C at each corner point, we find C is minimized at corner point (6, 10) where C = 6,600,000. The solution is 6 chambers of type A and 10 chambers of type B.

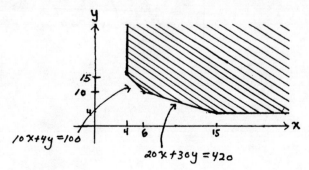

21. (a) A builds x km of highway and y km of expressway, so
    B builds (200 - x) km of highway and (100 - y) km of
    expressway. Thus
        D = 1x + 6y + 2(200 - x) + 5(100 - y = 900 - x + y.

    (b) The first constraint is company A's construction
    limit. The second constraint is company B's
    construction limit, which arises as follows:
        (200 - x) + (100 - y) ≦ 150,
                    300 - x - y ≦ 150,
                        -x - y ≦ -150,
                        x + y ≧ 150.
    The third constraint is the minimum contract for A.
    The fourth constraint is the minimum contract for B,
    which arises as follows:
        1(200 - x) + 5(100 - y) ≧ 250,
                700 - x - 5y ≧ 250,
                    -x - 5y ≧ -450
                    x + 5y ≦ 450.

    (c) The feasible region (see below) is bounded. The corner
    points are (75, 75), (125, 25), (187.5, 12.5) and
    (137.5, 62.5). Evaluating D at each corner point, we
    find that D is maximized at point (75, 75), where D =
    900. That is, D is maximized if x = y = 75.

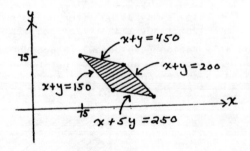

EXERCISE 7.3

1. The feasible region (see below) is unbounded. Z is minimized at corner points (2, 3) and (5, 2), where its value is 33. Z is also minimized at all points on the line segment joining (2, 3) and (5, 2), so the solution is
$$Z = 33 \text{ when } x = (1 - t)(2) + 5t = 2 + 3t,$$
$$y = (1 - t)(3) + 2t = 3 - t,$$
$$\text{and } 0 \leq t \leq 1.$$

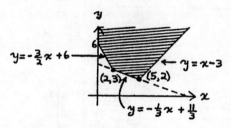

3. The feasible region appears below. The corner points are (0, 0), (0, 4), (3, 2), and (4, 0). Z is maximized at (3, 2) and (4, 0), where its value is 72. Thus Z is also maximized at all points on the line segment joining (3, 2) and (4,0). The solution is
$$Z = 72 \text{ when } x = (1 - t)(3) + 4t = 3 + t,$$
$$y = (1 - t)(2) + 0t = 2 - 2t,$$
$$\text{and } 0 \leq t \leq 1$$

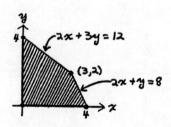

EXERCISE 7.4

1.

	x_1	x_2	s_1	s_2	Z		
s_1	2	1	1	0	0	8	8
s_2	2	[3]	0	1	0	12	4
Z	-1	-2	0	0	1	0	

	x_1	x_2	s_1	s_2	Z	
s_1	4/3	0	1	-1/3	0	4
x_2	2/3	1	0	1/3	0	4
Z	1/3	0	0	2/3	1	8

The solution is $Z = 8$ when $x_1 = 0$, $x_2 = 4$.

3.

	x_1	x_2	s_1	s_2	Z		
s_1	1	1	1	0	0	6	6
s_2	-1	[1]	0	1	0	4	4
Z	1	-3	0	0	1	0	

	x_1	x_2	s_1	s_2	Z		
s_1	[2]	0	1	-1	0	2	1
x_2	-1	1	0	1	0	4	
Z	-2	0	0	3	1	12	

	x_1	x_2	s_1	s_2	Z	
x_1	1	0	1/2	-1/2	0	1
x_2	0	1	1/2	1/2	0	5
Z	0	0	1	2	1	14

The solution is $Z = 14$ when $x_1 = 1$, $x_2 = 5$.

5.

	x_1	x_2	s_1	s_2	s_3	Z		
s_1	[1]	-1	1	0	0	0	1	1
s_2	1	2	0	1	0	0	8	8
s_3	1	1	0	0	1	0	5	5
Z	-8	-2	0	0	0	1	0	

	x_1	x_2	s_1	s_2	s_3	Z		
x_1	1	-1	1	0	0	0	1	
s_2	0	3	-1	1	0	0	7	7/3
s_3	0	[2]	-1	0	1	0	4	2
Z	0	-10	8	0	0	1	8	

$$
\begin{array}{c}
\begin{array}{ccccccc}
x_1 & x_2 & s_1 & s_2 & s_3 & Z &
\end{array}\\
\begin{array}{c}
x_1 \\ s_2 \\ x_2 \\ Z
\end{array}
\left[
\begin{array}{cccccc|c}
1 & 0 & 1/2 & 0 & 1/2 & 0 & 3 \\
0 & 0 & 1/2 & 1 & -3/2 & 0 & 1 \\
0 & 1 & -1/2 & 0 & 1/2 & 0 & 2 \\
\hline
0 & 0 & 3 & 0 & 5 & 1 & 28
\end{array}
\right]
\end{array}
$$

The solution is $Z = 28$ when $x_1 = 3$, $x_2 = 2$.

7.
$$
\begin{array}{c}
\begin{array}{cccccc}
x_1 & x_2 & x_3 & s_1 & s_2 & Z
\end{array}\\
\begin{array}{c}
s_1 \\ s_2 \\ Z
\end{array}
\left[
\begin{array}{cccccc|c}
1 & 2 & 0 & 1 & 0 & 0 & 10 \\
2 & \boxed{2} & 1 & 0 & 1 & 0 & 10 \\
\hline
-3 & -4 & -3/2 & 0 & 0 & 1 & 0
\end{array}
\right]
\begin{array}{l}
5 \\ 5 \\
\end{array}
\end{array}
$$
choosing s_2 as dep. var.

$$
\begin{array}{c}
\begin{array}{c}
s_1 \\ x_2 \\ Z
\end{array}
\left[
\begin{array}{cccccc|c}
-1 & 0 & -1 & 1 & -1 & 0 & 0 \\
1 & 1 & 1/2 & 0 & 1/2 & 0 & 5 \\
\hline
1 & 0 & 1/2 & 0 & 2 & 1 & 20
\end{array}
\right]
\end{array}
$$

The solution is $Z = 20$ when $x_1 = 0$, $x_2 = 5$, $x_3 = 0$

9. To obtain a standard linear programming problem, we write
the second constraint as $-x_1 + 2x_2 + x_3 \leqq 2$.

$$
\begin{array}{c}
\begin{array}{cccccc}
x_1 & x_2 & x_3 & s_1 & s_2 & Z
\end{array}\\
\begin{array}{c}
s_1 \\ s_2 \\ Z
\end{array}
\left[
\begin{array}{cccccc|c}
\boxed{1} & 1 & 0 & 1 & 0 & 0 & 1 \\
-1 & 2 & 1 & 0 & 1 & 0 & 2 \\
\hline
-2 & -1 & 1 & 0 & 0 & 1 & 0
\end{array}
\right]
\begin{array}{l}
1 \\
\end{array}
\end{array}
$$

$$
\begin{array}{c}
\begin{array}{c}
x_1 \\ s_2 \\ Z
\end{array}
\left[
\begin{array}{cccccc|c}
1 & 1 & 0 & 1 & 0 & 0 & 1 \\
0 & 3 & 1 & 1 & 1 & 0 & 3 \\
\hline
0 & 1 & 1 & 2 & 0 & 1 & 2
\end{array}
\right]
\end{array}
$$

The solution is $Z = 2$ when $x_1 = 1$, $x_2 = 0$, $x_3 = 0$.

11.
$$
\begin{array}{c}
\begin{array}{ccccccc}
x_1 & x_2 & s_1 & s_2 & s_3 & s_4 & Z
\end{array}\\
\begin{array}{c}
s_1 \\ s_2 \\ s_3 \\ s_4 \\ Z
\end{array}
\left[
\begin{array}{ccccccc|c}
1 & -1 & 1 & 0 & 0 & 0 & 0 & 4 \\
-1 & 1 & 0 & 1 & 0 & 0 & 0 & 4 \\
8 & 5 & 0 & 0 & 1 & 0 & 0 & 40 \\
\boxed{2} & 1 & 0 & 0 & 0 & 1 & 0 & 6 \\
\hline
-1 & -1 & 0 & 0 & 0 & 0 & 1 & 0
\end{array}
\right]
\begin{array}{l}
4 \\ \\ 5 \\ 3 \\
\end{array}
\end{array}
$$
choosing x_1 as ent. var.

	x_1	x_2	s_1	s_2	s_3	s_4	Z		
s_1	0	-3/2	1	0	0	-1/2	0	1	
s_2	0	3/2	0	1	0	1/2	0	7	14/3
s_3	0	1	0	0	1	-4	0	16	16
x_1	1	1/2	0	0	0	1/2	0	3	6
Z	0	-1/2	0	0	0	1/2	1	3	

	x_1	x_2	s_1	s_2	s_3	s_4	Z	
s_1	0	0	1	1	0	0	0	8
x_2	0	1	0	2/3	0	1/3	0	14/3
s_3	0	0	0	-2/3	1	-13/3	0	34/3
x_1	1	0	0	-1/3	0	1/3	0	2/3
Z	0	0	0	1/3	0	2/3	1	16/3

Thus the maximum value of Z is 16/3, when $x_1 = 2/3$, $x_2 = 14/3$. Choosing x_2 as the entering variable, we have:

	x_1	x_2	s_1	s_2	s_3	s_4	Z		
s_1	1	-1	1	0	0	0	0	4	
s_2	-1	1	0	1	0	0	0	4	4
s_3	8	5	0	0	1	0	0	40	8
s_4	2	1	0	0	0	1	0	6	6
Z	-1	-1	0	0	0	0	1	0	

	x_1	x_2	s_1	s_2	s_3	s_4	Z		
s_1	0	0	1	1	0	0	0	8	
x_2	-1	1	0	1	0	0	0	4	
s_3	13	0	0	-5	1	0	0	20	20/13
s_4	3	0	0	-1	0	1	0	2	2/3
Z	-2	0	0	1	0	0	1	4	

	x_1	x_2	s_1	s_2	s_3	s_4	Z	
s_1	0	0	1	1	0	0	0	8
x_2	0	1	0	2/3	0	1/3	0	14/3
s_3	0	0	0	-2/3	1	-13/3	0	34/3
x_1	1	0	0	-1/3	0	1/3	0	2/3
Z	0	0	0	1/3	0	2/3	1	16/3

The solution is Z = 16/3 when $x_1 = 2/3$, $x_2 = 14/3$.

13. To obtain a standard linear programming problem, we write the second constraint as $x_1 - x_2 + x_3 \le 2$ and the third constraint as $x_1 - x_2 - x_3 \le 1$.

	x_1	x_2	x_3	s_1	s_2	s_3	W		
s_1	4	3	-1	1	0	0	0	1	
s_2	-1	-1	[1]	0	1	0	0	2	2
s_3	1	-1	-1	0	0	1	0	1	
W	-1	12	-4	0	0	0	1	0	
s_1	[3]	2	0	1	1	0	0	3	1
x_3	-1	-1	1	0	1	0	0	2	
s_3	0	-2	0	0	1	1	0	3	
W	-5	8	0	0	4	0	1	8	
x_1	1	2/3	0	1/3	1/3	0	0	1	
x_3	0	-1/3	1	1/3	4/3	0	0	3	
s_3	0	-2	0	0	1	1	0	3	
W	0	34/3	0	5/3	17/3	0	1	13	

The solution is W = 13 when $x_1 = 1$, $x_2 = 0$, $x_3 = 3$.

		x_1	x_2	x_3	x_4	s_1	s_2	s_3	s_4	Z		
15.	s_1	1	-2	0	0	1	0	0	0	0	2	
	s_2	1	1	0	0	0	1	0	0	0	5	
	s_3	0	0	[1]	1	0	0	1	0	0	4	4
	s_4	0	0	1	-2	0	0	0	1	0	7	7
	Z	-60	0	-90	0	0	0	0	0	1	0	
	s_1	[1]	-2	0	0	1	0	0	0	0	2	2
	s_2	1	1	0	0	0	1	0	0	0	5	5
	x_3	0	0	1	1	0	0	1	0	0	4	
	s_4	0	0	0	-3	0	0	-1	1	0	3	
	Z	-60	0	0	90	0	0	90	0	1	360	
	x_1	1	-2	0	0	1	0	0	0	0	2	
	s_2	0	[3]	0	0	-1	1	0	0	0	3	1
	x_3	0	0	1	1	0	0	1	0	0	4	
	s_4	0	0	0	-3	0	0	-1	1	0	3	
	Z	0	-120	0	90	60	0	90	0	1	480	
	x_1	1	0	0	0	1/3	2/3	0	0	0	4	
	x_2	0	1	0	0	-1/3	1/3	0	0	0	1	
	x_3	0	0	1	1	0	0	1	0	0	4	
	s_4	0	0	0	-3	0	0	-1	1	0	3	
	Z	0	0	0	90	20	40	90	0	1	600	

The solution is Z = 600 for $x_1 = 4$, $x_2 = 1$, $x_3 = 4$, $x_4 = 0$.

17. Let x_1 and x_2 denote the numbers of boxes transported from A and B, respectively. The revenue received is $R = 0.75x_1 + 0.50x_2$. We want to maximize R subject to

$$2x_1 + x_2 \leq 2400 \quad \text{(volume)},$$
$$3x_1 + 5x_2 \leq 9200 \quad \text{(weight)},$$
$$x_1, x_2 \geq 0.$$

	x_1	x_2	s_1	s_2	R		
s_1	[2]	1	1	0	0	2400	1200
s_2	3	5	0	1	0	9200	$3066\frac{2}{3}$
R	-3/4	-1/2	0	0	1	0	
x_1	1	1/2	1/2	0	0	1200	2400
s_2	0	[7/2]	-3/2	1	0	5600	1600
R	0	-1/8	3/8	0	1	900	
x_1	1	0	5/7	-1/7	0	400	
x_2	0	1	-3/7	2/7	0	1600	
R	0	0	9/28	1/28	1	1100	

Thus 400 boxes from A and 1600 from B give a maximum revenue of $1100.

19. Let x_1, x_2, and x_3 denote the numbers of chairs, rockers, and chaise lounges produced, respectively. We want to maximize $R = 7x_1 + 8x_2 + 12x_3$ subject to

$$x_1 + x_2 + x_3 \leq 400,$$
$$x_1 + x_2 + 2x_3 \leq 500,$$
$$2x_1 + 3x_2 + 5x_3 \leq 1450,$$
$$x_1, x_2, x_3 \geq 0.$$

	x_1	x_2	x_3	s_1	s_2	s_3	R		
s_1	1	1	1	1	0	0	0	400	400
s_2	1	1	[2]	0	1	0	0	500	250
s_3	2	3	5	0	0	1	0	1450	290
R	-7	-8	-12	0	0	0	1	0	

	x_1	x_2	x_3	s_1	s_2	s_3	R		
s_1	1/2	⌐1/2¬	0	1	-1/2	0	0	150	300
x_3	1/2	1/2	1	0	1/2	0	0	250	500
s_3	-1/2	1/2	0	0	-5/2	1	0	200	400
R	-1	-2	0	0	6	0	1	3000	
x_2	1	1	0	2	-1	0	0	300	
x_3	0	0	1	-1	1	0	0	100	
s_3	-1	0	0	-1	-2	1	0	50	
R	1	0	0	4	4	0	1	3600	

The production order of 0 chairs, 300 rockers, 100 chaise
lounges give maximum revenue of $3600.

EXERCISE 7.5

1. Yes; for the tableau, x_2 is the entering variable and the
   quotients 6/2 and 3/1 tie for being the smallest.

	x_1	x_2	s_1	s_2	s_3	Z	
3. s_1	4	-3	1	0	0	0	4
s_2	3	-1	0	1	0	0	6
s_3	5	0	0	0	1	0	8
Z	-2	-7	0	0	0	1	0

The entering variable is x_2. Since no quotients exist,
the problem has an unbounded solution. Thus, no optimum
solution (unbounded).

	x_1	x_2	s_1	s_2	s_3	Z		
5. s_1	⌐1¬	-1	1	0	0	0	4	4
s_2	-1	1	0	1	0	0	4	
s_3	1	1	0	0	1	0	6	6
Z	-3	3	0	0	0	1	0	
x_1	1	-1	1	0	0	0	4	
s_2	0	0	1	1	0	0	8	
s_3	0	⌐2¬	-1	0	1	0	2	1
Z	0	0	3	0	0	1	12	

The maximum value of Z is 12, when $x_1 = 4$, $x_2 = 0$. Since x_2 is nonbasic for the last tableau and its indicator is 0, there may be multiple optimum solutions. Treating x_2 as an entering variable and continuing, we have

	x_1	x_2	s_1	s_2	s_3	Z	
x_1	1	0	1/2	0	1/2	0	5
s_2	0	0	1	1	0	0	8
x_2	0	1	-1/2	0	1/2	0	1
Z	0	0	3	0	0	1	12

Here Z = 12 when $x_1 = 5$, $x_2 = 1$. Thus multiple optimum solutions exist. Hence Z is maximum when

$$x_1 = (1-t)(4) + 5t = 4 + t,$$
$$x_2 = (1-t)(0) + (1)t = t,$$

and $0 \leq t \leq 1$. For the last tableau, s_3 is nonbasic and its indicator is 0. If we continue the process for determining other optimum solutions, we return to the second tableau.

<u>Ans.</u> Z = 12 when $x_1 = (1-t)(4) + 5t = 4 + t$,
$x_2 = (1-t)(0) + (1)t = t$, and $0 \leq t \leq 1$

	x_1	x_2	x_3	s_1	s_2	s_3	Z		
7. s_1	9	3	-2	1	0	0	0	5	5/3
s_2	4	[2]	-1	0	1	0	0	2	1
s_3	1	-4	1	0	0	1	0	3	
Z	-5	-6	-1	0	0	0	1	0	
s_1	3	0	-1/2	1	-3/2	0	0	2	
x_2	2	1	-1/2	0	1/2	0	0	1	
s_3	9	0	-1	0	2	1	0	7	
Z	7	0	-4	0	3	0	1	6	

For the last tableau, x_3 is the entering variable. Since no quotients exist, the problem has an unbounded solution.

<u>Ans.</u> no optimum solution (unbounded)

9. To obtain a standard linear programming problem, we write the second constraint as $4x_1 + x_2 \leq 6$.

$$
\begin{array}{c}
 \\
s_1 \\
s_2 \\
Z
\end{array}
\begin{array}{c}
x_1 \quad\; x_2 \quad\; x_3 \quad\; s_1 \quad\; s_2 \quad\; Z \\
\left[\begin{array}{cccccc|c}
2 & 1 & 1 & 1 & 0 & 0 & 7 \\
\boxed{4} & 1 & 0 & 0 & 1 & 0 & 6 \\
\hline
-6 & -2 & -1 & 0 & 0 & 1 & 0
\end{array}\right]
\begin{array}{c}
7/2 \\
3/2 \\

\end{array}
\end{array}
$$

$$
\begin{array}{c}
s_1 \\
x_1 \\
Z
\end{array}
\left[\begin{array}{cccccc|c}
0 & 1/2 & \boxed{1} & 1 & -1/2 & 0 & 4 \\
1 & 1/4 & 0 & 0 & 1/4 & 0 & 3/2 \\
\hline
0 & -1/2 & -1 & 0 & 3/2 & 1 & 9
\end{array}\right]
\begin{array}{c}
4 \\
 \\

\end{array}
$$

$$
\begin{array}{c}
x_3 \\
x_1 \\
Z
\end{array}
\left[\begin{array}{cccccc|c}
0 & 1/2 & 1 & 1 & -1/2 & 0 & 4 \\
1 & \boxed{1/4} & 0 & 0 & 1/4 & 0 & 3/2 \\
\hline
0 & 0 & 0 & 1 & 1 & 1 & 13
\end{array}\right]
\begin{array}{c}
8 \\
6 \\

\end{array}
$$

Z has a maximum value of 13 when $x_1 = 3/2$, $x_2 = 0$, $x_3 = 4$. Since x_2 is nonbasic for the last tableau and its indicator is 0, there may be multiple optimum solutions. Treating x_2 as an entering variable, we have

$$
\begin{array}{c}
x_3 \\
x_2 \\
Z
\end{array}
\begin{array}{c}
x_1 \quad\; x_2 \quad\; x_3 \quad\; s_1 \quad\; s_2 \quad\; Z \\
\left[\begin{array}{cccccc|c}
-2 & 0 & 1 & 1 & -1 & 0 & 1 \\
4 & 1 & 0 & 0 & 1 & 0 & 6 \\
\hline
0 & 0 & 0 & 1 & 1 & 1 & 13
\end{array}\right]
\end{array}
$$

Here $Z = 13$ when $x_1 = 0$, $x_2 = 6$, $x_3 = 1$. Thus multiple optimum solutions exist. Hence Z is maximum when
$$x_1 = (1-t)(3/2) + 0t = (3/2) - (3/2)t,$$
$$x_2 = (1-t)(0) + 6t = 6t,$$
$$x_3 = (1-t)(4) + (1)t = 4 - 3t,$$
and $0 \leq t \leq 1$. For the last tableau, x_1 is nonbasic and its indicator is 0. If we continue the process for determining other optimum solutions, we return to the third tableau.

<u>Ans.</u> $Z = 13$ when $x_1 = (1-t)(3/2) + 0t = (3/2) - (3/2)t$, $x_2 = (1-t)(0) + 6t = 6t$, $x_3 = (1-t)(4) + (1)t = 4 - 3t$, and $0 \leq t \leq 1$

11. Let x_1, x_2, and x_3 denote the numbers of chairs, rockers, and chaise lounges produced, respectively. We want to maximize $R = 6x_1 + 8x_2 + 12x_3$ subject to

$$x_1 + x_2 + x_3 \leq 400,$$
$$x_1 + x_2 + 2x_3 \leq 600,$$
$$2x_1 + 3x_2 + 5x_3 \leq 1500,$$
$$x_1, x_2, x_3 \geq 0.$$

	x_1	x_2	x_3	s_1	s_2	s_3	R			
s_1	1	1	1	1	0	0	0	400	400	
s_2	1	1	2	0	1	0	0	600	300	choosing
s_3	2	3	[5]	0	0	1	0	1500	300	s_3 as dep. var.
R	-6	-8	-12	0	0	0	1	0		

	x_1	x_2	x_3	s_1	s_2	s_3	R		
s_1	3/5	2/5	0	1	0	-1/5	0	100	500/3
s_2	[1/5]	-1/5	0	0	1	-2/5	0	0	0
x_3	2/5	3/5	1	0	0	1/5	0	300	750
R	-6/5	-4/5	0	0	0	12/5	1	3600	

	x_1	x_2	x_3	s_1	s_2	s_3	R		
s_1	0	[1]	0	1	-3	1	0	100	100
x_1	1	-1	0	0	5	-2	0	0	0
x_3	0	1	1	0	-2	1	0	300	300
R	0	-2	0	0	6	0	1	3600	

	x_1	x_2	x_3	s_1	s_2	s_3	R		
x_2	0	1	0	1	-3	1	0	100	
x_1	1	0	0	1	[2]	-1	0	100	50
x_3	0	0	1	-1	1	0	0	200	200
R	0	0	0	2	0	2	1	3800	

The maximum value of R is 3800 when $x_1 = 100$, $x_2 = 100$, $x_3 = 200$. Since s_2 is nonbasic for the last tableau and its indicator is 0, there may be multiple optimum solutions. Treating s_2 as an entering variable, we have

	x_1	x_2	x_3	s_1	s_2	s_3	R	
x_2	3/2	1	0	5/2	0	-1/2	0	250
s_2	1/2	0	0	1/2	1	-1/2	0	50
x_3	-1/2	0	1	-3/2	0	1/2	0	150
R	0	0	0	2	0	2	1	3800

Here $R = 3800$ when $x_1 = 0$, $x_2 = 250$, $x_3 = 150$. Thus multiple optimum solutions exist. Hence R is maximum when

$$x_1 = (1-t)(100) + 0t = 100 - 100t,$$

$$x_2 = (1-t)(100) + 250t = 100 + 150t,$$

$$x_3 = (1-t)(200) + 150t = 200 - 50t,$$

and $0 \leqq t \leqq 1$. For the last tableau, x_1 is nonbasic and its indicator is 0. If we continue the process for determining other optimum solutions, we return to the fourth tableau. If we were to initially choose s_2 as the departing variable, then

	x_1	x_2	x_3	s_1	s_2	s_3	R		
s_1	1	1	1	1	0	0	0	400	400
s_2	1	1	②	0	1	0	0	600	300
s_3	2	3	5	0	0	1	0	1500	300
R	-6	-8	-12	0	0	0	1	0	
s_1	1/2	1/2	0	1	-1/2	0	0	100	200
x_3	1/2	1/2	1	0	1/2	0	0	300	600
s_3	-1/2	1/2	0	0	-5/2	1	0	0	0
R	0	-2	0	0	6	0	1	3600	
s_1	1	0	0	1	②	-1	0	100	50
x_3	1	0	1	0	3	-1	0	300	100
x_2	-1	1	0	0	-5	2	0	0	
R	-2	0	0	0	-4	4	1	3600	
s_2	1/2	0	0	1/2	1	-1/2	0	50	100
x_3	-1/2	0	1	-3/2	0	1/2	0	150	
x_2	3/2	1	0	5/2	0	-1/2	0	250	500/3
R	0	0	0	2	0	2	1	3800	

The maximum value of R is 3800 when $x_1 = 0$, $x_2 = 250$, $x_3 = 150$. For the last tableau, x_1 is nonbasic and its indicator is 0. Treating x_1 as an entering variable, we have

	x_1	x_2	x_3	s_1	s_2	s_3	R	
x_1	1	0	0	1	2	-1	0	100
x_3	0	0	1	-1	1	0	0	200
x_2	0	1	0	1	-3	1	0	100
R	0	0	0	2	0	2	1	3800

Here R = 3800 when x_1 = 100, x_2 = 100, x_3 = 200. Hence R
is maximum when

$$x_1 = (1-t)(100) + 0t = 100 - 100t,$$
$$x_2 = (1-t)(100) + 250t = 100 + 150t,$$
$$x_3 = (1-t)(200) + 150t = 200 - 50t,$$

and $0 \leq t \leq 1$. For the last tableau, s_2 is nonbasic and
its indicator is 0. If we continue the process of
determining other optimum solutions, we return to the
tableau corresponding to the solution x_1 = 0, x_2 = 250,
x_3 = 150.

<u>Ans.</u> $3800; if x_1, x_2, and x_3 denote the numbers of
chairs, rockers, and chaise lounges produced, respectively, then x_1 = (1-t)(100) + 0t = 100 - 100t,
x_2 = (1-t)(100) + 250t = 100 + 150t,
x_3 = (1-t)(200) + 150t = 200 - 50t, and $0 \leq t \leq 1$

<u>EXERCISE 7.6</u>

1.

	x_1	x_2	s_1	s_2	t_2	W	
	1	1	1	0	0	0	6
	-1	1	0	-1	1	0	4
	-2	-1	0	0	M	1	0

	x_1	x_2	s_1	s_2	t_2	W		
s_1	1	1	1	0	0	0	6	6
t_2	-1	☐1	0	-1	1	0	4	4
W	-2+M	-1-M	0	M	0	1	-4M	

	x_1	x_2	s_1	s_2	t_2	W		
s_1	☐2	0	1	1	-1	0	2	1
x_2	-1	1	0	-1	1	0	4	
W	-3	0	0	-1	M+1	1	4	

	x_1	x_2	s_1	s_2	Z	
x_1	1	0	1/2	1/2	0	1
x_2	0	1	1/2	-1/2	0	5
Z	0	0	3/2	1/2	1	7

<u>Ans.</u> Z = 7 when x_1 = 1, x_2 = 5

3.

	x_1	x_2	x_3	s_1	s_2	t_2	W	
	1	2	1	1	0	0	0	5
	-1	1	1	0	-1	1	0	1
	-2	-1	1	0	0	M	1	0

	x_1	x_2	x_3	s_1	s_2	t_2	W		
s_1	1	2	1	1	0	0	0	5	5/2
t_2	-1	☐1	1	0	-1	1	0	1	1
W	-2+M	-1-M	1-M	0	M	0	1	-M	

	x_1	x_2	x_3	s_1	s_2	t_2	W		
s_1	☐3	0	-1	1	2	-2	0	3	1
x_2	-1	1	1	0	-1	1	0	1	
W	-3	0	2	0	-1	1+M	1	1	

	x_1	x_2	x_3	s_1	s_2	Z	
x_1	1	0	-1/3	1/3	2/3	0	1
x_2	0	1	2/3	1/3	-1/3	0	2
Z	0	0	1	1	1	1	4

<u>Ans.</u> $Z = 4$ when $x_1 = 1$, $x_2 = 2$, $x_3 = 0$

5.

	x_1	x_2	x_3	s_1	t_2	W	
	2	1	3	1	0	0	10
	1	-1	1	0	1	0	4
	-4	-1	-2	0	M	1	0

	x_1	x_2	x_3	s_1	t_2	W		
s_1	2	1	3	1	0	0	10	5
t_2	☐1	-1	1	0	1	0	4	4
W	-4-M	-1+M	-2-M	0	0	1	-4M	

	x_1	x_2	x_3	s_1	t_2	W		
s_1	0	☐3	1	1	-2	0	2	2/3
x_1	1	-1	1	0	1	0	4	
W	0	-5	2	0	4+M	1	16	

	x_1	x_2	x_3	s_1	Z	
x_2	0	1	1/3	1/3	0	2/3
x_1	1	0	4/3	1/3	0	14/3
Z	0	0	11/3	5/3	1	58/3

<u>Ans.</u> $Z = 58/3$ when $x_1 = 14/3$, $x_2 = 2/3$, $x_3 = 0$

7.

	x_1	x_2	s_1	s_2	s_3	t_3	W	
	1	-1	1	0	0	0	0	1
	1	2	0	1	0	0	0	8
	1	1	0	0	-1	1	0	5
	-1	10	0	0	0	M	1	0

	x_1	x_2	s_1	s_2	s_3	t_3	W		
s_1	[1]	-1	1	0	0	0	0	1	1
s_2	1	2	0	1	0	0	0	8	8
t_3	1	1	0	0	-1	1	0	5	5
W	-1-M	10-M	0	0	0	M	0	1	-5M

	x_1	x_2	s_1	s_2	s_3	t_3	W		
x_1	1	-1	1	0	0	0	0	1	
s_2	0	3	-1	1	0	0	0	7	7/3
t_3	0	[2]	-1	0	-1	1	0	4	2
W	0	9-2M	1+M	0	M	0	1	1-4M	

	x_1	x_2	s_1	s_2	s_3	t_3	W	
x_1	1	0	1/2	0	-1/2	1/2	0	3
s_2	0	0	1/2	1	3/2	-3/2	0	1
x_2	0	1	-1/2	0	-1/2	1/2	0	2
W	0	0	11/2	0	9/2	$-\frac{9}{2}$+M	1	-17

For the above tableau, $t_3 = 0$. Thus W = Z.

<u>Ans.</u> Z = -17 when x_1 = 3, x_2 = 2

9. We first write the third constraint as $-x_1 + x_2 + x_3 \geq 6$.

	x_1	x_2	x_3	s_1	s_2	s_3	t_2	t_3	W	
	1	1	1	1	0	0	0	0	0	1
	1	-1	1	0	-1	0	1	0	0	2
	-1	1	1	0	0	-1	0	1	0	6
	-3	2	-1	0	0	0	M	M	1	0

	x_1	x_2	x_3	s_1	s_2	s_3	t_2	t_3	W		
s_1	1	1	[1]	1	0	0	0	0	0	1	1
t_2	1	-1	1	0	-1	0	1	0	0	2	2
t_3	-1	1	1	0	0	-1	0	1	0	6	6
W	-3	2	-1-2M	0	M	M	0	0	1	-8M	

	x_1	x_2	x_3	s_1	s_2	s_3	t_2	t_3	W	
x_3	1	1	1	1	0	0	0	0	0	1
t_2	0	-2	0	-1	-1	0	1	0	0	1
t_3	-2	0	0	-1	0	-1	0	1	0	5
W	-2+2M	3+2M	0	1+2M	M	M	0	0	1	1-6M

<u>Ans.</u> no solution (empty feasible region)

	x_1	x_2	s_1	s_3	t_2	t_3	W	
11.	1	-1	1	0	0	0	0	4
	-1	1	0	0	1	0	0	4
	1	0	0	-1	0	1	0	6
	3	-2	0	0	M	M	1	0

	x_1	x_2	s_1	s_3	t_2	t_3	W		
s_1	1	-1	1	0	0	0	0	4	
t_2	-1	[1]	0	0	1	0	0	4	4
t_3	1	0	0	-1	0	1	0	6	
W	3	-2-M	0	M	0	0	1	-10M	

	x_1	x_2	s_1	s_3	t_2	t_3	W		
s_1	0	0	1	0	1	0	0	8	
x_2	-1	1	0	0	1	0	0	4	
t_3	[1]	0	0	-1	0	1	0	6	6
W	1-M	0	0	M	2+M	0	1	8-6M	

	x_1	x_2	s_1	s_3	t_2	t_3	W	
s_1	0	0	1	0	1	0	0	8
x_2	0	1	0	-1	1	1	0	10
x_1	1	0	0	-1	0	1	0	6
W	0	0	0	1	2+M	-1+M	1	2

For the above tableau, $t_2 = t_3 = 0$. Thus $W = Z$.

<u>Ans.</u> $Z = 2$ when $x_1 = 6$, $x_2 = 10$

13. Let x_1 and x_2 denote the numbers of Standard and Executive bookcases produced, respectively, each week. We want to maximize the profit function $P = 10x_1 + 12x_2$ subject to

$$x_1 + 2x_2 \leqq 400,$$
$$2x_1 + 3x_2 \leqq 510,$$
$$2x_1 + 3x_2 \geqq 240,$$
$$x_1, x_2 \geqq 0.$$

The artificial objective function is $W = P - Mt_3$.

	x_1	x_2	s_1	s_2	s_3	t_3	W	
	1	2	1	0	0	0	0	400
	2	3	0	1	0	0	0	510
	2	3	0	0	-1	1	0	240
	-10	-12	0	0	0	M	1	0

	x_1	x_2	s_1	s_2	s_3	t_3	W		
s_1	1	2	1	0	0	0	0	400	200
s_2	2	3	0	1	0	0	0	510	170
t_3	2	③	0	0	-1	1	0	240	80
W	-10-2M	-12-3M	0	0	M	0	1	-240M	

	x_1	x_2	s_1	s_2	s_3	t_3	W		
s_1	-1/3	0	1	0	2/3	-2/3	0	240	360
s_2	0	0	0	1	①	-1	0	270	270
x_2	2/3	1	0	0	-1/3	1/3	0	80	
W	-2	0	0	0	-4	4+M	1	960	

	x_1	x_2	s_1	s_2	s_3	P		
s_1	-1/3	0	1	-2/3	0	0	60	
s_3	0	0	0	1	1	0	270	
x_2	⌊2/3⌋	1	0	1/3	0	0	170	255
P	-2	0	0	4	0	1	2040	

	x_1	x_2	s_1	s_2	s_3	P	
s_1	0	1/2	1	-1/2	0	0	145
s_3	0	0	0	1	1	0	270
x_1	1	3/2	0	1/2	0	0	255
P	0	3	0	5	0	1	2550

<u>Ans.</u> 255 Standard, 0 Executive bookcases

15. Suppose I is the total investment. Let x_1, x_2, and x_3 be the proportions invested in A, AA, and AAA bonds, respectively. If Z is the total annual yield expressed as a proportion of I, then $ZI = 0.08x_1I + 0.07x_2I + 0.06x_3I$, or equivalently, $Z = 0.08x_1 + 0.07x_2 + 0.06x_3$. We want to maximize Z subject to

$$x_1 + x_2 + x_3 = 1,$$
$$x_2 + x_3 \geq 0.50,$$
$$x_1 + x_2 \leq 0.30,$$
$$x_1,\ x_2,\ x_3 \geq 0.$$

The artificial objective function is $W = Z - Mt_1 - Mt_2$.

	x_1	x_2	x_3	s_2	s_3	t_1	t_2	W		
	1	1	1	0	0	1	0	0|	1	
	0	1	1	-1	0	0	1	0|	.5	
	1	1	0	0	1	0	0	0|	.3	
	-.08	-.07	-.06	0	0	M	M	1|	0	

	x_1	x_2	x_3	s_2	s_3	t_1	t_2	W		
t_1	1	1	1	0	0	1	0	0|	1	1
t_2	0	1	1	-1	0	0	1	0|	.5	.5
s_3	1	☐1	0	0	1	0	0	0|	.3	.3
W	-.08-M	-.07-2M	-.06-2M	M	0	0	0	1|	-1.5M	

	x_1	x_2	x_3	s_2	s_3	t_1	t_2	W		
t_1	0	0	1	0	-1	1	0	0|	.7	.7
t_2	-1	0	☐1	-1	-1	0	1	0|	.2	.2
x_2	1	1	0	0	1	0	0	0|	.3	
W	-.01+M	0	-.06-2M	M	.07+2M	0	0	1|	.021-.9M	

	x_1	x_2	x_3	s_2	s_3	t_1	t_2	W		
t_1	1	0	0	1	0	1	-1	0|	.5	.5
x_3	-1	0	1	-1	-1	0	1	0|	.2	
x_2	☐1	1	0	0	1	0	0	0|	.3	.3
W	-.07-M	0	0	-.06-M	.01	0	.06+2M	1|	.033-.5M	

	x_1	x_2	x_3	s_2	s_3	t_1	t_2	W		
t_1	0	-1	0	☐1	-1	1	-1	0|	.2	.2
x_3	0	1	1	-1	0	0	1	0|	.5	
x_1	1	1	0	0	1	0	0	0|	.3	
W	0	.07+M	0	-.06-M	.08+M	0	.06+2M	1|	.054-.2M	

	x_1	x_2	x_3	s_2	s_3	t_1	t_2	W	
s_2	0	-1	0	1	-1	1	-1	0|	.2
x_3	0	0	1	0	-1	1	0	0|	.7
x_1	1	1	0	0	1	0	0	0|	.3
W	0	.01	0	0	.02	.06+M	M	1|	.066

For the above tableau, $t_1 = t_2 = 0$. Thus W = Z.

<u>Ans.</u> 30% in A, 0% in AA, 70% in AAA; 6.6%

EXERCISE 7.7

1.

	x_1	x_2	s_1	s_2	t_1	t_2	W	
	-1	1	-1	0	1	0	0 |	6
	1	1	0	-1	0	1	0 |	10
	3	6	0	0	M	M	1 |	0

	x_1	x_2	s_1	s_2	t_1	t_2	W		
t_1	-1	[1]	-1	0	1	0	0	6	6
t_2	1	1	0	-1	0	1	0	10	10
W	3	6-2M	M	M	0	0	1	-16M	
x_2	-1	1	-1	0	1	0	0	6	
t_2	[2]	0	1	-1	-1	1	0	4	2
W	9-2M	0	6-M	M	-6+2M	0	1	-36-4M	
x_2	0	1	-1/2	-1/2	1/2	1/2	0	8	
x_1	1	0	1/2	-1/2	-1/2	1/2	0	2	
W	0	0	3/2	9/2	$-\frac{3}{2}+M$	$-\frac{9}{2}+M$	1	-54	

<u>Ans.</u> Z = 54 when x_1 = 2, x_2 = 8

3.

	x_1	x_2	x_3	s	t	W		
	1	-1	-1	-1	1	0	9	
	4	2	1	0	M	1	0	
t	[1]	-1	-1	-1	1	0	9	9
W	4-M	2+M	1+M	M	0	1	-9M	
x_1	1	-1	-1	-1	1	0	9	
W	0	6	5	4	-4+M	1	-36	

<u>Ans.</u> Z = 36 when x_1 = 9, x_2 = 0, x_3 = 0

5. We write the second constraint as $-x_1 + x_3 \geq 4$.

	x_1	x_2	x_3	s_1	s_2	s_3	t_2	W		
	1	1	1	1	0	0	0	0	6	
	-1	0	1	0	-1	0	1	0	4	
	0	1	1	0	0	1	0	0	5	
	2	3	1	0	0	0	M	1	0	
s_1	1	1	1	1	0	0	0	0	6	6
t_2	-1	0	[1]	0	-1	0	1	0	4	4
s_3	0	1	1	0	0	1	0	0	5	5
W	2+M	3	1-M	0	M	0	0	1	-4M	

	x_1	x_2	x_3	s_1	s_2	s_3	t_2	W	
s_1	2	1	0	1	1	0	-1	0	2
x_3	-1	0	1	0	-1	0	1	0	4
s_3	1	1	0	0	1	1	-1	0	1
W	3	3	0	0	1	0	-1+M	1	-4

<u>Ans.</u> Z = 4 when x_1 = 0, x_2 = 0, x_3 = 4

7.

	x_1	x_2	x_3	s_3	t_1	t_2	W		
	1	2	1	0	1	0	0	4	
	0	1	1	0	0	1	0	1	
	1	1	0	1	0	0	0	6	
W	1	-1	-3	0	M	M	1	0	

	x_1	x_2	x_3	s_3	t_1	t_2	W		
t_1	1	2	1	0	1	0	0	4	2
t_2	0	[1]	1	0	0	1	0	1	1
s_3	1	1	0	1	0	0	0	6	6
W	1-M	-1-3M	-3-2M	0	0	0	1	-5M	

	x_1	x_2	x_3	s_3	t_1	t_2	W		
t_1	[1]	0	-1	0	1	-2	0	2	2
x_2	0	1	1	0	0	1	0	1	
s_3	1	0	-1	1	0	-1	0	5	5
W	1-M	0	-2+M	0	0	1+3M	1	1-2M	

	x_1	x_2	x_3	s_3	t_1	t_2	W		
x_1	1	0	-1	0	1	-2	0	2	
x_2	0	1	[1]	0	0	1	0	1	1
s_3	0	0	0	1	-1	1	0	3	
W	0	0	-1	0	-1+M	3+M	1	-1	

	x_1	x_2	x_3	s_3	-Z	
x_1	1	1	0	0	0	3
x_3	0	1	1	0	0	1
s_3	0	0	0	1	0	3
-Z	0	1	0	0	1	0

<u>Ans.</u> Z = 0 when x_1 = 3, x_2 = 0, x_3 = 1

9.

	x_1	x_2	x_3	s_1	s_2	t_1	t_2	W	
	1	1	1	-1	0	1	0	0	8
	-1	2	1	0	-1	0	1	0	2
	1	8	5	0	0	M	M	1	0

	x_1	x_2	x_3	s_1	s_2	t_1	t_2	W		
t_1	1	1	1	-1	0	1	0	0	8	8
t_2	-1	[2]	1	0	-1	0	1	0	2	1
W	1	$8-3M$	$5-2M$	M	M	0	0	1	$-10M$	

	x_1	x_2	x_3	s_1	s_2	t_1	t_2	W		
x_2	[3/2]	0	1/2	-1	1/2	1	-1/2	0	7	14/3
t_2	-1/2	1	1/2	0	-1/2	0	1/2	0	1	
W	$5-\frac{3}{2}M$	0	$1-\frac{1}{2}M$	M	$4-\frac{1}{2}M$	0	$-4+\frac{3}{2}M$	1	$-8-7M$	

	x_1	x_2	x_3	s_1	s_2	t_1	t_2	W		
x_1	1	0	1/3	-2/3	1/3	2/3	-1/3	0	14/3	14
t_2	0	1	[2/3]	-1/3	-1/3	1/3	1/3	0	10/3	5
W	0	0	-2/3	10/3	7/3	$-\frac{10}{3}+M$	$-\frac{7}{3}+M$	1	$-94/3$	

	x_1	x_2	x_3	s_1	s_2	t_1	t_2	W	
x_1	1	-1/2	0	-1/2	1/2	1/2	-1/2	0	3
x_3	0	3/2	1	-1/2	-1/2	1/2	1/2	0	5
W	0	1	0	3	2	$-3+M$	$-2+M$	1	-28

<u>Ans.</u> Z = 28 when $x_1 = 3$, $x_2 = 0$, $x_3 = 5$

11. Let x_1, x_2, and x_3 denote the annual numbers of barrels of cement produced in kilns that use device A, device B, and no device, respectively. We want to minimize the annual emission control cost C (C in dollars) where

$$C = \tfrac{1}{4}x_1 + \tfrac{2}{5}x_2 + 0x_3 \text{ subject to}$$

$$x_1 + x_2 + x_3 = 3{,}300{,}000,$$

$$\tfrac{1}{2}x_1 + \tfrac{1}{4}x_2 + 2x_3 \leq 1{,}000{,}000,$$

$$x_1, x_2, x_3 \geq 0.$$

	x_1	x_2	x_3	s_2	t_1	W	
	1	1	1	0	1	0	3,300,000
	1/2	1/4	2	1	0	0	1,000,000
	1/4	2/5	0	0	M	1	0

	x_1	x_2	x_3	s_2	t_1	W		
t_1	1	1	1	0	1	0	3,300,000	3,300,000
s_2	1/2	1/4	[2]	1	0	0	1,000,000	500,000
W	$\tfrac{1}{4}-M$	$\tfrac{2}{5}-M$	$-M$	0	0	1	$-3{,}300{,}000M$	

	x_1	x_2	x_3	s_2	t_1	W		
t_1	3/4	[7/8]	0	-1/2	1	0	2,800,000	3,200,000
x_3	1/4	1/8	1	1/2	0	0	500,000	4,000,000
W	$\tfrac{1}{4}-\tfrac{3}{4}M$	$\tfrac{2}{5}-\tfrac{7}{8}M$	0	$\tfrac{1}{2}M$	0	1	$-2{,}800{,}000M$	

	x_1	x_2	x_3	s_2	t_1	W		
x_2	6/7	1	0	-4/7	8/7	0	3,200,000	$\frac{11,200,000}{3}$
x_3	$\boxed{1/7}$	0	1	4/7	-1/7	0	100,000	700,000
W	$-\frac{13}{140}$	0	0	$\frac{8}{35}$	$-\frac{16}{35}+M$	1	-1,280,000	

	x_1	x_2	x_3	s_2	-C	
x_2	0	1	-6	-4	0	2,600,000
x_1	1	0	7	4	0	700,000
-C	0	0	$\frac{13}{20}$	$\frac{3}{5}$	1	-1,215,000

Thus the minimum value of C is 1,215,000 when x_1 = 700,000,
x_2 = 2,600,000, x_3 = 0.

Ans. Install device A on kilns producing 700,000 barrels
annually, and device B on kilns producing 2,600,000
barrels annually.

13. Let x_A = number of refrigerators shipped from A to Exton,
    x_B = number of refrigerators shipped from B to Exton,
    y_A = number of refrigerators shipped from A to Whyton,
    y_B = number of refrigerators shipped from B to Whyton.

We want to minimize C = $15x_A$ + $11x_B$ + $13y_A$ + $12y_B$
subject to
$$x_A + x_B = 30,$$
$$y_A + y_B = 30,$$
$$x_A + y_A \leq 50,$$
$$x_B + y_B \leq 20,$$
$$x_A, \ x_B, \ y_A, \ y_B \geq 0.$$

x_A	x_B	y_A	y_B	s_3	s_4	t_1	t_2	W	
1	1	0	0	0	0	1	0	0	30
0	0	1	1	0	0	0	1	0	30
1	0	1	0	1	0	0	0	0	50
0	1	0	1	0	1	0	0	0	20
15	11	13	12	0	0	M	M	1	0

	x_A	x_B	y_A	y_B	s_3	s_4	t_1	t_2	W		
t_1	1	1	0	0	0	0	1	0	0 |	30	30
t_2	0	0	1	1	0	0	0	1	0 |	30	
s_3	1	0	1	0	1	0	0	0	0 |	50	
s_4	0	[1]	0	1	0	1	0	0	0 |	20	20
W	15-M	11-M	13-M	12-M	0	0	0	0	1 |	-60M	

	x_A	x_B	y_A	y_B	s_3	s_4	t_1	t_2	W		
t_1	1	0	0	-1	0	-1	1	0	0 |	10	
t_2	0	0	[1]	1	0	0	0	1	0 |	30	30
s_3	1	0	1	0	1	0	0	0	0 |	50	50
x_B	0	1	0	1	0	1	0	0	0 |	20	
W	15-M	0	13-M	1	0	-11+M	0	0	1 |	-220-40M	

	x_A	x_B	y_A	y_B	s_3	s_4	t_1	t_2	W		
t_1	[1]	0	0	-1	0	-1	1	0	0 |	10	10
y_A	0	0	1	1	0	0	0	1	0 |	30	
s_3	1	0	0	-1	1	0	0	-1	0 |	20	20
x_B	0	1	0	1	0	1	0	0	0 |	20	
W	15-M	0	0	-12+M	0	-11+M	0	-13+M	1 |	-610-10M	

	x_A	x_B	y_A	y_B	s_3	s_4	t_1	t_2	W	
x_A	1	0	0	-1	0	-1	1	0	0 |	10
y_A	0	0	1	1	0	0	0	1	0 |	30
s_3	0	0	0	0	1	1	-1	-1	0 |	10
x_B	0	1	0	1	0	1	0	0	0 |	20
W	0	0	0	3	0	4	-15+M	-13+M	1 |	-760

<u>Ans.</u> to Exton, 10 from A and 20 from B; to Whyton, 30
from A; $760

15. Roll width $\begin{cases}15" \\ 10"\end{cases}$

3	2	1	0
0	1	3	4

Trim loss   3   8   3   8

We want to minimize $3x_1 + 8x_2 + 3x_3 + 8x_4$ subject to

$$3x_1 + 2x_2 + x_3 \geq 50,$$
$$x_2 + 3x_3 + 4x_4 \geq 60,$$
$$x_1, x_2, x_3, x_4 \geq 0.$$

	x_1	x_2	x_3	x_4	s_1	s_2	t_1	t_2	W		
	3	2	1	0	-1	0	1	0	0	50	
	0	1	3	4	0	-1	0	1	0	60	
	3	8	3	8	0	0	M	M	1	0	
t_1	3	2	1	0	-1	0	1	0	0	50	50
t_2	0	1	③	4	0	-1	0	1	0	60	20
W	3-3M	8-3M	3-4M	8-4M	M	M	0	0	1	-110M	
t_1	③	5/3	0	-4/3	-1	1/3	1	-1/3	0	30	10
x_3	0	1/3	1	4/3	0	-1/3	0	1/3	0	20	60
W	3-3M	$7-\frac{5}{3}M$	0	$4+\frac{4}{3}M$	M	$1-\frac{1}{3}M$	0	$-1+\frac{4}{3}M$	1	-60-30M	
x_1	1	5/9	0	-4/9	-1/3	1/9	1/3	-1/9	0	10	
x_3	0	1/3	1	4/3	0	-1/3	0	1/3	0	20	
W	0	$\frac{16}{3}$	0	$\frac{16}{3}$	1	$\frac{2}{3}$	-1+M	$-\frac{2}{3}+M$	1	-90	

Ans. (a) Column 3: 1, 3, 3; Column 4: 0, 4, 8.

(b) $x_1 = 10$, $x_2 = 0$, $x_3 = 20$, $x_4 = 0$.   (c) 90 in.

EXERCISE 7.8

1. Minimize $W = 6y_1 + 4y_2$
   subject to

$$y_1 - y_2 \geq 2,$$
$$y_1 + y_2 \geq 3,$$
$$y_1, y_2 \geq 0.$$

3. Maximize $W = 8y_1 + 2y_2$
   subject to

$$y_1 - y_2 \leq 1,$$
$$y_1 + 2y_2 \leq 8,$$
$$y_1 + y_2 \leq 5,$$
$$y_1, y_2 \geq 0.$$

5. Min. $W = 13y_1 - 3y_2 - 11y_3$

    subject to

$$-y_1 + y_2 - y_3 \geq 1,$$
$$2y_1 - y_2 - y_3 \geq -1,$$
$$y_1, y_2, y_3 \geq 0.$$

7. Maximize $W = -3y_1 + 3y_2$

    subject to

$$-y_1 + y_2 \leq 4,$$
$$y_1 - y_2 \leq 4,$$
$$y_1 + y_2 \leq 6,$$
$$y_1, y_2 \geq 0.$$

9. Dual is: Maximize $W = y_1 + 2y_2$ subject to

$$y_1 - y_2 \leq 4,$$
$$-y_1 + y_2 \leq 4,$$
$$y_1 + y_2 \leq 6,$$
$$y_1, y_2 \geq 0.$$

	y_1	y_2	s_1	s_2	s_3	W		
s_1	1	-1	1	0	0	0	4	
s_2	-1	[1]	0	1	0	0	4	4
s_3	1	1	0	0	1	0	6	6
W	-1	-2	0	0	0	1	0	
s_1	0	0	1	1	0	0	8	
y_2	-1	1	0	1	0	0	4	
s_3	[2]	0	0	-1	1	0	2	1
W	-3	0	0	2	0	1	8	
s_1	0	0	1	1	0	0	8	
y_2	0	1	0	1/2	1/2	0	5	
y_1	1	0	0	-1/2	1/2	0	1	
W	0	0	0	1/2	3/2	1	11	

<u>Ans.</u> $Z = 11$ when $x_1 = 0$, $x_2 = 1/2$, $x_3 = 3/2$

11. Dual is: Minimize $W = 8y_1 + 12y_2$ subject to

$$y_1 + y_2 \geq 3,$$
$$2y_1 + 6y_2 \geq 8,$$
$$y_1, y_2 \geq 0.$$

	y_1	y_2	s_1	s_2	t_1	t_2	U		
	1	1	-1	0	1	0	0	3	
	2	6	0	-1	0	1	0	8	
	8	12	0	0	M	M	1	0	

	y_1	y_2	s_1	s_2	t_1	t_2	U		
t_1	1	1	-1	0	1	0	0	3	3
t_2	2	6	0	-1	0	1	0	8	4/3
U	8-3M	12-7M	M	M	0	0	1	-11M	

	y_1	y_2	s_1	s_2	t_1	t_2	U		
t_1	2/3	0	-1	1/6	1	-1/6	0	5/3	5/2
y_2	1/3	1	0	-1/6	0	1/6	0	4/3	4
U	$4-\frac{2}{3}M$	0	M	$2-\frac{1}{6}M$	0	$-2+\frac{7}{6}M$	1	$-16-\frac{5}{3}M$	

	y_1	y_2	s_1	s_2	t_1	t_2	U	
y_1	1	0	-3/2	1/4	3/2	-1/4	0	5/2
y_2	0	1	1/2	-1/4	-1/2	1/4	0	1/2
U	0	0	6	1	-6+M	-1+M	1	-26

<u>Ans.</u> $Z = 26$ when $x_1 = 6$, $x_2 = 1$

13. Dual is: Maximize $W = -y_1 + 3y_2$ subject to

$$y_1 + y_2 \leq 6,$$
$$-y_1 + y_2 \leq 4,$$
$$y_1, y_2 \geq 0.$$

	y_1	y_2	s_1	s_2	W		
s_1	1	1	1	0	0	6	6
s_2	-1	1	0	1	0	4	4
W	1	-3	0	0	1	0	

	y_1	y_2	s_1	s_2	W		
s_1	2	0	1	-1	0	2	1
y_2	-1	1	0	1	0	4	
W	-2	0	0	3	1	12	

	y_1	y_2	s_1	s_2	W	
y_1	1	0	1/2	-1/2	0	1
y_2	0	1	1/2	1/2	0	5
W	0	0	1	2	1	14

<u>Ans.</u> $Z = 14$ when $x_1 = 1$, $x_2 = 2$

15. Let x_1 = amount spent on newspaper advertising,

$\quad\quad x_2$ = amount spent on radio advertising.

We want to minimize $C = x_1 + x_2$ subject to

$$40x_1 + 50x_2 \geq 8000,$$
$$100x_1 + 25x_2 \geq 6000,$$
$$x_1, x_2 \geq 0.$$

The dual is:

$\quad\quad$ Maximize $W = 8000y_1 + 6000y_2$ subject to

$$40y_1 + 100y_2 \leq 1,$$
$$50y_1 + 25y_2 \leq 1,$$
$$y_1, y_2 \geq 0.$$

$$
\begin{array}{c}
\begin{array}{ccccc}
y_1 & y_2 & s_1 & s_2 & W
\end{array} \\
\begin{array}{c} s_1 \\ s_2 \\ W \end{array}
\left[
\begin{array}{ccccc|c}
40 & 100 & 1 & 0 & 0 & 1 \\
\boxed{50} & 25 & 0 & 1 & 0 & 1 \\
\hline
-8000 & -6000 & 0 & 0 & 1 & 0
\end{array}
\right]
\begin{array}{c} 1/40 \\ 1/50 \\ \\ \end{array}
\end{array}
$$

$$
\begin{array}{c}
\begin{array}{c} s_1 \\ y_1 \\ W \end{array}
\left[
\begin{array}{ccccc|c}
0 & \boxed{80} & 1 & -\frac{4}{5} & 0 & \frac{1}{5} \\
1 & \frac{1}{2} & 0 & \frac{1}{50} & 0 & \frac{1}{50} \\
\hline
0 & -2000 & 0 & 160 & 1 & 160
\end{array}
\right]
\begin{array}{c} \frac{1}{400} \\ \frac{1}{25} \\ \\ \end{array}
\end{array}
$$

$$
\begin{array}{c}
\begin{array}{c} y_2 \\ y_1 \\ W \end{array}
\left[
\begin{array}{ccccc|c}
0 & 1 & \frac{1}{80} & -\frac{1}{100} & 0 & \frac{1}{400} \\
1 & 0 & -\frac{1}{160} & \frac{1}{40} & 0 & \frac{3}{160} \\
\hline
0 & 0 & 25 & 140 & 1 & 165
\end{array}
\right]
\end{array}
$$

Ans. $25 on newspaper advertising, $140 on radio advertising; $165

17. Let y_1 = number of shipping clerk apprentices,

$\quad\quad y_2$ = number of shipping clerks,

$\quad\quad y_3$ = number of semiskilled workers,

$\quad\quad y_4$ = number of skilled workers.

We want to minimize $W = 2y_1 + 5y_2 + 4y_3 + 7y_4$ subject to

$$y_1 + y_2 \geq 60,$$
$$-2y_1 + y_2 \geq 0,$$
$$y_3 + y_4 \geq 90,$$
$$y_3 - 2y_4 \geq 0,$$
$$y_1, y_2, y_3, y_4 \geq 0.$$

The dual is: Maximize $Z = 60x_1 + 0x_2 + 90x_3 + 0x_4$

subject to

$$x_1 - 2x_2 \leq 2,$$
$$x_1 + x_2 \leq 5,$$
$$x_3 + x_4 \leq 4,$$
$$x_3 - 2x_4 \leq 7,$$
$$x_1, x_2, x_3, x_4 \geq 0.$$

	x_1	x_2	x_3	x_4	s_1	s_2	s_3	s_4	Z		
s_1	1	-2	0	0	1	0	0	0	0	2	
s_2	1	1	0	0	0	1	0	0	0	5	
s_3	0	0	[1]	1	0	0	1	0	0	4	4
s_4	0	0	1	-2	0	0	0	1	0	7	7
Z	-60	0	-90	0	0	0	0	0	1	0	

	x_1	x_2	x_3	x_4	s_1	s_2	s_3	s_4	Z		
s_1	[1]	-2	0	0	1	0	0	0	0	2	2
s_2	1	1	0	0	0	1	0	0	0	5	5
x_3	0	0	1	1	0	0	1	0	0	4	
s_4	0	0	0	-3	0	0	-1	1	0	3	
Z	-60	0	0	90	0	0	90	0	1	360	

	x_1	x_2	x_3	x_4	s_1	s_2	s_3	s_4	Z		
x_1	1	-2	0	0	1	0	0	0	0	2	
s_2	0	[3]	0	0	-1	1	0	0	0	3	1
x_3	0	0	1	1	0	0	1	0	0	4	
s_4	0	0	0	-3	0	0	-1	1	0	3	
Z	0	-120	0	90	60	0	90	0	1	480	

	x_1	x_2	x_3	x_4	s_1	s_2	s_3	s_4	Z	
x_1	1	0	0	0	1/3	2/3	0	0	0	4
x_2	0	1	0	0	-1/3	1/3	0	0	0	1
x_3	0	0	1	1	0	0	1	0	0	4
s_4	0	0	0	-3	0	0	-1	1	0	3
Z	0	0	0	90	20	40	90	0	1	600

Ans. 20 shipping clerk apprentices, 40 shipping clerks, 90 semiskilled workers, 0 skilled workers; $600

CHAPTER 7 - REVIEW PROBLEMS

1.

3.

5.

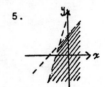

7.

9.

11. Feasible region appears below. Corner points are (0, 0),
(0, 2), (1, 3), (3, 1), (3, 0). Z is maximized at (3, 0)
where its value is 3. Thus Z = 3 when x = 3 and y = 0.

    11.   13.

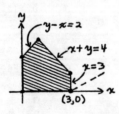

13. Feasible region (see above) is unbounded. Z is minimized
at the corner point (0, 2) where its value is -2. Thus
Z = -2 when x = 0 and y = 2.

15. Feasible region (see below) is empty. Thus there is no optimum solution.

15. 17.

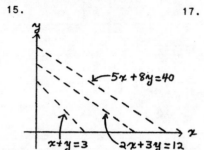

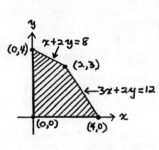

17. Feasible region appears above. The corner points are (0, 0), (0, 4), (2, 3), and (4, 0). Z is maximized at (2, 3) and (4, 0) where its value is 36. Thus Z is maximized at all points on the line segment joining (2, 3) and (4, 0). The solution is

$$Z = 36 \text{ when } x = (1 - t)(2) + 4t = 2 + 2t,$$
$$y = (1 - t)(3) + 0t = 3 - 3t, \text{ and } 0 \le t \le 1$$

	x_1	x_2	s_1	s_2	Z			
19. s_1	1	⑥	1	0	0		12	2
s_2	1	2	0	1	0		8	4
Z	-4	-5	0	0	1		0	
x_2	1/6	1	1/6	0	0		2	12
s_2	⌊2/3⌋	0	-1/3	1	0		4	6
Z	-19/6	0	5/6	0	1		10	
x_2	0	1	⌊1/4⌋	-1/4	0		1	4
x_1	1	0	-1/2	3/2	0		6	
Z	0	0	-3/4	19/4	1		29	
s_1	0	4	1	-1	0		4	
x_1	1	2	0	1	0		8	
Z	0	3	0	4	1		32	

Thus $Z = 32$ when $x_1 = 8$ and $x_2 = 0$.

21.

	x_1	x_2	x_3	s	t	W	
	1	2	3	-1	1	0	6
	2	3	1	0	M	1	0

	x_1	x_2	x_3	s	t	W		
t	1	2	③	-1	1	0	6	2
W	2-M	3-2M	1-3M	M	0	1	-6M

	x_1	x_2	x_3	s	t	W	
x_3	1/3	2/3	1	-1/3	1/3	0	2
W	5/3	7/3	0	1/3	$-\frac{1}{3}$+M	1	-2

Thus $Z = 2$ when $x_1 = 0$, $x_2 = 0$ and $x_3 = 2$.

23.

	x_1	x_2	s_1	s_2	s_3	t_2	W	
	1	1	1	0	0	0	0	12
	1	1	0	-1	0	1	0	5
	1	0	0	0	1	0	0	10
	-1	-2	0	0	0	M	1	0

	x_1	x_2	s_1	s_2	s_3	t_2	W		
s_1	1	1	1	0	0	0	0	12	12
t_2	1	①	0	-1	0	1	0	5	5
s_3	1	0	0	0	1	0	0	10	
W	-1-M	-2-M	0	M	0	0	1	-5M	

	x_1	x_2	s_1	s_2	s_3	t_2	W		
s_1	0	0	1	①	0	-1	0	7	7
x_2	1	1	0	-1	0	1	0	5	
s_3	1	0	0	0	1	0	0	10	
W	1	0	0	-2	0	2+M	1	10	

	x_1	x_2	s_1	s_2	s_3	Z	
s_2	0	0	1	1	0	0	7
x_2	1	1	1	0	0	0	12
s_3	1	0	0	0	1	0	10
Z	1	0	2	0	0	1	24

Thus $Z = 24$ when $x_1 = 0$ and $x_2 = 12$.

25. We write the first constraint as $-x_1 + x_2 + x_3 \geq 1$.

$$
\begin{array}{ccccccc}
x_1 & x_2 & x_3 & s_1 & t_1 & t_2 & W
\end{array}
$$

$$
\left[\begin{array}{ccccccc|c}
-1 & 1 & 1 & -1 & 1 & 0 & 0 & 1 \\
6 & 3 & 2 & 0 & 0 & 1 & 0 & 12 \\
\hline
1 & 2 & 1 & 0 & M & M & 1 & 0
\end{array}\right]
$$

$$
\begin{array}{c}
t_1 \\
t_2 \\
W
\end{array}
\left[\begin{array}{ccccccc|c}
-1 & 1 & 1 & -1 & 1 & 0 & 0 & 1 \\
\boxed{6} & 3 & 2 & 0 & 0 & 1 & 0 & 12 \\
\hline
1-5M & 2-4M & 1-3M & M & 0 & 0 & 1 & -13M
\end{array}\right] \begin{array}{c} \\ 2 \\ \end{array}
$$

$$
\begin{array}{c}
t_1 \\
x_1 \\
W
\end{array}
\left[\begin{array}{ccccccc|c}
0 & \boxed{3/2} & 4/3 & -1 & 1 & 1/6 & 0 & 3 \\
1 & 1/2 & 1/3 & 0 & 0 & 1/6 & 0 & 2 \\
\hline
0 & \frac{3}{2}-\frac{3}{2}M & \frac{2}{3}-\frac{4}{3}M & M & 0 & -\frac{1}{6}+\frac{5}{6}M & 1 & -2-3M
\end{array}\right] \begin{array}{c} 2 \\ 4 \\ \end{array}
$$

$$
\begin{array}{c}
x_2 \\
x_1 \\
W
\end{array}
\left[\begin{array}{ccccccc|c}
0 & 1 & \boxed{8/9} & -2/3 & 2/3 & 1/9 & 0 & 2 \\
1 & 0 & -1/9 & 1/3 & -1/3 & 1/9 & 0 & 1 \\
\hline
0 & 0 & -2/3 & 1 & -1+M & -\frac{1}{3}+M & 1 & -5
\end{array}\right] \begin{array}{c} 9/4 \\ \\ \end{array}
$$

$$
\begin{array}{ccccc}
x_1 & x_2 & x_3 & s_1 & -Z
\end{array}
$$

$$
\begin{array}{c}
x_3 \\
x_1 \\
-Z
\end{array}
\left[\begin{array}{ccccc|c}
0 & 9/8 & 1 & -3/4 & 0 & 9/4 \\
1 & 1/8 & 0 & 1/4 & 0 & 5/4 \\
\hline
0 & 3/4 & 0 & 1/2 & 1 & -7/2
\end{array}\right]
$$

Thus $Z = 7/2$ when $x_1 = 5/4$, $x_2 = 0$, and $x_3 = 9/4$.

$$
\begin{array}{cccccc}
x_1 & x_2 & x_3 & s_1 & s_2 & Z
\end{array}
$$

27. $\begin{array}{c}
s_1 \\
s_2 \\
Z
\end{array}
\left[\begin{array}{cccccc|c}
4 & -1 & 0 & 1 & 0 & 0 & 2 \\
-10 & \boxed{1} & 3 & 0 & 1 & 0 & 1 \\
\hline
-1 & -4 & -2 & 0 & 0 & 1 & 0
\end{array}\right] \begin{array}{c} \\ 1 \\ \end{array}$

$\begin{array}{c}
x_2 \\
s_2 \\
Z
\end{array}
\left[\begin{array}{cccccc|c}
-6 & 0 & 3 & 1 & 1 & 0 & 3 \\
-10 & 1 & 3 & 0 & 0 & 0 & 1 \\
\hline
-41 & 0 & 10 & 0 & 4 & 1 & 4
\end{array}\right]$

For the last tableau, x_1 is the entering variable. Since
no quotients exist, the problem has an unbounded solution.
That is, no optimum solution (unbounded).

29. Dual is: Maximize $W = 35y_1 + 25y_2$ subject to
$$y_1 + y_2 \leq 2,$$
$$2y_1 + y_2 \leq 7,$$
$$3y_1 + y_2 \leq 8,$$
$$y_1, y_2 \geq 0.$$

	y_1	y_2	s_1	s_2	s_3	W		
s_1	☐1	1	1	0	0	0	2	2
s_2	2	1	0	1	0	0	7	7/2
s_3	3	1	0	0	1	0	8	8/3
W	-35	-25	0	0	0	1	0	
y_1	1	1	1	0	0	0	2	
s_2	0	-1	-2	1	0	0	3	
s_3	0	-2	-3	0	1	0	2	
W	0	10	35	0	0	1	70	

Thus $Z = 70$ when $x_1 = 35$, $x_2 = 0$, and $x_3 = 0$.

31. Let x, y, and z denote the numbers of units of X, Y, and Z produced weekly, respectively. If P is the total profit obtained, we want to maximize $P = 10x + 15y + 22z$ subject to
$$x + 2y + 2z \leq 40,$$
$$x + y + 2z \leq 34,$$
$$x, y, z \geq 0.$$

	x	y	z	s_1	s_2	P		
s_1	1	2	2	1	0	0	40	20
s_2	1	1	☐2	0	1	0	34	17
P	-10	-15	-22	0	0	1	0	
s_1	0	☐1	0	1	-1	0	6	6
z	1/2	1/2	1	0	1/2	0	17	34
P	1	-4	0	0	11	1	374	
y	0	1	0	1	-1	0	6	
z	1/2	0	1	-1/2	1	0	14	
P	1	0	0	4	7	1	398	

Thus 0 units of X, 6 units of Y, and 14 units of Z give a maximum profit of $398.

33. Let x_{AC}, x_{AD}, x_{BC}, and x_{BD} denote the amounts (in hundreds of thousands of gallons) transported from A to C, A to D, B to C, and B to D, respectively. If c is the total transportation cost in thousands of dollars, we want to minimize $c = x_{AC} + 2x_{AD} + 2x_{BC} + 4x_{BD}$ subject to

$$x_{AC} + x_{AD} \le 6,$$
$$x_{BC} + x_{BD} \le 6,$$
$$x_{AC} + x_{BC} = 5,$$
$$x_{AD} + x_{BD} = 5,$$
$$x_{AC}, \ x_{AD}, \ x_{BC}, \ x_{BD} \ge 0.$$

	x_{AC}	x_{AD}	x_{BC}	x_{BD}	s_1	s_2	t_3	t_4	W		
	1	1	0	0	1	0	0	0	0	6	
	0	0	1	1	0	1	0	0	0	6	
	1	0	1	0	0	0	1	0	0	5	
	0	1	0	1	0	0	0	1	0	5	
	1	2	2	4	0	0	M	M	1	0	
s_1	1	1	0	0	1	0	0	0	0	6	6
s_2	0	0	1	1	0	1	0	0	0	6	
t_3	[1]	0	1	0	0	0	1	0	0	5	5
t_4	0	1	0	1	0	0	0	1	0	5	
W	1-M	2-M	2-M	4-M	0	0	0	0	1	-10M	
s_1	0	[1]	-1	0	1	0	-1	0	0	1	1
s_2	0	0	1	1	0	1	0	0	0	6	
x_{AC}	1	0	1	0	0	1	1	0	0	5	
t_4	0	1	0	1	0	0	0	1	0	5	5
W	0	2-M	1	4-M	0	0	-1+M	0	1	-5-5M	
x_{AD}	0	1	-1	0	1	0	-1	0	0	1	
s_2	0	0	1	1	0	1	0	0	0	6	6
x_{AC}	1	0	1	0	0	0	1	0	0	5	5
t_4	0	0	[1]	1	-1	0	1	1	0	4	4
W	0	0	3-M	4-M	-2+M	0	1	0	1	-7-4M	
x_{AD}	0	1	0	1	0	0	0	1	0	5	
s_2	0	0	0	0	1	1	-1	-1	0	2	
x_{AC}	1	0	0	-1	1	0	0	-1	0	1	
x_{BC}	0	0	1	1	-1	0	1	1	0	4	
W	0	0	0	1	1	1	-2+M	-3+M	1	-19	

The minimum value of c is 19, when $x_{AC} = 1$, $x_{AD} = 5$, $x_{BC} = 4$, and $x_{BD} = 0$. Thus 100,000 gal from A to C, 500,000 gal from A to D, and 400,000 gal from B to C give a minimum cost of $19,000.

35. Let x (y) represent daily consumption of food A (B) in ounces. We want to minimize $C = 4x + 11y$ subject to the constraints

$$8x + 4y \geq 176,$$
$$16x + 32y \geq 1024,$$
$$2x + 5y \geq 200.$$
$$x \geq 0,$$
$$y \geq 0,$$

The feasible region (see region below) is unbounded with corner points (100, 0), (5/2, 39) and (0, 44). C has a minimum value at (100, 0). Thus the animals should be fed 100 ounces of food A each day.

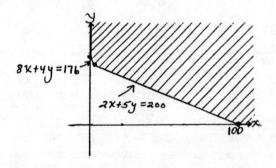

CHAPTER 7 — MATHEMATICAL SNAPSHOT

1.

	CURATIVE UNITS	TOXIC UNITS	RELATIVE DISCOMFORT
Drug (per ounce)	500	400	1
Radiation (per min.)	1000	600	1
Requirement	≥ 2000	≤ 1400	

Let x_1 = number of ounces of drug and let x_2 = number of minutes of radiation. We want to minimize the discomfort D, where $D = x_1 + x_2$, subject to

$$500x_1 + 1000x_2 \geq 2000,$$
$$400x_1 + 600x_2 \leq 1400,$$
$$x_1, x_2 \geq 0.$$

The feasible region is indicated below.

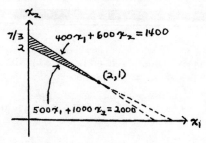

The corner points are $(0,2)$, $\left(0, \frac{7}{3}\right)$, and $(2,1)$. At $(0,2)$, $D = 0 + 2 = 2$; at $\left(0, \frac{7}{3}\right)$, $D = 0 + \frac{7}{3} = \frac{7}{3}$; at $(2,1)$, $D = 2 + 1 = 3$. Thus D is minimum at $(0,2)$.

<u>Ans.</u> 0 ounce of drug, 2 minutes of radiation

8

Mathematics of Finance

1. (a) $6000(1.08)^8 \approx 6000(1.850930) \approx \$11,105.58$
   (b) $11,105.58 - 6000 = \$5105.58$

3. $\left(1 + \frac{0.08}{4}\right)^4 - 1 = (1.02)^4 - 1 \approx 0.082432$ or 8.24%

5. $\left(1 + \frac{0.08}{365}\right)^{365} - 1 \approx 0.0832776$ or 8.32776%

7. (a) A nominal rate compounded yearly is the same as the effective rate, so the effective rate is 10%.

   (b) $\left(1 + \frac{0.10}{2}\right)^2 - 1 \approx 0.1025$ or 10.25%

   (c) $\left(1 + \frac{0.10}{4}\right)^4 - 1 \approx 0.10381$ or 10.381%

   (d) $\left(1 + \frac{0.10}{12}\right)^{12} - 1 \approx 0.10471$ or 10.471%

   (e) $\left(1 + \frac{0.10}{365}\right)^{365} - 1 \approx 0.10516$ or 10.516%

9. Let r_e be the *effective* rate. Then $2000(1 + r_e)^5 = 2950$,

$$(1 + r_e)^5 = \frac{2950}{2000}, \quad 1 + r_e = \sqrt[5]{\frac{2950}{2000}}, \quad r_e = \sqrt[5]{\frac{2950}{2000}} - 1,$$

$$r_e \approx 8.08\%$$

11. From Example 6, the number of years, n, is given by

$$n = \frac{0.69315}{\ln(1.08)} \approx \frac{0.69315}{0.07696} \approx 9.0 \text{ years}$$

13. $6000(1.08)^7 \approx 6000(1.713824) \approx \$10,282.94$

15. $16,000(1.06)^{10} \approx 16,000(1.790848) \approx \$28,653.57$

17. (a) $(0.015)(12) = 0.18$ or 18%

    (b) $(1.015)^{12} - 1 \approx 0.195618$ or 19.56%

19. The compound amount after the first four years is $2000(1.06)^4$. After the next four years the compound amount is

$$[2000(1.06)^4](1.03)^8 \approx 2000(1.262477)(1.266770)$$
$$\approx \$3198.54$$

21. 7.8% compounded semiannually is equivalent to an effective rate of $(1.039)^2 - 1 = 0.079521$ or 7.9521%. Thus 8% compounded annually, which is the effective rate, is the better rate.

23. (a) $\left(1 + \frac{0.0525}{360}\right)^{365} - 1 \approx 0.0547$ or 5.47%

    (b) $\left(1 + \frac{0.0525}{365}\right)^{365} - 1 \approx 0.0539$ or 5.39%

25. Let r_e = effective rate.

$$300,000 = 100,000(1 + r_e)^{10}, \quad (1 + r_e)^{10} = 3, \quad 1 + r_e = \sqrt[10]{3},$$

$$r_e = \sqrt[10]{3} - 1 \approx 0.1161 \text{ or } 11.61\%$$

27. Let r = the required nominal rate

$$220\left(1 + \frac{r}{2}\right)^{28} = 1000, \quad \left(1 + \frac{r}{2}\right)^{28} = \frac{1000}{220} = \frac{50}{11},$$

$$1 + \frac{r}{2} = \sqrt[28]{\frac{50}{11}}, \quad r = 2\left[\sqrt[28]{\frac{50}{11}} - 1\right] \approx 0.1111 \text{ or } 11.11\%$$

EXERCISE 8.2

1. $6000(1.05)^{-20} \approx 6000(0.376889) \approx \$2261.33$

3. $4000(1.035)^{-24} \approx 4000(0.437957) \approx \$1751.83$

5. $2000(1.0075)^{-30} \approx 2000(0.799187) \approx \$1598.37$

7. $8000\left(1 + \frac{0.06}{4}\right)^{-30} \approx 8000(1.015)^{-30} \approx 8000(0.639762)$
$$\approx \$5118.10$$

9. $8000\left(1 + \frac{0.10}{12}\right)^{-60} \approx \$4862.31$

11. $10,000\left(1 + \frac{0.095}{365}\right)^{-4(365)} \approx \$6838.95$

13. $10,000(1 + 0.01)^{-12} \approx 10,000(0.887449) = \$8874.49$

15. $27,000(1.03)^{-22} \approx 27,000(0.521893) \approx \$14,091.11$

17. Let x be the payment 2 years from now. The equation of value at year 2 is
$$x = 600(1.04)^{-2} + 800(1.04)^{-4}$$
$$\approx 600(0.924556) + 800(0.854804)$$
$$\approx \$1238.58.$$

19. Let x be the payment at the end of 6 years. The equation of value at year 6 is
$$2000(1.07)^{4} + 4000(1.07)^{2} + x = 5000(1.07) + 5000(1.07)^{-4}$$
Thus

$$x = 5000(1.07) + 5000(1.07)^{-4} - 2000(1.07)^4 - 4000(1.07)^2$$
$$\approx 5000(1.07) + 5000(0.762895) - 2000(1.310796)$$
$$- 4000(1.144900)$$
$$\approx \$1963.28.$$

21. (a) NPV
$$= 8000(1.025)^{-6} + 10,000(1.025)^{-8} + 14,000(1.025)^{-12} - 25,000$$
$$\approx 8000(0.862297) + 10,000(0.820747) + 14,000(0.743556) - 25,000$$
$$= \$515.63.$$

(b) Since NPV > 0, the investment is profitable.

23. We consider the value of each investment at the end of eight years. The savings account has a value of
$$10,000(1.03)^{16} \approx 1000(1.604706) = \$16,047.06.$$
The business investment has a value of \$16,000. Thus the better choice is the savings account.

25. $1000\left(1 + \frac{0.115}{4}\right)^{-80} \approx \$103.56$

27. Let r be the nominal discount rate, compounded quarterly. Then $4700 = 10,000\left(1 + \frac{r}{4}\right)^{-32}$, $4700 = \dfrac{10,000}{\left(1 + \frac{r}{4}\right)^{32}}$,

$\left(1 + \frac{r}{4}\right)^{32} = \frac{10,000}{4700} = \frac{100}{47}$, $1 + \frac{r}{4} = \sqrt[32]{\frac{100}{47}}$,

$r = 4\left[\sqrt[32]{\frac{100}{47}} - 1\right] \approx 0.0955$ or 9.55%

EXERCISE 8.3

1. 64, $64\left(\frac{1}{2}\right) = 32$, $64\left(\frac{1}{2}\right)^2 = 16$, $64\left(\frac{1}{2}\right)^3 = 8$, $64\left(\frac{1}{2}\right)^4 = 4$

3. 100, $100(1.02) = 102$, $100(1.02)^2 = 104.04$

5. $s = \dfrac{\frac{2}{3}\left[1 - \left(\frac{2}{3}\right)^5\right]}{1 - \frac{2}{3}} = \dfrac{\frac{2}{3}\left[\frac{211}{243}\right]}{\frac{1}{3}} = \dfrac{422}{243}$

7. $s = \dfrac{1[1 - (0.1)^6]}{1 - 0.1} = \dfrac{0.999999}{0.9} = 1.11111$

9. $a_{\overline{35}|.04} \approx 18.664613$

11. $s_{\overline{8}|.0075} \approx 8.213180$

13. $500a_{\overline{5}|.07} \approx 500(4.100197) \approx \$2050.10$

15. $2000a_{\overline{18}|.02} \approx 2000(14.992031) \approx \$29,984.06$

17. $800 + 800a_{\overline{11}|.035} \approx 800 + 800(9.001551) \approx \$8001.24$

19. $2000s_{\overline{36}|.0125} \approx 2000(45.115505) = \$90,231.01$

21. $5000s_{\overline{20}|.07} \approx 5000(40.995492) = \$204,977.46$

23. $1200(s_{\overline{13}|.08} - 1) \approx 1200(21.495297 - 1) \approx \$24,594.36$

25. $75a_{\overline{30}|.005} - 25a_{\overline{6}|.005} \approx 75(27.794054) - 25(5.896384)$
$\approx \$1937.14$

27. $R = \dfrac{5000}{a_{\overline{12}|.015}} \approx \dfrac{5000}{10.907505} \approx \$458.40$

29. (a) $\left(50s_{\overline{48}|.005}\right)(1.005)^{24} \approx 50(54.097832)(1.127160)$
$\approx \$3048.85$

    (b) $3048.85 - 48(50) = \$648.85$

31. $R = \dfrac{48,000}{s_{\overline{10}|.07}} \approx \dfrac{48,000}{13.816448} \approx \$3474.12$

33. The original annual payment is $25,000/s_{\overline{10}|.06}$. After six years the value of the fund is
$$\dfrac{25,000}{s_{\overline{10}|.06}}s_{\overline{6}|.06}.$$
This accumulates to

$$\left[\frac{25,000}{s_{\overline{10}|.06}} s_{\overline{6}|.06}\right](1.07)^4.$$

Let x be the amount of the new payment.

$$x s_{\overline{4}|.07} = 25,000 - \left[\frac{25,000}{s_{\overline{10}|.06}} s_{\overline{6}|.06}(1.07)^4\right]$$

$$x = \frac{25,000 - \left[\frac{25,000}{s_{\overline{10}|.06}} s_{\overline{6}|.06}(1.07)^4\right]}{s_{\overline{4}|.07}}$$

$$x \approx \frac{25,000 - \left[\frac{25,000}{13.180795}(6.975319)(1.310796)\right]}{4.439943}$$

$$x \approx \$1725$$

35. $s_{\overline{60}|.017} = \frac{(1.017)^{60} - 1}{0.017} \approx 102.91305$

37. $700 a_{\overline{360}|.0125} = 700\left[\frac{1 - (1.0125)^{-360}}{0.0125}\right] \approx 55,360.30$

39. $R = \frac{3000}{s_{\overline{20}|.01375}} = \frac{3000(0.01375)}{(1.01375)^{20} - 1} \approx \$131.34$

41. $50,000 + 50,000 a_{\overline{19}|.12}$

$$= 50,000 + 50,000\left[\frac{1 - (1.12)^{-19}}{0.12}\right] \approx \$418,288.84$$

EXERCISE 8.4

1. $R = \frac{2000}{a_{\overline{36}|.0125}} \approx \frac{2000}{28.847267} \approx \$69.33$

3. $R = \frac{8000}{a_{\overline{36}|.01}} \approx \frac{8000}{30.107505} \approx \$265.71$

   Finance charge = $36(265.71) - 8000 = \$1565.56$

5. (a) $R = \frac{7500}{a_{\overline{36}|.01}} \approx \frac{7500}{30.107505} \approx \$249.11$

   (b) $7500(.01) = \$75$

   (c) $249.11 - 75 = \$174.11$

7. $R = \dfrac{5000}{a_{\overline{4}|.07}} \approx \dfrac{5000}{3.387211} \approx \$1476.14$

The interest for the first period is $(0.07)(5000) = \$350$,
so the principal repaid at the end of that period is
$1476.14 - 350 = \$1126.14$. The principal outstanding at
the beginning of period 2 is $5000 - 1126.14 = \$3873.86$.
The interest for period 2 is $(0.07)(3873.86) = \$271.17$,
so the principal repaid at the end of that period is
$1476.14 - 271.17 = \$1204.97$. The principal outstanding
at beginning of period 3 is $3873.86 - 1204.97 = \$2668.89$.
Continuing in this manner, we construct the following
amortization schedule.

Period	Prin. Outs. at beginning	Int. for period	Pmt. at end	Prin. repaid at end
1	5000.00	350.00	1476.14	1126.14
2	3873.86	271.17	1476.14	1204.97
3	2668.89	186.82	1476.14	1289.32
4	1379.57	96.57	1476.14	1379.57
		904.56	5904.56	5000.00

9. $R = \dfrac{900}{a_{\overline{5}|.025}} \approx \dfrac{900}{4.645828} \approx \$193.72$

The interest for period 1 is $(0.025)(900) = \$22.50$,
so the principal repaid at the end of that period is
$193.72 - 22.50 = \$171.22$. The principal outstanding at
the beginning of period 2 is $900 - 171.22 = \$728.78$. The
interest for that period is $(0.025)(728.78) = \$18.22$,
so the principal repaid at the end of that period is
$193.72 - 18.22 = \$175.50$. The principal outstanding at
the beginning of period 3 is $728.78 - 175.50 = \$553.28$.
Continuing in this manner, we obtain the following
amortization schedule. Note the adjustment of the final
payment.

Period	Prin. Outs. at beginning	Int. for period	Pmt. at end	Prin. repaid at end
1	900.00	22.50	193.72	171.22
2	728.78	18.22	193.72	175.50
3	553.28	13.83	193.72	179.89
4	373.39	9.33	193.72	184.39
5	189.00	4.73	193.73	189.00
		68.61	968.61	900.00

11. From Eq.(1),

$$n = \frac{\ln\left[\frac{100}{100 - 1000(0.02)}\right]}{\ln(1.02)} = \frac{\ln(1.25)}{\ln(1.02)} \approx \frac{0.22314}{0.01980} \approx 11.270.$$

Thus the number of full payments is 11.

13. Each of the original payments is $\frac{10,000}{a_{\overline{10}|}.04}$. After two
    years the value of the remaining payments is
    $\left(\frac{10,000}{a_{\overline{10}|}.04}\right)a_{\overline{6}|}.04$. Thus the new annual payment is

$$\frac{10,000 a_{\overline{6}|}.04}{a_{\overline{10}|}.04} \cdot \frac{1}{a_{\overline{6}|}.05} \approx \frac{10,000(5.242137)}{8.110896} \cdot \frac{1}{5.075692} \approx \$1273.$$

15. (a) Monthly interest rate is $0.102/12 = 0.0085$. Monthly
        payment is

$$\frac{45,000}{a_{\overline{300}|}.0085} = 45,000\left[\frac{0.0085}{1 - (1.0085)^{-300}}\right] \approx \$415.28$$

    (b) $45,000(0.0085) = \$382.50$
    (c) $415.28 - 382.50 = \$32.78$
    (d) $300(415.28) - 45,000 = \$79,584$

17. $n = \frac{\ln\left[\frac{100}{100 - 2000(0.015)}\right]}{\ln 1.015} \approx 23.956$. Thus the number of
    full payments is 23.

19. Present value of mortgage payments is

$$600 a_{\overline{360}|}.126/12 = 600\left[\frac{1 - \left(1 + \frac{0.126}{12}\right)^{-360}}{0.126/12}\right] = \$55,812.75$$

    This amount is 75% of the purchase price x.
    $$0.75x = 55,812.75$$
    $$x = \$74,417$$

21. $\dfrac{25,000}{a_{\overline{60}|}.0125} - \dfrac{25,000}{a_{\overline{60}|}.01} = 25,000\left[\dfrac{1}{a_{\overline{60}|}.0125} - \dfrac{1}{a_{\overline{60}|}.01}\right]$

$$= 25,000\left[\frac{0.0125}{1 - (1.0125)^{-60}} - \frac{0.01}{1 - (1.01)^{-60}}\right] \approx \$38.64$$

CHAPTER 8 - REVIEW PROBLEMS

1. $s = 2 + 1 + \frac{1}{2} + \ldots + 2\left(\frac{1}{2}\right)^5 = \dfrac{2\left[1 - \left(\frac{1}{2}\right)^6\right]}{1 - \frac{1}{2}} = \dfrac{2\left[\frac{63}{64}\right]}{\frac{1}{2}} = \frac{63}{16}$

3. 8.2% compounded semiannually corresponds to an effective rate of $(1.041)^2 - 1 \approx 0.083681$ or 8.37%. Thus the better choice is 8.5% compounded annually.

5. Let x be the payment at the end of 2 years. The equation of value at the end of year 2 is
$$1000(1.04)^4 + x = 1200(1.04)^{-4} + 1000(1.04)^{-8}.$$
Thus
$$x = 1200(1.04)^{-4} + 1000(1.04)^{-8} - 1000(1.04)^4$$
$$\approx 1200(0.854804) + 1000(0.730690) - 1000(1.169859)$$
$$\approx \$586.60$$

7. (a) $A = 200a_{\overline{13}|.04} \approx 200(9.985648) \approx \$1997.13$
   (b) $S = 200s_{\overline{13}|.04} \approx 200(16.626838) \approx \$3325.37$

9. $100s_{\overline{9}|.035} \approx 100(10.368496) \approx \$1036.85$

11. $5000/s_{\overline{5}|.06} \approx 5000/5.637093 \approx \$886.98$

13. Let x be the first payment. An equation of value now is
$$x + 2x(1.07)^{-3} = 500(1.05)^{-3} + 500(1.03)^{-8}$$
$$x[1 + 2(1.07)^{-3}] = 500(1.05)^{-3} + 500(1.03)^{-8}$$
$$x = \frac{500(1.05)^{-3} + 500(1.03)^{-8}}{1 + 2(1.07)^{-3}}$$
$$x \approx \frac{500(0.863838) + 500(0.789409)}{1 + 2(0.816298)}$$
$$\approx \$314.00$$

15. $R = 15,000/a_{\overline{5}|.0075} \approx 15,000/4.889440 \approx \$3067.84$
    The interest for period 1 is $(0.0075)(15,000) = \$112.50$, so the principal repaid at the end of that period is $3067.84 - 112.50 = \$2955.34$. The principal outstanding at beginning of period 2 is $15,000 - 2955.34 = \$12,044.66$.

The interest for period 2 is 0.0075(12,044.66) = $90.33,
so the principal repaid at the end of that period is
3067.84 - 90.33 = $2977.51. Principal outstanding at the
beginning of period 3 is 12,044.66 - 2977.51 = $9067.15.
Continuing we obtain the following amortization schedule.
Note the adjustment of the final payment.

Period	Prin. Outs. at beginning	Int. for period	Pmt. at end	Prin. repaid at end
1	15,000.00	112.50	3067.84	2955.34
2	12,044.66	90.33	3067.84	2977.51
3	9067.15	68.00	3067.84	2999.84
4	6067.31	45.50	3067.84	3022.34
5	3044.97	22.84	3067.81	3044.97
		339.17	15,339.17	15,000.00

17. The monthly payment is

$$\frac{11,000}{a_{\overline{48}|.01125}} = 11,000\left[\frac{0.01125}{1 - (1.01125)^{-48}}\right] \approx \$297.84$$

The finance charge is 48(297.84) - 11,000 = $3296.32

MATHEMATICAL SNAPSHOT - CHAPTER 6

1. $\frac{12 + 11 + 10 + 9}{78}\cdot 150 = \$80.77$

3. Monthly payment $= \frac{7500}{a_{\overline{36}|.01}} = \frac{7500}{30.107505} = \$249.11$

Finance charge = 36(249.11) - 7500 = $1467.96

We have $1 + 2 + \cdots + 36 = \frac{36(37)}{2} = 666$, so the interest for

the first month is $\frac{36}{666}(1467.96) = \$79.35$. Thus the principal
repaid is 249.11 - 79.35 = $169.76. The amount owed at the
beginning of the second month is 7500 - 169.76 = $7330.24.
The interest for the second month is $\frac{35}{666}(1467.96) = \$77.15$,
so the principal repaid is 249.11 - 77.15 = $171.96. Thus
the amount owed at the beginning of the second month is
7330.24 - 171.96 = $7158.28, which is the payoff amount.

9

Introduction to Probability and Statistics

1.

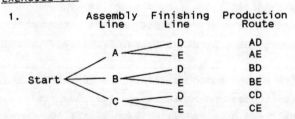

	Assembly Line	Finishing Line	Production Route

6 possible production routes

3.

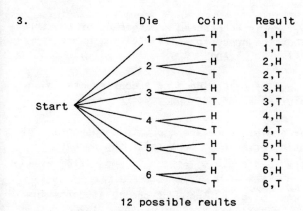

	Die	Coin	Result
	1	H	1,H
		T	1,T
	2	H	2,H
		T	2,T
	3	H	3,H
Start		T	3,T
	4	H	4,H
		T	4,T
	5	H	5,H
		T	5,T
	6	H	6,H
		T	6,T

12 possible reults

5. There are 5 science courses and 4 humanities. By the
basic counting principle, the number of selections is
5·4 = 20.   <u>Ans.</u> 20

7. There are 2 appetizers, 4 entrees, 4 desserts, and 3 bev-
erages. By the basic counting principle, the number of
possible complete dinners is 2·4·4·3 = 96.   <u>Ans.</u> 96

9. For each of the 10 questions, there are 2 choices. By
the basic counting principle, the number of ways to
answer the examination is $2·2·...·2 = 2^{10} = 1024$.
<u>Ans.</u> 1024

11. $_5P_2 = 5·4 = 20$

13. $_6P_6 = \frac{6!}{(6-6)!} = \frac{6!}{0!} = \frac{6·5·4·3·2·1}{1} = 720$

15. $_4P_2 \cdot _5P_3 = (4·3)(5·4·3) = (12)(60) = 720$

17. $\frac{1000!}{999!} = \frac{1000·999!}{999!} = 1000$. For most calculators, attempt-
ing to evaluate 1000!/999! results in an error message
(because of the magnitude of the numbers involved).

19. A name for the firm is an ordered arrangement of the three last names. Thus the number of possible firm names is $_3P_3 = 3! = 3 \cdot 2 \cdot 1 = 6$. Ans. 6

21. The number of ways of selecting 3 of 8 contestants in an order is $_8P_3 = 8 \cdot 7 \cdot 6 = 336$. Ans. 336

23. On each roll of a die, there are 6 possible outcomes. By the basic counting principle, on 3 rolls the number of possible results is $6 \cdot 6 \cdot 6 = 6^3 = 216$. Ans. 216

25. The number of ways of selecting 3 of the 12 students in an order is $_{12}P_3 = 12 \cdot 11 \cdot 10 = 1320$. Ans. 1320

27. The number of ways an employee can choose 3 of the 8 items in an order is $_8P_3 = 8 \cdot 7 \cdot 6 = 336$. Ans. 336

29. The number of ways to select six of the six different letters in the word WINDOW in an order is $_6P_6 = 6! = 6 \cdot 5 \cdot 4 \cdot 3 \cdot 2 \cdot 1 = 720$. Ans. 720

31. For an arrangement of books, order is important. The number of ways to arrange 5 of 7 books is $_7P_5 = 7 \cdot 6 \cdot 5 \cdot 4 \cdot 3 = 2520$. All 7 books can be arranged in $_7P_7 = 7! = 5040$ ways. Ans. 2520; 5040

33. After a "four of a kind" hand is dealt, the cards can be arranged so that the first four have the same face value, and order is not important. There are 13 possibilities for the first four cards (all 2's, all 3's,..., all aces). The fifth card can be any one of the 48 cards that remain. By the basic counting principle, the number of "four of a kind" hands is $13 \cdot 48 = 624$. Ans. 624

35. The number of ways the waitress can place four of the four different sandwiches (and order is important) is $_4P_4 = 4! = 4 \cdot 3 \cdot 2 \cdot 1 = 24$. Ans. 24

37. a. To fill the four offices by different people, 4 of 12
       members must be selected, and order is important. This
       can be done in $_{12}P_4$ = 12·11·10·9 = 11,880 ways.

    b. If the president and vice president must be different
       members, then there are 12 choices for president, 11
       for vice president, 12 for secretary, and 12 for
       treasurer. By the basic counting principle, the
       offices can be filled in 12·11·12·12 = 19,008 ways.

   Ans. (a) 11,880; (b) 19,008

39. There are 2 choices for the center position. After that
    choice is made, to fill the remaining four positions (and
    order is important), there are $_4P_4$ ways. By the basic
    counting principle, to assign positions to the five-member
    team there are $2 \cdot _4P_4$ = 2(4!) = 2(24) = 48 ways. **Ans.** 48

41. There are $_4P_4$ ways to select the first four batters (order
    is important) and there are $_5P_5$ ways to select the
    remaining batters. By the basic counting principle, the
    number of possible batting orders is $_4P_4 \cdot _5P_5$ = 4!·5! =
    24·120 = 2880.   **Ans.** 2880

EXERCISE 9.2

1. $_6C_4 = \frac{6!}{4!(6-4)!} = \frac{6!}{4! \cdot 2!} = \frac{6 \cdot 5 \cdot 4!}{4!(2 \cdot 1)} = \frac{6 \cdot 5}{2 \cdot 1} = 15$

3. $_{100}C_{100} = \frac{100!}{100!(100-100)!} = \frac{1}{0!} = \frac{1}{1} = 1$

5. $_3P_2 \cdot _3C_2 = (3 \cdot 2)\frac{3!}{2!(3-2)!} = (3 \cdot 2)\frac{3 \cdot 2!}{2! \cdot 1!} = (3 \cdot 2) \cdot 3 = 18$

7. $_nC_r = \frac{n!}{r!(n-r)!}.$ $_nC_{n-r} = \frac{n!}{(n-r)![n-(n-r)]!} = \frac{n!}{(n-r)!r!}.$
   Thus $_nC_r = _nC_{n-r}.$

9. The number of ways of selecting 5 of 15 people so that
order is not important is $_{15}C_5 = \frac{15!}{5!(15-5)!} = \frac{15!}{5! \cdot 10!} = \frac{15 \cdot 14 \cdot 13 \cdot 12 \cdot 11 \cdot 10!}{5 \cdot 4 \cdot 3 \cdot 2 \cdot 1(10!)} = 3003.$ **Ans.**   3003

11. The number of ways of selecting 10 of 12 questions (without regard to order) is $_{12}C_{10} = \frac{12!}{10!(12-10)!} = \frac{12!}{10! \cdot 2!} = \frac{12 \cdot 11 \cdot 10!}{10! \cdot 2 \cdot 1} = 66.$ **Ans.**   66

13. The order of selecting 10 of the 74 dresses is of no concern. Thus the number of possible samples $_{74}C_{10} = \frac{74!}{10! \cdot (74-10)!} = \frac{74!}{10! \cdot 64!}.$ **Ans.** $\frac{74!}{10! \cdot 64!}$

15. To score 80, 90, or 100, exactly 8, 9, or 10 questions must be correct, respectively. The number of ways in which 8 of 10 questions are correct is $_{10}C_8 = \frac{10!}{8!(10-8)!} = \frac{10!}{8! \cdot 2!} = \frac{10 \cdot 9 \cdot 8!}{8! \cdot 2 \cdot 1} = 45.$ For 9 of 10 questions, the number of ways is $_{10}C_9 = \frac{10!}{9!(10-9)!} = \frac{10!}{9! \cdot 1!} = \frac{10 \cdot 9!}{9! \cdot 1} = 10,$ and for 10 of 10 questions, it is $_{10}C_{10} = \frac{10!}{10!(10-10)!} = \frac{10!}{10! \cdot 0!} = 1.$ Thus the number of ways to score 80 or better is $45 + 10 + 1 = 56.$ **Ans.**   56

17. The word REMEMBER has 8 letters with repetition: two R's, three E's, two M's, and one B. Thus the number of distinguishable horizontal arrangements is $\frac{8!}{2! \cdot 3! \cdot 2! \cdot 1!} = \frac{8 \cdot 7 \cdot 6 \cdot 5 \cdot 4 \cdot 3!}{(2)3!(2)} = 1680.$ **Ans.**   1680

19. The number of ways 4 heads and 2 tails can occur in 6 tosses of a coin is the same as the number of distinguishable permutations in the "word" HHHHTT, which is $\frac{6!}{4! \cdot 2!} = \frac{6 \cdot 5 \cdot 4!}{4!(2)} = 15.$ **Ans.**   15

21. Since the order in which the calls are made is important, the number of possible schedules for the 6 calls is $_6P_6$ = 6! = 720.   Ans. 720

23. The number of ways to assign 9 scientists so that 3 work on project A, 3 work on B, and 3 work on C is $\frac{9!}{3! \cdot 3! \cdot 3!}$ = 1680.   Ans. 1680

25. A response to the true-false questions can be considered an ordered arrangement of 8 letter, 4 of which are T's and 4 of which are F's. The number of different responses is $\frac{8!}{4! \cdot 4!}$ = $\frac{8 \cdot 7 \cdot 6 \cdot 5 \cdot 4!}{4!(4 \cdot 3 \cdot 2 \cdot 1)}$ = 70.   Ans. 70

27. The number of ways to assign 15 clients to 3 caseworkers (cells) with 5 clients to each caseworker is $\frac{15!}{5! \cdot 5! \cdot 5!}$ = 756,756.   Ans. 756,756

29. (a) Six flags must be arranged: two are red (type 1), two are green (type 2), and two are yellow (type 3). Thus the number of distinguishable arrangements (messages) is $\frac{6!}{2! \cdot 2! \cdot 2!}$ = 90.

    (b) If exactly one yellow flag is used, then five flags are involved and the number of different messages is $\frac{5!}{2! \cdot 2! \cdot 1!}$ = 30. If exactly two yellow flags are used, then six flags are involved and the number of different messages is $\frac{6!}{2! \cdot 2! \cdot 2!}$ = 90. If all three yellow flags are used, then seven flags are involved and the number of different messages is $\frac{7!}{2! \cdot 2! \cdot 3!}$ = 210. Thus if at least one yellow flag is used, the number of different messages is 30 + 90 + 210 = 330.
    Ans. (a) 90; (b) 330

31. The order in which the securities go into the portfolio is not important. The number of ways to select 8 of 12 stocks is $_{12}C_8$. The number of ways to select 4 of 7 bonds is

$_7C_4$. By the basic counting principle, the numbevo bays

to create the portfolio is $_{12}C_8 \cdot _7C_4 = \dfrac{12!}{8!(12-8)!} \cdot \dfrac{7!}{4!(7-4)!} =$

$\dfrac{12!}{8! \cdot 4!} \cdot \dfrac{7!}{4! \cdot 3!} = \dfrac{12 \cdot 11 \cdot 10 \cdot 9 \cdot 8!}{8! \cdot 4 \cdot 3 \cdot 2 \cdot 1} \cdot \dfrac{7 \cdot 6 \cdot 5 \cdot 4!}{4! \cdot 3 \cdot 2 \cdot 1} = 495 \cdot 35 = 17,325.$

<u>Ans.</u> 17,325

33. (a) Selecting 3 of the 3 males can be done in only 1 way.

    (b) Selecting 4 of the 4 females can be done in only 1 way.

    (c) Selecting 2 males and 2 females can be considered as a two-stage process. In the first stage, 2 of the 3 males are selected (and order is not important), which can be done in $_3C_2$ ways. In the second stage, 2 of the 4 females are selected, which can be done in $_4C_2$ ways. By the basic counting principle, the ways of selecting the subcommittee is $_3C_2 \cdot _4C_2 =$

$$\dfrac{3!}{2!(3-2)!} \cdot \dfrac{4!}{2!(4-2)!} = \dfrac{3!}{2! \cdot 1!} \cdot \dfrac{4!}{2! \cdot 2!} = 3 \cdot 6 = 18.$$

<u>Ans.</u> (a) 1; (b) 1; (c) 18

35. There are 4 cards of a given denomination and the number of ways of selecting 3 cards of that denomination is $_4C_3$.

Since there are 13 denominations, the number of ways of selecting 3 cards of one denomination is $13 \cdot _4C_3$. After that selection is made, the 2 other cards must be of the same denomination (of which 12 denominations remain). Thus for the remaining 2 cards there are $12 \cdot _4C_2$ selections. By the basic counting principle, the number of possible full-house hands is $13 \cdot _4C_3 \cdot 12 \cdot _4C_2 =$

$13 \cdot \dfrac{4!}{3! \cdot 1!} \cdot 12 \cdot \dfrac{4!}{2! \cdot 2!} = 13 \cdot 4 \cdot 12 \cdot 6 = 3744.$ <u>Ans.</u> 3744

37. This situation can be considered as placing 18 tourists into 3 cells: 6 tourist go to the 6-passenger tram, 8 go to the 8-passenger tram, and 4 tourists remain at the bottom of the mountain. This can be done in $\dfrac{18!}{6! \cdot 8! \cdot 4!} =$ 9,189,180 ways.   <u>Ans.</u> 9,189,180

EXERCISE 9.3

1. {9D,9H,9C,9S}

3. {1H,1T,2H,2T,3H,3T,4H,4T,5H,5T,6H,6T}

5. {lo,lv,le,ov,oe,ve,ol,vl,el,vo,eo,ev}

7. (a) {RR,RW,RB,WR,WW,WB,BR,BW,BB};
   (b) {RW,RB,WR,WB,BR,BW}

9. Sample space consists of ordered sets of six elements
   and each element is H or T. Since there are two
   possibilities for each toss (H or T), and there are six
   tosses, by the basic counting principle, the number of
   sample points is $2 \cdot 2 \cdot 2 \cdot 2 \cdot 2 \cdot 2 = 2^6 = 64$.

11. Sample space consists of ordered pairs where the first
    element indicates the card drawn (52 possibilities) and
    the second element indicates the number on the die (6
    possibilities). By the basic counting principle, the
    number of sample points is $52 \cdot 6 = 312$.

13. Sample space consists of combinations of 52 cards taken
    13 at a time. Thus the number of sample points is $_{52}C_{13}$.

15. The sample points that are either in E, or in F, or in
    both E and F are 1, 3, 5, 7, and 9. Thus E $\cup$ F =
    {1,3,5,7,9}. <u>Ans.</u> {1,3,5,7,9}

17. The sample points common to both E and F are 3 and 5.
    Thus E $\cap$ F = {3,5}. <u>Ans.</u> {3,5}

19. The sample points in S that are not in F are 1, 2, 4, 6,
    8, and 10. Thus F' = {1,2,4,6,8,10}.
    <u>Ans.</u> {1,2,4,6,8,10}

21. (F $\cap$ G)' = $\emptyset$' = S

23. $E_1 \cap E_2 \neq \emptyset$; $E_1 \cap E_3 \neq \emptyset$; $E_1 \cap E_4 = \emptyset$; $E_2 \cap E_3 = \emptyset$; $E_2 \cap E_4 \neq \emptyset$; $E_3 \cap E_4 = \emptyset$. Thus E_1 and E_4, E_2 and E_3, and E_3 and E_4 are mutually exclusive.

Ans. E_1 and E_4, E_2 and E_3, E_3 and E_4

25. $E \cap F \neq \emptyset$, $E \cap G \neq \emptyset$, $E \cap H = \emptyset$, $E \cap I \neq \emptyset$, $F \cap G \neq \emptyset$, $F \cap H \neq \emptyset$, $F \cap I \neq \emptyset$, $G \cap H = \emptyset$, $G \cap I \neq \emptyset$, $H \cap I = \emptyset$.

Ans. E and H, G and H, H and I

27. (a) $S = \{HHH,HHT,HTH,HTT,THH,THT,TTH,TTT\}$;
    (b) $E_1 = \{HHH,HHT,HTH,HTT,THH,THT,TTH\}$;
    (c) $E_2 = \{HHT,HTH,HTT,THH,THT,TTH,TTT\}$;
    (d) $E_1 \cup E_2 = \{HHH,HHT,HTH,HTT,THH,THT,TTH,TTT\} = S$;
    (e) $E_1 \cap E_2 = \{HHT,HTH,HTT,THH,THT,TTH\}$;
    (f) $(E_1 \cup E_2)' = S' = \emptyset$;
    (g) $(E_1 \cap E_2)' = \{HHT,HTH,HTT,THH,THT,TTH\}' = \{HHH,TTT\}$

29. (a) $\{ABC,ACB,BAC,BCA,CAB,CBA\}$;
    (b) $\{ABC,ACB\}$;   (c) $\{BAC,BCA,CAB,CBA\}$

31. $(E \cap F) \cap (E \cap F')$
        $= (E \cap F \cap E) \cap F'$ [property 15]
        $= (E \cap E \cap F) \cap F'$ [property 11]
        $= (E \cap E) \cap (F \cap F')$ [property 15]
        $= E \cap \emptyset$ [property 5]
        $= \emptyset$ [property 9].
    Thus $(E \cap F) \cap (E \cap F') = \emptyset$, so $E \cap F$ and $E \cap F'$ are mutually exclusive.

EXERCISE 9.4

1. $2000P(E) = 2000(0.3) = 600$
   $= 3000(0.55) = 1650$

3. (a) $P(E') = 1 - P(E) = 1 - 0.2 = 0.8$
   (b) $P(E \cup F) = P(E) + P(F) - P(E \cap F)$
   $= 0.2 + 0.3 - 0.1 = 0.4$

5. If E and F are mutually exclusive, then $E \cap F = \emptyset$. Thus $P(E \cap F) = P(\emptyset) = 0$. Since it is given that $P(E \cap F) = 0.831 \neq 0$, E and F are not mutually exclusive. <u>Ans.</u> no

7. (a) $E_8 = \{(2,6),(3,5),(4,4),(5,3),(6,2)\}$.

   $P(E_8) = \frac{n(E_8)}{n(S)} = \frac{5}{36}$.

   (b) $E_{2or3} = \{(1,1),(1,2),(2,1)\}$.

   $P(E_{2or3}) = \frac{n(E_{2or3})}{n(S)} = \frac{3}{36} = \frac{1}{12}$.

   (c) $E_{3,4,or5} = \left\{ \begin{array}{l} (1,2),(2,1),(1,3),(2,2), \\ (3,1),(1,4),(2,3),(3,2),(4,1) \end{array} \right\}$.

   $P(E_{3,4,or5}) = \frac{n(E_{3,4,or5})}{n(S)} = \frac{9}{36} = \frac{1}{4}$.

   (d) $E_{12or13} = E_{12}$, since E_{13} is an impossible event.

   $E_{12} = \{(6,6)\}$. $P(E_{12or13}) = \frac{n(E_{12or13})}{n(S)} = \frac{1}{36}$.

   (e) $E_2 = \{(1,1)\}$, $E_4 = \{(1,3),(2,2),(3,1)\}$,

   $E_6 = \{(1,5),(2,4),(3,3),(4,2),(5,1)\}$,

   $E_8 = \{(2,6),(3,5),(4,4),(5,3),(6,2)\}$,

   $E_{10} = \{(4,6),(5,5),(6,4)\}$, $E_{12} = \{(6,6)\}$.

   $P(E_{even}) = P(E_2)+P(E_4)+P(E_6)+P(E_8)+P(E_{10})+P(E_{12})$

   $= \frac{1}{36} + \frac{3}{36} + \frac{5}{36} + \frac{5}{36} + \frac{3}{36} + \frac{1}{36} = \frac{18}{36} = \frac{1}{2}$.

   (f) $P(E_{odd}) = 1 - P(E_{even}) = 1 - \frac{1}{2} = \frac{1}{2}$.

   (g) $E'_{less\ than\ 10} = E_{10} \cup E_{11} \cup E_{12} =$
   $\{(4,6),(5,5),(6,4)\} \cup \{(5,6),(6,5)\} \cup \{(6,6)\} =$
   $\{(4,6),(5,5),(6,4),(5,6),(6,5),(6,6)\}$.

   $P(E_{less\ than\ 10}) = 1 - P(E'_{less\ than\ 10}) = 1 - \frac{6}{36} = \frac{5}{6}$.

9. $n(S) = 52$.

   (a) $P(king\ of\ hearts) = \frac{n(E_{king\ of\ hearts})}{n(S)} = \frac{1}{52}$

   (b) $P(diamond) = \frac{n(E_{diamond})}{n(S)} = \frac{13}{52} = \frac{1}{4}$

   (c) $P(jack) = \frac{n(E_{jack})}{n(S)} = \frac{4}{52} = \frac{1}{13}$

(d) $P(\text{red}) = \frac{n(E_{red})}{n(S)} = \frac{26}{52} = \frac{1}{2}$

(e) Because a heart is not a club, $E_{heart} \cap E_{club} = \emptyset$.

Thus $P(E_{heart \text{ or club}}) = P(E_{heart} \cup E_{club}) =$

$P(E_{heart}) + P(E_{club}) = \frac{n(E_{heart})}{n(S)} + \frac{n(E_{club})}{n(S)} = \frac{13}{52} + \frac{13}{52} = \frac{26}{52} = \frac{1}{2}.$

(f) $E_{club \text{ and } 4} = \{4C\}$.

$P(E_{club \text{ and } 4}) = \frac{n(E_{club \text{ and } 4})}{n(S)} = \frac{1}{52}.$

(g) $P(\text{club or 4}) = P(\text{club}) + P(4) - P(\text{club and 4}) = \frac{13}{52} + \frac{4}{52} - \frac{1}{52} = \frac{16}{52} = \frac{4}{13}$

(h) $E_{red \text{ and king}} = \{KH, KD\}$.

$P(\text{red and king}) = \frac{n(E_{red \text{ and king}})}{n(S)} = \frac{2}{52} = \frac{1}{26}.$

(i) $E_{spade \text{ and heart}} = \emptyset$. Thus $P(\text{spade and heart}) = 0$.

11. $n(S) = 2 \cdot 6 \cdot 52 = 624$.

(a) $P(\text{tail,3,queen of hearts}) = \frac{n(E_{T,3,QH})}{n(S)} = \frac{1 \cdot 1 \cdot 1}{624} = \frac{1}{624}$

(b) $P(\text{tail,3,queen}) = \frac{n(E_{T,3,Q})}{n(S)} = \frac{1 \cdot 1 \cdot 4}{624} = \frac{1}{156}$

(c) $P(\text{head,2or3,queen}) = \frac{n(E_{H,2or3,Q})}{n(S)} = \frac{1 \cdot 2 \cdot 4}{624} = \frac{1}{78}$

(d) $P(\text{head,even,diamond}) = \frac{n(E_{H,E,D})}{n(S)} = \frac{1 \cdot 3 \cdot 13}{624} = \frac{1}{16}$

13. $n(S) = 52 \cdot 51 = 2652$.

(a) $P(\text{both kings}) = \frac{n(E_{both \text{ kings}})}{n(S)} = \frac{4 \cdot 3}{2652} = \frac{1}{221}$

(b) The number of ways that first card is a diamond and second card is a heart is $13 \cdot 13$; for first card a heart and second card a diamond, the number of ways is $13 \cdot 13$. Thus

$P(E_{one \text{ card diamond \& other is heart}}) = \frac{13 \cdot 13 + 13 \cdot 13}{2652} = \frac{338}{2652} = \frac{13}{102}.$

15. $n(S) = 2 \cdot 2 \cdot 2 = 8$.

   (a) $E_{3 \text{ girls}} = \{GGG\}$. $P(3 \text{ girls}) = \frac{n(E_{3 \text{ girls}})}{n(S)} = \frac{1}{8}$.

   (b) $E_{1 \text{ boy}} = \{BGG, GBG, GGB\}$. $P(1 \text{ boy}) = \frac{n(E_{1 \text{ boy}})}{n(S)} = \frac{3}{8}$.

   (c) $E_{\text{no girl}} = \{BBB\}$. $P(\text{no girl}) = \frac{n(E_{\text{no girl}})}{n(S)} = \frac{1}{8}$.

   (d) $P(\text{at least 1 girl}) = 1 - P(\text{no girl}) = 1 - \frac{1}{8} = \frac{7}{8}$.

17. The sample space consists of 60 stocks. Thus $n(S) = 60$.

   (a) $P(10\% \text{ or more}) = \frac{n(E_{10\% \text{ or more}})}{n(S)} = \frac{48}{60} = \frac{4}{5}$

   (b) $P(\text{less than } 10\%) = 1 - P(10\% \text{ or more}) = 1 - \frac{4}{5} = \frac{1}{5}$

19. $n(S) = 40$. Of the 40 students, 4 received an A, 10 a B, 14 a C, 10 a D, and 2 an F.

   (a) $P(A) = \frac{n(E_A)}{n(S)} = = \frac{4}{40} = \frac{1}{10} = 0.1$

   (b) $P(A \text{ or } B) = \frac{n(E_{A \text{ or } B})}{n(S)} = \frac{4+10}{40} = \frac{14}{40} = 0.35$

   (c) $P(\text{neither D nor F}) = P(A,B,\text{or}C) = \frac{n(E_{A,B,\text{or}C})}{n(S)} =$
   $\frac{4+10+14}{40} = \frac{28}{40} = 0.7$

   (d) $P(\text{no F}) = 1 - P(F) = 1 - \frac{n(E_F)}{n(S)} = 1 - \frac{2}{40} = \frac{38}{40} = 0.95$

   (e) Let N = number of students. Then $n(S) = N$. Of the N students, 0.10N received an A, 0.25N a B, 0.35N a C, 0.25N a D, 0.05N an F.

   $P(A) = \frac{0.10N}{N} = 0.1$;

   $P(A \text{ or } B) = \frac{0.10N + 0.25N}{N} = \frac{0.35N}{N} = 0.35$;

   $P(\text{neither D nor F}) = P(A,B,\text{or}C)$
   $= \frac{0.10N + 0.25N + 0.35N}{N}$
   $= \frac{0.70N}{N} = 0.7$;

   $P(\text{no F}) = 1 - P(F) = 1 - \frac{0.05N}{N} = 1 - 0.05 = 0.95$.

21. The sample space consists of combinations of 2 people
    selected from 5. Thus $n(S) = {}_5C_2 = \frac{5!}{2! \cdot 3!} = \frac{5 \cdot 4}{2} = 10$.
    Because there are only 2 women in the group, the number
    of possible 2 women committees is 1. Thus P(2 women) =
    $\frac{n(E_{2 \text{ women}})}{n(S)} = \frac{1}{10}$.

23. Number of ways to answer exam is $2^{10} = 1024 = n(S)$.
    (a) There is only one way to achieve 100 points, namely,
        to answer each question correctly. Thus
        $$P(100 \text{ points}) = \frac{n(E_{100 \text{ points}})}{n(S)} = \frac{1}{1024}.$$
    (b) Number of ways to score 90 points = number of ways
        that exactly one question is answered incorrectly =
        10. Thus P(90 or more points) = P(90 points) +
        $$P(100 \text{ points}) = \frac{10}{1024} + \frac{1}{1024} = \frac{11}{1024}.$$

25. A poker hand is a 5 card deal from 52 cards. Thus
    $n(S) = {}_{52}C_5$. In 52 cards, there are 4 cards of a partic-
    ticular denomination. Thus, for a full house, the number
    of ways of selecting 3 of 4 cards of a particular
    denomination is ${}_4C_3$. Since there are 13 denominations, 3
    cards of the same denomination can be dealt in $13 \cdot {}_4C_3$
    ways. For the two remaining cards in the full house,
    there are 12 denominations that are possible, and for each
    denomination there are ${}_4C_2$ ways of dealing a pair. Thus
    $$P(\text{full house}) = \frac{n(E_{\text{full house}})}{n(S)} = \frac{13 \cdot {}_4C_3 \cdot 12 \cdot {}_4C_2}{{}_{52}C_5}.$$

27. $n(S) = {}_{100}C_3 = \frac{100!}{3! \cdot 97!} = 161,700$.
    (a) $n(E_{3 \text{ females}}) = {}_{35}C_3 = \frac{35!}{3! \cdot 32!} = 6545$.
        $$P(E_{3 \text{ females}}) = \frac{n(E_{3 \text{ females}})}{n(S)} = \frac{6545}{161,700} \approx 0.040.$$
    (b) The number of ways of selecting one professor is 15;
        the number of ways of selecting two associate
        professors is ${}_{24}C_2$. Thus $n(E_{1 \text{ prof \& 2 assoc profs}}) =$

$15 \cdot \dfrac{24!}{2! \cdot 22!} = 15 \cdot 276 = 4140.$ Therefore,

$P(E_{1 \text{ prof \& 2 assoc profs}}) = \dfrac{4140}{161,700} \approx 0.026.$

29. Let $p = P(1) = P(3) = P(5)$. Then $2p = P(2) = P(4) = P(6)$.
    Since $P(S) = 1$, then $3(p) + 3(2p) = 1$, $9p = 1$, $p = 1/9$.
    <u>Ans.</u> 1/9

31. (a) Of the 100 voters, 51 favor the tax increase. Thus
    $P(\text{favors tax increase}) = \dfrac{51}{100} = 0.51.$

    (b) Of the 100 voters, 44 oppose the tax increase. Thus
    $P(\text{opposes tax increase}) = \dfrac{44}{100} = 0.44.$

    (c) Of the 100 voters, 3 are Republican with no opinion.
    Thus $P(\text{is a Republican with no opinion}) = \dfrac{3}{100} = 0.03.$

33. $\dfrac{P(E)}{P(E')} = \dfrac{P(E)}{1 - P(E)} = \dfrac{4/5}{1 - (4/5)} = \dfrac{\frac{4}{5}}{\frac{1}{5}} = \dfrac{4}{1}.$ <u>Ans.</u> 4:1

35. $\dfrac{P(E)}{P(E')} = \dfrac{P(E)}{1 - P(E)} = \dfrac{0.3}{1 - 0.3} = \dfrac{0.3}{0.7} = \dfrac{3}{7}.$ <u>Ans.</u> 3:7

37. $P(E) = \dfrac{5}{5 + 4} = \dfrac{5}{9}$

39. $P(E) = \dfrac{4}{4 + 10} = \dfrac{4}{14} = \dfrac{2}{7}$

41. $P(\text{rain}) = \dfrac{3}{3 + 1} = \dfrac{3}{4}$

EXERCISE 9.5

1. (a) $P(E|F) = \dfrac{n(E \cap F)}{n(F)} = \dfrac{2}{5}.$

   (b) Using the result of part (a),
   $$P(E'|F) = 1 - P(E|F) = 1 - \dfrac{2}{5} = \dfrac{3}{5}.$$

   (c) $F' = \{3, 7, 8\}.$ $P(E|F') = \dfrac{n(E \cap F')}{n(F')} = \dfrac{1}{3}.$

(d) $P(F|E) = \frac{n(F \cap E)}{n(E)} = \frac{2}{3}$.

(e) $F \cap G = \{2, 4, 5\}$. $P(E|(F \cap G)) = \frac{n(E \cap (F \cap G))}{n(F \cap G)} = \frac{1}{3}$.

3. $P(E|E) = \frac{P(E \cap E)}{P(E)} = \frac{P(E)}{P(E)} = 1$.

5. $P(E'|F) = 1 - P(E|F) = 1 - 0.63 = 0.37$.

7. (a) $P(E|F) = \frac{P(E \cap F)}{P(F)} = \frac{1/6}{1/3} = \frac{1}{2}$.

   (b) $P(F|E) = \frac{P(F \cap E)}{P(E)} = \frac{1/6}{1/4} = \frac{2}{3}$.

9. (a) $P(F|E) = \frac{P(F \cap E)}{P(E)} = \frac{1/5}{1/3} = \frac{3}{5}$.

   (b) $P(E \cup F) = P(E) + P(F) - P(E \cap F)$, $\frac{8}{15} = \frac{1}{3} + P(F) - \frac{1}{5}$.

   Thus $P(F) = \frac{8}{15} - \frac{1}{3} + \frac{1}{5} = \frac{2}{5}$.

   (c) From part (b), $P(F) = \frac{2}{5}$. Then

   $$P(E|F) = \frac{P(E \cap F)}{P(F)} = \frac{1/5}{2/5} = \frac{1}{2}.$$

   (d) $P(E) = P(E \cap F) + P(E \cap F')$, $\frac{1}{3} = \frac{1}{5} + P(E \cap F')$, so

   $P(E \cap F') = \frac{1}{3} - \frac{1}{5} = \frac{2}{15}$. Then

   $$P(E|F') = \frac{P(E \cap F')}{P(F')} = \frac{2/15}{1 - (2/5)} = \frac{2/15}{3/5} = \frac{2}{9}.$$

11. (a) $P(F) = \frac{125}{200} = \frac{5}{8}$.

   (b) $P(F|II) = \frac{n(F \cap II)}{n(II)} = \frac{35}{58}$.

   (c) $P(O|I) = \frac{n(O \cap I)}{n(I)} = \frac{22}{78} = \frac{11}{39}$.

   (d) $P(III) = \frac{64}{200} = \frac{8}{25}$.

   (e) $P(III|O) = \frac{n(III \cap O)}{n(O)} = \frac{10}{47}$.

   (f) $P(II|N') = \frac{n(II \cap N')}{n(N')} = \frac{35 + 15}{125 + 47} = \frac{50}{172} = \frac{25}{86}$.

13. (a) $P(A|B) = \frac{P(A \cap B)}{P(B)} = \frac{0.20}{0.40} = \frac{1}{2}$.

   (b) $P(B|A) = \frac{P(B \cap A)}{P(A)} = \frac{0.20}{0.45} = \frac{4}{9}$.

15. S = {BB, BG, GG, GB}.
    Let E = {at least one girl} = {BG, GG, GB},
         F = {at least one boy} = {BB, BG, GB}.
    *(In each listing, the order of the letters indicates the
    order in which the children were born.)*
    $$P(E|F) = \frac{n(E \cap F)}{n(F)} = \frac{2}{3}.$$

17. S = {HHH, HHT, HTH, HTT, THH, THT, TTH, TTT}.
    Let E = {exactly two tails} = {HTT, THT, TTH},
         F = {second toss is a tail} = {HTH, HTT, TTH, TTT},
         G = {second toss is a head} = {HHH, HHT, THH, THT}.
    (a) $P(E|F) = \frac{n(E \cap F)}{n(F)} = \frac{2}{4} = \frac{1}{2}.$
    (b) $P(E|G) = \frac{n(E \cap G)}{n(G)} = \frac{1}{4}.$

19. $P(< 4 | odd) = \frac{n(< 4 \cap odd)}{n(odd)} = \frac{n\{1,3\}}{n\{1,3,5\}} = \frac{2}{3}.$

21. *Method 1.* The usual sample space has 36 outcomes, where
    the event "two 6's" is {(6,6)}. Note that
                  {at least one 6}' = {no 6's},
    and the event "no 6's" occurs in 5·5 = 25 ways. Thus
    $$P(\text{two 6's}|\text{at least one 6}) = \frac{n(\text{two 6's} \cap \text{at least one 6})}{n(\text{at least one 6})}$$
    $$= \frac{n\{(6,6)\}}{36 - 25} = \frac{1}{11}.$$

    Method 2. From the usual sample space, we find that the
    reduced sample space for "at least one 6" (which has 11
    outcomes) is
        {(6,1),(6,2),(6,3),(6,4),(6,5),(6,6),(1,6),(2,6),(3,6),
                                              (4,6),(5,6)}.
    Thus P(two 6's|at least one 6) = $\frac{1}{11}$.

23. The usual sample space consists of ordered pairs (R,G),
    where R = no. on red die and G = no. on green die. Now,
    n(green is even) = 6·3 = 18, because the red die can show
    any of six numbers and the green any of three: 2,4,or 6.
    Also,
        n(total of 7 ∩ green even) = n{(5,2),(3,4),(1,6)} = 3.
    Thus

$$P(\text{total of 7|green even}) = \frac{n(\text{total of 7} \cap \text{green even})}{n(\text{green even})}$$

$$= \frac{3}{18} = \frac{1}{6}.$$

25. The usual sample space consists of 36 ordered pairs.
Let E = {total > 5} and F = {first toss < 4}. Then
$n(F) = 3 \cdot 6 = 18$, and

$n(E \cap F) = n\{(1,5),(1,6),(2,4),(2,5),(2,6),(3,3),$
$(3,4),(3,5),(3,6)\} = 9.$

Thus $P(E|F) = \frac{n(E \cap F)}{n(F)} = \frac{9}{18} = \frac{1}{2}.$

27. $P(K|H) = \frac{n(K \cap H)}{n(H)} = \frac{1}{13}.$

29. Let E = {second card is not a face card} and F = {first
card is a face card}.

$$P(E|F) = \frac{n(E \cap F)}{n(F)} = \frac{12 \cdot (51 - 11)}{12 \cdot 51} = \frac{40}{51}.$$

31. $P(K_1 \cap Q_2 \cap J_3) = P(K_1)P(Q_2|K_1)P(J_3|(K_1 \cap Q_2))$

$$= \frac{4}{52} \cdot \frac{4}{51} \cdot \frac{4}{50} = \frac{8}{16,575}.$$

33. $P(D_1 \cap D_2 \cap D_3) = P(D_1)P(D_2|D_1)P(D_3|(D_1 \cap D_2))$

$$= \frac{13}{52} \cdot \frac{12}{51} \cdot \frac{11}{50} = \frac{11}{850}.$$

35. Let D = {two diamonds} and R = {first card red}. We have
$D \cap R = \{\text{two diamonds}\} = D$ and $P(D) = \frac{13}{52} \cdot \frac{12}{51}.$ Thus

$$P(D|R) = \frac{P(D \cap R)}{P(R)} = \frac{\frac{13 \cdot 12}{52 \cdot 51}}{\frac{26}{52}} = \frac{2}{17}.$$

37. (a) $P(U) = P(F \cap U) + P(O \cap U) + P(N \cap U)$
$= P(F)P(U|F) + P(O)P(U|O) + P(N)P(U|N)$
$= (0.60)(0.45) + (0.30)(0.55) + (0.10)(0.35)$
$= 0.47 = \frac{47}{100}.$

(b) $P(F|U) = \frac{P(F \cap U)}{P(U)} = \frac{(0.60)(0.45)}{0.47} = \frac{27}{47}.$

39. (a) After the first draw, if the marble drawn is red, then 4 marbles remain, 3 of which are yellow.

P(second is yellow|first is red) = $\frac{3}{4}$.

(b) After red marble is replaced, 5 marbles remain, 3 of which are yellow.

P(second is yellow|first is red) = $\frac{3}{5}$.

41. P(W) = P(Urn 1 ∩ W) + P(Urn 2 ∩ W)
= P(Urn 1)P(W|Urn 1) + P(Urn 2)P(W|Urn 2)
= $\frac{1}{2} \cdot \frac{2}{5} + \frac{1}{2} \cdot \frac{2}{4} = \frac{9}{20}$.

43. P(W$_3$) = P(U1 ∩ G$_2$ ∩ W$_3$) + P(U1 ∩ R$_2$ ∩ W$_3$) + P(U2 ∩ W$_2$ ∩ W$_3$)
= $\frac{1}{2} \cdot \frac{1}{2} \cdot \frac{1}{3} + \frac{1}{2} \cdot \frac{1}{2} \cdot \frac{1}{3} + \frac{1}{2} \cdot \frac{1}{2} \cdot \frac{1}{3} = \frac{1}{4}$.

45. P(Def) = P(A ∩ Def) + P(B ∩ Def)
= P(A)P(Def|A) + P(B)P(Def|B)
= $\frac{200}{600} \cdot \frac{2}{100} + \frac{400}{600} \cdot \frac{5}{100} = \frac{1}{25}$.

47. P(Def) = P(A ∩ Def) + P(B ∩ Def) + P(C ∩ Def)
= P(A)P(Def|A) + P(B)P(Def|B) + P(C)P(Def|C)
= (0.10)(0.06) + (0.20)(0.04) + (0.70)(0.05) = 0.049

49. (a) P(D ∩ V) = P(D)P(V|D) = (0.40)(0.15) = 0.06.
(b) P(V) = P(D ∩ V) + P(R ∩ V) + P(I ∩ V)
= P(D)P(V|D) + P(R)P(V|R) + P(I)P(V|I)
= (0.40)(0.15) + (0.35)(0.20) + (0.25)(0.10)
= 0.155

51. P(3 Fem|at least one Fem)

$= \frac{P(3\ Fem\ \cap\ at\ least\ one\ Fem)}{P(at\ least\ one\ Fem)} = \frac{P(3\ Fem)}{1 - P(no\ Fem)}$

$= \frac{\frac{_5C_3}{_9C_3}}{1 - \frac{_4C_3}{_9C_3}} = \frac{\frac{5}{42}}{1 - \frac{1}{21}} = \frac{1}{8}$.

EXERCISE 9.6

1. (a) $P(E \cap F) = P(E)P(F) = \frac{1}{3} \cdot \frac{3}{4} = \frac{1}{4}$.

   (b) $P(E \cup F) = P(E) + P(F) - P(E \cap F) = \frac{1}{3} + \frac{3}{4} - \frac{1}{4} = \frac{5}{6}$.

   (c) $P(E|F) = \frac{P(E \cap F)}{P(F)} = \frac{1/4}{3/4} = \frac{1}{3}$.

   (d) $P(E'|F) = 1 - P(E|F) = 1 - \frac{1}{3} = \frac{2}{3}$.

   (e) $P(E \cap F') = P(E)P(F') = \frac{1}{3} \cdot \frac{1}{4} = \frac{1}{12}$.

   (f) $P(E \cup F') = P(E) + P(F') - P(E \cap F') = \frac{1}{3} + \frac{1}{4} - \frac{1}{12} = \frac{1}{2}$.

   (g) $P(E|F') = \frac{P(E \cap F')}{P(F')} = \frac{1/12}{1/4} = \frac{1}{3}$.

3. $P(E \cap F) = P(E)P(F)$, $\quad \frac{1}{3} = \frac{2}{5} \cdot P(F) \Rightarrow P(F) = \frac{1}{3} \cdot \frac{5}{2} = \frac{5}{6}$.

5. $P(E)P(F) = \frac{3}{4} \cdot \frac{8}{9} = \frac{2}{3} = P(E \cap F)$. Since $P(E)P(F) = P(E \cap F)$, events E and F are independent.

7. Let F = {full service} and I = {increase in value}.

   $P(F) = \frac{400}{600} = \frac{2}{3}$ and $P(F|I) = \frac{n(F \cap I)}{n(I)} = \frac{320}{480} = \frac{2}{3}$.

   Since $P(F|I) = P(F)$, events F and I are independent.

9. Let S be the usual sample space consisting of ordered pairs of the form (R, G), where the first component of each pair represents the number showing on the red die. Then $n(S) = 6 \cdot 6 = 36$. For E, any of four numbers can occur on the red die, and any of six numbers can appear on the green die. Thus $n(E) = 4 \cdot 6 = 24$. For F we have F = {(1,4), (2,3), (3,2), (4,1)}, so n(F) = 4. Also, E $\cap$ F = {(1,4), (2,3), (3,2)}, so n(E $\cap$ F) = 3. Thus

   $P(E)P(F) = \frac{24}{36} \cdot \frac{4}{36} = \frac{2}{27}$ and $P(E \cap F) = \frac{3}{36} = \frac{1}{12}$.

   Since $P(E)P(F) \neq P(E \cap F)$, events E and F are dependent.

11. S = {HH, HT, TH, TT}, E = {HT, TH, TT}, F = {HT, TH}, and E $\cap$ F = {HT, TH}. Thus $P(E) = \frac{3}{4}$, $P(F) = \frac{2}{4} = \frac{1}{2}$, and $P(E \cap F) = \frac{2}{4} = \frac{1}{2}$. We have $P(E)P(F) = \frac{3}{4} \cdot \frac{1}{2} = \frac{3}{8} \neq P(E \cap F)$, so events E and F are dependent.

13. Let S be the set of ordered pairs those first *(second)*
    component represents the number on the first *(second)* chip.
    Then n(S) = 6·6 = 36, n(E) = 1·6 = 6, and n(F) = 6·1 = 6.
    For G, if the first chip is 1, 3 or 5, then the second
    chip must be 2, 4 or 6; if the first chip is 2, 4 or 6,
    the second must be 1, 3 or 5. Thus n(G) = 3·3 + 3·3 = 18.
    (a) E ∩ F = {(3,3)}, so P(E ∩ F) = $\frac{1}{36}$. Since

    $$P(E)P(F) = \frac{6}{36} \cdot \frac{6}{36} = \frac{1}{36} = P(E \cap F),$$

    events E and F are independent.
    (b) E ∩ G = {(3,2), (3,4), (3,6)}, so P(E ∩ G) = $\frac{3}{36} = \frac{1}{12}$.
    Since

    $$P(E)P(G) = \frac{6}{36} \cdot \frac{18}{36} = \frac{1}{12} = P(E \cap G),$$

    events E and G are independent.
    (c) F ∩ G = {(2,3), (4,3), (6,3)} so P(F ∩ G) = $\frac{3}{36} = \frac{1}{12}$.
    Since

    $$P(F)P(G) = \frac{6}{36} \cdot \frac{18}{36} = \frac{1}{12} = P(F \cap G),$$

    events F and G are independent.
    (d) E ∩ F ∩ G = ∅, so P(E ∩ F ∩ G) = 0. However,
    $$P(E)P(F)P(G) \neq 0 = P(E \cap F \cap G),$$
    so events E, F a,d G are not independent.

15. P(E ∩ F) = P(E)P(F|E) ⟹ P(E) = $\frac{P(E \cap F)}{P(F|E)} = \frac{0.28}{0.4} = 0.7$.
    Since P(E) = 0.7 ≠ 0.6 = P(E|F), E and F are dependent.

17. Let E = {red 4} and F = {green > 4}. Assume E and F are
    independent. P(E ∩ F) = P(E)P(F) = $\frac{1}{6} \cdot \frac{1}{3} = \frac{1}{18}$.

19. Let F = {first person attends regularly} and
    S = {second person attends regularly}.
    Then P(F ∩ S) = P(F)P(S) = $\frac{1}{5} \cdot \frac{1}{5} = \frac{1}{25}$.

21. Because of replacements, assume the cards selected on the
    draws are independent events.
    P(ace, then face card, then spade)
    = P(ace)·P(face card)·P(spade) = $\frac{4}{52} \cdot \frac{12}{52} \cdot \frac{13}{52} = \frac{3}{676}$.

23. (a)   P(Bill gets A ∩ Jim gets A ∩ Linda gets A)
       = P(Bill gets A)·P(Jim gets A)·P(Linda gets A)
       = $\frac{3}{4}\cdot\frac{1}{2}\cdot\frac{4}{5} = \frac{3}{10}$.
    (b)   P(Bill no A ∩ Jim no A ∩ Linda no A)
       = P(Bill no A)·P(Jim no A)·P(Linda no A) = $\frac{1}{4}\cdot\frac{1}{2}\cdot\frac{1}{5} = \frac{1}{40}$.
    (c)   P(Bill no A ∩ Jim no A ∩ Linda gets A)
       = P(Bill no A)·P(Jim no A)·P(Linda gets A)
       = $\frac{1}{4}\cdot\frac{1}{2}\cdot\frac{4}{5} = \frac{1}{10}$.

25. Let A = {A survives 15 more years}, B = {B survives 15 more years}.
    (a) P(A ∩ B) = P(A)P(B) = $\frac{2}{3}\cdot\frac{3}{5} = \frac{2}{5}$.
    (b) P(A' ∩ B) = P(A')P(B) = $\frac{1}{3}\cdot\frac{3}{5} = \frac{1}{5}$.
    (c)   P[(A ∩ B') ∪ (A' ∩ B)]
       = P(A)P(F') + P(A')P(B)   (A ∩ B' and A' ∩ B are mutually exclusive)
       = $\frac{2}{3}\cdot\frac{2}{5} + \frac{1}{3}\cdot\frac{3}{5} = \frac{7}{15}$.
    (d) From parts (c) and (a),
          P(at least one survives)
        = P(exactly one survives) + P(both survive) =
        = $\frac{7}{15} + \frac{2}{5} = \frac{13}{15}$.
    (e) P(neither survives) = 1 - P(at least one survives)
                            = $1 - \frac{13}{15} = \frac{2}{15}$.

27. Assume the colors selected on the draws are independent events.
    (a) P(W$_1$ ∩ G$_2$) = P(W$_1$)P(G$_2$) = $\frac{6}{15}\cdot\frac{5}{15} = \frac{2}{15}$.
    (b) P[(W$_1$ ∩ G$_2$) ∪ (G$_1$ ∩ W$_2$)] = P(W$_1$)P(G$_2$) + P(G$_1$)P(G$_2$)
                            = $\frac{6}{15}\cdot\frac{5}{15} + \frac{5}{15}\cdot\frac{6}{15} = \frac{4}{15}$.

29. Assume that the selections are independent.
          P(both red ∪ both white ∪ both green)
        = P(both red) + P(both white) + P(both green)
        = $\frac{3}{18}\cdot\frac{3}{18} + \frac{6}{18}\cdot\frac{6}{18} + \frac{9}{18}\cdot\frac{9}{18} = \frac{7}{18}$.

31. Assume that the draws are independent.

P(particular 1st ticket $\cap$ particular 2nd ticket)

$$= \frac{1}{20} \cdot \frac{1}{20} = \frac{1}{400}.$$

P(sum is 35) = P{(20,15),(19,16),(18,17),(17,18),(16,19),
(15,20)}

$$= 6\left(\frac{1}{400}\right) = \frac{3}{200}.$$

33. (a) $\frac{1}{12} \cdot \frac{1}{12} \cdot \frac{1}{12} = \frac{1}{1728}.$

(b) To get exactly one even, there are $_3C_1 = 3$ ways.

P(one even and two odd)
= 3[P(even 1st spin)·P(odd 2nd spin)·P(odd 3rd spin)]

$$= 3\left(\frac{6}{12} \cdot \frac{6}{12} \cdot \frac{6}{12}\right) = \frac{3}{8}.$$

35. (a) The number of ways of getting exactly four correct
answers out of five is $_5C_4 = 5$. Each of these ways

has a probability of $\frac{1}{4} \cdot \frac{1}{4} \cdot \frac{1}{4} \cdot \frac{1}{4} \cdot \frac{3}{4} = \frac{3}{1024}$. Thus

$$P(\text{exactly 4 correct}) = 5 \cdot \frac{3}{1024} = \frac{15}{1024}.$$

(b) P(at least 4 correct) = P(exactly 4) + P(exactly 5)

$$= \frac{15}{1024} + \frac{1}{4} \cdot \frac{1}{4} \cdot \frac{1}{4} \cdot \frac{1}{4} \cdot \frac{1}{4} = \frac{1}{64}.$$

(c) The number of ways of getting exactly three correct
answers out of five is $_5C_3 = 10$. Each of these ways

has a probability of $\frac{1}{4} \cdot \frac{1}{4} \cdot \frac{1}{4} \cdot \frac{3}{4} \cdot \frac{3}{4} = \frac{9}{1024}$, so

$$P(\text{exactly 3 correct}) = 10 \cdot \frac{9}{1024} = \frac{45}{512}.$$

Thus

P(3 or more correct) = P(exactly 3) + P(at least 4)

$$= \frac{45}{512} + \frac{1}{64} = \frac{53}{512}.$$

37. A wrong majority decision can occur in one of two mutually
exclusive ways: exactly two wrong recommendations, or three
wrong recommendations. Exactly two wrong recommendations
can occur in $_3C_2 = 3$ mutually exclusive ways. Thus

P(wrong majority decision)
= [(0.05)(0.05)(0.9)+(0.05)(0.95)(0.1)+(0.95)(0.05)(0.1)] +
(0.05)(0.05)(0.1)

= 0.012.

EXERCISE 9.7

1. $P(E|D) = \dfrac{P(E)P(D|E)}{P(E)P(D|E) + P(F)P(D|F)} = \dfrac{\frac{2}{5} \cdot \frac{1}{10}}{\frac{2}{5} \cdot \frac{1}{10} + \frac{3}{5} \cdot \frac{1}{5}} = \dfrac{1}{4}.$

For the second part, $P(D'|F) = 1 - P(D|F) = 1 - \frac{1}{5} = \frac{4}{5}$, and

$P(D'|E) = 1 - P(D|E) = 1 - \frac{1}{10} = \frac{9}{10}.$ Then

$P(F|D') = \dfrac{P(F)P(D'|F)}{P(E)P(D'|E) + P(F)P(D'|F)} = \dfrac{\frac{3}{5} \cdot \frac{4}{5}}{\frac{2}{5} \cdot \frac{9}{10} + \frac{3}{5} \cdot \frac{4}{5}} = \dfrac{4}{7}.$

3. Let D = {is Democrat}, R = {is Republican},
   I = {is Independent}, V = {voted}.

$$P(D|V) = \dfrac{P(D)P(V|D)}{P(D)P(V|D) + P(R)P(V|R) + P(I)P(V|I)}$$

$$= \dfrac{(0.40)(0.15)}{(0.40)(0.15) + (0.35)(0.20) + (0.25)(0.10)}$$

$$= \tfrac{12}{31} \approx 0.387$$

5. D = {has the disease}, D' = {does not have the disease},
   R = {positive reaction}, N = {negative reaction} = R'.

(a) $P(D|R) = \dfrac{P(D)P(R|D)}{P(D)P(R|D) + P(D')P(R|D')}$

$= \dfrac{(0.05)(0.98)}{(0.05)(0.98) + (0.95)(0.06)} = \tfrac{49}{106} \approx 0.462$

(b) $P(D|N) = \dfrac{P(D)P(N|D)}{P(D)P(N|D) + P(D')P(N|D')}$

$= \dfrac{(0.05)(0.02)}{(0.05)(0.02) + (0.95)(0.94)} = \tfrac{1}{894} \approx 0.001$

7. U_1 = {first urn selected}, U_2 = {second urn selected},
   R = {red marble drawn}. $P(U_1) = P(U_2) = 1/2.$

$P(U_1|R) = \dfrac{P(U_1)P(R|U_1)}{P(U_1)P(R|U_1) + P(U_2)P(R|U_2)} = \dfrac{\frac{1}{2} \cdot \frac{4}{6}}{\frac{1}{2} \cdot \frac{4}{6} + \frac{1}{2} \cdot \frac{2}{5}} = \dfrac{5}{8}.$

9. A = {unit from line A}, B = {unit from line B},
   D = {defective unit}. $P(A) = \frac{200}{600} = \frac{1}{3}$, $P(B) = \frac{400}{600} = \frac{2}{3}.$

$P(A|D) = \dfrac{P(A)P(D|A)}{P(A)P(D|A) + P(B)P(D|B)} = \dfrac{\frac{1}{3} \cdot \frac{2}{100}}{\frac{1}{3} \cdot \frac{2}{100} + \frac{2}{3} \cdot \frac{5}{100}} = \dfrac{1}{6}.$

11. C = {call made}, T = {on time for meeting}.

$$P(C|T) = \frac{P(C)P(T|C)}{P(C)P(T|C) + P(C')P(T|C')}$$

$$= \frac{(0.9)(0.9)}{(0.9)(0.9) + (0.1)(0.8)} = \frac{81}{89} \approx 0.910$$

13. W = {walking reported}, B = {bicycling reported},
R = {running reported}, C = {completed requirement}.

$$P(W|C) = \frac{P(W)P(C|W)}{P(W)P(C|W) + P(B)P(C|B) + P(R)P(C|R)}$$

$$= \frac{\frac{1}{2} \cdot \frac{9}{10}}{\frac{1}{2} \cdot \frac{9}{10} + \frac{1}{4} \cdot \frac{4}{5} + \frac{1}{4} \cdot \frac{2}{3}} = \frac{27}{49} \approx 55.1\%$$

15. J = {had Japanese-made car}, E = {had European-made car},
A = {had American-made car}, B = {buy same make again}.

$$P(J|B) = \frac{P(J)P(B|J)}{P(J)P(B|J) + P(E)P(B|E) + P(A)P(B|A)}$$

$$= \frac{\frac{3}{5} \cdot \frac{85}{100}}{\frac{3}{5} \cdot \frac{85}{100} + \frac{1}{10} \cdot \frac{50}{100} + \frac{3}{10} \cdot \frac{40}{100}} = \frac{51}{68}.$$

17. P = {pass the exam}, A = {answer every question}.

$$P(A|P) = \frac{P(A)P(P|A)}{P(A)P(P|A) + P(A')P(P|A')}$$

$$= \frac{(0.75)(0.8)}{(0.75)(0.8) + (0.25)(0.50)} = \frac{24}{29} \approx 0.828$$

19. S = {signals sent}, D = {signals detected}.

$$P(S|D) = \frac{P(S)P(D|S)}{P(S)P(D|S) + P(S')P(D|S')} = \frac{\frac{2}{5} \cdot \frac{3}{5}}{\frac{2}{5} \cdot \frac{3}{5} + \frac{3}{5} \cdot \frac{1}{10}} = \frac{4}{5}.$$

21. S = {movie is a success}, U = {"Two Thumbs Up"}.

$$P(S|U) = \frac{P(S)P(U|S)}{P(S)P(U|S) + P(S')P(U|S')}$$

$$= \frac{\frac{7}{10} \cdot \frac{80}{100}}{\frac{7}{10} \cdot \frac{80}{100} + \frac{3}{10} \cdot \frac{10}{100}} = \frac{56}{59} \approx 0.949$$

23. S = {is substandard request}, C = {is considered
substandard request by Risky}.

(a) $P(C) = P(S)P(C|S) + P(S')P(C|S')$

$$= (0.15)(0.8) + (0.85)(0.10) = 0.205 = \frac{41}{200}$$

(b) $P(S|C) = \dfrac{P(S)P(C|S)}{P(S)P(C|S) + P(S')P(C|S')}$

$= \dfrac{(0.15)(0.8)}{0.205} = \dfrac{0.12}{0.205} = \dfrac{120}{205} = \dfrac{24}{41} \approx 0.585$

(c) $P(Error) = P(C \cap S') + P(C' \cap S)$

$= P(S)P(C'|S) + P(S')P(C|S')$

$= (0.15)(0.2) + (0.85)(0.10) = 0.115 = \dfrac{23}{200}$

25. (a) $P(L|E) = \dfrac{P(L)P(E|L)}{P(L)P(E|L) + P(M)P(E|M) + P(H)P(E|H)}$

$= \dfrac{(0.25)(0.49)}{(0.25)(0.49) + (0.25)(0.64) + (0.5)(0.81)}$

≈ 0.18

(b) $P(M|E) = \dfrac{P(M)P(E|M)}{P(L)P(E|L) + P(M)P(E|M) + P(H)P(E|H)}$

$= \dfrac{(0.25)(0.64)}{(0.25)(0.49) + (0.25)(0.64) + (0.5)(0.81)}$

≈ 0.23

(c) $P(H|E) = \dfrac{P(H)P(E|H)}{P(L)P(E|L) + P(M)P(E|M) + P(H)P(E|H)}$

$= \dfrac{(0.5)(0.81)}{(0.25)(0.49) + (0.25)(0.64) + (0.5)(0.81)}$

$\approx 0.59.$

(d) High quality

27. F = {fair weather}, I = {inclement weather}, W = {predict fair weather}.

$P(F|W) = \dfrac{P(F)P(W|F)}{P(F)P(W|F) + P(I)P(W|I)}$

$= \dfrac{(0.8)(0.7)}{(0.8)(0.7) + (0.2)(0.3)} = \dfrac{28}{31} \approx 0.90$

CHAPTER 9 - REVIEW PROBLEMS

1. $_8P_3 = 8 \cdot 7 \cdot 6 = 336$

3. $_9C_7 = \dfrac{9!}{7!(9-7)!} = \dfrac{9!}{7! \cdot 2!} = \dfrac{9 \cdot 8 \cdot 7!}{7! \cdot 2 \cdot 1} = \dfrac{9 \cdot 8}{2} = 36$

5. For each of the first two symbols there are 26 choices, for the third there are 9 choices, and for each of the last two symbols there are 10 choices. By the basic

counting principle, the number of license plates that are
possible is 26·26·9·10·10 = 608,400.   <u>Ans.</u> 608,400

7. Each of the five switches has 2 possible positions. By
the basic counting principle, the number of different
codes is 2·2·2·2·2 = 2^5 = 32.   <u>Ans.</u> 32

9. A possibility for first, second, and third place is a
selection of three of the seven teams so that order is
important. Thus the number of ways the season can end is
$_7P_3$ = 7·6·5 = 210.   <u>Ans.</u> 210

11. A group is any five of the eight people, and the order of
the five is not important. Thus the number of groups is
$_8C_5 = \frac{8!}{5!\cdot3!} = \frac{8\cdot7\cdot6\cdot5!}{5!\cdot3\cdot2\cdot1} = 56$.   <u>Ans.</u> 56

13. (a) Three bulbs are selected from 24, and the order of
selection is not important. Thus the number of
possible selections is $_{24}C_3 = \frac{24!}{3!(24-3)!} = \frac{24!}{3!\cdot21!} =$
$\frac{24\cdot23\cdot22\cdot21!}{3\cdot2\cdot1\cdot21!} = \frac{24\cdot23\cdot22}{3\cdot2\cdot1} = 2024$.
(b) Only one bulb is defective and that bulb must be
included in the selection. The other two bulbs must
be selected from the 23 remaining bulbs and there are
$_{23}C_2$ such selections possible. Thus the number of
ways of selecting three bulbs such that one is
defective is $1\cdot_{23}C_2 = _{23}C_2 = \frac{23!}{2!(23-2)!} = \frac{23!}{2!\cdot21!} =$
$\frac{23\cdot22\cdot21!}{2\cdot1\cdot21!} = \frac{23\cdot22}{2\cdot1} = 253$.
<u>Ans.</u> (a) 2024; (b) 253

15. In the word MISSISSIPPI, there are 11 letters with
repeition: 1 M, 4 I's, 4 S's, and 2 P's. Thus the number
of distinguishable horizontal arrangements is
$\frac{11!}{1!\cdot4!\cdot4!\cdot2!} = 34,650$.   <u>Ans.</u> 34,650

17. Of the eight nurses, three go to hospital A (cell A), two go to hospital B (cell B), and three are not assigned (cell C). The number of possible assignments is

$$\frac{8!}{3! \cdot 2! \cdot 3!} = 560.$$ Ans. 560

19. (a) $\{1,2,3,4,5,6,7\}$;   (b) $\{4,5,6\}$;
    (c) $E_1' \cup E_2 = \{7,8\} \cup \{4,5,6,7\} = \{4,5,6,7,8\}$;
    (d) The intersection of any event and its complement is $\emptyset$. Thus $E_1 \cap E_1' = \emptyset$.
    (e) $(E_1 \cap E_2')' = (\{1,2,3,4,5,6\} \cap \{1,2,3,8\})' = \{1,2,3\}' = \{4,5,6,7,8\}$;
    (f) From (b), $E_1 \cap E_2 \neq \emptyset$. Thus E_1 and E_2 are not mutually exclusive.

21. (a) $\left\{ \begin{array}{l} R_1R_2R_3, \ R_1R_2G_3, \ R_1G_2R_3, \ R_1G_2G_3, \\ G_1R_2R_3, \ G_1R_2G_3, \ G_1G_2R_3, \ G_1G_2G_3 \end{array} \right\}$;
    (b) $\{R_1R_2G_3, \ R_1G_2R_3, \ G_1R_2R_3\}$;
    (c) $\{R_1R_2R_3, \ G_1G_2G_3\}$

23. $n(S) = {}_{10}C_2 = \frac{10!}{2! \cdot 8!} = \frac{10 \cdot 9 \cdot 8!}{2 \cdot 1 \cdot 8!} = \frac{10 \cdot 9}{2 \cdot 1} = 45.$ Let E be the event that box is rejected. If box is rejected, the one defective chip must be in the two-chip sample and there are nine possibilities for the other chip. Thus $n(E) = 9$ and $P(E) = \frac{n(E)}{n(S)} = \frac{9}{45} = \frac{1}{5} = 0.2.$ Ans. 0.2

25. Number of ways to answer exam is $4^5 = 1024 = n(S)$. Let E = {exactly two questions are incorrect}. The number of ways of selecting two of the five questions that are incorrect is ${}_5C_2 = \frac{5!}{2! \cdot 3!} = 10.$ However, there are three ways to answer a question incorrectly. Since two questions are incorrect, $n(E) = 10 \cdot 3 \cdot 3 = 90.$ Thus $P(E) = \frac{n(E)}{n(S)} = \frac{90}{1024} = \frac{45}{512}.$ Ans. $\frac{45}{512}$

27. (a) There are 10 marbles in the urn. $n(S) = 10 \cdot 10 = 100$.
$n(E_{both\ red}) = 4 \cdot 4 = 16$. Thus $P(E_{both\ red}) =$
$$\frac{n(E_{both\ red})}{n(S)} = \frac{16}{100} = \frac{4}{25}.$$
(b) $n(S) = 10 \cdot 9 = 90$. $n(E_{both\ red}) = 4 \cdot 3 = 12$. Thus
$$P(E_{both\ red}) = \frac{12}{90} = \frac{2}{15}.$$
Ans. (a) 4/25; (b) 2/15

29. $n(S) = 52 \cdot 52$.
(a) There are 26 red cards in a deck. Thus
$n(E_{both\ red}) = 26 \cdot 26$ and $P(E_{both\ red}) = \frac{26 \cdot 26}{52 \cdot 52} = \frac{1}{4}$.
(b) There are 13 clubs in a deck, none of which are red.
If E = event that one card is red and the other is a
club, then E occurs if the first card is red and the
second is a club, or vice versa. Thus $n(E) = 26 \cdot 13 +$
$13 \cdot 26 = 2 \cdot 26 \cdot 13$ and $P(E) = \frac{2 \cdot 26 \cdot 13}{52 \cdot 52} = \frac{1}{4}$.
Ans. (a) 1/4; (b) 1/4

31. $\frac{P(E)}{P(E')} = \frac{3/8}{1 - (3/8)} = \frac{3/8}{5/8} = \frac{3}{5}$. Ans. 3:5

33. $P(E) = \frac{6}{6 + 1} = \frac{6}{7}$

35. $P(F|H') = \frac{P(F \cap H')}{P(H')} = \frac{9/52}{39/52} = \frac{3}{13}$

37. $P(S \cap M) = P(S)P(M|S) = (0.6)(0.7) = 0.42$

39. (a) The reduced sample space consists of
    (4,1),(4,2),(4,3),(4,4),(4,5),(4,6),
    (1,4),(2,4),(3,4),(5,4),(6,4).
    In two of these 11 points, the sum of the components
    is 7. Thus $P(\text{sum} = 7 \mid \text{a 4 shows}) = \frac{2}{11}$.
(b) Out of 36 sample points, the event {getting a total
    of 7 and having a 4 show} is {(4,3),(3,4)}. Thus the
    probability of this event is $\frac{2}{36} = \frac{1}{18}$.

41. The second number must be a 5 or 6, so the reduced sample
    space has $6 \cdot 2 = 12$ sample points. Of these, the event
    {first number $\geq$ second number} consists of (5,5), (6, 5),
    and (6,6). Thus the conditional probability is $\frac{3}{12} = \frac{1}{4}$.

43. (a) $P(L'|F) = \frac{n(L' \cap F)}{n(F)} = \frac{160}{480} = \frac{1}{3}$

    (b) $P(L) = \frac{400}{600} = \frac{2}{3}$ and $P(L|M) = \frac{n(L \cap M)}{n(M)} = \frac{80}{120} = \frac{2}{3}$.
    Since $P(L|M) = P(L)$, events L and M are independent.

45. P = {attend public college}, M = {from mid-cls family}.
    $P(P) = \frac{125}{175} = \frac{5}{7}$, $P(P|M) = \frac{n(P \cap M)}{n(M)}$ $\frac{55}{80} = \frac{11}{16}$.
    Since $P(P|C) \neq P(P)$, events P and C are dependent.

47. (a) P(none takes root) = $(0.3)(0.3)(0.3)(0.3) = 0.0081$.

    (b) The probability that a particular two shrubs take root
    and the remaining two do not is $(0.7)(0.7)(0.3)(0.3)$.
    The number of ways the two that take root can be
    chosen from the four shrubs is $_4C_2$. Thus

    P(exactly two take root) = $_4C_2(0.7)^2(0.3)^2 = 0.2646$.

    (c) For at most two shrubs to take root, either none does,
    exactly one does, or exactly two do.
    P(none) + P(exactly one) + P(exactly two)
    $$= 0.0081 + {}_4C_1(0.7)(0.3)^3 + 0.2646$$
    $$= 0.0081 + 0.0756 + 0.2646$$
    $$= 0.3483$$

49. $P(R_{II}) = P(G_I)P(R_{II}|G_I) + P(R_I)P(R_{II}|R_I) = \frac{3}{5} \cdot \frac{4}{9} + \frac{2}{5} \cdot \frac{5}{9} = \frac{22}{45}$

51. $P(G|A) = \frac{P(G \cap A)}{P(A)} = \frac{0.1}{0.4} = \frac{1}{4}$

53. (a) F = {produced by first shift}, S = {produced by
    second shift}, D = {defective}.
    $$P(D) = P(F)P(D|F) + P(S)P(D|S)$$
    $$= \frac{300}{500} \cdot (0.01) + \frac{200}{500} \cdot (0.02)$$
    $$= 0.006 + 0.008 = 0.014$$

    (b) $P(S|D) = \frac{P(S)P(D|S)}{P(F)P(D|F) + P(S)P(D|S)} = \frac{0.008}{0.014} = \frac{4}{7} \approx 0.57$

10

Additional Topics in Probability

1. $\mu = \sum\limits_{x} xf(x) = 0(0.1) + 1(0.4) + 2(0.2) + 3(0.3)$

   $\qquad = 0 + 0.4 + 0.4 + 0.9 = 1.7.$

   $Var(X) = \sum\limits_{x} x^2 f(x) - \mu^2$

   $\qquad = [0^2(0.1) + 1^2(0.4) + 2^2(0.2) + 3^2(0.3)] - (1.7)^2$

   $\qquad = [0 + 0.4 + 0.8 + 2.7] - 2.89 = 3.9 - 2.89 = 1.01.$

   $\sigma = \sqrt{Var(X)} = \sqrt{1.01} = 1.00.$

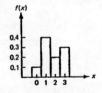

3. $\mu = \sum_x xf(x) = 1\left(\frac{1}{4}\right) + 2\left(\frac{1}{4}\right) + 3\left(\frac{1}{2}\right) = \frac{9}{4} = 2.25.$

   $Var(X) = \sum_x x^2 f(x) - \mu^2 = \left[1^2\left(\frac{1}{4}\right) + 2^2\left(\frac{1}{4}\right) + 3^2\left(\frac{1}{2}\right)\right] - \left(\frac{9}{4}\right)^2$

   $= \frac{23}{4} - \frac{81}{16} = \frac{11}{16} = 0.6875.$

   $\sigma = \sqrt{\frac{11}{16}} = \frac{\sqrt{11}}{4} = 0.83.$

5. (a) $P(2) = 1 - [P(4) + P(7)] = 1 - 0.5 - 0.4 = 0.1.$

   (b) $\mu = \sum_x xf(x) = 2(0.1) + 4(0.5) + 7(0.4) = 5.$

   (c) $\sigma^2 = \sum_x x^2 f(x) - \mu^2$

   $= [2^2(0.1) + 4^2(0.5) + 7^2(0.4)] - 5^2 = 3.$

7. Distribution of X:

   $$f(0) = \frac{1}{8}, \quad f(1) = \frac{3}{8}, \quad f(2) = \frac{3}{8}, \quad f(3) = \frac{1}{8}.$$

   $E(X) = \sum_x xf(x) = 0\left(\frac{1}{8}\right) + 1\left(\frac{3}{8}\right) + 2\left(\frac{3}{8}\right) + 3\left(\frac{1}{8}\right) = \frac{12}{8} = \frac{3}{2} = 1.5;$

   $\sigma^2 = Var(X) = \sum_x x^2 f(x) - [E(x)]^2$

   $= \left[0^2\left(\frac{1}{8}\right) + 1^2\left(\frac{3}{8}\right) + 2^2\left(\frac{3}{8}\right) + 3^2\left(\frac{1}{8}\right)\right] - \left(\frac{3}{2}\right)^2$

   $= \frac{24}{8} - \frac{9}{4} = \frac{6}{8} = \frac{3}{4} = 0.75;$

   $\sigma = \sqrt{\frac{3}{4}} = \frac{\sqrt{3}}{2} = 0.87.$

9. The number of outcomes in the sample space is $_5C_2 = 10.$

   Distribution of X:

   $$f(0) = \frac{_3C_2}{10} = \frac{3}{10}; \quad f(1) = \frac{_2C_1 \cdot _3C_1}{10} = \frac{3}{5}; \quad f(2) = \frac{_2C_2}{10} = \frac{1}{10}.$$

   $E(X) = \sum_x xf(x) = 0\left(\frac{3}{10}\right) + 1\left(\frac{3}{5}\right) + 2\left(\frac{1}{10}\right) = \frac{4}{5} = 0.8;$

   $\sigma^2 = \sum_x x^2 f(x) - [E(x)]^2 = \left[0^2\left(\frac{3}{10}\right) + 1^2\left(\frac{3}{5}\right) + 2^2\left(\frac{1}{10}\right)\right] - \left(\frac{4}{5}\right)^2$

   $= 1 - \frac{16}{25} = \frac{9}{25} = 0.36;$

   $\sigma = \sqrt{0.36} = 0.6.$

11. $f(0) = P(X = 0) = \frac{_2C_2}{_5C_2} = \frac{1}{10}$;

     $f(1) = P(X = 1) = \frac{_3C_1 \cdot _2C_1}{_5C_2} = \frac{6}{10} = \frac{3}{5}$;

     $f(2) = P(X = 2) = \frac{_3C_2}{_5C_2} = \frac{3}{10}$.

13. (a) If X is the gain (in dollars), then X = -1 or 8499.
     Distribution of X:

$$f(-1) = \frac{9999}{10,000}, \quad f(8499) = \frac{1}{10,000}.$$

$$E(X) = \sum_X xf(x) = -1 \cdot \frac{9999}{10,000} + 8499 \cdot \frac{1}{10,000}$$

$$= -\frac{1500}{10,000} = -\$0.15 \text{ (a loss)}.$$

     (b) Here X = -2 or 8498. Distribution of X:

$$f(-2) = \frac{9998}{10,000}, \quad f(8499) = \frac{2}{10,000}.$$

$$E(X) = \sum_X xf(x) = -2 \cdot \frac{9998}{10,000} + 8498 \cdot \frac{2}{10,000}$$

$$= -\$0.30 \text{ (a loss)}.$$

15. Let X = daily earnings (in dollars).
     Distribution of X:

$$f(200) = \frac{4}{7}, \quad f(-30) = \frac{3}{7}.$$

$$E(X) = \sum_X xf(x) = 200 \cdot \frac{4}{7} + (-30) \cdot \frac{3}{7} = \frac{710}{7} = \$101.43.$$

17. The probability that a person in the group is not
hospitalized is
     $1 - (0.001 + 0.002 + 0.003 + 0.004 + 0.008)$
     $= 1 - 0.018 = 0.982.$
Let X = gain (in dollars) to the company from a policy.
Distribution of X:
     $f(10) = 0.982, \quad\quad f(-90) = 0.001, \quad f(-190) = 0.002,$
     $f(-290) = 0.003, \quad\quad f(-390) = 0.004, \quad f(-490) = 0.008.$

$$E(X) = 10(0.982) + (-90)(0.001) + (-190)(0.002) +$$
$$(-290)(0.003) + (-390)(0.004) + (-490)(0.008)$$
$$= 9.82 - 0.09 - 0.38 - 0.87 - 1.56 - 3.92 = \$3.00.$$

19. Let p = the annual premium (in dollars) per policy. If X = gain (in dollars) to the company from a policy, then X = p or X = -(80,000 - p). Thus we want

$$E(X) = -(80,000 - p)(0.0002) + p(0.9998) = 50,$$
$$-16 + 0.0002p + 0.9998p = 50,$$
$$-16 + p = 50,$$
$$p = \$66.$$

21. Let X = gain (in dollars) on a play.
    If 0 heads show, then X = 0 - 1.25 = -5/4;
    if exactly 1 head shows, then X = 1.00 - 1.25 = -1/4;
    if 2 heads show, then X = 2.00 - 1.25 = 3/4.
Distribution of X:

$$f(-5/4) = 1/4, \quad f(-1/4) = 1/2, \quad f(3/4) = 1/4.$$

$$E(X) = \left(-\tfrac{5}{4}\right)\left(\tfrac{1}{4}\right) + \left(-\tfrac{1}{4}\right)\left(\tfrac{1}{2}\right) + \left(\tfrac{3}{4}\right)\left(\tfrac{1}{4}\right) = -\tfrac{1}{4} = -0.25.$$

Thus there is an expected loss of \$0.25 on each play.

For a fair game, let p = amount (in dollars) paid to play.
Distribution of X:

$$f(-p) = 1/4, \quad f(1 - p) = 1/2, \quad f(2 - p) = 1/4.$$

Thus we want

$$E(X) = (-p)\tfrac{1}{4} + (1 - p)\tfrac{1}{2} + (2 - p)\tfrac{1}{4} = 0.$$
$$-\tfrac{p}{4} + \tfrac{1}{2} - \tfrac{p}{2} + \tfrac{1}{2} - \tfrac{p}{4} = 0,$$
$$1 - p = 0,$$
$$p = 1.$$

Thus you should pay \$1 for a fair game.

EXERCISE 10.2

1. $f(0) = {}_2C_0\left(\tfrac{1}{4}\right)^0\left(\tfrac{3}{4}\right)^2 = \tfrac{2!}{0!\cdot 2!}\cdot 1\cdot\tfrac{9}{16} = 1\cdot 1\cdot\tfrac{9}{16} = \tfrac{9}{16};$

   $f(1) = {}_2C_1\left(\tfrac{1}{4}\right)^1\left(\tfrac{3}{4}\right)^1 = \tfrac{2!}{1!\cdot 1!}\cdot\tfrac{1}{4}\cdot\tfrac{3}{4} = 2\cdot\tfrac{1}{4}\cdot\tfrac{3}{4} = \tfrac{3}{8};$

   $f(2) = {}_2C_2\left(\tfrac{1}{4}\right)^2\left(\tfrac{3}{4}\right)^0 = \tfrac{2!}{2!\cdot 0!}\cdot\tfrac{1}{16}\cdot 1 = 1\cdot\tfrac{1}{16}\cdot 1 = \tfrac{1}{16}.$

   $\mu = np = 2\cdot\tfrac{1}{4} = \tfrac{1}{2}; \quad \sigma = \sqrt{npq} = \sqrt{2\cdot\tfrac{1}{4}\cdot\tfrac{3}{4}} = \sqrt{\tfrac{6}{16}} = \tfrac{\sqrt{6}}{4}.$

3. $f(0) = {}_3C_0\left(\frac{2}{3}\right)^0\left(\frac{1}{3}\right)^3 = \frac{3!}{0!\cdot3!}\cdot1\cdot\frac{1}{27} = 1\cdot1\cdot\frac{1}{27} = \frac{1}{27}$;

$f(1) = {}_3C_1\left(\frac{2}{3}\right)^1\left(\frac{1}{3}\right)^2 = \frac{3!}{1!\cdot2!}\cdot\frac{2}{3}\cdot\frac{1}{9} = 3\cdot\frac{2}{3}\cdot\frac{1}{9} = \frac{2}{9}$;

$f(2) = {}_3C_2\left(\frac{2}{3}\right)^2\left(\frac{1}{3}\right)^1 = \frac{3!}{2!\cdot1!}\cdot\frac{4}{9}\cdot\frac{1}{3} = 3\cdot\frac{4}{9}\cdot\frac{1}{3} = \frac{4}{9}$;

$f(3) = {}_3C_3\left(\frac{2}{3}\right)^3\left(\frac{1}{3}\right)^0 = \frac{3!}{3!\cdot0!}\cdot\frac{8}{27}\cdot1 = 1\cdot\frac{8}{27}\cdot1 = \frac{8}{27}$.

$\mu = np = 3\cdot\frac{2}{3} = 2$; $\sigma = \sqrt{npq} = \sqrt{3\cdot\frac{2}{3}\cdot\frac{1}{3}} = \sqrt{\frac{2}{3}} = \frac{\sqrt{6}}{3}$.

5. $P(X=5) = {}_6C_5(0.2)^5(0.8)^1 = 6(0.00032)(0.8) = 0.001536$

7. $P(X=2) = {}_4C_2\left(\frac{4}{5}\right)^2\left(\frac{1}{5}\right)^2 = 6\cdot\frac{16}{25}\cdot\frac{1}{25} = \frac{96}{625} = 0.1536$

9. $P(X<2) = P(X=0) + P(X=1) = {}_5C_0\left(\frac{1}{2}\right)^0\left(\frac{1}{2}\right)^5 + {}_5C_1\left(\frac{1}{2}\right)^1\left(\frac{1}{2}\right)^4$

$= 1\cdot1\cdot\frac{1}{32} + 5\cdot\frac{1}{2}\cdot\frac{1}{16} = \frac{6}{32} = \frac{3}{16}$

11. Let X = number of heads that occurs. $p = 1/2$, $n = 10$.

$P(X=8) = {}_{10}C_8\left(\frac{1}{2}\right)^8\left(\frac{1}{2}\right)^2 = 45\cdot\frac{1}{256}\cdot\frac{1}{4} = \frac{45}{1024} \approx 0.044$.

13. Let X = number of green marbles drawn. The probability of selecting a green marble on any draw is $p = \frac{6}{10} = \frac{3}{5}$. $n = 4$.

$P(X=1) = {}_4C_1\left(\frac{3}{5}\right)^1\left(\frac{2}{5}\right)^3 = 4\cdot\frac{3}{5}\cdot\frac{8}{125} = \frac{96}{625} = 0.1536$.

15. Let X = number of defective switches selected. The probability that a switch is defective is $p = 0.02$. $n = 4$.

$P(X=2) = {}_4C_2(0.02)^2(0.98)^2 = 6(0.0004)(0.9604) \approx 0.002$.

17. Let X = number of heads that occurs. $p = 1/4$, $n = 3$.

(a) $P(X=2) = {}_3C_2\left(\frac{1}{4}\right)^2\left(\frac{3}{4}\right)^1 = 3\cdot\frac{1}{16}\cdot\frac{3}{4} = \frac{9}{64}$.

(b) $P(X=3) = {}_3C_3\left(\frac{1}{4}\right)^3\left(\frac{3}{4}\right)^0 = 1\cdot\frac{1}{64}\cdot1 = \frac{1}{64}$.

Thus $P(X=2) + P(X=3) = \frac{9}{64} + \frac{1}{64} = \frac{10}{64} = \frac{5}{32}$.

<u>Ans.</u> (a) 9/64; (b) 5/32

19. Let X = number of defective in sample. p = 1/3, n = 4.

$P(X{\leq}1) = P(X{=}0) + P(X{=}1) = {}_4C_0\left(\frac{1}{3}\right)^0\left(\frac{2}{3}\right)^4 + {}_4C_1\left(\frac{1}{3}\right)^1\left(\frac{2}{3}\right)^3 =$

$1\cdot1\cdot\frac{16}{81} + 4\cdot\frac{1}{3}\cdot\frac{8}{27} = \frac{48}{81} = \frac{16}{27} \approx 0.593.$

21. Let X = number of hits in four bats. p = 0.300, n = 4.

$P(X{\geq}1) = 1 - P(X{=}0) = 1 - {}_4C_0(0.300)^0(0.700)^4 =$

$1 - 1\cdot1\cdot(0.2401) = 0.7599.$

23. Let X = number of girls. The probability that a child is a
    girl is p = 1/2. Here n = 5. We must find $P(X{\geq}2) =$
    $1 - P(X{<}2) = 1 - [P(X{=}0) + P(X{=}1)]$.

$P(X{=}0) = {}_5C_0\left(\frac{1}{2}\right)^0\left(\frac{1}{2}\right)^5 = 1\cdot1\cdot\frac{1}{32} = \frac{1}{32}.$

$P(X{=}1) = {}_5C_1\left(\frac{1}{2}\right)^1\left(\frac{1}{2}\right)^4 = 5\cdot\frac{1}{2}\cdot\frac{1}{16} = \frac{5}{32}.$

Thus $P(X{\geq}2) = 1 - [P(X{=}0) + P(X{=}1)] = 1 - \left[\frac{1}{32} + \frac{5}{32}\right] =$

$1 - \frac{3}{16} = \frac{13}{16}.$ Ans. $\frac{13}{16}$

25. $\mu = 2$, $\sigma^2 = \frac{3}{2}$. Since $\mu = np$, then np = 2. Since $\sigma^2 = npq$,

then $(np)q = \frac{3}{2}$, or $2q = \frac{3}{2}$, or $q = \frac{3}{4}$. Thus $p = 1 - q =$

$1 - \frac{3}{4} = \frac{1}{4}$. Since np = 2, then $n\cdot\frac{1}{4} = 2$, or n = 8. Thus

$P(X{=}1) = {}_8C_1\left(\frac{1}{4}\right)^1\left(\frac{3}{4}\right)^7 = 8\cdot\frac{3^7}{4^8} = \frac{2187}{8192} \approx 0.267.$

Ans. $\frac{2187}{8192} \approx 0.267$

EXERCISE 10.3

1. $\begin{bmatrix} \frac{1}{2} & 0 \\ \frac{2}{3} & \frac{1}{3} \end{bmatrix}$. The sum of the entries in row 1 is $\frac{1}{2} + 0 = \frac{1}{2} \neq 1$.
   Thus matrix cannot be a transition matrix.
   Ans. no

3. $\begin{bmatrix} \frac{1}{2} & -\frac{1}{4} & \frac{3}{4} \\ \frac{1}{8} & \frac{5}{8} & \frac{1}{4} \\ \frac{1}{3} & \frac{1}{3} & \frac{1}{3} \end{bmatrix}$. Entry in row 1 and column 2 is negative. Thus matrix cannot be a transition matrix. <u>Ans.</u> no

5. $\begin{bmatrix} 0.4 & 0.2 & 0.4 \\ 0 & 0.1 & 0.9 \\ 0.5 & 0.3 & 0.2 \end{bmatrix}$. All entries are nonnegative and sum of entries in each row is 1. <u>Ans.</u> yes

7. $\begin{bmatrix} \frac{2}{3} & a \\ b & \frac{1}{4} \end{bmatrix}$. $a + \frac{2}{3} = 1$, so $a = \frac{1}{3}$. $b + \frac{1}{4} = 1$, so $b = \frac{3}{4}$. <u>Ans.</u> $a = \frac{1}{3}$, $b = \frac{3}{4}$

9. $\begin{bmatrix} 0.4 & a & 0.2 \\ a & 0.1 & b \\ a & b & c \end{bmatrix}$. $0.4 + a + 0.2 = 1$, so $a = 0.4$. $a + 0.1 + b = 1$, $0.4 + 0.1 + b = 1$, so $b = 0.5$.
$a + b + c = 1$, $0.4 + 0.5 + c = 1$, so $c = 0.1$.
<u>Ans.</u> $a = 0.4$, $b = 0.5$, $c = 0.1$

11. $[0.4 \quad 0.6]$. All entries are nonnegative and their sum is 1.
<u>Ans.</u> yes

13. $[0.2 \quad 0.7 \quad 0.5]$. The sum of the entries is not 1.
<u>Ans.</u> no

15. $X_1 = X_0 T = \begin{bmatrix} \frac{1}{4} & \frac{3}{4} \end{bmatrix} \begin{bmatrix} \frac{2}{3} & \frac{1}{3} \\ 1 & 0 \end{bmatrix} = \begin{bmatrix} \frac{11}{12} & \frac{1}{12} \end{bmatrix}$;

$X_2 = X_1 T = \begin{bmatrix} \frac{11}{12} & \frac{1}{12} \end{bmatrix} \begin{bmatrix} \frac{2}{3} & \frac{1}{3} \\ 1 & 0 \end{bmatrix} = \begin{bmatrix} \frac{25}{36} & \frac{11}{36} \end{bmatrix}$;

$X_3 = X_2 T = \begin{bmatrix} \frac{25}{36} & \frac{11}{36} \end{bmatrix} \begin{bmatrix} \frac{2}{3} & \frac{1}{3} \\ 1 & 0 \end{bmatrix} = \begin{bmatrix} \frac{83}{108} & \frac{25}{108} \end{bmatrix}$.

17. $X_1 = X_0 T = [0.4 \quad 0.6] \begin{bmatrix} 0.5 & 0.5 \\ 0.5 & 0.5 \end{bmatrix} = [0.5 \quad 0.5]$;

$$X_2 = X_1T = \begin{bmatrix} 0.5 & 0.5 \end{bmatrix} \begin{bmatrix} 0.5 & 0.5 \\ 0.5 & 0.5 \end{bmatrix} = \begin{bmatrix} 0.5 & 0.5 \end{bmatrix};$$

$$X_3 = X_2T = \begin{bmatrix} 0.5 & 0.5 \end{bmatrix} \begin{bmatrix} 0.5 & 0.5 \\ 0.5 & 0.5 \end{bmatrix} = \begin{bmatrix} 0.5 & 0.5 \end{bmatrix}.$$

19. $X_1 = X_0T = \begin{bmatrix} 0.2 & 0 & 0.8 \end{bmatrix} \begin{bmatrix} 0.1 & 0.2 & 0.7 \\ 0 & 0.4 & 0.6 \\ 0.3 & 0.3 & 0.4 \end{bmatrix} = \begin{bmatrix} 0.26 & 0.28 & 0.4 \end{bmatrix}$

$X_2 = X_1T = \begin{bmatrix} 0.26 & 0.28 & 0.46 \end{bmatrix} \begin{bmatrix} 0.1 & 0.2 & 0.7 \\ 0 & 0.4 & 0.6 \\ 0.3 & 0.3 & 0.4 \end{bmatrix}$

$= \begin{bmatrix} 0.164 & 0.302 & 0.534 \end{bmatrix};$

$X_3 = X_2T = \begin{bmatrix} 0.164 & 0.3020 & 0.534 \end{bmatrix} \begin{bmatrix} 0.1 & 0.2 & 0.7 \\ 0 & 0.4 & 0.6 \\ 0.3 & 0.3 & 0.4 \end{bmatrix}$

$= \begin{bmatrix} 0.1766 & 0.3138 & 0.5096 \end{bmatrix}.$

21. (a) $T^2 = \begin{bmatrix} \frac{1}{4} & \frac{3}{4} \\ \frac{3}{4} & \frac{1}{4} \end{bmatrix} \begin{bmatrix} \frac{1}{4} & \frac{3}{4} \\ \frac{3}{4} & \frac{1}{4} \end{bmatrix} = \begin{bmatrix} \frac{5}{8} & \frac{3}{8} \\ \frac{3}{8} & \frac{5}{8} \end{bmatrix};$

$T^3 = T^2T = \begin{bmatrix} \frac{5}{8} & \frac{3}{8} \\ \frac{3}{8} & \frac{5}{8} \end{bmatrix} \begin{bmatrix} \frac{1}{4} & \frac{3}{4} \\ \frac{3}{4} & \frac{1}{4} \end{bmatrix} = \begin{bmatrix} \frac{7}{16} & \frac{9}{16} \\ \frac{9}{16} & \frac{7}{16} \end{bmatrix}.$

(b) Entry in row 1, column 2, of T^2 is 3/8.

(c) Entry in row 2, column 1 of T^3 is 9/16.

23. (a) $T^2 = \begin{bmatrix} 0 & 1 & 0 \\ 0.5 & 0.4 & 0.1 \\ 0.3 & 0.3 & 0.4 \end{bmatrix} \begin{bmatrix} 0 & 1 & 0 \\ 0.5 & 0.4 & 0.1 \\ 0.3 & 0.3 & 0.4 \end{bmatrix} = \begin{bmatrix} 0.50 & 0.40 & 0.1 \\ 0.23 & 0.69 & 0.0 \\ 0.27 & 0.54 & 0.1 \end{bmatrix}$

$T^3 = T^2T = \begin{bmatrix} 0.50 & 0.40 & 0.10 \\ 0.23 & 0.69 & 0.08 \\ 0.27 & 0.54 & 0.19 \end{bmatrix} \begin{bmatrix} 0 & 1 & 0 \\ 0.5 & 0.4 & 0.1 \\ 0.3 & 0.3 & 0.4 \end{bmatrix} =$

$= \begin{bmatrix} 0.230 & 0.690 & 0.080 \\ 0.369 & 0.530 & 0.101 \\ 0.327 & 0.543 & 0.130 \end{bmatrix}.$

(b) Entry in row 1, column 2, of T^2 is 0.40.

(c) Entry in row 2, column 1 of T^3 is 0.369.

25. $T^T - I = \begin{bmatrix} 1/2 & 3/4 \\ 1/2 & 1/4 \end{bmatrix} - \begin{bmatrix} 1 & 0 \\ 0 & 1 \end{bmatrix} = \begin{bmatrix} -1/2 & 3/4 \\ 1/2 & -3/4 \end{bmatrix}.$

$\begin{bmatrix} -1/2 & 3/4 & | & 0 \\ 1/2 & -3/4 & | & 0 \\ 1 & 1 & | & 1 \end{bmatrix} \longrightarrow \cdots \longrightarrow \begin{bmatrix} 1 & 0 & | & 3/5 \\ 0 & 1 & | & 2/5 \\ 0 & 0 & | & 0 \end{bmatrix}.$

Ans. $Q = [3/5 \quad 2/5]$

27. $T^T - I = \begin{bmatrix} 1/5 & 3/5 \\ 4/5 & 2/5 \end{bmatrix} - \begin{bmatrix} 1 & 0 \\ 0 & 1 \end{bmatrix} = \begin{bmatrix} -4/5 & 3/5 \\ 4/5 & -3/5 \end{bmatrix}.$

$\begin{bmatrix} -4/5 & 3/5 & | & 0 \\ 4/5 & -3/5 & | & 0 \\ 1 & 1 & | & 1 \end{bmatrix} \longrightarrow \cdots \longrightarrow \begin{bmatrix} 1 & 0 & | & 3/7 \\ 0 & 1 & | & 4/7 \\ 0 & 0 & | & 0 \end{bmatrix}.$

Ans. $Q = [3/7 \quad 4/7]$

29. $T^T - I = \begin{bmatrix} 0.4 & 0.6 & 0.6 \\ 0.3 & 0.3 & 0.1 \\ 0.3 & 0.1 & 0.3 \end{bmatrix} - \begin{bmatrix} 1 & 0 & 0 \\ 0 & 1 & 0 \\ 0 & 0 & 1 \end{bmatrix} = \begin{bmatrix} -0.6 & 0.6 & 0.6 \\ 0.3 & -0.7 & 0.1 \\ 0.3 & 0.1 & -0.7 \end{bmatrix}.$

$\begin{bmatrix} -0.6 & 0.6 & 0.6 & | & 0 \\ 0.3 & -0.7 & 0.1 & | & 0 \\ 0.3 & 0.1 & -0.7 & | & 0 \\ 1 & 1 & 1 & | & 1 \end{bmatrix} \longrightarrow \cdots \longrightarrow \begin{bmatrix} 1 & 0 & 0 & | & 1/2 \\ 0 & 1 & 0 & | & 1/2 \\ 0 & 0 & 1 & | & 1/4 \\ 0 & 0 & 0 & | & 0 \end{bmatrix}.$

Ans. $Q = [1/2 \quad 1/4 \quad 1/4]$

31. (a) $T = \begin{array}{c} \\ \text{Flu} \\ \text{No flu} \end{array} \begin{array}{cc} \text{Flu} & \text{No flu} \\ \begin{bmatrix} 0.1 & 0.9 \\ 0.2 & 0.8 \end{bmatrix} \end{array}$

(b) $X_0 = [120/200 \quad 80/200] = [0.6 \quad 0.4].$

If a period is 4 days, then 8 days corresponds to 2 periods, and 12 days corresponds to 3 periods.
The state vector corresponding to 8 days from now is
$X_2 = X_0 T^2 = [0.6 \quad 0.4] \begin{bmatrix} 0.19 & 0.81 \\ 0.18 & 0.82 \end{bmatrix} = [0.186 \quad 0.914].$
Thus $0.186(200) \approx 37$ students can be expected to have the flu 8 days from now.
The state vector corresponding to 12 days from now is
$X_3 = X_0 T^3 = [0.6 \quad 0.4] \begin{bmatrix} 0.181 & 0.819 \\ 0.182 & 0.818 \end{bmatrix} = [0.1814 \quad 0.8186].$
Thus $0.1814(200) \approx 36$ students can be expected to have the flu 12 days from now.   Ans. 37; 36

33. (a) $T = \begin{matrix} A \\ B \end{matrix} \begin{bmatrix} \overset{A}{0.9} & \overset{B}{0.1} \\ 0.3 & 0.7 \end{bmatrix}$.

(b) Thursday corresponds to step 3.

$T^2 = \begin{bmatrix} 0.84 & 0.16 \\ 0.48 & 0.52 \end{bmatrix}$. $T^3 = \begin{bmatrix} 0.804 & 0.196 \\ 0.588 & 0.412 \end{bmatrix}$.

Ans. 0.804

35. (a) $T = \begin{matrix} D \\ R \\ O \end{matrix} \begin{bmatrix} \overset{D}{0.8} & \overset{R}{0.1} & \overset{O}{0.1} \\ 0.1 & 0.8 & 0.1 \\ 0.3 & 0.1 & 0.6 \end{bmatrix}$.

(b) $T^2 = \begin{bmatrix} 0.68 & 0.17 & 0.15 \\ 0.19 & 0.66 & 0.15 \\ 0.43 & 0.17 & 0.40 \end{bmatrix}$.   Ans. 0.19

(c) $X_1 = X_0T = [0.40 \quad 0.40 \quad 0.20] \begin{bmatrix} 0.8 & 0.1 & 0.1 \\ 0.1 & 0.8 & 0.1 \\ 0.3 & 0.1 & 0.6 \end{bmatrix}$

$= [\ 0.42 \quad 0.38 \quad 0.20]$.   Ans. 38%

37. (a) $T = \begin{matrix} A \\ Compet. \end{matrix} \begin{bmatrix} \overset{A}{0.8} & \overset{Compet.}{0.2} \\ 0.3 & 0.7 \end{bmatrix}$.

(b) $X_1 = X_0T = [0.70 \quad 0.30] \begin{bmatrix} 0.8 & 0.2 \\ 0.3 & 0.7 \end{bmatrix} = [0.65 \quad 0.35]$.

Ans. 65%

(c) $T^T - I = \begin{bmatrix} 0.8 & 0.3 \\ 0.2 & 0.7 \end{bmatrix} - \begin{bmatrix} 1 & 0 \\ 0 & 1 \end{bmatrix} = \begin{bmatrix} -0.2 & 0.3 \\ 0.2 & -0.3 \end{bmatrix}$.

$\begin{bmatrix} -0.2 & 0.3 & | & 0 \\ 0.2 & -0.3 & | & 0 \\ 1 & 1 & | & 1 \end{bmatrix} \longrightarrow \cdots \longrightarrow \begin{bmatrix} 1 & 0 & | & 3/5 \\ 0 & 1 & | & 2/5 \\ 0 & 0 & | & 0 \end{bmatrix}$.

$Q = [3/5 \quad 2/5] = [0.60 \quad 0.20]$.   Ans. 60%

39. (a) $T = \begin{matrix} 1 \\ 2 \end{matrix} \begin{bmatrix} \overset{1}{5/7} & \overset{2}{2/7} \\ 3/7 & 4/7 \end{bmatrix}$.

(b) $X_2 = X_0 T^2 = [1/2 \quad 1/2]\begin{bmatrix} 31/49 & 18/49 \\ 27/49 & 22/49 \end{bmatrix} = [58/98 \quad 40/98]$

$\approx [0.5918 \quad 0.4082]$.

Ans. 59.18% in compartment 1, 40.82% in compartment 2

(c) $T^T - I = \begin{bmatrix} 5/7 & 3/7 \\ 2/7 & 4/7 \end{bmatrix} - \begin{bmatrix} 1 & 0 \\ 0 & 1 \end{bmatrix} = \begin{bmatrix} -2/7 & 3/7 \\ 2/7 & -3/7 \end{bmatrix}$.

$\begin{bmatrix} -2/7 & 3/7 & | & 0 \\ 2/7 & -3/7 & | & 0 \\ 1 & 1 & | & 1 \end{bmatrix} \longrightarrow \cdots \longrightarrow \begin{bmatrix} 1 & 0 & | & 3/5 \\ 0 & 1 & | & 2/5 \\ 0 & 0 & | & 0 \end{bmatrix}$.

$Q = [3/5 \quad 2/5] = [0.60 \quad 0.40]$.

Ans. 60% in compartment 1, 40% in compartment 2

41. (a) $T^T - I = \begin{bmatrix} 3/4 & 1/2 \\ 1/4 & 1/2 \end{bmatrix} - \begin{bmatrix} 1 & 0 \\ 0 & 1 \end{bmatrix} = \begin{bmatrix} -1/4 & 1/2 \\ 1/4 & -1/2 \end{bmatrix}$.

$\begin{bmatrix} -1/4 & 1/2 & | & 0 \\ 1/4 & -1/2 & | & 0 \\ 1 & 1 & | & 1 \end{bmatrix} \longrightarrow \cdots \longrightarrow \begin{bmatrix} 1 & 0 & | & 2/3 \\ 0 & 1 & | & 1/3 \\ 0 & 0 & | & 0 \end{bmatrix}$.

Ans. $Q = [2/3 \quad 1/3]$

(b) Presently, A accounts for 50% of sales and in long run A will account for $\frac{2}{3}$, or $66\frac{2}{3}$%, of sales. Thus the percentage increase in sales above the present level is

$$\frac{66\frac{2}{3} - 50}{50} \cdot 100\% = \frac{16\frac{2}{3}}{50} \cdot 100\% = 33\frac{1}{3}\% \qquad \underline{\text{Ans.}} \quad 33\frac{1}{3}\%$$

43. $T^2 = TT = \begin{bmatrix} 1/2 & 1/2 \\ 1 & 0 \end{bmatrix}\begin{bmatrix} 1/2 & 1/2 \\ 1 & 0 \end{bmatrix} = \begin{bmatrix} 3/4 & 1/4 \\ 1/2 & 1/2 \end{bmatrix}$. Since all entries of T^2 are positive, T is regular.

CHAPTER 10 - REVIEW PROBLEMS

1. $\mu = \Sigma_x x f(x) = 1 \cdot f(1) + 2 \cdot f(2) + 3 \cdot f(3)$

   $= 1(0.7)\ 2(0.1) + 3(0.2) = 0.7 + 0.2 + 0.6 = 1.5.$

   $Var(X) = \Sigma_x x^2 f(x) - \mu^2$

   $\qquad = [1^2(0.7) + 2^2(0.1) + 3^2(0.2)] - (1.5)^2$

   $\qquad = [0.7 + 0.4 + 1.8] - 2.25 = 2.9 - 2.25 = 0.65.$

   $\sigma = \sqrt{Var(X)} = \sqrt{0.65} \approx 0.81.$

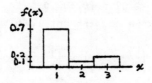

3. (a) $n(S) = 2 \cdot 6 = 12.$ $E_1 = \{T1\},$ $E_2 = \{T2, H1\},$

   $E_3 = \{T3, H2\},$ $E_4 = \{T4, H3\},$ $E_5 = \{T5, H4\},$

   $E_6 = \{T6, H5\},$ $E_7 = \{H6\}.$

   $f(1) = P(E_1) = \dfrac{n(E_1)}{n(S)} = \dfrac{1}{12},$ $f(2) = P(E_2) = \dfrac{n(E_2)}{n(S)} =$

   $\dfrac{2}{12} = \dfrac{1}{6}.$ Similarly, $f(3),\ f(4),\ f(5),$ and $f(6)$ equal

   $1/6.$ $f(7) = P(E_7) = \dfrac{n(E_7)}{n(S)} = \dfrac{1}{12}.$

   (b) $E(X) = \Sigma_x x f(x) = 1 \cdot \dfrac{1}{12} + \dfrac{2+3+4+5+6}{6} + 7 \cdot \dfrac{1}{12} =$

   $\dfrac{1}{12} + \dfrac{20}{6} + \dfrac{7}{12} = \dfrac{48}{12} = 4.$

   Ans.   (a) $f(1) = 1/12,\ f(2) = f(3) = f(4) = f(5) = f(6) =$
   $1/6,\ f(7) = 1/12;$ (b) 4

5. Let X = gain (in dollars) on a play. If no 10 appears,
   then $X = 0 - (1/4) = -1/4;$ if exactly one 10 appears, then
   $X = 1 - (1/4) = 3/4;$ if two 10's appear, then $X = 2 - (1/4) =$
   $7/4.$ $n(S) = 52 \cdot 52.$ In a deck, there are 4 10's and 48
   non-10's. Thus $n(E_{no\ 10}) = 48 \cdot 48.$ The event $E_{one\ 10}$
   occurs if the first card is a 10 and the second is a
   non-10, or vice versa. Thus $n(E_{one\ 10}) = 4 \cdot 48 + 48 \cdot 4 =$

2·4·48. $n(E_{two \ 10's}) = 4·4$.

Dist. of X: $f\left(-\frac{1}{4}\right) = \frac{48·48}{52·52} = \frac{144}{169}$, $f\left(\frac{3}{4}\right) = \frac{2·4·48}{52·52} = \frac{24}{169}$,

$f\left(\frac{7}{4}\right) = \frac{4·4}{52·52} = \frac{1}{169}$.

$E(X) = -\frac{1}{4}·\frac{144}{169} + \frac{3}{4}·\frac{24}{169} + \frac{7}{4}·\frac{1}{169} = \frac{-144+72+7}{4·169} = -\frac{65}{676} =$

$-\frac{5}{52} \approx -0.10$. <u>Ans.</u> loss of $0.10 per play

7. (a) Let X = gain (in dollars) on each unit shipped. Then
   $P(X=-100) = 0.08$ and $P(X=200) = 1 - 0.08 = 0.92$.
   $E(X) = -100f(-100) + 200f(200)$
   $= -100(0.08) + 200(0.92) = \$176$ per unit.
   (b) Since the expected gain per unit is $176 and 4000 units
   are shipped per year, then then expected annual profit
   is $4000(176) = \$704,000$.
   <u>Ans.</u> (a) $176; (b) $704,000

9. $f(0) = {}_3C_0(0.1)^0(0.9)^3 = \frac{3!}{0!·3!}·1(0.729) = 1·1(0.729)$
   $= 0.729$,

   $f(1) = {}_3C_1(0.1)^1(0.9)^2 = \frac{3!}{1!·2!}(0.1)(0.81) = 3(0.1)(0.81)$
   $= 0.243$,

   $f(2) = {}_3C_2(0.1)^2(0.9)^1 = \frac{3!}{2!·1!}(0.01)(0.9) = 3(0.01)(0.9)$
   $= 0.027$,

   $f(3) = {}_3C_3(0.1)^3(0.9)^0 = \frac{3!}{3!·0!}(0.001)·1 = 1·(0.001)·1$
   $= 0.001$;

   $\mu = np = 3(0.1) = 0.3$;

   $\sigma = \sqrt{npq} = \sqrt{3(0.1)(0.9)} = \sqrt{0.27} \approx 0.52$

11. $P(X\leq 1) = P(X=0) + P(X=1) = {}_5C_0\left(\frac{3}{4}\right)^0\left(\frac{1}{4}\right)^5 + {}_5C_1\left(\frac{3}{4}\right)^1\left(\frac{1}{4}\right)^4$

    $= 1·1·\frac{1}{1024} + 5·\frac{3}{4}·\frac{1}{256} = \frac{16}{1024} = \frac{1}{64}$

13. The probability that a 2 or 3 results on one roll is $2/6 = 1/3$. Let X = number of 2's or 3's that appear on 4 rolls.
    Then X is binomial with $p = 1/3$ and $n = 4$.

    $P(X=3) = {}_4C_3\left(\frac{1}{3}\right)^3\left(\frac{2}{3}\right)^1 = 4·\frac{1}{27}·\frac{2}{3} = \frac{8}{81}$.   <u>Ans.</u> $\frac{8}{81}$

15. Let X = number of heads that occur. Then X is binomial.

$P(X=0) = {}_4C_0\left(\frac{1}{3}\right)^0\left(\frac{2}{3}\right)^4 = \frac{4!}{0!\cdot 4!}(1)\left(\frac{16}{81}\right) = 1(1)\left(\frac{16}{81}\right) = \frac{16}{81},$

$P(X=1) = {}_4C_1\left(\frac{1}{3}\right)^1\left(\frac{2}{3}\right)^3 = \frac{4!}{1!\cdot 3!}\left(\frac{1}{3}\right)\left(\frac{8}{27}\right) = 4\left(\frac{1}{3}\right)\left(\frac{8}{27}\right) = \frac{32}{81}.$

$P(X\geq 2) = 1 - [P(X=0) + P(X=1)] = 1 - \left[\frac{16}{81} + \frac{32}{81}\right] =$

$1 - \frac{48}{81} = \frac{33}{81}.$ **Ans.** $\frac{33}{81}$

17. From row 1, 0.1 + a + 0.6 = 1, so a = 0.3. From row 2,
2a + b + b = 1, so 2b = 1 - 2a, or $b = \frac{1 - 2a}{2} = \frac{1 - 2(0.3)}{2} =$
0.2. From row 3, a + b + c = 1, so c = 1 - a - b, or
c = 1 - 0.3 - 0.2 = 0.5.   **Ans.** a = 0.3, b = 0.2, c = 0.5

19. $X_1 = X_0T = [0.5 \quad 0 \quad 0.5]\begin{bmatrix} 0.1 & 0.2 & 0.7 \\ 0.3 & 0.4 & 0.3 \\ 0.1 & 0.1 & 0.8 \end{bmatrix} = [0.10 \quad 0.15 \quad 0.75]$

$X_2 = X_1T = [0.10 \quad 0.15 \quad 0.75]\begin{bmatrix} 0.1 & 0.2 & 0.7 \\ 0.3 & 0.4 & 0.3 \\ 0.1 & 0.1 & 0.8 \end{bmatrix}$

$= [0.130 \quad 0.155 \quad 0.715].$

21. (a) $T^2 = TT = \begin{bmatrix} \frac{1}{7} & \frac{6}{7} \\ \frac{3}{7} & \frac{4}{7} \end{bmatrix}\begin{bmatrix} \frac{1}{7} & \frac{6}{7} \\ \frac{3}{7} & \frac{4}{7} \end{bmatrix} = \begin{bmatrix} \frac{19}{49} & \frac{30}{49} \\ \frac{15}{49} & \frac{34}{49} \end{bmatrix},$

$T^3 = T^2T = \begin{bmatrix} \frac{19}{49} & \frac{30}{49} \\ \frac{15}{49} & \frac{34}{49} \end{bmatrix}\begin{bmatrix} \frac{1}{7} & \frac{6}{7} \\ \frac{3}{7} & \frac{4}{7} \end{bmatrix} = \begin{bmatrix} \frac{109}{343} & \frac{234}{343} \\ \frac{117}{343} & \frac{226}{343} \end{bmatrix}.$

(b) From T^2, entry in row 1, column 2, is 30/49.

(c) From T^3, entry in row 2, column 1, is 117/343.

23. $T^T - I = \begin{bmatrix} 1/3 & 2/3 \\ 2/3 & 1/3 \end{bmatrix} - \begin{bmatrix} 1 & 0 \\ 0 & 1 \end{bmatrix} = \begin{bmatrix} -2/3 & 2/3 \\ 2/3 & -2/3 \end{bmatrix}.$

$\begin{bmatrix} -2/3 & 2/3 & | & 0 \\ 2/3 & -2/3 & | & 0 \\ 1 & 1 & | & 1 \end{bmatrix} \longrightarrow \cdots \longrightarrow \begin{bmatrix} 1 & 0 & | & 1/2 \\ 0 & 1 & | & 1/2 \\ 0 & 0 & | & 0 \end{bmatrix}.$

Ans. Q = [1/2 1/2]

25. T =

	Japanese	Non-Japanese
Japanese	0.8	0.2
Non-Japanese	0.6	0.4

$T = \begin{bmatrix} 0.8 & 0.2 \\ 0.6 & 0.4 \end{bmatrix}$.

(a) $T^2 = \begin{bmatrix} 0.8 & 0.2 \\ 0.6 & 0.4 \end{bmatrix} \begin{bmatrix} 0.8 & 0.2 \\ 0.6 & 0.4 \end{bmatrix} = \begin{bmatrix} 0.76 & 0.24 \\ 0.72 & 0.28 \end{bmatrix}$.

From row 1, column 1, the probability that a person who presently owns a Japanese car will buy a Japanese car two cars later is 0.76. Ans. 76%

(b) $X_2 = X_0 T^2 = [0.6 \quad 0.4] \begin{bmatrix} 0.76 & 0.24 \\ 0.72 & 0.28 \end{bmatrix} = [0.744 \quad 0.256]$.

Ans. 74.4% Japanese, 25.6% non-Japanese

(c) $T^T - I = \begin{bmatrix} 0.8 & 0.6 \\ 0.2 & 0.4 \end{bmatrix} - \begin{bmatrix} 1 & 0 \\ 0 & 1 \end{bmatrix} = \begin{bmatrix} -0.2 & 0.6 \\ 0.2 & -0.6 \end{bmatrix}$.

$$\begin{bmatrix} -0.2 & 0.6 & | & 0 \\ 0.2 & -0.6 & | & 0 \\ 1 & 1 & | & 1 \end{bmatrix} \longrightarrow \cdots \longrightarrow \begin{bmatrix} 1 & 0 & | & 0.75 \\ 0 & 1 & | & 0.25 \\ 0 & 0 & | & 0 \end{bmatrix}.$$

$Q = [0.75 \quad 0.25]$.

Ans. 75% Japanese, 25% non-Japanese

CHAPTER 10 — MATHEMATICAL SNAPSHOT

1. If X = number of correct predictions on first part of game, then X has a binomial distribution with n = 3 and p = 1/2. Thus $E(X) = np = 3 \cdot \frac{1}{2} = \frac{3}{2}$. If P(B) is the probability of selecting the shell with the ball under it, then $P(B) = \frac{\text{avg. no. of correct predictions}}{\text{no. of shells}} = \frac{3/2}{4} = \frac{3}{8}$.

Ans. $\frac{3}{8}$

11

Limits
and Continuity

1. (a) 1; (b) 0; (c) 1

3. (a) 1; (b) does not exist; (c) 3

5.   f(0.9) = 2.8   f(1.1) = 3.2
     f(0.99) = 2.98   f(1.01) = 3.02
     f(0.999) = 2.998   f(1.001) = 3.002
              estimate of limit: 3

7.   f(-0.1) = 0.9516   f(0.1) = 1.0517
     f(-0.01) = 0.9950   f(0.01) = 1.0050
     f(-0.001) = 0.9950   f(0.001) = 1.0005
              estimate of limit: 1

9. $\lim_{x \to 2} 16 = 16$

11. $\lim_{t \to -5} (t^2 - 5) = (-5)^2 - 5 = 25 - 5 = 20$

13. $\lim\limits_{x\to-1} (x^3-3x^2-2x+1) = (-1)^3-3(-1)^2-2(-1)+1 = -1-3+2+1 = -1$

15. $\lim\limits_{t\to-3} \dfrac{t-2}{t+5} = \dfrac{\lim\limits_{t\to-3} (t-2)}{\lim\limits_{t\to-3} (t+5)} = \dfrac{-3-2}{-3+5} = \dfrac{-5}{2} = -\dfrac{5}{2}$

17. $\lim\limits_{h\to0} \dfrac{h}{h^2-7h+1} = \dfrac{\lim\limits_{h\to0} h}{\lim\limits_{h\to0} (h^2-7h+1)} = \dfrac{0}{0^2-7(0)+1} = 0$

19. $\lim\limits_{p\to4} \sqrt{p^2+p+5} = \sqrt{\lim\limits_{p\to4} (p^2+p+5)} = \sqrt{4^2+4+5} = \sqrt{25} = 5$

21. $\lim\limits_{x\to-2} \dfrac{x^2+2x}{x+2} = \lim\limits_{x\to-2} \dfrac{x(x+2)}{x+2} = \lim\limits_{x\to-2} x = -2$

23. $\lim\limits_{x\to2} \dfrac{x^2-x-2}{x-2} = \lim\limits_{x\to2} \dfrac{(x-2)(x+1)}{x-2} = \lim\limits_{x\to2} (x+1) = 3$

25. $\lim\limits_{x\to-1} \dfrac{x^2+2x+1}{x+1} = \lim\limits_{x\to-1} \dfrac{(x+1)^2}{x+1} = \lim\limits_{x\to-1} (x+1) = 0$

27. $\lim\limits_{x\to3} \dfrac{x-3}{x^2-9} = \lim\limits_{x\to3} \dfrac{x-3}{(x+3)(x-3)} = \lim\limits_{x\to3} \dfrac{1}{x+3} = \dfrac{1}{6}$

29. $\lim\limits_{x\to4} \dfrac{x^2-9x+20}{x^2-3x-4} = \lim\limits_{x\to4} \dfrac{(x-4)(x-5)}{(x-4)(x+1)} = \lim\limits_{x\to4} \dfrac{x-5}{x+1} = -\dfrac{1}{5}$

31. $\lim\limits_{x\to2} \dfrac{3x^2-x-10}{x^2+5x-14} = \lim\limits_{x\to2} \dfrac{(3x+5)(x-2)}{(x+7)(x-2)} = \lim\limits_{x\to2} \dfrac{3x+5}{x+7} = \dfrac{11}{9}$

33. $\lim\limits_{h\to0} \dfrac{(2+h)^2-2^2}{h} = \lim\limits_{h\to0} \dfrac{[4+4h+h^2]-4}{h} = \lim\limits_{h\to0} \dfrac{4h+h^2}{h} =$

$\lim\limits_{h\to0} \dfrac{h(4+h)}{h} = \lim\limits_{h\to0} (4+h) = 4$

35. $\lim\limits_{h\to0} \dfrac{(x+h)^2-x^2}{h} = \lim\limits_{h\to0} \dfrac{2xh+h^2}{h} = \lim\limits_{x\to0} (2x+h) = 2x$

37. $\lim\limits_{h\to 0} \dfrac{f(x+h)-f(x)}{h} = \lim\limits_{h\to 0} \dfrac{[4-(x+h)]-(4-x)}{h} = \lim\limits_{h\to 0} \dfrac{-h}{h} =$

$\lim\limits_{h\to 0} -1 = -1$

39. $\lim\limits_{h\to 0} \dfrac{f(x+h)-f(x)}{h} = \lim\limits_{h\to 0} \dfrac{[(x+h)^2-3]-(x^2-3)}{h} =$

$\lim\limits_{h\to 0} \dfrac{x^2+2xh+h^2-3-(x^2-3)}{h} = \lim\limits_{h\to 0} \dfrac{2xh+h^2}{h} = \lim\limits_{h\to 0} (2x+h) = 2x$

41. $\lim\limits_{h\to 0} \dfrac{f(x+h)-f(x)}{h} = \lim\limits_{h\to 0} \dfrac{[2(x+h)^2-3(x+h)]-(2x^2-3x)}{h} =$

$\lim\limits_{h\to 0} \dfrac{2x^2+4xh+2h^2-3x-3h-(2x^2-3x)}{h} = \lim\limits_{h\to 0} \dfrac{4xh+2h^2-3h}{h} =$

$\lim\limits_{h\to 0} \dfrac{h(4x+2h-3)}{h} = \lim\limits_{h\to 0} (4x+2h-3) = 4x-3$

43. $\lim\limits_{x\to 6} \dfrac{\sqrt{x-2} - 2}{x-6} = \lim\limits_{x\to 6} \dfrac{(\sqrt{x-2} - 2)(\sqrt{x-2} + 2)}{(x-6)(\sqrt{x-2} + 2)} =$

$\lim\limits_{x\to 6} \dfrac{(x-2) - 4}{(x-6)(\sqrt{x-2} + 2)} = \lim\limits_{x\to 6} \dfrac{x-6}{(x-6)(\sqrt{x-2} + 2)} =$

$\lim\limits_{x\to 6} \dfrac{1}{\sqrt{x-2} + 2} = \dfrac{1}{4}$

45. (a) $\lim\limits_{T_c\to 0} \dfrac{T_h-T_c}{T_h} = \dfrac{T_h-0}{T_h} = \dfrac{T_h}{T_h} = 1$

(b) $\lim\limits_{T_c\to T_h} \dfrac{T_h-T_c}{T_h} = \dfrac{T_h-T_h}{T_h} = \dfrac{0}{T_h} = 0$

47. (a) $\lim\limits_{R\to 0} i = \lim\limits_{R\to 0} \dfrac{10(5 + R)}{10 + 7R} = \dfrac{10(5)}{10} = 5$ amperes

(b) $\lim\limits_{R\to 2} i = \lim\limits_{R\to 2} \dfrac{10(5 + R)}{10 + 7R} = \dfrac{10(7)}{10 + 14} = \dfrac{70}{24} = \dfrac{35}{12}$ amperes

49. 11.00

51. -7.00

EXERCISE 11.2

1. a. 2   b. 3   c. does not exist   d. $-\infty$ e. ∞ f. ∞
   g. ∞ h. 0   i. 1   j. 1   k. 1

3. $\lim\limits_{x\to 3^+} (x-2)$. As $x \to 3^+$, then $x-2 \to 1$.   Ans. 1

5. $\lim\limits_{x\to -\infty} 5x$. As x becomes very negative, so does 5x. Thus
   $\lim\limits_{x\to -\infty} 5x = -\infty$.   Ans. $-\infty$

7. $\lim\limits_{x\to 0^-} \frac{6x}{x^4} = \lim\limits_{x\to 0^-} \frac{6}{x^3} = -\infty$ since x^3 is negative and close to
   0 for $x\to 0^-$.   Ans. $-\infty$

9. $\lim\limits_{x\to -\infty} x^2 = \infty$ since x^2 is positive for $x\to -\infty$.   Ans. ∞

11. $\lim\limits_{h\to 0^+} \sqrt{h}$. If h is positive and close to 0, then $\sqrt{h}$ is
    close to 0.   Ans. 0

13. $\lim\limits_{x\to 5^-} \frac{3}{x-5} = -\infty$; $\lim\limits_{x\to 5^+} \frac{3}{x-5} = \infty$. So $\lim\limits_{x\to 5} \frac{3}{x-5}$ does not exist.
    Ans. does not exist

15. $\lim\limits_{x\to 1^+} (4\sqrt{x-1})$. As $x \to 1^+$, then x-1 approaches 0 through
    positive values. So $\sqrt{x-1} \to 0$. Thus
    $$\lim\limits_{x\to 1^+} (4\sqrt{x-1}) = 4\cdot \lim\limits_{x\to 1^+} \sqrt{x-1} = 4\cdot 0 = 0.$$

17. $\lim\limits_{x\to \infty} \sqrt{x+10}$. As x becomes very large, so does x+10.
    Because square roots of very large numbers are very large,
    $\lim\limits_{x\to \infty} \sqrt{x+10} = \infty$.   Ans. ∞

19. $\lim\limits_{x\to\infty} \dfrac{3}{\sqrt{x}} = 3 \lim\limits_{x\to\infty} \dfrac{1}{x^{1/2}} = 3\cdot 0 = 0$

21. $\lim\limits_{x\to\infty} \dfrac{x+2}{x+3} = \lim\limits_{x\to\infty} \dfrac{x}{x} = \lim\limits_{x\to\infty} 1 = 1$

23. $\lim\limits_{x\to-\infty} \dfrac{x^2-1}{x^3+4x-3} = \lim\limits_{x\to-\infty} \dfrac{x^2}{x^3} = \lim\limits_{x\to-\infty} \dfrac{1}{x} = 0$

25. $\lim\limits_{t\to\infty} \dfrac{5t^2+2t+1}{4t+7} = \lim\limits_{t\to\infty} \dfrac{5t^2}{4t} = \lim\limits_{t\to\infty} \dfrac{5t}{4} = \lim\limits_{t\to\infty} \left(\dfrac{5}{4}t\right) = \infty$

27. $\lim\limits_{x\to\infty} \dfrac{7}{2x+1} = \lim\limits_{x\to\infty} \dfrac{7}{2x} = \dfrac{7}{2}\cdot\lim\limits_{x\to\infty} \dfrac{1}{x} = \dfrac{7}{2}\cdot 0 = 0$

29. $\lim\limits_{x\to\infty} \dfrac{3-4x-2x^3}{5x^3-8x+1} = \lim\limits_{x\to\infty} \dfrac{-2x^3}{5x^3} = \lim\limits_{x\to\infty} \dfrac{-2}{5} = -\dfrac{2}{5}$

31. $\lim\limits_{x\to 3^-} \dfrac{x+3}{x^2-9} = \lim\limits_{x\to 3^-} \dfrac{x+3}{(x+3)(x-3)} = \lim\limits_{x\to 3^-} \dfrac{1}{x-3} = -\infty$

33. $\lim\limits_{w\to\infty} \dfrac{2w^2-3w+4}{5w^2+7w-1} = \lim\limits_{w\to\infty} \dfrac{2w^2}{5w^2} = \lim\limits_{w\to\infty} \dfrac{2}{5} = \dfrac{2}{5}$

35. $\lim\limits_{x\to\infty} \dfrac{6-4x^2+x^3}{4+5x-7x^2} = \lim\limits_{x\to\infty} \dfrac{x^3}{-7x^2} = \lim\limits_{x\to\infty} \dfrac{-x}{7} = -\infty$

37. $\lim\limits_{x\to-5} \dfrac{2x^2+9x-5}{x^2+5x} = \lim\limits_{x\to-5} \dfrac{(2x-1)(x+5)}{x(x+5)} = \lim\limits_{x\to-5} \dfrac{2x-1}{x} = \dfrac{11}{5}$

39. $\lim\limits_{x\to 1} \dfrac{x^2-3x+1}{x^2+1} = \dfrac{\lim\limits_{x\to 1}(x^2-3x+1)}{\lim\limits_{x\to 1}(x^2+1)} = \dfrac{-1}{2} = -\dfrac{1}{2}$

41. As $x\to 1^+$, then $\dfrac{1}{x-1} \to \infty$. Thus $\lim\limits_{x\to 1^+}\left[1 + \dfrac{1}{x-1}\right] = \infty$. <u>Ans.</u> ∞

43. $\lim\limits_{x \to -7^-} \dfrac{x^2+1}{\sqrt{x^2-49}}$.   As $x \to -7^-$, then $x^2+1 \to 50$ and $\sqrt{x^2-49}$

approaches 0 through positive values.   Thus $\dfrac{x^2+1}{\sqrt{x^2-49}} \to \infty$.

Ans. ∞

45. As $x \to 0^+$, $x+x^2$ approaches 0 through positive values.   Thus
$\dfrac{2}{x+x^2} \to \infty$.   Ans. ∞

47. $\lim\limits_{x \to 1^-} \dfrac{x}{x-1} = -\infty$; $\lim\limits_{x \to 1^+} \dfrac{x}{x-1} = \infty$.   Ans. does not exist

49. As $x \to 0^+$, $\dfrac{3}{x} \to \infty$.   Thus $-\dfrac{3}{x} \to -\infty$.   Ans. $-\infty$

51. $\lim\limits_{x \to 0^+} |x| = \lim\limits_{x \to 0^+} x = 0$; $\lim\limits_{x \to 0^-} |x| = \lim\limits_{x \to 0^-} (-x) = 0$.   Thus
$\lim\limits_{x \to 0} |x| = 0$.   Ans. 0

53. $\lim\limits_{x \to -\infty} \dfrac{x+1}{x} = \lim\limits_{x \to -\infty} \dfrac{x}{x} = \lim\limits_{x \to -\infty} 1 = 1$

55. $f(x)$ a. 1   b. 2   c. does not exist
                              d. 1   e. 2

57. $g(x)$ a. 0   b. 0   c. 0   d. $-\infty$ e. $-\infty$

59. $\lim\limits_{q\to\infty} \bar{c} = \lim\limits_{q\to\infty} \left(\dfrac{5000}{q} + 6\right)$

$= 0 + 6 = 6$

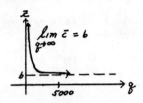

61. $\lim\limits_{t\to\infty} \left[20,000 + \dfrac{10,000}{(t+2)^2}\right] = 20,000 + 0 = 20,000.$ <u>Ans.</u>   20,000

63. $\lim\limits_{x\to\infty} y = \lim\limits_{x\to\infty} \dfrac{900x}{10 + 45x} = \lim\limits_{x\to\infty} \dfrac{900x}{45x} = \lim\limits_{x\to\infty} 20 = 20.$ <u>Ans.</u>   20

65. 1, 0.5, 0.525, 0.631, 0.912, 0.986, 0.998; conclude limit is 1

67. 0

69. (a) 11;   (b) 9;   (c) does not exist

EXERCISE 11.3

1. $S = 4000e^{0.055(6)} = 4000e^{0.33} \approx \$5564;$
5564 - 4000 = \$1564

3. $P = 2500e^{-0.0675(8)} = 2500e^{-0.54} \approx \$1456.88$

5. $e^{0.04} - 1 \approx 0.0408.$ <u>Ans.</u>   4.08%

7. $e^{0.10} - 1 \approx 0.1052.$ <u>Ans.</u>   10.52%

9. $S = 100e^{0.055(2)} = 100e^{0.11} \approx \$111.63$

11. $P = 1,000,000e^{-0.08(5)} = 1,000,000e^{-0.40} \approx \$670,320$

13. (a) $10,000(1 + 0.1)^{20} \approx \$67,275$
(b) $P = 67,275e^{-(0.12)(20)} = 67,275e^{-2.4} \approx \$6103$

15. Effective rate = $e^r - 1$. Thus $0.05 = e^r - 1$, $e^r = 1.05$, $r = \ln 1.05 \approx 0.04879 \approx 0.0488$. Ans. 4.88%

17. $100\dfrac{1-e^{-0.09(20)}}{0.09} = 100\dfrac{1-e^{-1.8}}{0.09} \approx \$927$

19. $3P = Pe^{0.07t}$, $3 = e^{0.07t}$, $0.07t = \ln 3$, $t = \dfrac{\ln 3}{0.07} \approx 16$.
    Ans. 16 years

21. The accumulated amounts under each option are:
    a: $1000e^{(0.1)(2)} = 1000e^{0.2} \approx \$1221.40$
    b: $1050(1.05)^4 \approx \$1276.28$
    c: $500e^{(0.1)(2)} + 500(1.05)^4 \approx 610.70 + 607.75 = \$1218.45$

23. (a) $9000(1.025)^4 \approx \$9934.32$
    (b) After one year the accumulated amount of the invest-
        ment is $10,000e^{0.11} \approx \$11,162.78$. The payoff for the
        loan (including interest) is $1000 + 1000(0.16) =$
        $\$1160$. The net return is
                $11,162.78 - 1160 = \$10,002.78$.
        Thus this strategy is better by
                $10,002.78 - 9934.32 = \$68.46$.

EXERCISE 11.4

1. $f(x) = x^3-5x$; $x = 2$. (i) f is defined at $x = 2$: $f(2) =$
   -2. (ii) $\lim\limits_{x\to2} f(x) = \lim\limits_{x\to2} (x^3-5x) = 2^3-5(2) = -2$, which
   exists. (iii) $\lim\limits_{x\to2} f(x) = -2 = f(2)$. Thus f is
   continuous at $x = 2$.

3. $g(x) = \sqrt{2-3x}$; $x = 0$. (i) g is defined at $x = 0$; $g(0) =$
   $\sqrt{2}$. (ii) $\lim\limits_{x\to0} g(x) = \lim\limits_{x\to0} \sqrt{2-3x} = \sqrt{2}$, which exists.
   (iii) $\lim\limits_{x\to0} g(x) = \sqrt{2} = g(0)$. Thus g is continuous at $x = 0$.

5. $h(x) = \frac{x-4}{x+4}$; $x = 4$. (i) h is defined at $x = 4$; $h(4) = 0$.

   (ii) $\lim\limits_{x \to 4} h(x) = \lim\limits_{x \to 4} \frac{x-4}{x+4} = \frac{0}{8} = 0$, which exists.

   (iii) $\lim\limits_{x \to 4} h(x) = 0 = h(4)$. Thus h is continuous at $x = 4$.

7. continuous at -2 and 0 because f is a rational function
   and at neither point is the denominator zero.

9. discontinuous at 3 and -3 because at both points the
   denominator of this rational function is zero.

11. $F(x) = \begin{cases} x+2, & \text{if } x \geq 2, \\ x^2, & \text{if } x < 2. \end{cases}$ F is defined at $x = 2$ and $x = 0$;
    $F(2) = 4$, $F(0) = 0$. Because $\lim\limits_{x \to 2^+} F(x) = \lim\limits_{x \to 2^+} (x+2) = 4$ and

    $\lim\limits_{x \to 2^-} F(x) = \lim\limits_{x \to 2^-} x^2 = 4$, we have $\lim\limits_{x \to 2} F(x) = 4$. In addition,

    $\lim\limits_{x \to 0} F(x) = \lim\limits_{x \to 0} x^2 = 0$. Since $\lim\limits_{x \to 2} F(x) = 4 = F(2)$ and
    $\lim\limits_{x \to 0} F(x) = 0 = F(0)$, F is continuous at both 2 and 0.
    <u>Ans.</u> continuous at 2 and 0

13. f is a polynomial function.

15. f is a rational function and the denominator is never zero.

17. none, because f is a polynomial function.

19. The denominator of this rational function is zero only
    when $x = 4$. Thus f is discontinuous only for $x = 4$.

21. none, because g is a polynomial function.

23. $x^2+2x-15 = 0$, $(x+5)(x-3) = 0$, $x = -5$ or 3. Discontinuous
    at -5 and 3.

25. $x^3-x = 0$, $x(x^2-1) = 0$, $x(x+1)(x-1) = 0$, $x = 0, \pm 1$.
    Discontinuous at 0, ± 1.

27. $x^2+1 = 0$ has no real roots, so no discontinuity exists.

29. $f(x) = \begin{cases} 1, & \text{if } x \geq 0, \\ -1, & \text{if } x < 0. \end{cases}$ For $x < 0$, $f(x) = -1$, which is a

polynomial and hence continuous. For $x > 0$, $f(x) = 1$,
which is a polynomial and hence continuous. Because
$$\lim_{x \to 0^-} f(x) = \lim_{x \to 0^-} (-1) = -1 \text{ and } \lim_{x \to 0^+} f(x) = \lim_{x \to 0^+} 1 = 1,$$
$\lim_{x \to 0} f(x)$ does not exist. Thus f is discontinuous at $x = 0$.
Ans. 0

31. $f(x) = \begin{cases} 0, & \text{if } x \leq 1, \\ x-1, & \text{if } x > 1. \end{cases}$ For $x < 1$, $f(x) = 0$, which is a

polynomial and hence continuous. For $x > 1$, $f(x) = x-1$,
which is a polynomial and hence continuous. For $x = 1$, f
is defined [f(1) = 0]. Because $\lim_{x \to 1^-} f(x) = \lim_{x \to 1^-} 0 = 0$ and
$\lim_{x \to 1^+} f(x) = \lim_{x \to 1^+} (x-1) = 0$, then $\lim_{x \to 1} f(x) = 0$. Since
$\lim_{x \to 1} f(x) = 0 = f(0)$, f is continuous at $x = 1$. Ans. none

33. $f(x) = \begin{cases} x^2, & \text{if } x > 2, \\ x-1, & \text{if } x < 2. \end{cases}$ For $x < 2$, $f(x) = x-1$, which is a

polynomial and hence continuous. For $x > 2$, $f(x) = x^2$,
which is a polynomial and hence continuous. Because f is
not defined at $x = 2$, it is discontinuous there.
Ans. 2

35.

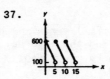

Discontinuities
at 1, 2, 3, 4.

37.

f is continuous at 2.
f is discontinuous at 5.
f is discontinuous at 10.
Ans. yes, no, no

EXERCISE 11.5

1. $x^2-3x-4 > 0$. $f(x) = x^2-3x-4 = (x+1)(x-4)$ has zeros -1 and 4. By considering the intervals $(-\infty,-1)$, $(-1,4)$, and $(4,\infty)$, we find $f(x) > 0$ on $(-\infty,-1)$ and $(4,\infty)$.
   Ans. $(-\infty, -1)$, $(4, \infty)$

3. $x^2-5x+6 \leqq 0$. $f(x) = x^2-5x+6 = (x-2)(x-3)$ has zeros 2 and 3. By considering the intervals $(-\infty,2)$, $(2,3)$, and $(3,\infty)$, we find $f(x) < 0$ on $(2,3)$. Ans. $[2, 3]$

5. $2x^2+11x+14 < 0$. $f(x)= 2x^2+11x+14 = (2x+7)(x+2)$ has zeros -7/2 and -2. By considering the intervals $(-\infty,-7/2)$, $(-7/2,-2)$, and $(-2,\infty)$, we find $f(x) < 0$ on $(-7/2,-2)$.
   Ans. $(-7/2, -2)$

7. $x^2+4 < 0$. Since x^2+4 is always positive, the inequality $x^2+4 < 0$ has no solution. Ans. no solution

9. $(x+2)(x-3)(x+6) \leqq 0$. $f(x) = (x+2)(x-3)(x+6)$ has zeros -2, 3, and -6. By considering the intervals $(-\infty,-6)$, $(-6,-2)$, $(-2,3)$, and $(3,\infty)$, we find $f(x) < 0$ on $(-\infty,-6)$ and $(-2,3)$.
   Ans. $(-\infty, -6]$, $[-2, 3]$

11. $-x(x-5)(x+4) > 0$, or equivalently, $x(x-5)(x+4) < 0$. $f(x) = x(x-5)(x+4)$ has zeros 0, 5, and -4. By considering the intervals $(-\infty,-4)$, $(-4,0)$, $(0,5)$, and $(5,\infty)$, we find $f(x) < 0$ on $(-\infty,-4)$ and $(0,5)$. Ans. $(-\infty, -4)$, $(0, 5)$

13. $x^3+4x \geqq 0$. $f(x) = x(x^2+4)$ has 0 as the only (real) zero. By considering the intervals $(-\infty,0)$ and $(0,\infty)$, we find $f(x) > 0$ on $(0,\infty)$. Ans. $[0, \infty)$

15. $x^3+2x^2-3x > 0$. $f(x) = x(x+3)(x-1)$ has zeros 0, -3, and 1. By considering the intervals $(-\infty,-3)$, $(-3,0)$, $(0,1)$, and $(1,\infty)$, we find $f(x) > 0$ on $(-3,0)$ and $(1,\infty)$.
    Ans. $(-3, 0)$, $(1, \infty)$

17. $\frac{x}{x^2-1} < 0$. $f(x) = \frac{x}{x^2-1}$ is discontinuous when x = ±1; f has
    0 as a zero. By considering the intervals $(-\infty,-1)$,
    $(-1,0)$, $(0,1)$, and $(1,\infty)$, we find $f(x) < 0$ on $(-\infty,-1)$ and
    $(0,1)$.   Ans. $(-\infty, -1)$, $(0, 1)$

19. $\frac{4}{x-1} \geq 0$. $f(x) = \frac{4}{x-1}$ is discontinuous when x = 1, and
    $f(x) = 0$ has no root. By considering the intervals $(-\infty,1)$
    and $(1,\infty)$, we find $f(x) > 0$ on $(1,\infty)$. Note also that
    $f(x) \neq 0$ for any x.   Ans. $(1, \infty)$

21. $\frac{x^2-x-6}{x^2+4x-5} \geq 0$. $f(x) = \frac{x^2-x-6}{x^2+4x-5} = \frac{(x-3)(x+2)}{(x+5)(x-1)}$ is discontinuous
    at x = -5 and x = 1; f has zeros 3 and -2. By considering
    the intervals $(-\infty,-5)$, $(-5,-2)$, $(-2,1)$, $(1,3)$, and $(3,\infty)$,
    we find $f(x) > 0$ on $(-\infty,-5)$, $(-2,1)$, and $(3,\infty)$.
    Ans. $(-\infty, -5)$, $[-2, 1)$, $[3, \infty)$

23. $\frac{3}{x^2+6x+8} \leq 0$. $f(x) = \frac{3}{x^2+6x+8} = \frac{3}{(x+4)(x+2)}$ is never zero,
    but is discontinuous at x = -4 and x = -2. By considering
    the intervals $(-\infty,-4)$, $(-4,-2)$, and $(-2,\infty)$, we find that
    $f(x) < 0$ on $(-4,-2)$.   Ans. $(-4, -2)$

25. $x^2+2x \geq 2$, or equivalently, $x^2+2x-2 \geq 0$. $f(x) = x^2+2x-2$
    has zeros $-1\pm\sqrt{3}$. By considering the intervals
    $(-\infty, -1-\sqrt{3})$, $(-1-\sqrt{3}, -1+\sqrt{3})$, and $(-1+\sqrt{3}, \infty)$, we find
    $f(x) > 0$ on $(-\infty, -1-\sqrt{3})$ and $(-1+\sqrt{3}, \infty)$.
    Ans. $(-\infty, -1-\sqrt{3}]$, $[-1+\sqrt{3}, \infty)$

27. Revenue = (no. of units)(price per unit). We want
    $$q(20 - 0.1q) \geq 750$$
    $$0.1q^2 - 20q + 750 \leq 0$$
    $$q^2 - 200q + 7500 \leq 0$$
    $$(q - 50)(q - 150) \leq 0.$$
    Solving gives $50 \leq q \leq 150$.
    Ans. between 50 and 150 units, inclusive

29.

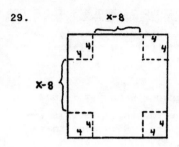

If x is the length of the piece of aluminum, then the box will be 4 by x - 8 by x - 8.

$$4(x - 8)^2 \geqq 324$$
$$(x - 8)^2 \geqq 81$$
$$x^2 - 16x - 17 \geqq 0$$
$$(x - 17)(x + 1) \geqq 0.$$

Solving gives $x \leqq -1$ or $x \geqq 17$. Since x must be positive, we have $x \geqq 17$.

<u>Ans.</u> 17 in. by 17 in.

31. $(-\infty, -7.72]$

CHAPTER 11 - REVIEW PROBLEMS

1. $\lim\limits_{x \to -1} (2x^2+6x-1) = 2(-1)^2+6(-1)-1 = -5$

3. $\lim\limits_{x \to 3} \dfrac{x^2-9}{x^2-3x} = \lim\limits_{x \to 3} \dfrac{(x+3)(x-3)}{x(x-3)} = \lim\limits_{x \to 3} \dfrac{x+3}{x} = \dfrac{6}{3} = 2$

5. $\lim\limits_{h \to 0} (x+h) = x+0 = x$

7. $\lim\limits_{x \to -4} \dfrac{x^3+4x^2}{x^2+2x-8} = \lim\limits_{x \to -4} \dfrac{x^2(x+4)}{(x+4)(x-2)} = \lim\limits_{x \to -4} \dfrac{x^2}{x-2} = \dfrac{16}{-6} = -\dfrac{8}{3}$

9. As $x \to \infty$, $x+1 \to \infty$. Thus $\lim\limits_{x \to \infty} \dfrac{2}{x+1} = 0$.   <u>Ans.</u> 0

11. $\lim\limits_{x \to \infty} \dfrac{3x-2}{5x+3} = \lim\limits_{x \to \infty} \dfrac{3x}{5x} = \lim\limits_{x \to \infty} \dfrac{3}{5} = \dfrac{3}{5}$

13. $\lim\limits_{t \to 3^-} \dfrac{2t-3}{t-3} = -\infty$ and $\lim\limits_{t \to 3^+} \dfrac{2t-3}{t-3} = \infty$. Thus $\lim\limits_{t \to 3} \dfrac{2t-3}{t-3}$ does not exist.   <u>Ans.</u> limit does not exist

15. $\lim\limits_{x\to-\infty} \frac{x+3}{1-x} = \lim\limits_{x\to-\infty} \frac{x}{-x} = \lim\limits_{x\to-\infty} (-1) = -1$

17. $\lim\limits_{x\to\infty} \frac{x^2-1}{(3x+2)^2} = \lim\limits_{x\to\infty} \frac{x^2-1}{9x^2+12x+4} = \lim\limits_{x\to\infty} \frac{x^2}{9x^2} = \lim\limits_{x\to\infty} \frac{1}{9} = \frac{1}{9}$

19. $\lim\limits_{x\to3^-} \frac{x+3}{x^2-9} = \lim\limits_{x\to3^-} \frac{x+3}{(x+3)(x-3)} = \lim\limits_{x\to3^-} \frac{1}{x-3} = -\infty$

21. $\lim\limits_{x\to\infty} \sqrt{3x}$. As x becomes large, so does 3x. Because the square roots of large numbers are also large,

$\lim\limits_{x\to\infty} \sqrt{3x} = \infty$.   <u>Ans.</u> ∞

23. $\lim\limits_{x\to\infty} \dfrac{x^{100} + \frac{1}{x^2}}{e - x^{98}} = \lim\limits_{x\to\infty} \dfrac{x^2\left(x^{100} + \frac{1}{x^2}\right)}{x^2(e - x^{98})} = \lim\limits_{x\to\infty} \dfrac{x^{102} + 1}{ex^2 - x^{100}} =$

$\lim\limits_{x\to\infty} \dfrac{x^{102}}{-x^{100}} = \lim\limits_{x\to\infty} (-x^2) = -\infty$

25. $\lim\limits_{x\to1^-} f(x) = \lim\limits_{x\to1^-} x^2 = 1$; $\lim\limits_{x\to1^+} f(x) = \lim\limits_{x\to1^+} x = 1$. Thus

$\lim\limits_{x\to1} f(x) = 1$.   <u>Ans.</u> 1

27. $\lim\limits_{x\to4^+} \dfrac{\sqrt{x^2-16}}{4-x} = \lim\limits_{x\to4^+} \dfrac{\sqrt{x-4}\sqrt{x+4}}{-(x-4)} = \lim\limits_{x\to4^+} - \dfrac{\sqrt{x+4}}{\sqrt{x-4}}$. As $x\to4^+$,

$\sqrt{x-4}$ approaches 0 through positive values and $\sqrt{x+4} \to \sqrt{8}$.

Thus $- \dfrac{\sqrt{x+4}}{\sqrt{x-4}} \to -\infty$.   <u>Ans.</u> $-\infty$

29. $\lim\limits_{h\to0} \dfrac{f(x+h) - f(x)}{h} = \lim\limits_{h\to0} \dfrac{[8(x+h)-2] - [8x-2]}{h} = \lim\limits_{h\to0} \dfrac{8h}{h} =$

$\lim\limits_{h\to0} 8 = 8$

31. $y = 11\left(1 - \frac{1}{1+2x}\right)$. Considering $\frac{1}{1+2x}$, we have $\lim\limits_{x\to\infty} \frac{1}{1+2x} =$

$\lim\limits_{x\to\infty} \frac{1}{2x} = \frac{1}{2}\cdot\lim\limits_{x\to\infty} \frac{1}{x} = \frac{1}{2}\cdot 0 = 0$. Thus

$$\lim_{x \to \infty} y = \lim_{x \to \infty} \left[11\left(1 - \frac{1}{1+2x}\right) \right] = 11(1 - 0) = 11.$$

Ans. 11

33. (a) $S = 2500e^{0.07(14)} = 2500e^{0.98} \approx \$6661.25$

(b) $P = 2500e^{-0.07(14)} = 2500e^{-0.98} \approx \$938.28$

35. effective rate = $e^{0.06} - 1 \approx 0.0618$.   Ans. 6.18%

37. $S = Pe^{rt}$, $2P = Pe^{0.1t}$, $2 = e^{0.1t}$, $0.1t = \ln 2$,

$t = \frac{\ln 2}{0.1} = 10 \ln 2$.   Ans. $10 \ln 2$

39. $f(x) = x+5$; $x = 7$. (i) f is defined at $x = 7$; $f(7) = 12$.

(ii) $\lim_{x \to 7} f(x) = \lim_{x \to 7} (x+5) = 7+5 = 12$, which exists.

(iii) $\lim_{x \to 7} f(x) = 12 = f(7)$. Thus f is continuous at $x = 7$.

41. Since $f(x) = \frac{1}{4}x$ is a polynomial function, it is continuous everywhere.

43. $f(x) = \frac{x^2}{x+3}$ is a rational function and the denominator is zero at $x = -3$. Thus f is discontinuous at $x = -3$.
Ans. $x = -3$

45. Since $f(x) = \frac{x-1}{2x^2+3}$ is a rational function whose denominator is never zero, f is continuous everywhere.   Ans. none

47. $f(x) = \frac{4-x^2}{x^2+3x-4} = \frac{4-x^2}{(x+4)(x-1)}$ is a rational function and the denominator is zero only when $x = -4$ or $x = 1$, so f is discontinuous there.   Ans. $x = -4, 1$

49. $f(x) = \begin{cases} x+4, & \text{if } x > -2, \\ 3x+6, & \text{if } x \le -2. \end{cases}$ For $x < -2$, $f(x) = 3x+6$, which is a polynomial and hence continuous. For $x > -2$, $f(x) =$

x+4, which is a polynomial and hence continuous. Because

$$\lim_{x \to -2^-} f(x) = \lim_{x \to -2^-} (3x+6) = 0 \text{ and } \lim_{x \to -2^+} f(x) =$$
$$\lim_{x \to -2^+} (x+4) = 2, \lim_{x \to -2} f(x) \text{ does not exist. Thus f is}$$
discontinuous at x = -2. **Ans.** x = -2

51. $x^2+4x-12 > 0$. $f(x) = x^2+4x-12 = (x+6)(x-2)$ has zeros -6 and 2. By considering the intervals $(-\infty,-6)$, $(-6,2)$, and $(2,\infty)$, we find $f(x) > 0$ on $(-\infty,-6)$ and $(2,\infty)$.
    Ans. $(-\infty, -6)$, $(2, \infty)$

53. $x^3 \geq 2x^2$, $x^3-2x^2 \geq 0$. $f(x) = x^3-2x^2 = x^2(x-2)$ has zeros 0 and 2. By considering the intervals $(-\infty,0)$, $(0,2)$, and $(2,\infty)$, we find $f(x) > 0$ on $(2,\infty)$.
    Ans. $[2, \infty)$, x = 0

55. $\frac{x+5}{x^2-1} < 0$. $f(x) = \frac{x+5}{(x+1)(x-1)}$ is discontinuous when x = ±1, and f has -5 as a zero. By considering the intervals $(-\infty,-5)$, $(-5,-1)$, $(-1,1)$, and $(1,\infty)$, we find $f(x) < 0$ on $(-\infty,-5)$ and $(-1,1)$. **Ans.** $(-\infty, -5)$, $(-1, 1)$

57. $\frac{x^2+3x}{x^2+2x-8} \geq 0$, or $\frac{x(x+3)}{(x+4)(x-2)} \geq 0$. We consider the intervals determined by x = 0, -3, -4, and 2, namely $(-\infty,-4)$, $(-4,-3)$, $(-3,0)$, $(0,2)$, and $(2,\infty)$. We find $f(x) > 0$ on $(-\infty,-4)$, $(-3,0)$, and $(2,\infty)$.
    Ans. $(-\infty, -4)$, $[-3, 0]$, $(2, \infty)$

59. 1.00   61. 0   63. $[2.00, \infty)$

CHAPTER 11 — MATHEMATICAL SNAPSHOT

1. $D = 351.9e^{-rt}$. For 1994, t = 1 and D = 300. Thus
   $300 = 351.9e^{-r}$, $e^{-r} = \frac{300}{351.9}$, $-r = \ln \frac{300}{351.9}$,
   $r = -\ln \frac{300}{351.9} \approx 0.16$.   **Ans.** 16%

12

Differentiation

1. (a) $f(x) = x^3 + 3$, $P = (2, 11)$

   To begin, if $x = 3$, then $m_{PQ} = \frac{(3^3 + 3) - 11}{3 - 2} = 19$.

   If $x = 2.5$, then $m_{PQ} = \frac{[(2.5)^3 + 3] - 11}{3 - 2.5} = 15.25$.

   Continuing in this manner, we complete the table:

x-value of Q	3	2.5	2.2	2.1	2.01	2.001
m_{PQ}	19	15.25	13.24	12.61	12.0601	12.0060

   (b) We estimate that m_{tan} at P is 12.

3. $f(x) = x$.

   $f'(x) = \lim_{h \to 0} \frac{f(x + h) - f(x)}{h} = \lim_{h \to 0} \frac{(x + h) - x}{h}$

   $= \lim_{h \to 0} \frac{h}{h} = \lim_{h \to 0} 1 = 1$.

-262-

5. $y = 3x + 7$. Let $y = f(x)$.

$$\frac{dy}{dx} = \lim_{h \to 0} \frac{f(x + h) - f(x)}{h} = \lim_{h \to 0} \frac{[3(x + h) + 7] - [3x + 7]}{h}$$

$$= \lim_{h \to 0} \frac{3x + 3h + 7 - 3x - 7}{h} = \lim_{h \to 0} \frac{3h}{h} = \lim_{h \to 0} 3 = 3.$$

7. Let $f(x) = 5 - 4x$.

$$\frac{d}{dx}(5 - 4x) = \lim_{h \to 0} \frac{f(x + h) - f(x)}{h}$$

$$= \lim_{h \to 0} \frac{[5 - 4(x + h)] - [5 - 4x]}{h}$$

$$= \lim_{h \to 0} \frac{-4h}{h} = \lim_{h \to 0} (-4) = -4.$$

9. $f(x) = 3$.

$$f'(x) = \lim_{h \to 0} \frac{f(x + h) - f(x)}{h}$$

$$= \lim_{h \to 0} \frac{3 - 3}{h} = \lim_{h \to 0} \frac{0}{h} = \lim_{h \to 0} 0 = 0.$$

11. $f(x) = x^2 + 4x - 8$.

$$\frac{d}{dx}(x^2 + 4 - 8)$$

$$= \lim_{h \to 0} \frac{f(x + h) - f(x)}{h}$$

$$= \lim_{h \to 0} \frac{[(x + h)^2 + 4(x + h) - 8] - [x^2 + 4x - 8]}{h}$$

$$= \lim_{h \to 0} \frac{x^2 + 2xh + h^2 + 4x + 4h - 8 - x^2 - 4x + 8}{h}$$

$$= \lim_{h \to 0} \frac{2xh + h^2 + 4h}{h}$$

$$= \lim_{h \to 0} (2x + h + 4) = 2x + 0 + 4 = 2x + 4.$$

13. $p = f(q) = 2q^2 + 5q - 1$.

$$\frac{dp}{dq} = \lim_{h \to 0} \frac{f(q + h) - f(q)}{h}$$

$$= \lim_{h \to 0} \frac{[2(q + h)^2 + 5(q + h) - 1] - [2q^2 + 5q - 1]}{h}$$

$$= \lim_{h \to 0} \frac{4qh + 2h^2 + 5h}{h}$$

$$= \lim_{h \to 0} (4q + 2h + 5) = 4q + 0 + 5 = 4q + 5.$$

15. $y = f(x) = \frac{1}{x}$.

$$y' = \lim_{h \to 0} \frac{f(x + h) - f(x)}{h} = \lim_{h \to 0} \frac{\frac{1}{x + h} - \frac{1}{x}}{h}.$$

Multiplying the numerator and denominator by $x(x + h)$ gives

$$y' = \lim_{h \to 0} \frac{x - (x + h)}{h(x)(x + h)} = \lim_{h \to 0} \frac{-h}{h(x)(x + h)}$$

$$= \lim_{h \to 0} \left[- \frac{1}{x(x + h)} \right] = - \frac{1}{x(x + 0)} = - \frac{1}{x^2}.$$

17. $f(x) = \sqrt{x + 2}$.

$$f'(x) = \lim_{h \to 0} \frac{f(x + h) - f(x)}{h} = \lim_{h \to 0} \frac{\sqrt{x + h + 2} - \sqrt{x + 2}}{h}.$$

Rationalizing the numerator gives

$$\frac{\sqrt{x + h + 2} - \sqrt{x + 2}}{h}$$

$$= \frac{\sqrt{x + h + 2} - \sqrt{x + 2}}{h} \cdot \frac{\sqrt{x + h + 2} + \sqrt{x + 2}}{\sqrt{x + h + 2} + \sqrt{x + 2}}$$

$$= \frac{(x + h + 2) - (x + 2)}{h(\sqrt{x + h + 2} + \sqrt{x + 2})} = \frac{1}{\sqrt{x + h + 2} + \sqrt{x + 2}}.$$

Thus $f'(x) = \lim_{h \to 0} \frac{1}{\sqrt{x + h + 2} + \sqrt{x + 2}} = \frac{1}{2\sqrt{x + 2}}.$

19. $y = f(x) = x^2 + 4.$

$$y' = \lim_{h \to 0} \frac{f(x + h) - f(x)}{h}$$

$$= \lim_{h \to 0} \frac{[(x + h)^2 + 4] - [x^2 + 4]}{h}$$

$$= \lim_{h \to 0} \frac{2xh + h^2}{h} = \lim_{h \to 0} (2x + h) = 2x + 0 = 2x.$$

The slope at $(-2, 8)$ is $y'(-2) = 2(-2) = -4.$

21. $y = 4x^2 - 5.$

$$y' = \lim_{h \to 0} \frac{[4(x + h)^2 - 5] - [4x^2 - 5]}{h}$$

$$= \lim_{h \to 0} \frac{8xh + 4h^2}{h} = \lim_{h \to 0} (8x + 4h) = 8x.$$

The slope when $x = 0$ is $y'(0) = 8(0) = 0.$

23. $y = x + 4.$

$$y' = \lim_{h \to 0} \frac{[(x + h) + 4] - [x + 4]}{h} = \lim_{h \to 0} \frac{h}{h} = 1.$$

If $x = 3$, then $y' = 1$. The tangent line at the point
$(3,7)$ is $y - 7 = 1(x - 3)$, or $y = x + 4.$

25. $y = 3x^2 + 3x - 4.$

$$y' = \lim_{h \to 0} \frac{[3(x + h)^2 + 3(x + h) - 4] - [3x^2 + 3x - 4]}{h}$$

$$= \lim_{h \to 0} \frac{6xh + 3h^2 + 3h}{h}$$

$$= \lim_{h \to 0} (6x + 3h + 3) = 6x + 3.$$

If $x = -1$, then $y' = 6(-1) + 3 = -3$. The tangent line at
the point $(-1,-4)$ is $y + 4 = -3(x + 1)$, or $y = -3x - 7.$

27. $y = \frac{3}{x + 1}$.

$y' = \lim\limits_{h \to 0} \dfrac{\frac{3}{(x + h) + 1} - \frac{3}{x + 1}}{h}$

$= \lim\limits_{h \to 0} \dfrac{\frac{3(x + 1) - 3(x + h + 1)}{(x + h + 1)(x + 1)}}{h}$

$= \lim\limits_{h \to 0} \dfrac{-3h}{h(x + h + 1)(x + 1)} = \lim\limits_{h \to 0} \dfrac{-3}{(x + h + 1)(x + 1)}$

$= -\dfrac{3}{(x + 1)^2}$.

If $x = 2$, then $y' = -\frac{3}{9} = -\frac{1}{3}$. The tangent line at
$(2, 1)$ is $y - 1 = -\frac{1}{3}(x - 2)$, or $y = -\frac{1}{3}x + \frac{5}{3}$.

29. $r = \left(\frac{\eta}{1 + \eta}\right)\left(r_L - \frac{dC}{dD}\right)$, $(1 + \eta)r = \eta\left(r_L - \frac{dC}{dD}\right)$,

$r + \eta r = \eta\left(r_L - \frac{dC}{dD}\right)$, $r = \eta\left(r_L - \frac{dC}{dD}\right) - \eta r$,

$r = \eta\left(r_L - \frac{dC}{dD} - r\right)$, $\eta = \dfrac{r}{r_L - r - \frac{dC}{dD}}$.

31. For the x-values of the points where the tangent to the
graph of f is horizontal, the corresponding values of
f'(x) are 0. This is expected because the slope of a
horizontal line is zero and the derivative gives the
slope of the tangent line.

EXERCISE 12.2

1. $f(x) = 5$ is a constant function, so $f'(x) = 0$

3. $y = x^6$, $y' = 6x^{6-1} = 6x^5$

5. $y = x^{80}$, $\frac{dy}{dx} = 80x^{80-1} = 80x^{79}$

7. $f(x) = 9x^2$, $f'(x) = 9(2x^{2-1}) = 18x$

9. $g(w) = 4w^5$, $g'(w) = 4(5w^{5-1}) = 20w^4$

11. $y = \frac{2}{3}x^4$, $y' = \frac{2}{3}\left(4x^{3-1}\right) = \frac{8}{3}x^3$

13. $f(t) = \frac{1}{18}t^9$, $f'(t) = \frac{1}{18}(9t^{9-1}) = \frac{1}{2}t^8$

15. $f(x) = x + 3$, $f'(x) = 1 + 0 = 1$

17. $f'(x) = 3(2x) - 2(1) - 0 = 6x - 2$

19. $g'(p) = 4p^{4-1} - 3(3p^{3-1}) - 0 = 4p^3 - 9p^2$

21. $y' = -(8x^{8-1}) + 5x^{5-1} = -8x^7 + 5x^4$

23. $y' = -13(3x^2) + 14(2x) - 2(1) + 0 = -39x^2 + 28x - 2$

25. $f'(x) = 2(0 - 4x^3) = -8x^3$

27. $g(x) = \frac{1}{3}(13 - x^4)$, $g'(x) = \frac{1}{3}(0 - 4x^3) = -\frac{4}{3}x^3$

29. $h(x) = 4x^4 + x^3 - \frac{9}{2}x^2 + 9x$,

     $h'(x) = 4(4x^3) + 3x^2 - \frac{9}{2}(2x) + 9(1) = 16x^3 + 3x^2 - 9x + 9$

31. $f(x) = \frac{3}{2}x^4 + \frac{7}{3}x^3$, $f'(x) = \frac{3}{2}(4x^3) + \frac{7}{3}(3x^2) = 6x^3 + 7x^2$

33. $f'(x) = \frac{7}{2}x^{(7/2)-1} = \frac{7}{2}x^{5/2}$

35. $y' = \frac{3}{4}x^{(3/4)-1} + \frac{5}{3}x^{(5/3)-1} = \frac{3}{4}x^{-1/4} + \frac{5}{3}x^{2/3}$

37. $f(x) = \sqrt{x} = x^{1/2}$,

     $f'(x) = \frac{1}{2}x^{(1/2)-1} = \frac{1}{2}x^{-1/2} = \frac{1}{2\sqrt{x}}$

39. $f(r) = 6r^{1/3}$, $f'(r) = 6\left(\frac{1}{3}r^{-2/3}\right) = 2r^{-2/3}$

41. $f'(x) = -4x^{-4-1} = -4x^{-5}$

43. $f'(x) = -3x^{-3-1} + (-5x^{-5-1}) - 2(-6x^{-6-1})$
$$= -3x^{-4} - 5x^{-6} + 12x^{-7}$$

45. $y = \frac{1}{x} = x^{-1}, \quad \frac{dy}{dx} = -1 \cdot x^{-1-1} = -x^{-2} = -\frac{1}{x^2}$

47. $y = \frac{3}{x^5} = 3x^{-5}, \quad y' = 3(-5x^{-6}) = -15x^{-6}$

49. $g(x) = \frac{4}{3x^3} = \frac{4}{3}x^{-3}, \quad g'(x) = \frac{4}{3}(-3x^{-4}) = -4x^{-4}$

51. $f(t) = \frac{1}{2}\left(\frac{1}{t}\right) = \frac{1}{2}t^{-1}, \quad f'(t) = \frac{1}{2}(-1 \cdot t^{-2}) = -\frac{1}{2}t^{-2}$

53. $f(x) = \frac{1}{7}x + 7x^{-1}.$
$$f'(x) = \frac{1}{7}(1) + 7(-1x^{-2}) = \frac{1}{7} - 7x^{-2}$$

55. $f'(x) = -9\left(\frac{1}{3}x^{-2/3}\right) + 5\left(-\frac{2}{5}x^{-7/5}\right) = -3x^{-2/3} - 2x^{-7/5}$

57. $q(x) = \frac{1}{\sqrt[5]{x}} = \frac{1}{x^{1/5}} = x^{-1/5}, \quad q'(x) = -\frac{1}{5}x^{-6/5}$

59. $y = \frac{2}{x^{1/2}} = 2x^{-1/2}, \quad y' = 2\left(-\frac{1}{2}x^{-1/2}\right) = -x^{-3/2}$

61. $y = x^2\sqrt{x} = x^2(x^{1/2}) = x^{2+(1/2)} = x^{5/2}, \quad y' = \frac{5}{2}x^{3/2}$

63. $f(x) = x(3x^2 - 7x + 7) = 3x^3 - 7x^2 + 7x$
$$f'(x) = 9x^2 - 14x + 7$$

65. $f(x) = x^3(3x)^2 = x^3(9x^2) = 9x^5, \quad f'(x) = 45x^4$

67. $v(x) = x^{-2/3}(x + 5) = x^{1/3} + 5x^{-2/3}$
    $v'(x) = \frac{1}{3}x^{-2/3} - \frac{10}{3}x^{-5/3} = \frac{1}{3}x^{-5/3}(x - 10)$

69. $f(q) = \frac{4q^3 + 7q - 4}{q} = \frac{4q^3}{q} + \frac{7q}{q} - \frac{4}{q} = 4q^2 + 7 - 4q^{-1}$
    $f'(q) = 8q + 4q^{-2} = 8q + \frac{4}{q^2}$

71. $f(x) = (x + 1)(x + 3) = x^2 + 4x + 3$
    $f'(x) = 2x + 4 = 2(x + 2)$

73. $w(x) = \frac{x^2 + x^3}{x^2} = \frac{x^2}{x^2} + \frac{x^3}{x^2} = 1 + x$, $w'(x) = 0 + 1 = 1$

75. $y' = 6x + 4$. $y'\big|_{x=0} = 4$, $y'\big|_{x=2} = 16$, $y'\big|_{x=-3} = -14$.

77. y is a constant, so $y' = 0$ for all x.

79. $y = 4x^2 + 5x + 2$. $y' = 8x + 5$. $y'\big|_{x=1} = 13$.
    An equation of the tangent line is $y - 11 = 13(x - 1)$,
    or $y = 13x - 2$.

81. $y = \frac{2}{x^2} = 2x^{-2}$. $y' = 2(-2x^{-3}) = \frac{-4}{x^3}$. $y'\big|_{x=1} = -4$.
    An equation of the tangent line is $y - 2 = -4(x - 1)$,
    or $y = -4x + 6$.

83. $y = 3 + x - 5x^2 + x^4$. $y' = 1 - 10x + 4x^3$.
    When $x = 0$, then $y = 3$ and $y' = 1$. Thus an equation of
    the tangent line is $y - 3 = 1(x - 0)$, or $y = x + 3$.

85. $y = \frac{1}{3}x^3 - x^2$. $y' = x^2 - 2x$. A horizontal tangent line
    has slope 0, so we set $x^2 - 2x = 0$. Then $x(x - 2) = 0$,
    $x = 0$ or 2. If $x = 0$, then $y = 0$. If $x = 2$, then
    $y = -\frac{4}{3}$. This gives the points $(0, 0)$ and $\left(2, -\frac{4}{3}\right)$.

87. $y = x^2 - 5x + 3.$ $y' = 2x - 5.$ Setting $2x - 5 = 1$ gives
$2x = 6$, $x = 3$. When $x = 3$, then $y = -3$. This gives the
point $(3, -3)$.

89. $f(x) = \sqrt{x} + \dfrac{1}{\sqrt{x}} = x^{1/2} + x^{-1/2}$,

$f'(x) = \frac{1}{2}x^{-1/2} - \frac{1}{2}x^{-3/2} = \dfrac{1}{2\sqrt{x}} - \dfrac{1}{2x\sqrt{x}} = \dfrac{x-1}{2x\sqrt{x}}.$

Thus $\dfrac{x-1}{2x\sqrt{x}} - f'(x) = \dfrac{x-1}{2x\sqrt{x}} - \dfrac{x-1}{2x\sqrt{x}} = 0.$

91. $y = x^3 - 3x.$ $y'(x) = 3x^2 - 3.$ $y'\big|_{x=2} = 3(2^2) - 3 = 9.$
The tangent line at $(2, 2)$ is given by $y - 2 = 9(x - 2)$,
or $y = 9x - 16$.

EXERCISE 12.3

1. $s = f(t) = t^3 + t.$ If $\Delta t = 1$, then over $[1, 2]$ we have
$$\dfrac{\Delta s}{\Delta t} = \dfrac{f(2) - f(1)}{\Delta t} = \dfrac{(2^3 + 2) - (1 + 1)}{1} = 8.$$

If $\Delta t = 0.5$, then over $[1, 1.5]$ we have
$$\dfrac{\Delta s}{\Delta t} = \dfrac{f(1.5) - f(1)}{\Delta t} = \dfrac{[(1.5)^3 + 1.5] - 2}{0.5} = 5.75.$$

If $\Delta t = 0.2$, then over $[1, 1.2]$ we have
$$\dfrac{\Delta s}{\Delta t} = \dfrac{f(1.2) - f(1)}{\Delta t} = \dfrac{[(1.2)^3 + 1.2] - 2}{0.2} = 4.64.$$

Continuing in this way we obtain the following table:

Δt	1	0.5	0.2	0.1	0.01	0.001
$\Delta s/\Delta t$	8	5.75	4.64	4.31	4.0301	4.003001

We estimate the velocity when $t = 1$ to be 4.0000 m/s.
With differentiation we get
$$v = \dfrac{ds}{dt} = 3t^2 + 1. \quad \dfrac{ds}{dt}\big|_{t=1} = 3(1^2) + 1 = 4 \text{ m/s}$$

3. $s = f(t) = t^2 - 3t$.

  (a) When $t = 4$, then $s = 4^2 - 3(4) = 4$ m.

  (b) $\frac{\Delta s}{\Delta t} = \frac{f(4.5) - f(4)}{0.5} = \frac{[(4.5)^2 - 3(4.5)] - 4}{0.5} = 5.5$ m/s

  (c) $v = \frac{ds}{dt} = 2t - 3$. If $t = 4$, then $v = 2(4) - 3 = 5$ m/s

5. $s = 2t^3 + 6$.

  (a) When $t = 1$, $s = 2(1)^3 + 6 = 8$ m

  (b) $\frac{\Delta s}{\Delta t} = \frac{f(1.02) - f(1)}{0.02} = \frac{[2(1.02)^3 + 6] - 8}{0.02} = 6.1208$ m/s

  (c) $v = \frac{ds}{dt} = 6t^2$. If $t = 1$, then $v = 6(1)^2 = 6$ m/s

7. $s = t^4 - 2t^3 + t$

  (a) When $t = 2$, $s = 2^4 - 2(2^3) + 2 = 2$ m

  (b) $\frac{\Delta s}{\Delta t} = \frac{f(2.1) - f(2)}{0.1} = \frac{[(2.1)^4 - 2(2.1)^3 + 2.1] - 2}{0.1}$

                $= 10.261$ m/s

  (c) $v = 4t^3 - 6t^2 + 1$. $v\big|_{t=1} = 4(2^3) - 6(2^2) + 1 = 9$ m/s

9. $i = \sqrt{P} = P^{1/2}$. $\frac{di}{dP} = \frac{1}{2}P^{-1/2} = \frac{1}{2\sqrt{P}}$. When $P = 4$, then

  $\frac{di}{dP} = \frac{1}{2\sqrt{4}} = \frac{1}{4}$.

11. $\frac{dy}{dx} = 10x^{3/2}$. If $x = 9$, $\frac{dy}{dx} = 10(27) = 270$.

13. $\frac{dT}{dT_e} = 0 + 0.27(1 - 0) = 0.27$.

15. $c = 500 + 10q$, $\frac{dc}{dq} = 10$. When $q = 100$, $\frac{dc}{dq} = 10$.

17. $\frac{dc}{dq} = 0.3(2q) + 2 = 0.6q + 2$.

  If $q = 3$, then $\frac{dc}{dq} = 0.6(3) + 2 = 3.8$.

19. $\frac{dc}{dq}$ = 2q + 50. Evaluating when q = 15, 16 and 17 gives 80, 82 and 84, respectively.

21. $\bar{c}$ = 0.01q + 5 + $\frac{500}{q}$.   c = $\bar{c}$q = 0.01q^2 + 5q + 500.

    $\frac{dc}{dq}$ = 0.02q + 5, $\frac{dc}{dq}\Big|_{q=50}$ = 6, $\frac{dc}{dq}\Big|_{q=100}$ = 7.

23. c = $\bar{c}$q = 0.00002q^3 - 0.01q^2 + 6q + 20,000.

    $\frac{dc}{dq}$ = 0.00006q^2 - 0.02q + 6.   If q = 100, then $\frac{dc}{dq}$ = 4.6.

    If q = 500, then $\frac{dc}{dq}$ = 11.

25. $\frac{dr}{dq}$ = 0.7 for all q.

27. $\frac{dr}{dq}$ = 250 + 90q - 3q^2.   Evaluating when q = 5, 10 and 25 gives 625, 850 and 625, respectively.

29. $\frac{dc}{dq}$ = 6.750 - 0.000328(2q) = 6.750 - 0.000656q.

    $\frac{dc}{dq}\Big|_{q=5000}$ = 6.750 - 0.000656(5000) = 3.47.

31. PR$^{0.93}$ = 5,000,000.   P = 5,000,000R$^{-0.93}$.

    $\frac{dP}{dR}$ = -4,650,000R$^{-1.93}$.

33. y = 59.3 - 1.5x - 0.5x^2.

    (a) $\frac{dy}{dx}$ = -1.5 - x. $\frac{dy}{dx}\Big|_{x=6}$ = -1.5 - 6 = -7.5.

    (b) Setting -1.5 - x = -6 gives x = 4.5.

35. y = $\dot{f}$(x) = x + 4.   (a) y' = 1,   (b) $\frac{y'}{y}$ = $\frac{1}{x + 4}$,   (c) 1

    (d) $\frac{1}{5 + 4}$ = $\frac{1}{9}$ ≈ 0.111,   (e) 11.1%

37. y = 3x^2 + 6.   (a) y' = 6x,   (b) $\frac{y'}{y}$ = $\frac{6x}{3x^2 + 6}$ = $\frac{2x}{x^2 + 2}$,

    (c) 6(2) = 12,   (d) $\frac{4}{4 + 2}$ = $\frac{2}{3}$ ≈ 0.667,   (e) 66.7%

39. $y = 8 - x^3$.   (a) $y' = -3x^2$,   (b) $\frac{y'}{y} = \frac{-3x^2}{8 - x^3}$,   (c) -3,

    (d) $-\frac{3}{7} \approx -0.429$,   (e) -42.9%

41. $c = 0.2q^2 + 1.2q + 4$. $\frac{dc}{dq} = 0.4q + 1.2$.

    If $q = 5$, then $\frac{dc}{dq} = 0.14(5) + 1.2 = 3.2$.

    If $q = 5$, then $c = 15$ and $\frac{dc/dq}{c}(100) = \frac{3.2}{15}(100) \approx 21.3\%$

43. $r = 30q - 0.3q^2$.   (a) $dr/dq = 30 - 0.6q$.   (b) If $q = 10$,

    $\frac{r'}{r} = \frac{30 - 6}{300 - 30} = \frac{24}{270} = \frac{4}{45} \approx 0.089$.   (c) $8.9\% \approx 9\%$.

45. $\frac{W'}{W} = \frac{0.864t^{-0.568}}{2t^{0.432}} = \frac{0.432}{t}$

47. The cost of $q = 20$ bikes is $q\bar{c} = 29(150) = \$3000$. The
    marginal cost, \$125, is the approximate cost of one
    additional bike. Thus the approximate cost of producing
    21 bikes is \$3000 + \$125 = \$3125.

EXERCISE 12.5

1. $f'(x) = (4x + 1)(6) + (6x + 3)(4)$
          $= 24x + 6 + 24x + 12$
          $= 48x + 18 = 6(8x + 3)$

3. $s'(t) = (8 - 7t)(2t) + (t^2 - 2)(-7)$
          $= 16t - 14t^2 - 7t^2 + 14$
          $= 14 + 16t - 21t^2$

5. $f'(r) = (3r^2 - 4)(2r - 5) + (r^2 - 5r + 1)(6r)$
          $= 6r^3 - 15r^2 - 8r + 20 + 6r^3 - 30r^2 + 6r$
          $= 12r^3 - 45r^2 - 2r + 20$

7. *Without* the product rule we have

$f(x) = x^2(x^2 - 5) = x^4 - 5x^2.$ $f'(x) = 4x^3 - 10x$

Alternatively, *with* the product rule we have

$$f'(x) = x^2(2x) + (x^2 - 5)(2x)$$
$$= 2x^3 + 2x^3 - 10x = 4x^3 - 10x$$

9. $y' = (x^2 + 3x - 2)(4x - 1) + (2x^2 - x - 3)(2x + 3)$

$$= (4x^3 + 12x^2 - 8x - x^2 - 3x + 2) +$$
$$(4x^3 - 2x^2 - 6x + 6x^2 - 3x - 9)$$
$$= 8x^3 + 15x^2 - 20x - 7$$

11. $f'(w) = (8w^2 + 2w - 3)(15w^2) + (5w^3 + 2)(16w + 2)$

$$= 120w^4 + 30w^3 - 45w^2 + 80w^4 + 10w^3 + 32w + 4$$
$$= 200w^4 + 40w^3 - 45w^2 + 32w + 4$$

13. $y' = (x^2 - 1)(9x^2 - 6) + (3x^3 - 6x + 5)(2x) - 4(8x + 2)$

$$= 15x^4 - 27x^2 - 22x - 2$$

15. $f'(p) = \frac{3}{2}\left[(p^{1/2} - 4)(4) + (4p - 5)\left(\frac{1}{2}p^{-1/2}\right)\right]$

$$= \frac{3}{2}\left[4p^{1/2} - 16 + 2p^{1/2} - \frac{5}{2}p^{-1/2}\right]$$
$$= \frac{3}{2}\left[6p^{1/2} - 16 - \frac{5}{2}p^{-1/2}\right]$$
$$= \frac{3}{4}(12p^{1/2} - 5p^{-1/2} - 32)$$

17. $y = 7 \cdot \frac{2}{3}$ is a constant function, so $y' = 0$.

19. $y = (2x - 1)(3x + 4)(x + 7) = [(2x - 1)(3x + 4)](x + 7)$

$y' = [(2x - 1)(3x + 4)][1] +$
$$(x + 7)[(2x - 1)(3) + (3x + 4)(2)]$$
$$= 6x^2 + 5x - 4 + (x + 7)[12x + 5]$$
$$= 6x^2 + 5x - 4 + 12x^2 + 89x + 35$$
$$= 18x^2 + 94x + 31.$$

21. $f'(x) = \dfrac{(x - 1)(1) - (x)(1)}{(x - 1)^2} = -\dfrac{1}{(x - 1)^2}$

23. $f(x) = \dfrac{3}{2x^6} = \dfrac{3}{2}x^{-6}, \quad f'(x) = \dfrac{3}{2}(-6x^{-7}) = -\dfrac{9}{x^7}$

25. $y' = \dfrac{(x - 1)(1) - (x + 2)(1)}{(x - 1)^2} = \dfrac{x - 1 - x - 2}{(x - 1)^2} = \dfrac{-3}{(x - 1)^2}$

27. $h'(z) = \dfrac{(z^2 - 4)(-2) - (5 - 2z)(2z)}{(z^2 - 4)^2}$

$\qquad = \dfrac{-2z^2 + 8 - 10z + 4z^2}{(z^2 - 4)^2} = \dfrac{2z^2 - 10z + 8}{(z^2 - 4)^2}$

$\qquad = \dfrac{2(z^2 - 5z + 4)}{(z^2 - 4)^2} = \dfrac{2(z - 4)(z - 1)}{(z^2 - 4)^2}$

29. $y' = \dfrac{(x^2 - 5x)(16x - 2) - (8x^2 - 2x + 1)(2x - 5)}{(x^2 - 5x)^2}$

$\qquad = \dfrac{16x^3 - 82x^2 + 10x - (16x^3 - 44x^2 + 12x - 5)}{(x^2 - 5x)^2}$

$\qquad = \dfrac{-38x^2 - 2x + 5}{(x^2 - 5x)^2}.$

31. $y' = \dfrac{(2x^2 - 3x + 2)(2x - 4) - (x^2 - 4x + 3)(4x - 3)}{(2x^2 - 3x + 2)^2}$

$\qquad = \dfrac{4x^3 - 6x^2 + 4x - 8x^2 + 12x - 8 - (4x^3 - 16x^2 + 12x - 3x^2 + 12x - 9)}{(2x^2 - 3x + 2)^2}$

$\qquad = \dfrac{5x^2 - 8x + 1}{(2x^2 - 3x + 2)^2}$

33. $g'(x) = \dfrac{(x^{100} + 1)(0) - (1)(100x^{99})}{(x^{100} + 1)^2} = \dfrac{-100x^{99}}{(x^{100} + 1)^2}$

35. $u(v) = \dfrac{v^5 - 8}{v} = \dfrac{v^5}{v} - \dfrac{8}{v} = v^4 - 8v^{-1},$

$\qquad u'(v) = 4v^3 + 8v^{-2} = 4\left(v^3 + \dfrac{2}{v^2}\right) = \dfrac{4(v^5 + 2)}{v^2}$

37. $y = \dfrac{3x^2 - x - 1}{\sqrt[3]{x}} = \dfrac{3x^2 - x - 1}{x^{1/3}} = 3x^{5/3} - x^{2/3} - x^{-1/3}$

  $y' = 5x^{2/3} - \frac{2}{3}x^{-1/3} + \frac{1}{3}x^{-4/3} = 5x^{2/3} - \dfrac{2}{3x^{1/3}} + \dfrac{1}{3x^{4/3}}$

  $= \dfrac{15x^2 - 2x + 1}{3x^{4/3}}$

39. $y' = -\dfrac{(x-8)(0) - (4)(1)}{(x-8)^2} + \dfrac{(3x+1)(2) - (2x)(3)}{(3x+1)^2}$

  $= \dfrac{4}{(x-8)^2} + \dfrac{2}{(3x+1)^2}$

41. $y' = \dfrac{[(x+2)(x-4)](1) - (x-5)(2x-2)}{[(x+2)(x-4)]^2}$

  $= \dfrac{x^2 - 2x - 8 - (2x^2 - 12x + 10)}{[(x+2)(x-4)]^2} = \dfrac{-(x^2 - 10x + 18)}{[(x+2)(x-4)]^2}$

43. $s'(t)$

  $= \dfrac{[(t^2-1)(t^3+7)](2t+3) - (t^2+3t)(5t^4 - 3t^2 + 14t)}{[(t^2-1)(t^3+7)]^2}$

  $= \dfrac{-3t^6 - 12t^5 + t^4 + 6t^3 - 21t^2 - 14t - 21}{[(t^2-1)(t^3+7)]^2}$

45. $y = 3x - \dfrac{\frac{2}{x} - \frac{3}{x-1}}{x-2} = 3x - \dfrac{\frac{2(x-1) - 3x}{x(x-1)}}{x-2}$

  $= 3x + \dfrac{x+2}{x(x-1)(x-2)} = \dfrac{x+2}{x^3 - 3x^2 + 2x}$

  $y' = 3 + \dfrac{(x^3 - 3x^2 + 2x)[1] - (x+2)[3x^2 - 6x + 2]}{[x(x-1)(x-2)]^2}$

  $= 3 - \dfrac{2x^3 + 3x^2 - 12x + 4}{[x(x-1)(x-2)]^2}$

47. $f'(x) = \dfrac{(a+x)[-1] - (a-x)[1]}{(a+x)^2} = \dfrac{-2a}{(a+x)^2}$

49. $y = (4x^2 + 2x - 5)(x^3 + 7x + 4)$.

$y' = (4x^2 + 2x - 5)[3x^2 + 7] + (x^3 + 7x + 4)[8x + 2]$.

$y'(-1) = (-3)[10] + (-4)[-6] = -6$

51. $y = \frac{6}{x - 1}$, $y' = \frac{(x - 1)[0] - (6)[1]}{(x - 1)^2} = -\frac{6}{(x - 1)^2}$.

$y'(3) = -\frac{6}{2^2} = -\frac{3}{2}$. The tangent is $y - 3 = -\frac{3}{2}(x - 3)$,

or $y = -\frac{3}{2}x + \frac{15}{2}$.

53. $y = (2x + 3)[2(x^4 - 5x^2 + 4)]$

$y' = (2x + 3)[2(4x^3 - 10x)] + [2(x^4 - 5x^2 + 4)](2)$.

$y'(0) = (3)[0] + [2(4)](2) = 16$.

The tangent line is $y - 24 = 16(x - 0)$, or $y = 16x + 24$.

55. $y = \frac{x}{2x - 6}$. $y' = \frac{(2x - 6)[1] - x[2]}{(2x - 6)^2} = \frac{-6}{(2x - 6)^2}$.

If $x = 1$, then $y = \frac{1}{2 - 6} = -\frac{1}{4}$ and $y' = \frac{-6}{(-4)^2} = \frac{-6}{16} = -\frac{3}{8}$.

Thus $\frac{y'}{y} = \frac{-3/8}{-1/4} = \frac{3}{2} = 1.5$.

57. $s = \frac{2}{t^3 + 1}$. When $t = 1$, then $s = 1$ m

$v = \frac{(t^3 + 1)[0] - 2[3t^2]}{(t^3 + 1)^2} = -\frac{6t^2}{(t^3 + 1)^2}$.

If $t = 1$, then $v = -\frac{6}{4} = -1.5$ m/s

59. $F = \frac{kq(Q - q)}{r^2} = \frac{k}{r^2}[q(Q - q)$

Product Rule: $\frac{dF}{dq} = \frac{k}{r^2}[q(-1) + (Q - q)(1)] = \frac{k}{r^2}(Q - 2q)$

Alternatively, $F = \frac{k}{r^2}(qQ - q^2)$, $\frac{dF}{dq} = \frac{k}{r^2}(Q - 2q)$

61. $p = 25 - 0.02q$. $r = pq = 25q - 0.02q^2$. $\frac{dr}{dq} = 25 - 0.04q$

63. $p = \frac{108}{q + 2} - 3$. $r = pq = \frac{108q}{q + 2} - 3q$.

$\frac{dr}{dq} = \frac{(q + 2)[108] - (108q)[1]}{(q + 2)^2} - 3 = \frac{216}{(q + 2)^2} - 3$

65. $\frac{dC}{dI} = 0.672$

67. $C = 2 + 2I^{1/2}$. $\frac{dC}{dI} = 0 + 2\left(\frac{1}{2}I^{-1/2}\right) = \frac{1}{\sqrt{I}}$.

When $I = 9$, then $\frac{dC}{dI} = \frac{1}{\sqrt{9}} = \frac{1}{3}$ and $\frac{dS}{dI} = 1 - \frac{1}{3} = \frac{2}{3}$

69. $\frac{dC}{dI} = \dfrac{(\sqrt{I}+4)\left[\frac{8}{\sqrt{I}} + 1.2\sqrt{I} - .2\right] - (16\sqrt{I}+.8\sqrt{I^3}-.2I)\left[\frac{1}{2\sqrt{I}}\right]}{(\sqrt{I}+4)^2}$

$\left.\frac{dC}{dI}\right|_{I=36} = 0.615$, so $\frac{dS}{dI} = 1 - 0.615 = 0.385$ when $I = 36$.

71. Simplifying gives $C = 10 + 0.7I - 0.2I^{1/2}$.

(a) $\frac{dC}{dI} = 0.7 - 0.1I^{-1/2} = 0.7 - \frac{0.1}{\sqrt{I}}$.

$\frac{dS}{dI} = 1 - \frac{dC}{dI} = 0.3 + \frac{0.1}{\sqrt{I}}$.

$\left.\frac{dS}{dI}\right|_{I=25} = 0.3 + \frac{0.1}{5} = 0.32$

(b) $\frac{dC/dI}{C}$ when $I = 25$ is $\frac{0.7 - \frac{0.1}{5}}{10 + 0.7(25) - 0.2(5)} \approx 0.026$

73. $\frac{dc}{dq} = 5 \cdot \frac{(q + 3)(2q) - q^2(1)}{(q + 3)^2} = 5 \cdot \frac{q^2 + 6q}{(q + 3)^2} = \frac{5q(q + 6)}{(q + 3)^2}$

75. $y = \frac{900x}{10 + 45x}$, $\frac{dy}{dx} = \frac{(10 + 45x)(900) - (900x)(45)}{(10 + 45x)^2}$.

$\left.\frac{dy}{dx}\right|_{x=2} = \frac{(100)(900) - (1800)(45)}{(100)^2} = \frac{9}{10}$.

77. $y = \frac{0.7355x}{1 + 0.02744x}$.

$\frac{dy}{dx} = \frac{(1 + 0.02744x)(0.7355) - (0.7355x)(0.02744)}{(1 + 0.02744x)^2}$

$= \frac{0.7355}{(1 + 0.02744x)^2}$.

79. $\frac{d\bar{c}}{dq} = \frac{d}{dq}\left(\frac{c}{q}\right) = \frac{q \cdot \frac{dc}{dq} - c(1)}{q^2}$. When $q = 20$ we have

$\frac{\frac{d\bar{c}}{dq}}{\bar{c}} = \frac{\frac{q \cdot \frac{dc}{dq} - c}{q^2}}{\bar{c}} = \frac{\frac{20(125) - 20(150)}{(20)^2}}{150} = -\frac{1}{120}$

81. $\frac{dy}{dx}$

$= (2x - 1)(x - 2)(1) + (2x - 1)(1)(x + 3) + 2(x - 2)(x + 3)$

$= (2x^2 - 5x + 2) + (2x^2 + 5x - 3) + (2x^2 + 2x - 12)$

EXERCISE 12.6

1. $\frac{dy}{dx} = \frac{dy}{du} \cdot \frac{du}{dx} = (2u - 2)(2x - 1)$. Expressing the answer in terms of x, we have

$(2u - 2)(2x - 1) = [2(x^2 - x) - 2](2x - 1)$

$= (2x^2 - 2x - 2)(2x - 1)$

$= 4x^3 - 2x^2 - 4x^2 + 2x - 4x + 2$

$= 4x^3 - 6x^2 - 2x + 2.$

3. $y = w^{-2}$, so $\frac{dy}{dw} = -2w^{-3} = -\frac{2}{w^3}$. $\frac{dw}{dx} = -1$. Thus

$\frac{dy}{dx} = \frac{dy}{dw} \cdot \frac{dw}{dx} = \left(-\frac{2}{w^3}\right)(-1) = \frac{2}{w^3} = \frac{2}{(2 - x)^3}$

5. $\frac{dw}{dt} = \frac{dw}{du} \cdot \frac{du}{dt} = (2u)\left[\frac{(t - 1) - (t + 1)}{(t - 1)^2}\right] = 2u\left[\frac{-2}{(t - 1)^2}\right]$.

If $t = 3$, then $u = \frac{3 + 1}{3 - 1} = 2$, so $\left.\frac{dw}{dt}\right|_{t=3} = 2(2)\left[\frac{-2}{4}\right] = -2$.

7. $\frac{dy}{dx} = \frac{dy}{dw} \cdot \frac{dw}{dx} = (6w - 8)(6x)$. If $x = 0$, then $\frac{dy}{dx} = 0$.

9. $y' = 6(3x + 2)^5 \cdot \frac{d}{dx}(3x + 2) = 6(3x + 2)^5(3) = 18(3x + 2)^5$

11. $y' = 3(5 - x^2)^2 \cdot \frac{d}{dx}(5 - x^2) = 3(5 - x^2)^2(-2x) = -6x(5 - x^2)^2$

13. $y' = 3 \cdot 100(x^3 - 8x^2 + x)^{99} \cdot \frac{d}{dx}(x^3 - 8x^2 + x)$

   $= 300(x^3 - 8x^2 + x)^{99}(3x^2 - 16x + 1)$

   $= 300(3x^2 - 16x + 1)(x^3 - 8x^2 + x)^{99}$

15. $y' = -3(x^2 - 2)^{-4} \cdot \frac{d}{dx}(x^2 - 2)$

   $= -3(x^2 - 2)^{-4}(2x) = -6x(x^2 - 2)^{-4}$

17. $y' = \left(-\frac{10}{3}\right)(2x^2 - 3x - 1)^{-13/3}(4x - 3)$

   $= -\frac{10}{3}(4x - 3)(2x^2 - 3x + 1)^{-13/3}$

19. $y = \sqrt{5x^2 - x} = (5x^2 - x)^{1/2}$

   $y' = \frac{1}{2}(5x^2 - x)^{-1/2}(10x - 1) = \frac{1}{2}(10x - 1)(5x^2 - x)^{-1/2}$

21. $y = \sqrt[4]{2x - 1} = (2x - 1)^{1/4}$

   $y' = \frac{1}{4}(2x - 1)^{-3/4}(2) = \frac{1}{2}(2x - 1)^{-3/4}$

23. $y = 2\sqrt[5]{(x^3 + 1)^2} = 2(x^3 + 1)^{2/5}$

   $y' = 2\left(\frac{2}{5}\right)(x^3 + 1)^{-3/5}(3x^2) = \frac{12}{5}x^2(x^3 + 1)^{-3/5}$

25. $y = \dfrac{6}{2x^2 - x + 1} = 6(2x^2 - x + 1)^{-1}$

   $y' = 6(-1)(2x^2 - x + 1)^{-2}(4x - 1)$

   $= -6(4x - 1)(2x^2 - x + 1)^{-2}$

27. $y = \dfrac{1}{(x^2 - 3x)^2} = (x^2 - 3x)^{-2}$

$y' = -2(x^2 - 3x)^{-3}(2x - 3) = -2(2x - 3)(x^2 - 3x)^{-3}$

29. $y = \dfrac{2}{\sqrt{8x - 1}} = 2(8x - 1)^{-1/2}$

$y' = 2\left(-\dfrac{1}{2}\right)(8x - 1)^{-3/2}(8) = -8(8x - 1)^{-3/2}$

31. $y = \sqrt[3]{7x} + \sqrt[3]{7}x = (7x)^{1/3} + \sqrt[3]{7}x$

$y' = \dfrac{1}{3}(7x)^{-2/3}(7) + \sqrt[3]{7}(1) = \dfrac{7}{3}(7x)^{-2/3} + \sqrt[3]{7}$

33. By the product rule,
$$y' = x^2[5(x - 4)^4(1)] + (x - 4)^5(2x).$$
Factoring out $x(x - 4)^4$ from both terms gives
$$y' = x(x - 4)^4[5x + 2(x - 4)] = x(x - 4)^4(7x - 8)$$

35. $y = 2x\sqrt{6x - 1} = 2x(6x - 1)^{1/2}$. By the product rule,

$y' = 2x\left[\dfrac{1}{2}(6x - 1)^{-1/2}(6)\right] + \sqrt{6x - 1}(2)$

$\quad = 6x(6x - 1)^{-1/2} + 2\sqrt{6x - 1}$

37. By the product rule,

$y' = (x^2 + 2x - 1)^3(5) + (5x)[3(x^2 + 2x - 1)^2(2x + 2)]$

$\quad = 5(x^2 + 2x - 1)^2[(x^2 + 2x - 1) + 3x(2x + 2)]$

$\quad = 5(x^2 + 2x - 1)^2(7x^2 + 8x - 1)$

39. $y' = (8x - 1)^3[4(2x + 1)^3(2)] + (2x + 1)^4[3(8x - 1)^2(8)]$

$\quad = 8(8x - 1)^2(2x + 1)^3[(8x - 1) + 3(2x + 1)]$

$\quad = 8(8x - 1)^2(2x + 1)^3(14x + 2)$

$\quad = 16(8x - 1)^2(2x + 1)^3(7x + 1)$

41. $y' = 10\left(\dfrac{x - 7}{x + 4}\right)^9\left[\dfrac{(x + 4)(1) - (x - 7)(1)}{(x + 4)^2}\right]$

$\quad = 10\left(\dfrac{x - 7}{x + 4}\right)^9\left[\dfrac{11}{(x + 4)^2}\right] = \dfrac{110(x - 7)^9}{(x + 4)^9(x + 4)^2} = \dfrac{110(x - 7)^9}{(x + 4)^{11}}$

43. $y' = \frac{1}{2}\left(\frac{x-2}{x+3}\right)^{-1/2}\left[\frac{(x+3)(1)-(x-2)(1)}{(x+3)^2}\right]$

$= \frac{5}{2(x+3)^2}\left(\frac{x-2}{x+3}\right)^{-1/2} = \frac{5}{2(x+3)^2}\sqrt{\frac{x+3}{x-2}}$

45. $y' = \frac{(x^2+4)^3(2)-(2x-5)[3(x^2+4)^2(2x)]}{(x^2+4)^6}$.

Factoring out $(x^2+4)^2$ from the numerator and cancelling,

$y' = \frac{(x^2+4)^2\{(x^2+4)(2)-(2x-5)[3(2x)]\}}{(x^2+4)^6}$

$= \frac{(x^2+4)(2)-(2x-5)(6x)}{(x^2+4)^4}$

$= \frac{2x^2+8-12x^2+30x}{(x^2+4)^4} = \frac{-10x^2+30x+8}{(x^2+4)^4}$

$= \frac{-2(5x^2-15x-4)}{(x^2+4)^4}$

47. $y' = \frac{(3x-1)^3[5(8x-1)^4(8)]-(8x-1)^5[3(3x-1)^2(3)]}{(3x-1)^6}$

$= \frac{(3x-1)^2[(3x-1)(40)(8x-1)^4-(8x-1)^5(9)]}{(3x-1)^6}$

$= \frac{(3x-1)(40)(8x-1)^4-(8x-1)^5(9)]}{(3x-1)^4}$

$= \frac{(8x-1)^4[40(3x-1)-(8x-1)(9)]}{(3x-1)^4}$

$= \frac{(8x-1)^4(48x-31)}{(3x-1)^4}$

49. $y = 6(5x^2+2)\sqrt{x^4+5} = 6\left[(5x^2+2)(x^4+5)^{1/2}\right]$

$y' = 6\left[(5x^2+2)\cdot\frac{1}{2}(x^4+5)^{-1/2}(4x^3)+(x^4+5)^{1/2}(10x)\right]$

$= 6\left[(5x^2+2)(x^4+5)^{-1/2}(2x^3)+(x^4+5)^{1/2}(10x)\right]$.

Factoring out $2x(x^4+5)^{-1/2}$ gives

$y' = 12x(x^4+5)^{-1/2}[(5x^2+2)(x^2)+(x^4+5)(5)]$

$= 12x(x^4+5)^{-1/2}(10x^4+2x^2+25)$

51. $y' = 8 + \dfrac{(t + 4)(1) - (t - 1)(1)}{(t + 4)^2} - 2\left(\dfrac{8t - 7}{4}\right)\left[\dfrac{1}{4}\cdot 8\right]$

$= 8 + \dfrac{5}{(t + 4)^2} - (8t - 7) = 15 - 8t + \dfrac{5}{(t + 4)^2}$

53. $y' = \dfrac{\left[\begin{array}{l}(x^2-7)^4[(2x+1)(2)(3x-5)(3)+(3x-5)^2(2)]-\\ \qquad\qquad (2x+1)(3x-5)^2[4(x^2-7)^3(2x)]\end{array}\right]}{(x^2-7)^8}$

55. $\dfrac{dy}{dx} = \dfrac{dy}{du}\cdot\dfrac{du}{dx} = [3(5u + 6)^2(5)][4(x^2 + 1)^3(2x)]$.

When $x = 0$, then $\dfrac{dy}{dx} = 0$.

57. $y' = 3(x^2 - 7x - 8)^2(2x - 7)$.

If $x = 8$, then slope $= y' = 3(64 - 56 - 8)^2(16 - 7) = 0$.

59. $y = (x^2 - 8)^{2/3}$. $\quad y' = \dfrac{2}{3}(x^2 - 8)^{-1/3}(2x) = \dfrac{4x}{3(x^2 - 8)^{1/3}}$.

If $x = 3$, then $y' = \dfrac{12}{3(1)} = 4$. Thus the tangent is

$y - 1 = 4(x - 3)$, or $y = 4x - 11$.

61. $y' = \dfrac{(x + 1)\left(\dfrac{1}{2}\right)(7x + 2)^{-1/2}(7) - \sqrt{7x + 2}(1)}{(x + 1)^2}$.

$= \dfrac{(x + 1)\left(\dfrac{7}{2}\right)\dfrac{1}{\sqrt{7x + 2}} - \sqrt{7x + 2}}{(x + 1)^2}$.

If $x = 1$, then $y' = \dfrac{2\left(\dfrac{7}{2}\right)\left(\dfrac{1}{3}\right) - 3(1)}{4} = -\dfrac{1}{6}$. The tangent

line is $y - \dfrac{3}{2} = -\dfrac{1}{6}(x - 1)$, or $y = -\dfrac{1}{6}x + \dfrac{5}{3}$.

63. $y = (x^2 + 9)^3$ and $y' = 6x(x^2 + 9)^2$. When $x = 4$, then

$y = (25)^3$ and $y' = 6(4)(25)^2$, so

$$\dfrac{y'}{y}(100) = \dfrac{6(4)(25)^2}{(25)^3} = \dfrac{24}{25} = 96\%$$

65. Given $q = 2m$, $p = -0.5q + 20$; $m = 5$. $\frac{dr}{dm} = \frac{dr}{dq} \cdot \frac{dq}{dm}$.

Since $r = pq = -0.5q^2 + 20q$, we have $\frac{dr}{dq} = -q + 20$.

For $m = 5$, then $q = 2(5) = 10$, so $\frac{dr}{dq} = -10 + 20 = 10$.

Also, $\frac{dq}{dm} = 2$. Thus $\frac{dr}{dm}\Big|_{m=5} = (10)(2) = 20$.

67. $q = \frac{10m^2}{\sqrt{m^2 + 9}}$, $p = \frac{525}{q + 3}$; $m = 4$. $\frac{dr}{dm} = \frac{dr}{dq} \cdot \frac{dq}{dm}$.

$r = pq = \frac{525q}{q + 3}$, so $\frac{dr}{dq} = 525 \cdot \frac{(q + 3)(1) - q(1)}{(q + 3)^2} = \frac{1575}{(q + 3)^2}$.

If $m = 4$, then $q = 32$, so $\frac{dr}{dq} = \frac{1575}{1225}$.

$\frac{dq}{dm} = \frac{(m^2 + 9)^{1/2}(20m) - 10m^2 \cdot \frac{1}{2}(m^2 + 9)^{-1/2}(2m)}{m^2 + 9}$

$= \frac{(m^2 + 9)^{-1/2}[20m(m^2 + 9) - 10m^3]}{m^2 + 9} = \frac{10m^3 + 180m}{(m^2 + 9)^{3/2}}$

When $m = 4$, then $\frac{dq}{dm} = \frac{10(64) + 180(4)}{(25)^{3/2}} = \frac{1360}{125}$.

Thus $\frac{dr}{dm}\Big|_{m=4} = \frac{1575}{1225} \cdot \frac{1360}{125} = 13.99$ (approx).

69. $p = 100 - \sqrt{q^2 + 20}$

(a) $\frac{dp}{dq} = 0 - \frac{1}{2}(q^2 + 20)^{-1/2}(2q) = \frac{-q}{\sqrt{q^2 + 20}}$.

(b) $\frac{dp/dq}{p} = \frac{\frac{-q}{\sqrt{q^2 + 20}}}{100 - \sqrt{q^2 + 20}} = -\frac{q}{\sqrt{q^2 + 20}\left(100 - \sqrt{q^2 + 20}\right)}$

$= -\frac{q}{100\sqrt{q^2 + 20} - q^2 - 20}$

(c) $r = pq = 100q - q\sqrt{q^2 + 20}$.

$\frac{dr}{dq} = 100 - \left[q \cdot \frac{1}{2}(q^2 + 20)^{-1/2}(2q) + \sqrt{q^2 + 20}(1)\right]$

$= 100 - \frac{q^2}{\sqrt{q^2 + 20}} - \sqrt{q^2 + 20}$

71. $\frac{dc}{dp} = \frac{dc}{dq}\cdot\frac{dq}{dp} = (10 + 0.2q)(-2.5)$.

When $p = 80$, then $q = 600$, so $\frac{dc}{dp}\Big|_{p=80} = -325$.

73. $\frac{dc}{dq} = \dfrac{(q^2 + 3)^{1/2}[10q] - (5q^2)\left[\frac{1}{2}(q^2 + 3)^{-1/2}(2q)\right]}{q^2 + 3}$.

Multiplying numerator and denominator by $(q^2 + 3)^{1/2}$ gives

$\frac{dc}{dq} = \dfrac{(q^2 + 3)(10q) - 5q^2(q)}{(q^2 + 3)^{3/2}} = \dfrac{5q^3 + 30q}{(q^2 + 3)^{3/2}} = \dfrac{5q(q^2 + 6)}{(q^2 + 3)^{3/2}}$.

75. $\frac{dV}{dt} = \frac{dV}{dr}\cdot\frac{dr}{dt} = (4\pi r^2)[10^{-8}(2t) + 10^{-7}]$. When $t = 10$, then

$r = 10^{-8}(10^2) + 10^{-7}(10) = 10^{-6} + 10^{-6} = 2(10)^{-6}$. Thus

$\frac{dV}{dt}\Big|_{t=10} = 4\pi[2(10)^{-6}]^2[10^{-8}(2)(10) + 10^{-7}]$

$= 4\pi[4(10)^{-12}][3(10^{-7})] = 48\pi(10)^{-19}$

77. (a) $l_x = 2000\sqrt{100 - x}$.

$\frac{d}{dx}(l_x) = 2000\left(\frac{1}{2}\right)(100 - x)^{-1/2}(-1) = -\dfrac{1000}{\sqrt{100 - x}}$.

If $x = 36$, $\frac{d}{dx}(l_x) = -\dfrac{1000}{8} = -125$.

(b) If $x = 36$,

$\dfrac{\frac{d}{dx}(l_x)}{l_x} = \dfrac{-125}{2000(8)} = -\dfrac{1}{128}$.

79. $P = \varepsilon^2\cdot\dfrac{R}{(r + R)^2}$,

$\frac{dP}{dR} = \varepsilon^2\cdot\dfrac{(r + R)^2(1) - R(2)(r + R)(1)}{(r + R)^4}$

$= \varepsilon^2\cdot\dfrac{(r + R)[(r + R) - 2R]}{(r + R)^4} = \dfrac{\varepsilon^2(r - R)}{(r + R)^3}$

81. By the chain rule, $\frac{dc}{dp} = \frac{dc}{dq} \cdot \frac{dq}{dp}$. We are given that

$q = \frac{100}{p} = 100p^{-1}$, so $\frac{dq}{dp} = -100p^{-2} = \frac{-100}{p^2}$. Thus

$\frac{dc}{dp} = \frac{dc}{dq}\left[\frac{-100}{p^2}\right]$. When $q = 200$, then $p = \frac{100}{200} = \frac{1}{2}$ and

we are given that $dc/dq = 0.01$. Therefore

$$\frac{dc}{dp} = 0.01\left[\frac{-100}{(1/2)^2}\right] = -4$$

83. $\frac{dy}{dt} = \frac{dy}{dx} \cdot \frac{dx}{dt} = f'(x)g'(t)$. We are given that $g(2) = 3$, so

$x = 3$ when $t = 2$. Thus

$$\left.\frac{dy}{dt}\right|_{t=2} = \left.\frac{dy}{dx}\right|_{x=g(2)} \cdot \left.\frac{dx}{dt}\right|_{t=2} = f'(3)g'(2) = 10(4) = 40$$

CHAPTER 12 - REVIEW PROBLEMS

1. $f(x) = 2 - x^2$.

   $f'(x)$

   $= \lim_{h \to 0} \frac{f(x + h) - f(x)}{h} = \lim_{h \to 0} \frac{[2 - (x + h)^2] - (2 - x^2)}{h}$

   $= \lim_{h \to 0} \frac{[2 - x^2 - 2hx - h^2] - (2 - x^2)}{h} = \lim_{h \to 0} \frac{-2hx - h^2}{h}$

   $= \lim_{h \to 0} \frac{-h(2x + h)}{h} = \lim_{h \to 0} -(2x + h) = -2x$

3. $f(x) = \sqrt{3x}$.

   $f'(x)$

   $= \lim_{h \to 0} \frac{f(x + h) - f(x)}{h} = \lim_{h \to 0} \frac{\sqrt{3(x + h)} - \sqrt{3x}}{h}$

   $= \lim_{h \to 0} \frac{\sqrt{3(x + h)} - \sqrt{3x}}{h} \cdot \frac{\sqrt{3(x + h)} + \sqrt{3x}}{\sqrt{3(x + h)} + \sqrt{3x}}$

   $= \lim_{h \to 0} \frac{3(x + h) - 3x}{h(\sqrt{3(x + h)} + \sqrt{3x})} = \lim_{h \to 0} \frac{3h}{h(\sqrt{3(x + h)} + \sqrt{3x})}$

   $= \lim_{h \to 0} \frac{3}{\sqrt{3(x + h)} + \sqrt{3x}}$

   $= \frac{3}{\sqrt{3x} + \sqrt{3x}} = \frac{3}{2\sqrt{3x}} = \frac{\sqrt{3}}{2\sqrt{x}}$

5. y is a constant function, so y' = 0.

7. $y' = 7(4x^3) - 6(3x^2) + 5(2x) + 0$
   $= 28x^3 - 18x^2 + 10x = 2x(14x^2 - 9x + 5)$

9. $f(s) = s^2(s^2 + 2) = s^4 + 2s^2$
   $f'(s) = 4s^3 + 2(2s) = 4s^3 + 4s = 4s(s^2 + 1)$

11. $y = \frac{1}{5}(x^2 + 3).$ $y' = \frac{1}{5}(2x) = \frac{2x}{5}$

13. $y' = (x^2 + 6x)(3x^2 - 12x) + (x^3 - 6x^2 + 4)(2x + 6)$
    $= 3x^4 + 6x^3 - 72x^2 + (2x^4 + 6x^3 - 12x^3 - 36x^2 + 8x + 24)$
    $= 5x^4 - 108x^2 + 8x + 24$

15. $f'(x) = 100(2x^2 + 4x)^{99}(4x + 4)$
    $= 400(x + 1)[(2x)(x + 2)]^{99}$

17. $y = (2x + 1)^{-1}.$ $y' = (-1)(2x + 1)^{-2}(2) = -\dfrac{2}{(2x + 1)^2}$

19. $y' = (8 + 2x)[(4)(x^2 + 1)^3(2x)] + (x^2 + 1)^4[2]$
    $= 2(x^2 + 1)^3[4x(8 + 2x) + (x^2 + 1)]$
    $= 2(x^2 + 1)^3(32x + 8x^2 + x^2 + 1)$
    $= 2(x^2 + 1)^3(9x^2 + 32x + 1)$

21. $f'(z) = \dfrac{(z^2 + 1)(2z) - (z^2 - 1)(2z)}{(z^2 + 1)^2} = \dfrac{4z}{(z^2 + 1)^2}$

23. $y = (4x - 1)^{1/3}.$ $y' = \frac{1}{3}(4x - 1)^{-2/3}(4) = \frac{4}{3}(4x - 1)^{-2/3}$

25. $y = (1 - x)^{-1/2}.$ $y' = \left(-\frac{1}{2}\right)(1 - x)^{-3/2}(-1) = \frac{1}{2}(1 - x)^{-3/2}$

27. $h'(x) = (x - 6)^4[3(x + 5)^2] + (x + 5)^3[4(x - 6)^3]$
    $= (x - 6)^3(x + 5)^2[3(x - 6) + 4(x + 5)]$
    $= (x - 6)^3(x + 5)^2(7x + 2)$

29. $y' = \dfrac{(x + 1)(5) - (5x - 4)(1)}{(x + 1)^2} = \dfrac{9}{(x + 1)^2}$

31. $y' = 2\left(-\dfrac{3}{8}\right)x^{-11/8} + \left(-\dfrac{3}{8}\right)(2x)^{-11/8}(2)$

$= -\dfrac{3}{4}x^{-11/8} - \dfrac{3}{4}(2^{-11/8})x^{-11/8}$

$= -\dfrac{3}{4}x^{-11/8}(1 + 2^{-11/8}) = -\dfrac{3}{4}(1 + 2^{-11/8})x^{-11/8}$

33. $y' = \dfrac{(x^2 + 5)^{1/2}(2x) - (x^2 + 6)(1/2)(x^2 + 5)^{-1/2}(2x)}{x^2 + 5}$.

Multiplying the numerator and denominator by $(x^2 + 5)^{1/2}$,

$y' = \dfrac{(x^2 + 5)(2x) - x(x^2 + 6)}{(x^2 + 5)^{3/2}} = \dfrac{x^3 + 4x}{(x^2 + 5)^{3/2}} = \dfrac{x(x^2 + 4)}{(x^2 + 5)^{3/2}}$

35. $y' = \dfrac{3}{5}(x^3 + 6x^2 + 9)^{-2/5}(3x^2 + 12x)$

$= \dfrac{3}{5}(x^3 + 6x^2 + 9)^{-2/5}(3x)(x + 4)$

$= \dfrac{9}{5}x(x + 4)(x^3 + 6x + 9)^{-2/5}$

37. $g(z) = -7z(z - 1) = -7(z^2 - z)$
$g'(z) = -7(2z - 1) = 7(1 - 2z)$

39. $y = x^2 - 6x + 4$, $\quad y' = 2x - 6$.
When $x = 1$, then $y = -1$ and $y' = -4$. An equation of the tangent is $y - (-1) = -4(x - 1)$, or simply $y = -4x + 3$.

41. $y = x^{1/3}$, $\quad y' = \dfrac{1}{3}x^{-2/3}$.

When $x = 8$, then $y = 2$ and $y' = \dfrac{1}{12}$. An equation of the tangent line is $y - 2 = \dfrac{1}{12}(x - 8)$, or $y = \dfrac{1}{12}x + \dfrac{4}{3}$

43. $f(x) = 4x^2 + 2x + 8$. $\quad f'(x) = 8x + 2$.
$f(1) = 14$ and $f'(1) = 10$. The relative rate of change is $\dfrac{f'(1)}{f(1)} = \dfrac{10}{14} = \dfrac{5}{7} \approx 0.714$, so the percentage rate of change is 71.4%.

45. $r = q(20 - 0.1q) = 20q - 0.1q^2$. $\frac{dr}{dq} = 20 - 0.2q$.

47. $\frac{dC}{dI} = 0.6 - 0.25\left(\frac{1}{2}\right)I^{-1/2} = 0.6 - \frac{1}{8\sqrt{I}}$. $\left.\frac{dC}{dI}\right|_{I=16} = 0.569$.
    Thus the marginal propensity to consume is 0.569, so the marginal propensity to save is $1 - 0.569 = 0.431$.

49. Since $p = -0.5q + 450$, then $r = pq = -0.5q^2 + 450q$. Thus
    $\frac{dr}{dq} = 450 - q$.

51. $\frac{dc}{dq} = 0.125 + 0.00878q$. $\left.\frac{dc}{dq}\right|_{q=70} = 0.7396$.

53. $\frac{dy}{dx} = 42x^2 - 34x - 16$. $\left.\frac{dy}{dx}\right|_{x=2} = 84$.

55. (a) $\frac{dt}{dT}$ when $T = 38$ is
$$\left.\frac{T}{dT}\left[\frac{4}{3}T - \frac{175}{4}\right]\right|_{T=38} = \left.\frac{4}{3}\right|_{T=38} = \frac{4}{3}.$$
    (b) $\frac{dt}{dT}$ when $T = 35$ is
$$\left.\frac{d}{dT}\left[\frac{1}{24}T + \frac{11}{4}\right]\right|_{T=35} = \left.\frac{1}{24}\right|_{T=35} = \frac{1}{24}.$$

57. $B = k[x^{-1} - (d - x)^{-1}]$,
$$\frac{dB}{dx} = k[-1x^{-2} - (-1)(d - x)^{-2}(-1)] = -k\left[\frac{1}{x^2} - \frac{1}{(d - x)^2}\right]$$

59. $V' = \frac{1}{2}\pi d^2$. If $d = 4$ ft, then $V' = 8\pi$ ft^3/ft.

61. $c = \bar{c}q = 2q^2 + \frac{10,000}{q} = 2q^2 + 10,000q^{-1}$.
$$\frac{dc}{dq} = 4q - 10,000q^{-2} = 4q - \frac{10,000}{q^2}$$

13

Additional Differentiation Topics

1. $\frac{dy}{dx} = 4 \cdot \frac{d}{dx}(\ln x) = 4 \cdot \frac{1}{x} = \frac{4}{x}$

3. $\frac{dy}{dx} = \frac{1}{3x - 4}(3) = \frac{3}{3x - 4}$

5. $y = \ln x^2 = 2 \ln x. \quad \frac{dy}{dx} = 2 \cdot \frac{1}{x} = \frac{2}{x}$

7. $\frac{dy}{dx} = \frac{1}{1 - x^2}(-2x) = -\frac{2x}{1 - x^2}$

9. $f'(p) = \frac{1}{2p^3 + 3p}(6p^2 + 3) = \frac{6p^2 + 3}{2p^3 + 3p} = \frac{3(2p^2 + 1)}{p(2p^2 + 3)}$

11. $f'(t) = t\left(\frac{1}{t}\right) + (\ln t)(1) = 1 + \ln t$

13. $\frac{dy}{dx} = x^2\left[\frac{1}{4x+3}(4)\right] + [\ln(4x+3)](2x) = \frac{4x^2}{4x+3} + 2x \ln(4x+3)$

15. $y = \log_3(2x - 1) = \frac{\ln(2x - 1)}{\ln 3}$.

$$\frac{dy}{dx} = \frac{1}{\ln 3} \cdot \frac{d}{dx}[\ln(2x - 1)] = \frac{1}{\ln 3} \cdot \frac{1}{2x - 1}(2)$$

$$= \frac{2}{(2x - 1)(\ln 3)}$$

17. $y = x^2 + \log_2(x^2+4) = x^2 + \frac{\ln(x^2+4)}{\ln 2}$.

$$\frac{dy}{dx} = 2x + \frac{1}{\ln 2}\left[\frac{1}{x^2+4}(2x)\right] = 2x\left[1 + \frac{1}{(\ln 2)(x^2+4)}\right]$$

19. $f'(z) = \frac{z\left(\frac{1}{z}\right) - (\ln z)(1)}{z^2} = \frac{1 - \ln z}{z^2}$

21. $\frac{dy}{dx} = \frac{(\ln x)(2x) - (x^2 - 1)\left(\frac{1}{x}\right)}{\ln^2 x} = \frac{2x^2 \ln(x) - x^2 + 1}{x \ln^2 x}$

23. $y = \ln(x^2 + 4x + 5)^3 = 3 \ln(x^2 + 4x + 5)$.

$$\frac{dy}{dx} = 3 \cdot \frac{1}{x^2 + 4x + 5}(2x + 4) = \frac{3(2x + 4)}{x^2 + 4x + 5} = \frac{6(x + 2)}{x^2 + 4x + 5}$$

25. $y = \ln\sqrt{1 + x^2} = \frac{1}{2}\ln(1 + x^2)$. $\frac{dy}{dx} = \frac{1}{2} \cdot \frac{1}{1 + x^2}(2x) = \frac{x}{1 + x^2}$

27. $f(I) = \ln\left(\frac{1 + I}{1 - I}\right) = \ln(1 + I) - \ln(1 - I)$.

$$f'(I) = \frac{1}{1 + I} - \frac{1}{1 - I}(-1) = \frac{(1 - I) + (1 + I)}{(1 + I)(1 - I)} = \frac{2}{1 - I^2}$$

29. $y = \ln \sqrt[4]{\frac{1+x^2}{1-x^2}} = \frac{1}{4}[\ln(1+x^2) - \ln(1-x^2)]$.

$$\frac{dy}{dx} = \frac{1}{4}\left[\frac{2x}{1+x^2} - \frac{-2x}{1-x^2}\right] = \frac{1}{4}\left[\frac{2x(1-x^2)+2x(1+x^2)}{(1+x^2)(1-x^2)}\right] = \frac{x}{1-x^4}$$

31. $y = \ln[(x^2 + 2)^2(x^3 + x - 1)]$

$$= 2 \ln(x^2 + 2) + \ln(x^3 + x - 1).$$

$$\frac{dy}{dx} = 2 \cdot \frac{1}{x^2 + 2}(2x) + \frac{1}{x^3 + x - 1}(3x^2 + 1)$$

$$= \frac{4x}{x^2 + 2} + \frac{3x^2 + 1}{x^3 + x - 1}$$

33. $y = \ln(x\sqrt{2x+1}) = \ln x + \ln(2x+1)^{1/2} = \ln x + \frac{1}{2}\ln(2x+1)$.

$\frac{dy}{dx} = \frac{1}{x} + \frac{1}{2}\cdot\frac{1}{2x+1}(2) = \frac{1}{x} + \frac{1}{2x+1} = \frac{3x+1}{x(2x+1)}$

35. $\frac{dy}{dx} = (x^2+1)\left[\frac{1}{2x+1}(2)\right] + \ln(2x+1)\cdot[2x] = \frac{2(x^2+1)}{2x+1} + 2x \ln(2x+1)$

37. $y = \ln x^3 + \ln^3 x = 3 \ln x + (\ln x)^3$.

$\frac{dy}{dx} = 3\cdot\frac{1}{x} + 3(\ln x)^2\cdot\frac{1}{x} = \frac{3}{x} + \frac{3(\ln x)^2}{x} = \frac{3(1 + \ln^2 x)}{x}$

39. $y = \ln^4(ax) = [\ln(ax)]^4$. $\frac{dy}{dx} = 4[\ln(ax)]^3\left(\frac{1}{ax}\cdot a\right) = \frac{4 \ln^3(ax)}{x}$

41. $y = x \ln\sqrt{x - 1} = \frac{1}{2}x \ln(x - 1)$.   By the product rule,

$\frac{dy}{dx} = \frac{1}{2}\left[x\left(\frac{1}{x - 1}\right) + \ln(x-1)\cdot[1]\right] = \frac{x}{2(x - 1)} + \ln\sqrt{x - 1}$

43. $y = \sqrt{4 + \ln x} = (4 + \ln x)^{1/2}$.

$\frac{dy}{dx} = \frac{1}{2}(4 + \ln x)^{-1/2}\cdot\frac{1}{x} = \frac{1}{2x\sqrt{4 + \ln x}}$

45. $y = \ln(x^2 - 2x - 2)$, $y' = \frac{2x - 2}{x^2 - 2x - 2}$.   The slope of the tangent line at $x = 3$ is $y'(3) = \frac{6 - 2}{9 - 6 - 2} = 4$. Also, if $x = 3$, then $y = \ln(9 - 6 - 2) = \ln 1 = 0$. Thus an equation of the tangent line is $y - 0 = 4(x - 3)$, or $y = 4x - 12$.

47. $y = \frac{x}{\ln x}$, $y' = \frac{(\ln x)(1) - x\left(\frac{1}{x}\right)}{\ln^2 x} = \frac{\ln x - 1}{\ln^2 x}$. When $x = 2$ the slope is $y'(2) = \frac{(\ln 2) - 1}{\ln^2 2}$.

49. $c = 25 \ln(q+1) + 12$. $\frac{dc}{dq} = \frac{25}{q+1}$, so $\frac{dc}{dq}\Big|_{q=6} = \frac{25}{7}$.

51. $A = 6 \ln\left(\frac{T}{a-T} - a\right)$. Rate of change of A with respect to T:

$$\frac{dA}{dT} = 6 \cdot \frac{1}{\frac{T}{a-T} - a}\left[\frac{(a-T)(1) - T(-1)}{(a-T)^2}\right] = 6 \cdot \frac{1}{\frac{T - a(a-T)}{a-T}}\left[\frac{a}{(a-T)^2}\right]$$

$$= 6 \cdot \frac{a-T}{T-a^2+aT} \cdot \frac{a}{(a-T)^2} = \frac{6a}{(T-a^2+aT)(a-T)}$$

53. $\frac{d}{dx}(\log_b u) = \frac{d}{dx}\left(\frac{\ln u}{\ln b}\right) = \frac{1}{\ln b} \cdot \frac{d}{dx}(\ln u) = \frac{1}{\ln b}\left(\frac{1}{u} \cdot \frac{du}{dx}\right)$

$$= (\log_b e)\left(\frac{1}{u} \cdot \frac{du}{dx}\right) = \frac{1}{u}(\log_b e)\frac{du}{dx}.$$

EXERCISE 13.2

1. $y' = 7 \cdot \frac{d}{dx}(e^x) = 7e^x$

3. $y' = e^{x^2+1}(2x) = 2xe^{x^2+1}$

5. $y' = e^{3-5x} \cdot \frac{d}{dx}(e^{3-5x}) = e^{3-5x}(-5) = -5e^{3-5x}$

7. $f'(r) = e^{3r^2+4r+4}(6r + 4) = 2(3r + 2)e^{3r^2+4r+4}$

9. By the product rule, $y' = x(e^x) + e^x(1) = e^x(x + 1)$

11. $y' = x^2[e^{-x^2}(-2x)] + e^{-x^2}[2x] = 2xe^{-x^2}(1 - x^2)$

13. $y = \frac{1}{2}(e^x + e^{-x})$. $y' = \frac{1}{2}[e^x + e^{-x}(-1)] = \frac{e^x - e^{-x}}{2}$

15. $\frac{d}{dx}\left(4^{3x^2}\right) = \frac{d}{dx}[e^{(\ln 4)3x^2}] = e^{(\ln 4)3x^2}[(\ln 4)(6x)]$

$$= (6x)4^{3x^2} \ln 4$$

17. $f'(w) = \frac{w^2[e^{2w}(2)] - e^{2w}[2w]}{w^4} = \frac{2e^{2w}(w - 1)}{w^3}$

19. $y' = e^{1+\sqrt{x}}\left(\frac{1}{2}x^{-1/2}\right) = \frac{e^{1+\sqrt{x}}}{2\sqrt{x}}$

21. $y = x^3 - 3^x = x^3 - e^{(\ln 3)x}$.

$y' = 3x^2 - e^{(\ln 3)x}(\ln 3) = 3x^2 - 3^x \ln 3$

23. $\frac{dy}{dx} = \frac{(e^x + 1)[e^x] - (e^x - 1)[e^x]}{(e^x + 1)^2} = \frac{2e^x}{(e^x + 1)^2}$

25. $y = e^{\ln x} = x$ (by the property that $e^{\ln a} = a$). Thus $y' = 1$.

27. $y' = e^{x \ln x}\left[x\cdot\frac{1}{x} + (\ln x)(1)\right] = (1 + \ln x)e^{x \ln x}$

29. $f(x) = ee^x e^{x^2} = e^{1+x+x^2}$.

$f'(x) = e^{1+x+x^2}(1 + 2x) = (1 + 2x)e^{1+x+x^2}$.

$f'(-1) = [1 + 2(-1)]e^{1+(-1)+(-1)^2} = -e$

31. $y = e^x$, $y' = e^x$. When $x = 2$, then $y = e^2$ and $y' = e^2$. Thus an equation of the tangent is $y - e^2 = e^2(x - 2)$, or $y = e^2 x - e^2$.

33. $\frac{dp}{dq} = 15e^{-0.001q}(-0.001) = -0.015e^{-0.001q}$.

$\left.\frac{dp}{dq}\right|_{q=500} = -0.015e^{-0.5}$

35. $\bar{c} = \frac{7000e^{q/700}}{q}$), so $c = \bar{c}q = 7000e^{q/700}$. The marginal cost function is $\frac{dc}{dq} = 7000e^{q/700}\left(\frac{1}{700}\right) = 10e^{q/700}$. Thus $\left.\frac{dc}{dq}\right|_{q=350} = 10e^{0.5}$ and $\left.\frac{dc}{dq}\right|_{q=700} = 10e$.

37. $w = e^{x^3-4x} + x \ln(x-1)$ and $x = \frac{t+1}{t-1}$. By the chain rule,

$$\frac{dw}{dt} = \frac{dw}{dx} \cdot \frac{dx}{dt} =$$

$$\left[e^{x^3-4x}(3x^2-4)+x\left(\frac{1}{x-1}\right)+[\ln(x-1)](1) \right] \left[\frac{(t-1)(1)-(t+1)(1)}{(t-1)^2} \right] =$$

$$\left[(3x^2-4)e^{x^3-4x} + \frac{x}{x-1} + \ln(x-1) \right] \left[\frac{-2}{(t-1)^2} \right].$$

When $t = 3$, then $x = \frac{3+1}{3-1} = \frac{4}{2} = 2$ and

$$\frac{dw}{dt} = [8 + 2 + 0]\left[-\frac{1}{2} \right] = -5.$$

39. $\frac{d}{dx}(c^x - x^c) = \frac{d}{dx}\left[(e^{\ln c})^x - x^c \right] = \frac{d}{dx}\left[e^{(\ln c)x} - x^c \right]$

$$= (\ln c)e^{(\ln c)x} - cx^{c-1} = (\ln c)c^x - cx^{c-1}.$$

When $x = 1$, this derivative is 0. Thus $(\ln c)c - c = 0$, or $c[\ln(c) - 1] = 0$. Since $c > 0$, we must have $\ln(c) - 1 = 0$, $\ln c = 1$, or $c = e$.

41. $q = 500(1 - e^{-0.2t})$, $\frac{dq}{dt} = 500(-e^{-0.2t})(-0.2) = 100e^{-0.2t}$.

Thus $\left. \frac{dq}{dt} \right|_{t=10} = 100e^{-2}$.

43. $P = 20,000e^{0.03t}$.

$$\frac{dP}{dt} = 20,000e^{0.03t}(0.03) = P(0.03) = 0.03P = kP,$$

where $k = 0.03$.

45. Since $S = Pe^{rt}$, then $dS/dt = Pe^{rt}r = rPe^{rt}$. Thus

$$\frac{dS/dt}{S} = \frac{rPe^{rt}}{Pe^{rt}} = r.$$

47. $N = 10^A 10^{-bM} = 10^{A-bM} = e^{(\ln 10)(A-bM)}$.

$$\frac{dN}{dM} = e^{(\ln 10)(A-bM)}(\ln 10)(-b)$$

$$= 10^{A-bM}(\ln 10)(-b)$$

$$= -b(10^{A-bM}) \ln 10$$

49. $c(t) = c_0e^{-(r/V)t}$.

$\frac{dc}{dt} = c_0e^{-(r/V)t}\left(-\frac{r}{V}\right) = [c(t)]\left(-\frac{r}{V}\right) = -\left(\frac{r}{V}\right)c(t)$.

51. $f(t) = 1 - e^{-0.008t}$. $f'(t) = 0.008e^{-0.008t}$.

$f'(100) = 0.008e^{-0.8} \approx (0.008)(0.44933) \approx 0.0036$.

EXERCISE 13.3

1. $2x + 8yy' = 0$
$x + 4yy' = 0$,
$4yy' = -x$,
$y' = -\frac{x}{4y}$

3. $12y^3y' - 5 = 0$, $y' = 5/(12y^3)$

5. $\frac{1}{2}x^{-1/2} + \frac{1}{2}y^{-1/2}y' = 0$,
$x^{-1/2} + y^{-1/2}y' = 0$,
$y^{-1/2}y' = -x^{-1/2}$,
$y' = -\frac{x^{-1/2}}{y^{-1/2}} = -\frac{y^{1/2}}{x^{1/2}} = -\frac{\sqrt{y}}{\sqrt{x}} = -\sqrt{\frac{y}{x}}$.

7. $(3/4)x^{-1/4} + (3/4)y^{-1/4}y' = 0$, $y' = -y^{1/4}/x^{1/4} = -(y/x)^{1/4}$

9. By the product rule, $xy' + y(1) = 0$, $xy' = -y$, $y' = -\frac{y}{x}$

11. $xy' + y(1) - y' - 4 = 0$, $y'(x - 1) = 4 - y$, $y' = \frac{4 - y}{x - 1}$

13. $3x^2 + 3y^2y' - 12(xy' + y) = 0$, $3y^2y' - 12xy' = 12y - 3x^2$,
$y'(3y^2 - 12x) = 12y - 3x^2$, $y'(y^2 - 4x) = 4y - x^2$,
$y' = \frac{4y - x^2}{y^2 - 4x}$

15. $1 = \frac{1}{2}y^{-1/2}y' + \frac{1}{3}y^{-2/3}y'$, $\quad 6 = y'\left(\frac{3}{y^{1/2}} + \frac{2}{y^{2/3}}\right)$,

$6 = y'\left(\frac{3y^{1/6} + 2}{y^{2/3}}\right)$, $\quad y' = \frac{6y^{2/3}}{3y^{1/6} + 2}$

17. $3x^2(3y^2y') + y^3(6x) - 1 + y' = 0$,

$$9x^2y^2y' + y' = 1 - 6xy^3,$$
$$y'(9x^2y^2 + 1) = 1 - 6xy^3,$$
$$y' = \frac{1 - 6xy^3}{1 + 9x^2y^2}$$

19. $y\left(\frac{1}{x}\right) + (\ln x)y' = x(e^y y') + e^y(1)$,

$[\ln(x) - xe^y]y' = e^y - \frac{y}{x}$,

$[\ln(x) - xe^y]y' = \frac{xe^y - y}{x}$, $\quad y' = \frac{xe^y - y}{x[\ln(x) - xe^y]}$

21. $[x(e^y y') + e^y(1)] + y' = 0$, $\quad xe^y y' + e^y + y' = 0$,

$(xe^y + 1)y' = -e^y$, $\quad y' = -\frac{e^y}{xe^y + 1}$

23. $2(1 + e^{3x})(3e^{3x}) = \frac{1}{x + y}(1 + y')$,

$6e^{3x}(1 + e^{3x})(x + y) = 1 + y'$,

$y' = 6e^{3x}(1 + e^{3x})(x + y) - 1$

25. $1 + [xy' + y(1)] + 2yy' = 0$, $\quad xy' + 2yy' = -1 - y$,

$(x + 2y)y' = -(1 + y)$, $\quad y' = -\frac{1 + y}{x + 2y}$.

At the point $(1,2)$, $y' = -\frac{1 + 2}{1 + 4} = -\frac{3}{5}$

27. $8x + 18yy' = 0$, $\quad y' = -\frac{8x}{18y} = -\frac{4x}{9y}$.

Thus at $(0,1/3)$, $y' = 0$; at (x_0, y_0), $y' = -\frac{4x_0}{9y_0}$.

29. $3x^2 + 2yy' = 0$, $y' = -\frac{3x^2}{2y}$. At $(-1,2)$, $y' = -\frac{3}{4}$. The tangent is given by $y - 2 = -\frac{3}{4}(x + 1)$, or $y = -\frac{3}{4}x + \frac{5}{4}$.

31. $p = 100 - q^2$, $\frac{d}{dp}(p) = \frac{d}{dp}(100 - q^2)$, $1 = -2q \cdot \frac{dq}{dp}$, $\frac{dq}{dp} = -\frac{1}{2q}$

33. $p = \frac{20}{(q + 5)^2}$, $\frac{d}{dp}(p) = \frac{d}{dp}\left[\frac{20}{(q + 5)^2}\right]$,

$\frac{d}{dp}(p) = \frac{d}{dp}[20(q + 5)^{-2}]$,

$1 = -\frac{40}{(q + 5)^3} \cdot \frac{dq}{dp}$,

$\frac{dq}{dp} = -\frac{(q + 5)^3}{40}$

35. $\ln \frac{I}{I_0} = -\lambda t$, $\ln I - \ln I_0 = -\lambda t$, $\frac{1}{I}\frac{dI}{dt} = -\lambda$, $\frac{dI}{dt} = -\lambda I$

37. $1.5M = \log\left(\frac{E}{2.5 \times 10^{11}}\right)$, $1.5M = \log E - \log(2.5 \times 10^{11})$,

$\frac{d}{dM}(1.5M) = \frac{d}{dM}[\log E - \log(2.5 \times 10^{11})]$,

$\frac{d}{dM}(1.5M) = \frac{d}{dM}\left[\frac{\ln E}{\ln 10} - \log(2.5 \times 10^{11})\right]$,

$1.5 = \frac{1}{\ln 10}\left(\frac{1}{E} \cdot \frac{dE}{dM}\right)$, $\frac{dE}{dM} = 1.5E \ln 10$

39. $TV^{0.4} = 1500$. Differentiating implicitly with respect to T: $T\left(0.4V^{-0.6}\frac{dV}{dT}\right) + V^{0.4}(1) = 0$, $0.4TV^{-0.6}\frac{dV}{dT} = -V^{0.4}$,

$\frac{dV}{dT} = -\frac{V^{0.4}}{0.4TV^{-0.6}} = -\frac{V}{0.4T} = -2.5\frac{V}{T}$.

41. $S^2 + \frac{1}{4}I^2 = SI + I$. Differentiating implicitly with respect to I: $2S\frac{dS}{dI} + \frac{1}{2}I = \left[S(1) + I\frac{dS}{dI}\right] + 1$,

$2S\frac{dS}{dI} - I\frac{dS}{dI} = S + 1 - \frac{I}{2}$, $(2S - I)\frac{dS}{dI} = \frac{2S + 2 - I}{2}$,

$\frac{dS}{dI} = \frac{2S + 2 - I}{2(2S - I)}$.

Marginal propensity to consume $= \frac{dC}{dI} = 1 - \frac{dS}{dI}$. Thus

$$\frac{dC}{dI} = 1 - \frac{2S + 2 - I}{2(2S - I)}.$$

When I = 16 and S = 12,

$$\frac{dC}{dI} = 1 - \frac{24 + 2 - 16}{2(24 - 16)} = 1 - \frac{10}{16} = \frac{6}{16} = \frac{3}{8}.$$

EXERCISE 13.4

1. $y = (x + 1)^2(x - 1)(x^2 + 3)$. Taking natural logarithms of both sides gives

$$\ln y = \ln\left[(x + 1)^2(x - 1)(x^2 + 3)\right].$$

Using properties of logarithms on the right side gives

$$\ln y = 2 \ln(x + 1) + \ln(x - 1) + \ln(x^2 + 3).$$

Differentiating both sides with respect to x,

$$\frac{y'}{y} = \frac{2}{x + 1} + \frac{1}{x - 1} + \frac{2x}{x^2 + 3}.$$

Solving for y',

$$y' = y\left[\frac{2}{x + 1} + \frac{1}{x - 1} + \frac{2x}{x^2 + 3}\right].$$

Expressing y in terms of x,

$$y' = (x + 1)^2(x - 1)(x^2 + 3)\left[\frac{2}{x + 1} + \frac{1}{x - 1} + \frac{2x}{x^2 + 3}\right].$$

3. $\ln y = \ln\left[(3x^3-1)^2(2x+5)^3\right]$

$$= 2 \ln(3x^3-1) + 3 \ln(2x+5).$$

$$\frac{y'}{y} = 2 \cdot \frac{9x^2}{3x^3-1} + 3 \cdot \frac{2}{2x+5}, \quad y' = y\left[\frac{18x^2}{3x^3-1} + \frac{6}{2x+5}\right],$$

$$y' = (3x^3-1)^2(2x+5)^3\left[\frac{18x^2}{3x^3-1} + \frac{6}{2x+5}\right]$$

5. $y = \sqrt{x + 1}\sqrt{x^2 - 2}\sqrt{x + 4}$.

$$\ln y = \ln\left(\sqrt{x + 1}\sqrt{x^2 - 2}\sqrt{x + 4}\right),$$

$$\ln y = \frac{1}{2} \ln(x + 1) + \frac{1}{2} \ln(x^2 - 2) + \frac{1}{2} \ln(x + 4).$$

$$\frac{y'}{y} = \frac{1}{2}\left[\frac{1}{x + 1} + \frac{2x}{x^2 - 2} + \frac{1}{x + 4}\right].$$

Thus,

$$y' = \frac{y}{2}\left[\frac{1}{x+1} + \frac{2x}{x^2-2} + \frac{1}{x+4}\right]$$

$$= \frac{\sqrt{x+1}\sqrt{x^2-2}\sqrt{x+4}}{2}\left[\frac{1}{x+1} + \frac{2x}{x^2-2} + \frac{1}{x+4}\right]$$

7. $\ln y = \ln \sqrt{\frac{1-x^2}{1-2x}} = \frac{1}{2}\ln(1-x^2) - \ln(1-2x)$.

$\frac{y'}{y} = \frac{1}{2}\cdot\frac{-2x}{1-x^2} - \frac{-2}{1-2x}$, $\quad y' = y\left[-\frac{x}{1-x^2} + \frac{2}{1-2x}\right]$,

$y' = \sqrt{\frac{1-x^2}{1-2x}}\left[\frac{x}{x^2-1} + \frac{2}{1-2x}\right]$

9. $y = \dfrac{(2x^2+2)^2}{(x+1)^2(3x+2)}$

$\ln y = \ln\left[\dfrac{(2x^2+2)^2}{(x+1)^2(3x+2)}\right]$

$\quad = 2\ln(2x^2+2) - 2\ln(x+1) - \ln(3x+2)$.

$\frac{y'}{y} = 2\cdot\frac{4x}{2x^2+2} - 2\cdot\frac{1}{x+1} - \frac{3}{3x+2}$

$y' = y\left[\frac{8x}{2x^2+2} - \frac{2}{x+1} - \frac{3}{3x+2}\right]$

$\quad = \dfrac{(2x^2+2)^2}{(x+1)^2(3x+2)}\left[\frac{4x}{x^2+1} - \frac{2}{x+1} - \frac{3}{3x+2}\right]$

11. $\ln y = \ln\sqrt{\dfrac{(x-1)(x+1)}{3x-4}} = \frac{1}{2}\left[\ln(x-1) + \ln(x+1) - \ln(3x-4)\right]$.

$\frac{y'}{y} = \frac{1}{2}\left[\frac{1}{x-1} + \frac{1}{x+1} - \frac{3}{3x-4}\right]$, $\quad y' = \frac{y}{2}\left[\frac{1}{x-1} + \frac{1}{x+1} - \frac{3}{3x-4}\right]$,

$y' = \frac{1}{2}\sqrt{\dfrac{(x-1)(x+1)}{3x-4}}\left[\frac{1}{x-1} + \frac{1}{x+1} - \frac{3}{3x-4}\right]$

13. $y = x^{2x+1}$. Thus $\ln y = \ln x^{2x+1} = (2x+1)\ln x$.

$\frac{y'}{y} = (2x+1)\frac{1}{x} + (\ln x)(2)$, $\quad y' = y\left[\frac{2x+1}{x} + 2\ln x\right]$,

$y' = x^{2x+1}\left[\frac{2x+1}{x} + 2\ln x\right]$

15. $y = x^{1/x}$. Thus $\ln y = \frac{1}{x} \ln x = \frac{\ln x}{x}$.

$\frac{y'}{y} = \frac{x(1/x) - (\ln x)(1)}{x^2}$, $y' = y\left[\frac{1 - \ln x}{x^2}\right]$.

$y' = \frac{x^{1/x}(1 - \ln x)}{x^2}$

17. $y = (3x + 1)^{2x}$. Thus $\ln y = \ln[(3x + 1)^{2x}] = 2x \ln(3x+1)$.

$\frac{y'}{y} = 2\left\{x\left(\frac{3}{3x + 1}\right) + [\ln(3x + 1)](1)\right\}$

$y' = 2y\left[\frac{3x}{3x + 1} + \ln(3x + 1)\right]$

$\quad = 2(3x + 1)^{2x}\left[\frac{3x}{3x + 1} + \ln(3x + 1)\right]$

19. $y = e^x x^{3x}$.

Thus $\ln y = \ln(e^x x^{3x}) = \ln e^x + \ln x^{3x} = x + 3x \ln x$.

$\frac{y'}{y} = 1 + 3\left[x\left(\frac{1}{x}\right) + (\ln x)(1)\right]$, $y' = y(4 + 3 \ln x)$,

$y' = e^x x^{3x}(4 + 3 \ln x)$

21. $y = (4x-3)^{2x+1}$. $\ln y = \ln(4x-3)^{2x+1} = (2x+1)\ln(4x-3)$.

$\frac{y'}{y} = (2x+1)\left[\frac{4}{4x-3}\right] + [\ln(4x-3)](2)$,

$y' = y\left[\frac{4(2x+1)}{4x-3} + 2 \ln(4x-3)\right]$.

When $x = 1$, then $\frac{dy}{dx} = 1\left[\frac{12}{1} + 2 \ln(1)\right] = 12$.

23. $y = (x + 1)(x + 2)^2(x + 3)^2$.

$\ln y = \ln(x + 1) + 2 \ln(x + 2) + 2 \ln(x + 3)$

$\frac{y'}{y} = \frac{1}{x + 1} + \frac{2}{x + 2} + \frac{2}{x + 3}$

$y' = y\left[\frac{1}{x + 1} + \frac{2}{x + 2} + \frac{2}{x + 3}\right]$.

When $x = 0$, then $y = 36$ and $y' = 96$. Thus an equation of
the tangent line is $y - 36 = 96(x - 0)$, or $y = 96x + 36$.

25. $y = 3e^x(x^2-x+1)^x$. $\ln y = \ln 3 + \ln e^x + \ln(x^2-x+1)^x$

$\qquad\qquad\qquad\qquad = \ln(3) + x + x \ln(x^2-x+1)$.

$\frac{y'}{y} = 1 + \left[x\left(\frac{2x-1}{x^2-x+1}\right) + [\ln(x^2-x+1)](1)\right]$,

$y' = y\left[1 + \frac{x(2x-1)}{x^2-x+1} + \ln(x^2-x+1)\right]$. When $x = 1$, then $y = 3e$
and $y' = 3e[1 + 1 + \ln(1)] = 6e$. Thus an equation of the
tangent line is $y - 3e = 6e(x - 1)$, or $y = 6ex - 3e$.

27. $y = (3x)^{-2x}$. $\ln y = -2x \ln(3x)$.

$\frac{y'}{y} = -2\left\{x\left[\frac{1}{3x}(3)\right] + [\ln(3x)](1)\right\} = -2[1 + \ln(3x)]$.

$\frac{y'}{y}\cdot 100$ gives the percentage rate of change. Thus
$-2[1 + \ln(3x)](100) = 60$, $1 + \ln(3x) = -0.3$,

$\ln(3x) = -1.3$, $3x = e^{-1.3}$, $x = \frac{1}{3e^{1.3}}$.

EXERCISE 13.5

1. $y' = 12x^2 - 24x + 6$, $y'' = 24x - 24$, $y''' = 24$

3. $\frac{dy}{dx} = -1$, $\frac{d^2y}{dx^2} = 0$

5. $y' = 3x^2 + e^x$, $y'' = 6x + e^x$, $y''' = 6 + e^x$, $y^{(4)} = e^x$

7. $f(x) = x^2 \ln x$
     $f'(x) = x^2\left(\frac{1}{x}\right) + (\ln x)(2x) = x(1 + 2 \ln x)$
     $f''(x) = x\left(\frac{2}{x}\right) + (1 + 2 \ln x)(1) = 3 + 2 \ln x$

9. $f(p) = \frac{1}{6p^3} = \frac{1}{6}p^{-3}$, $f'(p) = -\frac{1}{2}p^{-4}$, $f''(p) = 2p^{-5}$,
     $f'''(p) = -10p^{-6} = -\frac{10}{p^6}$

11. $f(r) = \sqrt{1 - r} = (1 - r)^{1/2}$, $f'(r) = -\frac{1}{2}(1 - r)^{-1/2}$,
     $f''(r) = -\frac{1}{4}(1 - r)^{-3/2} = -\frac{1}{4(1 - r)^{3/2}}$

13. $y = \frac{1}{5x - 6} = (5x - 6)^{-1}$

$\frac{dy}{dx} = (-1)(5x - 6)^{-2}(5) = -5(5x - 6)^{-2}$

$\frac{d^2y}{dx^2} = -5(-2)(5x - 6)^{-3}(5) = 50(5x - 6)^{-3} = \frac{50}{(5x - 6)^3}$

15. $y = \frac{x + 1}{x - 1}$

$y' = \frac{(x - 1)(1) - (x + 1)(1)}{(x - 1)^2} = -\frac{2}{(x - 1)^2} = -2(x - 1)^{-2}$

$y'' = 4(x - 1)^{-3} = \frac{4}{(x - 1)^3}$

17. $y = \ln[x(x + 1)] = \ln(x) + \ln(x + 1)$

$y' = \frac{1}{x} + \frac{1}{x + 1} = x^{-1} + (x + 1)^{-1}$

$y'' = -x^{-2} + (-1)(x + 1)^{-2} = -\left[\frac{1}{x^2} + \frac{1}{(x + 1)^2}\right]$

19. $f(z) = z^2 e^z$

$f'(z) = z^2(e^z) + e^z(2z) = (ze^z)(z + 2).$

$f''(z) = (ze^z)(1) + (z + 2)[ze^z + e^z(1)] = e^z(z^2 + 4z + 2).$

21. $y = e^{2x}$, $\frac{dy}{dx} = 2e^{2x}$, $\frac{d^2y}{dx^2} = 4e^{2x}$, $\frac{d^3y}{dx^3} = 8e^{2x}$,

$\frac{d^4y}{dx^4} = 16e^{2x}$, $\frac{d^5y}{dx^5} = 32e^{2x}$. $\left.\frac{d^5y}{dx^5}\right|_{x=0} = 32e^0 = 32$

23. $x^2 + 4y^2 - 16 = 0.$

$2x + 8yy' = 0$, $8yy' = -2x$, $y' = -\frac{x}{4y}$.

$y'' = -\frac{4y(1) - x(4y')}{16y^2} = -\frac{4y - 4x\left(-\frac{x}{4y}\right)}{16y^2} = -\frac{4y^2 + x^2}{16y^3}$

$= -\frac{16}{16y^3} = -\frac{1}{y^3}.$

25. $y^2 = 4x.$

$2yy' = 4$, $y' = 2/y = 2y^{-1}.$

$y'' = -2y^{-2}y' = -2y^{-2}(2y^{-1}) = -4/y^3.$

27. $\sqrt{x} + 4\sqrt{y} = 4$, $x^{1/2} + 4y^{1/2} = 4$.

$\frac{1}{2}x^{-1/2} + 2y^{-1/2}y' = 0$, $2y^{-1/2}y' = -\frac{1}{2}x^{-1/2}$,

$y' = -\frac{1}{2} \cdot \frac{x^{-1/2}}{2y^{-1/2}} = -\frac{1}{4} \cdot \frac{y^{1/2}}{x^{1/2}}$.

$y'' = -\frac{1}{4}\left[\dfrac{x^{1/2}\left(\frac{1}{2}y^{-1/2}y'\right) - y^{1/2}\left(\frac{1}{2}x^{-1/2}\right)}{x}\right]$

$= -\frac{1}{8}\left[\dfrac{x^{1/2}\left(-\frac{y^{1/2}}{4x^{1/2}}\right) - \frac{y^{1/2}}{x^{1/2}}}{x}\right] = -\frac{1}{8}\left[\dfrac{-\frac{1}{4} - \frac{y^{1/2}}{x^{1/2}}}{x}\right]$

$= \frac{1}{8}\left[\dfrac{\frac{1}{4} + \frac{y^{1/2}}{x^{1/2}}}{x}\right] = \frac{1}{8}\left[\dfrac{x^{1/2} + 4y^{1/2}}{4x^{3/2}}\right] = \frac{1}{8}\left[\dfrac{4}{4x^{3/2}}\right] = \dfrac{1}{8x^{3/2}}$

29. $xy + y - x = 4$. $xy' + y(1) + y' - 1 = 0$,

$xy' + y' = 1 - y$, $(x + 1)y' = 1 - y$, $y' = \dfrac{1 - y}{1 + x}$.

$y'' = \dfrac{(1 + x)(-y') - (1 - y)(1)}{(1 + x)^2}$

$= \dfrac{(1 + x)[-(1 - y)/(1 + x)] - (1 - y)}{(1 + x)^2}$

$= \dfrac{-(1 - y) - (1 - y)}{(1 + x)^2} = \dfrac{-2(1 - y)}{(1 + x)^2} = \dfrac{2(y - 1)}{(1 + x)^2}$

31. $y = e^{x+y}$. $y' = e^{x+y}(1 + y')$, $y' - e^{x+y}y' = e^{x+y}$,

$y'(1 - e^{x+y}) = e^{x+y}$, $y' = \dfrac{e^{x+y}}{1 - e^{x+y}}$, $y' = \dfrac{y}{1 - y}$.

$y'' = \dfrac{(1 - y)y' - y(-y')}{(1 - y)^2} = \dfrac{y'}{(1 - y)^2} = \dfrac{\frac{y}{1 - y}}{(1 - y)^2} = \dfrac{y}{(1 - y)^3}$

33. $x^2 + 8y = y^2$. $2x + 8y' = 2yy'$, $x + 4y' = yy'$,

$x = yy' - 4y'$, $x = y'(y - 4)$, $y' = \dfrac{x}{y - 4}$.

$y'' = \dfrac{(y - 4)(1) - x(y')}{(y - 4)^2} = \dfrac{y - 4 - x\left(\frac{x}{y - 4}\right)}{(y - 4)^2} = \dfrac{(y - 4)^2 - x^2}{(y - 4)^3}$.

When $x = 3$ and $y = -1$, then $y'' = \dfrac{(-5)^2 - 3^2}{(-5)^3} = -\dfrac{16}{125}$.

35. $f(x) = (5x - 3)^4$, $f'(x) = 20(5x - 3)^3$, $f''(x) = 300(5x - 3)^2$

37. $\dfrac{dc}{dq} = 0.6q + 2$, $\dfrac{d^2c}{dq^2} = 0.6$, $\dfrac{d^2c}{dq^2}\bigg|_{q=100} = 0.6$

39. $f(x) = x^4 - 6x^2 + 5x - 6$, $f'(x) = 4x^3 - 12x + 5$,
$f''(x) = 12x^2 - 12 = 12(x + 1)(x - 1)$. Clearly $f''(x) = 0$
when $x = \pm 1$.

CHAPTER 13 - REVIEW PROBLEMS

1. $y = 2e^x + e^2 + e^{x^2}$, $y' = 2e^x + 0 + e^{x^2}(2x) = 2(e^x + xe^{x^2})$

3. $f'(r) = \dfrac{1}{r^2 + 5r}(2r + 5) = \dfrac{2r + 5}{r(r + 5)}$

5. $y = e^{x^2+4x+5}$, $y' = e^{x^2+4x+5}(2x + 4) = 2(x + 2)e^{x^2+4x+5}$

7. $y' = e^x(2x) + (x^2 + 2)e^x = e^x(x^2 + 2x + 2)$

9. $y = \sqrt{(x - 6)(x + 5)(9 - x)}$
$\ln y = \ln\sqrt{(x - 6)(x + 5)(9 - x)}$
$\quad = \dfrac{1}{2}\Big[\ln(x - 6) + \ln(x + 5) + \ln(9 - x)\Big]$.
$\dfrac{y'}{y} = \dfrac{1}{2}\Big[\dfrac{1}{x - 6} + \dfrac{1}{x + 5} + \dfrac{-1}{9 - x}\Big]$,
$y' = \dfrac{y}{2}\Big[\dfrac{1}{x - 6} + \dfrac{1}{x + 5} - \dfrac{1}{9 - x}\Big]$
$\quad = \dfrac{\sqrt{(x - 6)(x + 5)(9 - x)}}{2}\Big[\dfrac{1}{x - 6} + \dfrac{1}{x + 5} + \dfrac{1}{x - 9}\Big]$

11. $y' = \dfrac{e^x\left(\frac{1}{x}\right) - (\ln x)(e^x)}{e^{2x}} = \dfrac{e^x - xe^x \ln x}{e^{2x}} = \dfrac{1 - x \ln x}{xe^x}$

13. $f(q) = \ln\Big[(q + 1)^2(q + 2)^3\Big] = 2 \ln(q + 1) + 3 \ln(q + 2)$.
$f'(q) = \dfrac{2}{q + 1} + \dfrac{3}{q + 2}$

15. $y = e^{(2-7x)(\ln 10)}$.
$y' = e^{(2-7x)(\ln 10)}(-7 \ln 10) = - 7(\ln 10)10^{2-7x}$

17. $y = \dfrac{4e^{3x}}{xe^{x-1}} = \dfrac{4e^{2x+1}}{x}$.

$y' = 4 \cdot \dfrac{x[e^{2x+1}(2)] - e^{2x+1}[1]}{x^2} = \dfrac{4e^{2x+1}(2x - 1)}{x^2}$.

19. $y = \log_2(8x + 5)^2 = 2 \log_2(8x + 5) = 2 \cdot \dfrac{\ln(8x + 5)}{\ln 2}$.

$y' = 2 \cdot \dfrac{1}{\ln 2} \cdot \dfrac{8}{8x + 5} = \dfrac{16}{(8x + 5) \ln 2}$

21. $f(1) = \ln(1 + 1 + 1^2 + 1^3)$.

$f'(1) = \dfrac{1}{1 + 1 + 1^2 + 1^3}[1 + 21 + 31^2] = \dfrac{1 + 41 + 31^2}{1 + 1 + 1^2 + 1^3}$

23. $y = (x + 1)^{x+1}$,
$\ln y = (x + 1) \ln(x + 1)$.

$\dfrac{y'}{y} = (x + 1)\dfrac{1}{x + 1} + \ln(x + 1)[1] = 1 + \ln(x + 1)$,

$y' = y[1 + \ln(x + 1)] = (x + 1)^{x+1}[1 + \ln(x + 1)]$

25. $f(t) = \ln(t^2\sqrt{1 - t}) = 2 \ln t + (1/2) \ln(1 - t)$.

$f'(t) = 2\left(\dfrac{1}{t}\right) + \dfrac{1}{2}\left(\dfrac{1}{1 - t}\right)(-1) = \dfrac{2}{t} + \dfrac{1}{2(t - 1)} = \dfrac{5t - 4}{2t(t - 1)}$

27. $y = \dfrac{(x^2 + 2)^{3/2}(x^2 + 9)^{4/9}}{(x^3 + 6x)^{4/11}}$.

$\ln y = \dfrac{3}{2} \ln(x^2 + 2) + \dfrac{4}{9} \ln(x^2 + 9) - \dfrac{4}{11} \ln(x^3 + 6x)$.

$\dfrac{y'}{y} = \dfrac{3}{2}\left(\dfrac{1}{x^2 + 2}\right)(2x) + \dfrac{4}{9}\left(\dfrac{1}{x^2 + 9}\right)(2x) - \dfrac{4}{11}\left(\dfrac{1}{x^3 + 6x}\right)(3x^2 + 6)$,

$y' = y\left[\dfrac{3x}{x^2 + 2} + \dfrac{8x}{9(x^2 + 9)} - \dfrac{12(x^2 + 2)}{11x(x^2 + 6)}\right]$

$= \dfrac{(x^2+2)^{3/2}(x^2+9)^{4/9}}{(x^3+6x)^{4/11}}\left[\dfrac{3x}{x^2 + 2} + \dfrac{8x}{9(x^2 + 9)} - \dfrac{12(x^2 + 2)}{11x(x^2 + 6)}\right]$

29. $y = (x^x)^x = x^{x^2}$. $\ln y = \ln x^{x^2} = x^2 \ln x$.

$\dfrac{y'}{y} = x^2\left(\dfrac{1}{x}\right) + (\ln x)(2x) = x + 2x \ln x$,

$y' = y(x + 2x \ln x) = (x^x)^x(x + 2x \ln x)$

31. $y = (x + 1) \ln x^2 = 2(x + 1) \ln x$.

$y' = 2\left[(x + 1)\left(\frac{1}{x}\right) + (\ln x)(1)\right] = 2\left[\frac{x + 1}{x} + \ln x\right]$.

When $x = 1$, then $y' = 2\left[\frac{2}{1} + \ln 1\right] = 4$.

33. $y = e^{e + x \ln(1/x)} = e^{e - x \ln x}$.

$y' = e^{e - x \ln x}\left(-\left[x\left(\frac{1}{x}\right) + (\ln x)(1)\right]\right)$

$= -(1 + \ln x)e^{e - x \ln x}$.

When $x = e$, then $y' = -(1 + \ln e)e^{e - e \ln e} = -(2)e^0 = -2$.

35. $y = e^x$, $y' = e^x$.

If $x = \ln 2$, then $y = e^{\ln 2} = 2$ and $y' = e^{\ln 2} = 2$.

An equation of the tangent line is $y - 2 = 2(x - \ln 2)$,

$y = 2x + 2 - 2 \ln 2$, $y = 2x + 2(1 - \ln 2)$. Alternatively,

since $2 \ln 2 = \ln 2^2 = \ln 4$, the tangent line can be

written as $y = 2x + 2 - \ln 4$.

37. $y = x(2^{2-x^2})$. To find y' we shall use logarithmic
differentiation.

$\ln y = \ln[x(2^{2-x^2})] = \ln x + (2 - x^2) \ln 2$.

$\frac{y'}{y} = \frac{1}{x} + (-2x) \ln 2$, $y' = y\left[\frac{1}{x} - 2(\ln 2)x\right]$.

When $x = 1$, then $y = 2$ and $y' = 2(1 - 2 \ln 2)$.

Equation of tangent line is $y - 2 = 2(1 - 2 \ln 2)(x - 1)$.

The y-intercept of the tangent line corresponds to the

point where $x = 0$:

$y - 2 = 2(1 - 2 \ln 2)(-1) = -2 + 4 \ln 2$.

Thus $y = 4 \ln 2$ and the y-intercept is $(0, 4 \ln 2)$.

39. $y = e^{x^2-4}$. $y' = e^{x^2-4}[2x] = 2xe^{x^2-4}$.

$y'' = 2(xe^{x^2-4}[2x] + e^{x^2-4}\cdot 1) = 2e^{x^2-4}(2x^2 + 1)$.

At $(2,1)$, $y'' = 2e^0(9) = 18$.

41. $y = \ln(2x)$. $y' = \frac{1}{2x}(2) = x^{-1}$. $y'' = -1\cdot x^{-2} = -x^{-2}$.

$y''' = -(-2)x^{-3} = \frac{2}{x^3}$. At $(1, \ln 2)$, $y''' = \frac{2}{1^3} = 2$.

43. $2xy + y^2 = 6$, $2(xy' + y) + 2yy' = 0$, $2xy' + 2yy' = -2y$,
   $(x + y)y' = -y$, $y' = -\dfrac{y}{x + y}$

45. $\ln(xy^2) = xy$, $\ln x + 2 \ln y = xy$.
   $\dfrac{1}{x} + \dfrac{2}{y}y' = xy' + y$, $y + 2xy' = x^2yy' + xy^2$,
   $2xy' - x^2yy' = xy^2 - y$, $(2x - x^2y)y' = xy^2 - y$,
   $y' = \dfrac{xy^2 - y}{2x - x^2y}$

47. $x + xy + y = 5$, $1 + xy' + y(1) + y' = 0$, $(x + 1)y' = -1-y$
   $y' = -\dfrac{1 + y}{x + 1}$. $y'' = -\dfrac{(x + 1)y' - (1 + y)}{(x + 1)^2}$.
   At $(2,1)$, $y' = -\dfrac{1 + 1}{2 + 1} = -\dfrac{2}{3}$ and $y'' = -\dfrac{3(-2/3) - 2}{9} = \dfrac{4}{9}$

49. $e^y = (y + 1)e^x$. $e^y y' = (y + 1)e^x + e^x(y')$,
   $e^y y' - e^x y' = (y + 1)e^x$, $(e^y - e^x)y' = (y + 1)e^x$,
   $y' = \dfrac{(y + 1)e^x}{e^y - e^x} = \dfrac{(y + 1)\left(\dfrac{e^y}{y + 1}\right)}{e^y - \left(\dfrac{e^y}{y + 1}\right)} = \dfrac{e^y}{e^y - \dfrac{e^y}{y + 1}}$
   $= \dfrac{1}{1 - \dfrac{1}{y + 1}} = \dfrac{y + 1}{y}$.
   $y'' = \dfrac{y(y') - (y + 1)(y')}{y^2} = \dfrac{-y'}{y^2} = -\dfrac{\dfrac{y + 1}{y}}{y^2} = -\dfrac{y + 1}{y^3}$.

51. $f'(t) = -\left[0.8e^{-0.01t}(-0.01) + 0.2e^{-0.0002t}(-0.0002)\right]$
   $= 0.008e^{-0.01t} + 0.00004e^{-0.0002t}$

14

Curve Sketching

1. Decreasing on $(-\infty, -1)$ and $(3, \infty)$; increasing on $(-1, 3)$; rel. min. $(-1, -1)$; rel. max. $(3, 4)$.

3. Decreasing on $(-\infty, -2)$ and $(0, 2)$; increasing on $(-2, 0)$ and $(2, \infty)$; rel. min. $(-2, 1)$ and $(2, 1)$; no rel. max.

We denote the term *critical value* by CV.

5. $f'(x) = (x + 1)(x - 3)$. $f'(x) = 0$ when $x = -1, 3$.
   CV: $x = -1, 3$.

   Increasing on $(-\infty, -1)$ and $(3, \infty)$; decreasing on $(-1, 3)$; rel. max. when $x = -1$; rel. min. when $x = 3$.

7. $f'(x) = (x + 1)(x - 3)^2$. CV: $x = -1, 3$.

   Decreasing on $(-\infty, -1)$; increasing on $(-1, 3)$ and $(3, \infty)$; rel. min. when $x = -1$.

9. $y = x^2 + 2$. $y' = 2x = 0$ if $x = 0$.   CV: $x = 0$.

$$\diagdown \quad | \quad \diagup$$
$$0$$

Decreasing on $(-\infty, 0)$; increasing on $(0, \infty)$; rel. min. when $x = 0$.

11. $y = x - x^2 + 2$. $y' = 1 - 2x$.   CV: $x = 1/2$.

$$\diagup \quad | \quad \diagdown$$
$$1/2$$

Increasing on $(-\infty, 1/2)$; decreasing on $(1/2, \infty)$; rel. max. when $x = 1/2$.

13. $y = -(x^3/3) - 2x^2 + 5x - 2$.

$y' = -x^2 - 4x + 5 = -(x^2 + 4x - 5) = -(x + 5)(x - 1)$.
CV: $x = -5, 1$.

$$\diagdown \quad | \quad \diagup \quad | \quad \diagdown$$
$$-5 \qquad 1$$

Decreasing on $(-\infty, -5)$ and $(1, \infty)$; increasing on $(-5, 1)$; rel. min. when $x = -5$; rel. max. when $x = 1$.

15. $y = x^4 - 2x^2$.

$y' = 4x^3 - 4x = 4x(x^2 - 1) = 4x(x + 1)(x - 1)$.
CV: $x = 0, \pm 1$.

$$\diagdown \quad \diagup \quad | \quad \diagdown \quad | \quad \diagup$$
$$-1 \quad 0 \quad 1$$

Decreasing on $(-\infty, -1)$ and $(0, 1)$; increasing on $(-1, 0)$ and $(1, \infty)$; rel. max. when $x = 0$; rel. min. when $x = \pm 1$.

17. $y = x^3 - 6x^2 + 9x$.

$y' = 3x^2 - 12x + 9 = 3(x^2 - 4x + 3) = 3(x - 1)(x - 3)$.
CV: $x = 1, 3$.

$$\diagup \quad | \quad \diagdown \quad | \quad \diagup$$
$$1 \qquad 3$$

Increasing on $(-\infty, 1)$ and $(3, \infty)$; decreasing on $(1, 3)$; rel. max. when $x = 1$; rel. min. when $x = 3$.

19. $y = 2x^3 - \frac{11}{2}x^2 - 10x + 2$.

$y' = 6x^2 - 11x - 10 = (2x - 5)(3x + 2)$.   CV: $x = -\frac{2}{3}, \frac{5}{2}$.

$$\begin{array}{ccc} \nearrow & \searrow & \nearrow \\ \hline & \!\!+\!\!\quad\!\!+\!\! & \\ -2/3 & 5/2 & \end{array}$$

Increasing on $(-\infty, -\frac{2}{3})$ and $(\frac{5}{2}, \infty)$; decreasing on
$(-\frac{2}{3}, \frac{5}{2})$; rel. max. when $x = -\frac{2}{3}$; rel. min. when $x = \frac{5}{2}$.

21. $y = x^3 + 2x^2 - x - 1$. $y' = 3x^2 + 4x - 1$.   Setting $y' = 0$
gives $3x^2 + 4x - 1 = 0$. By the quadratic formula,

$x = \frac{-4 \pm 2\sqrt{7}}{6} = \frac{-2 \pm \sqrt{7}}{3}$.   CV: $x = \frac{-2 \pm \sqrt{7}}{3}$.

$$\begin{array}{ccc} \nearrow & \searrow & \nearrow \\ \hline & \!\!+\!\!\quad\!\!+\!\! & \\ \frac{-2-\sqrt{7}}{3} & \frac{-2+\sqrt{7}}{3} & \end{array}$$

Increasing on $\left(-\infty, \frac{-2 - \sqrt{7}}{3}\right)$ and $\left(\frac{-2 + \sqrt{7}}{3}, \infty\right)$; decreasing

on $\left(\frac{-2 - \sqrt{7}}{3}, \frac{-2 + \sqrt{7}}{3}\right)$; rel. max. when $x = \frac{-2 - \sqrt{7}}{3}$; rel.

min. when $x = \frac{-2 + \sqrt{7}}{3}$.

23. $y = 3x^5 - 5x^3$.

$y' = 15x^4 - 15x^2 = 15x^2(x + 1)(x - 1)$.   CV: $x = 0, \pm 1$.

$$\begin{array}{cccc} \nearrow & \searrow & \searrow & \nearrow \\ \hline & \!\!+\!\!\quad\!\!+\!\!\quad\!\!+\!\! & & \\ -1 & 0 & 1 & \end{array}$$

Increasing on $(-\infty, -1)$ and $(1, \infty)$; decreasing on $(-1, 0)$
and $(0, 1)$; rel. max. when $x = -1$; rel. min. when $x = 1$.

25. $y = -x^5 - 5x^4 + 200$.

$y' = -5x^4 - 20x^3 = -5x^3(x + 4)$.   CV: $x = 0, -4$.

$$\begin{array}{ccc} \searrow & \nearrow & \searrow \\ \hline & \!\!+\!\!\quad\!\!+\!\! & \\ -4 & 0 & \end{array}$$

Decreasing on $(-\infty, -4)$ and $(0, \infty)$; increasing on $(-4, 0)$;
rel. min. when $x = -4$; rel. max. when $x = 0$.

27. $y = 8x^4 - x^8$.

$y' = 32x^3 - 8x^7 = 8x^3(4 - x^4) = 8x^3(2 + x^2)(2 - x^2)$
$= 8x^3(2 + x^2)(\sqrt{2} - x)(\sqrt{2} + x)$. CV: $x = 0, \pm\sqrt{2}$.

$$\overset{\displaystyle \diagup \quad \diagdown \quad \diagup \quad \diagdown}{\underset{-\sqrt{2} \quad 0 \quad \sqrt{2}}{\rule{0pt}{0pt}\!+\!\!+\!\!+\!\!+\!}}$$

Increasing on $(-\infty, -\sqrt{2})$ and $(0, \sqrt{2})$; decreasing on $(-\sqrt{2}, 0)$ and $(\sqrt{2}, \infty)$; rel. max. when $x = \pm\sqrt{2}$, rel. min. when $x = 0$.

29. $y = (x^3 + 1)^3$.

$y' = 9x^2(x^3 + 1)^2 = 9x^2(x + 1)^2(x^2 - x + 1)^2$. CV: $0, -1$.

$$\overset{\displaystyle \diagup \quad \diagup \quad \diagup}{\underset{-1 \quad 0}{\rule{0pt}{0pt}\!+\!\!+\!\!+\!}}$$

Increasing on $(-\infty, -1)$, $(-1, 0)$, and $(0, \infty)$; never decreasing; no rel. extremum.

31. $y = \dfrac{1}{x - 1} = (x - 1)^{-1}$. $y' = -1(x - 1)^{-2} = -\dfrac{1}{(x - 1)^2}$.

CV: none, but $x = 1$ must be included in the sign chart because it is a point of discontinuity of y.

$$\overset{\displaystyle \diagdown \qquad \diagdown}{\underset{\boxed{1}}{\rule{0pt}{0pt}\!+\!\!+\!\!+\!}}$$

Decreasing on $(-\infty, 1)$ and $(1, \infty)$; no rel. max. or rel. min.

33. $y = \dfrac{10}{\sqrt{x}} = 10x^{-1/2}$. [Note: $x > 0$.]

$y' = -5x^{-3/2} = -\dfrac{5}{\sqrt{x^3}} < 0$ for $x > 0$.

Decreasing on $(0, \infty)$; no rel. max. or rel. min.

35. $y = \dfrac{x^2}{1 - x}$. $y' = \dfrac{x(2 - x)}{(1 - x)^2}$. CV: $x = 0, 2$, but $x = 1$ must be included in the sign chart because it is a point of

discontinuity of y.

Decreasing on $(-\infty, 0)$ and $(2, \infty)$; increasing on $(0, 1)$ and $(1, 2)$; rel. min. when $x = 0$; rel. max. when $x = 2$.

37. $y = \dfrac{x^2 - 3}{x + 2}$.

$y' = \dfrac{(x+2)(2x) - (x^2-3)(1)}{(x+2)^2} = \dfrac{x^2 + 4x + 3}{(x + 2)^2} = \dfrac{(x + 1)(x + 3)}{(x + 2)^2}$.

CV: $x = -3, -1$, but $x = -2$ must be included in the sign chart because it is a point of discontinuity of y.

Increasing on $(-\infty, -3)$ and $(-1, \infty)$; decreasing on $(-3, -2)$ and $(-2, -1)$; rel. max. when $x = -3$; rel. min. when $x = -1$.

39. $y = \dfrac{5x + 3}{x^2 + 1}$.

$y' = \dfrac{(x^2+1)(5) - (5x+3)(2x)}{(x^2+1)^2} = \dfrac{-5x^2 - 6x + 5}{(x^2 + 1)^2}$. y' is 0 when $-5x^2 - 6x + 5 = 0$; by the quadratic formula,

$x = \dfrac{-3 \pm \sqrt{34}}{5}$. CV: $x = \dfrac{-3 \pm \sqrt{34}}{5}$.

Decreasing on $\left(-\infty, \dfrac{-3 - \sqrt{34}}{5}\right)$ and $\left(\dfrac{-3 + \sqrt{34}}{5}, \infty\right)$;

increasing on $\left(\dfrac{-3 - \sqrt{34}}{5}, \dfrac{-3 + \sqrt{34}}{5}\right)$; rel. min. when

$x = \dfrac{-3 - \sqrt{34}}{5}$; rel. max. when $x = \dfrac{-3 + \sqrt{34}}{5}$.

41. $y = (x + 2)^3(x - 5)^2$.

$y' = (x + 2)^3[(2)(x - 5)] + (x - 5)^2[(3)(x + 2)^2]$

$= (x + 2)^2(x - 5)[2(x + 2) + 3(x - 5)]$. Simplifying,

$y' = (x + 2)^2(x - 5)(5x - 11)$.   CV: $x = -2, 5, 11/5$.

Increasing on $(-\infty, -2)$, $(-2, 11/5)$ and $(5, \infty)$; decreasing on $(11/5, 5)$; rel. max. when $x = 11/5$; rel. min. when $x = 5$

43. $y = x^3(x - 4)^4$.

$y' = x^3[4(x - 4)^3] + (x - 4)^4[3x^2]$
$= x^2(x - 4)^3[4x + 3(x - 4)] = x^2(x - 4)^3(7x - 12)$.

CV: $x = 0, 4, \frac{12}{7}$.

Increasing on $(-\infty, 0)$, $(0, \frac{12}{7})$, and $(4, \infty)$; decreasing on $(\frac{12}{7}, 4)$; rel. max. when $x = \frac{12}{7}$; rel. min. when $x = 4$.

45. $y = e^{-2x}$. $y' = -2e^{-2x} < 0$ for all x. Thus decreasing on $(-\infty, \infty)$; no rel. max. or rel. min.

47. $y = x^2 - 2 \ln x$. [Note: $x > 0$.]

$y' = 2x - \frac{2}{x} = \frac{2x^2 - 2}{x} = \frac{2(x^2 - 1)}{x} = \frac{2(x + 1)(x - 1)}{x}$.

CV: $x = 1$.

Decreasing on $(0, 1)$; increasing on $(1, \infty)$; rel. min. when $x = 1$.

49. $y = e^x + e^{-x}$. $y' = e^x - e^{-x}$. Setting $y' = 0$ gives
$e^x - e^{-x} = 0$, $e^x = e^{-x}$, $x = -x$, $x = 0$. CV: $x = 0$.

Decreasing on $(-\infty, 0)$; increasing on $(0, \infty)$; rel. min. when $x = 0$.

51. $y = x \ln x - x.$ [Note: $x > 0$.]

$y' = \left[x \cdot \frac{1}{x} + (\ln x)(1)\right] - 1 = \ln x.$ CV: $x = 1.$

Decreasing on $(0, 1)$; increasing on $(1, \infty)$; rel. min. when $x = 1$; no rel. max.

53. $y = x^2 - 6x - 7 = (x - 7)(x + 1).$
Intercepts $(7,0)$, $(-1,0)$, $(0,-7)$.
$y' = 2x - 6 = 2(x - 3).$ CV: $x = 3.$
Decreasing on $(-\infty, 3)$; increasing on $(3, \infty)$; rel. min. when $x = 3.$

55. $y = 3x - x^3 = x(\sqrt{3} + x)(\sqrt{3} - x).$
Intercepts $(0,0)$, $(\pm\sqrt{3},0)$.
Symmetric about origin.
$y' = 3 - 3x^2 = 3(1 + x)(1 - x).$
CV: $x = \pm 1.$ Decreasing on $(-\infty,-1)$
and $(1,\infty)$; increasing on $(-1,1)$;
rel.min. when $x = -1$;
rel. max. when $x = 1.$

57. $y = 2x^3 - 9x^2 + 12x = x(2x^2 - 9x + 12).$
Note that $2x^2 - 9x + 12 = 0$ has no
real roots. The only intercept is $(0,0)$.
$y' = 6x^2 - 18x + 12 = 6(x^2 - 3x + 2)$
$= 6(x - 2)(x - 1).$ CV: $x = 1, 2.$
Increasing on $(-\infty,1)$ and $(2,\infty)$;
decreasing on $(1,2)$; rel. max. when
$x = 1$; rel. min. when $x = 2.$

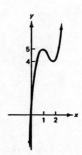

59. $y = x^4 + 4x^3 + 4x^2 = x^2(x + 2)^2$.
    Intercepts $(0,0)$, $(-2,0)$.
    $y' = 4x^3 + 12x^2 + 8x = 4x(x + 1)(x + 2)$.
    CV: $x = 0, -1, -2$. Increasing on
    $(-2,-1)$ and $(0,\infty)$; decreasing on
    $(-\infty,-2)$ and $(-1,0)$; rel. max. when
    $x = -1$; rel. min. when $x = -2$ or $x = 0$.

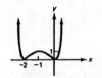

61. $y = (x - 1)^2(x + 2)^2$.
    Intercepts $(1,0)$, $(-2,0)$, $(0,4)$.
    $y' = (x-1)^2 \cdot 2(x+2) + (x+2)^2 \cdot 2(x-1)$
       $= 2(x-1)(x+2)[(x-1) + (x+2)]$
       $= 2(x-1)(x+2)(2x+1)$.
    CV: $x = 1, -2, -\frac{1}{2}$. Decreasing on

    $(-\infty,-2)$ and $\left(-\frac{1}{2}, 1\right)$; increasing on

    $\left(-2, -\frac{1}{2}\right)$ and $(1,\infty)$; rel. min. when

    $x = -2$ or $x = 1$; rel. max. when $x = -\frac{1}{2}$.

63. $y = 2\sqrt{x} - x = \sqrt{x}(2 - \sqrt{x})$. [Note: $x \geq 0$.]
    Intercepts $(0, 0)$, $(4,0)$.
    $y' = \frac{1}{\sqrt{x}} - 1 = \frac{1 - \sqrt{x}}{\sqrt{x}}$.
    CV: $x = 0, 1$. Increasing
    on $(0,1)$; decreasing on $(1,\infty)$;
    rel. max. when $x = 1$.

65.

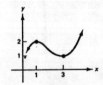

67. $c_f = 25{,}000$. $\bar{c}_f = \frac{c_f}{q} = \frac{25{,}000}{q}$.
    $\frac{d}{dq}(\bar{c}_f) = -\frac{25{,}000}{q^2} < 0$ for $q > 0$, so $\bar{c}_f$ is a decreasing
    function for $q > 0$.

69. $p = 400 - 2q$. Revenue is given by $r = pq = (400 - 2q)q = 400q - 2q^2$. Marginal revenue is $r' = 400 - 4q$. Marginal revenue is increasing when its derivative is positive. But $(r')' = -4 < 0$. Thus marginal revenue is never increasing.

71. $r = 240q + 57q^2 - q^3$.
$$r' = 240 + 114q - 3q^2 = 0$$
$$3(40 - q)(2 + q) = 0.$$
Since $q \geq 0$, we have $q = 40$ as the only CV. Since r is increasing on $(0,40)$ and decreasing on $(40,\infty)$, r is a maximum when output is 40.

73. $E = 0.71\left(1 - \frac{T_c}{T_h}\right)$. $\frac{dE}{dT_h} = 0.71\left(\frac{T_c}{T_h^2}\right) > 0$, so as T_h increases, E increases.

75. $C(k) = 100\left[100 + 9k + \frac{144}{k}\right]$, $1 \leq k \leq 100$.
   (a) $C(1) = 25,300$.
   (b) $C'(k) = 100\left[9 - \frac{144}{k^2}\right] = 100\left[\frac{9k^2 - 144}{k^2}\right]$
$$= 100\left[\frac{9(k + 4)(k - 4)}{k^2}\right].$$
   Since $k \geq 1$, the only critical value is $k = 4$. If $1 \leq k < 4$, then $C'(k) < 0$ and C is decreasing. If $4 < k \leq 100$, then $C'(k) > 0$ and C is increasing. Thus C has an absolute minimum for $k = 4$.
   (c) $C(4) = 17,200$

77. Rel. min.: $(-4.10, -3.40)$.

79. Rel. max.: $(2.74, 2.37)$; rel. min.: $(-2.74, -2.37)$.

EXERCISE 14.2

1. $f(x) = x^2 - 2x + 3$ and f is continuous over $[-1,2]$. $f'(x) = 2x - 2 = 2(x - 1)$. The only critical value on the interval $(-1,2)$ is $x = 1$. We evaluate f at this point and at the endpoints: $f(-1) = 6$, $f(1) = 2$, and $f(2) = 3$. Ab. maximum: $f(-1) = 6$; ab. minimum: $f(1) = 2$.

3. $f(x) = (1/3)x^3 - x^2 - 3x + 1$ and f is continuous over
[0,2]. $f'(x) = x^2 - 2x - 3 = (x + 1)(x - 3)$. There are
no critical values on (0,2), so we only have to evaluate f
at the endpoints: $f(0) = 1$ and $f(2) = -19/3$.
Ab. maximum: $f(0) = 1$; ab. minimum: $f(2) = -19/3$.

5. $f(x) = 4x^3 + 3x^2 - 18x + 3$ and f is continuous over $\left[\frac{1}{2},3\right]$.
$f'(x) = 12x^2 + 6x - 18 = 6(2x^2 + x - 3) = 6(2x + 3)(x - 1)$.
The only critical value on (1/2,3) is x = 1. We evaluate f
at this point and the endpoints: $f(1/2) = -19/4$, $f(1) = -8$,
$f(3) = 84$. Ab. maximum: $f(3) = 84$; ab. minimum: $f(1) = -8$.

7. $f(x) = -3x^5 + 5x^3$ and f is continuous over [-2,0].
$f'(x) = -15x^4 + 15x^2 = 15x^2(1 - x^2) = 15x^2(1 + x)(1 - x)$.
The only critical value on (-2,0) is x = -1. We have
$f(-2) = 56$, $f(-1) = -2$, and $f(0) = 0$.
Ab. maximum: $f(-2) = 56$; ab. minimum: $f(-1) = -2$.

9. $f(x) = 3x^4 - x^6$ and f is continuous over [-1,2].
$f'(x) = 12x^3 - 6x^5 = 6x^3(2 - x^2) = 6x^3(\sqrt{2} - x)(\sqrt{2} + x)$.
The only critical values on (-1,2) are x = 0, $\sqrt{2}$. We
have $f(-1) = 2$, $f(0) = 0$, $f(\sqrt{2}) = 4$, and $f(2) = -16$.
Ab. maximum: $f(\sqrt{2}) = 4$; ab. minimum: $f(2) = -16$.

11. $f(x) = x^4 - 9x^2 + 2$ and f is continuous over [-1,3].
$f'(x) = 4x^3 - 18x = 2x(2x^2 - 9) = 2x(\sqrt{2}x - 3)(\sqrt{2}x + 3)$.
The only critical values on (-1,3) are x = 0 and
$x = 3/\sqrt{2} = 3\sqrt{2}/2$. We have $f(-1) = -6$, $f(0) = 2$,
$f(3\sqrt{2}/2) = -73/4$, and $f(3) = 2$. Thus there is an
absolute maximum when x = 0 or x = 3, and an absolute
minimum when $x = 3\sqrt{2}/2$. Ab. maximum: $f(0) = f(3) = 2$;
ab. minimum: $f(3\sqrt{2}/2) = -73/4$.

13. $f(x) = x^{2/3}$ and f is continuous over [-2,3].
Ab. maximum: $f(3) = 2.08$; ab. minimum: $f(0) = 0$.

EXERCISE 14.3

1. $f(x) = x^4 - 3x^3 + 7x - 5$; $f''(x) = 6x(2x - 3)$.
   $f''(x)$ is 0 when $x = 0, 3/2$. Sign chart for f'':

   $$\begin{array}{ccccc} & + & - & + & \\ \hline & & | & | & \\ & & 0 & 3/2 & \end{array}$$

   Concave up on $(-\infty,0)$ and $(3/2,\infty)$; concave down on $(0,3/2)$.
   Inflection points when $x = 0, 3/2$.

3. $f(x) = \dfrac{2 + x - x^2}{x^2 - 2x + 1}$, $f''(x) = \dfrac{2(7 - x)}{(x - 1)^4}$.
   $f''(x)$ is 0 when $x = 7$. Although f'' is not defined when
   $x = 1$, f is not continuous at $x = 1$. So there is no
   inflection point when $x = 1$, but $x = 1$ must be considered
   in concavity analysis. Sign chart of f'':

   $$\begin{array}{ccccc} & + & + & - & \\ \hline & & | & | & \\ & & \boxed{1} & 7 & \end{array}$$

   Concave up on $(-\infty,1)$ and $(1,7)$; concave down on $(7,\infty)$.
   Inflection point when $x = 7$.

5. $f(x) = \dfrac{x^2 + 1}{x^2 - 2}$; $f''(x) = \dfrac{6(3x^2 + 2)}{(x^2 - 2)^3} = \dfrac{6(3x^2 + 2)}{[(x - \sqrt{2})(x + \sqrt{2})]^3}$.
   $f''(x)$ is never 0. Although f'' is not defined when
   $x = \pm\sqrt{2}$, f is not continuous at $x = \pm\sqrt{2}$. So there is no
   inflection point when $x = \pm\sqrt{2}$, but $x = \pm\sqrt{2}$ must be
   considered in concavity analysis. Sign chart of f'':

   $$\begin{array}{ccccc} & + & - & + & \\ \hline & & | & | & \\ & & \boxed{-\sqrt{2}} & \boxed{\sqrt{2}} & \end{array}$$

   Concave up on $(-\infty,-\sqrt{2})$ and $(\sqrt{2},\infty)$; concave down on
   $(-\sqrt{2},\sqrt{2})$. No inflection point.

7. $y = -2x^2 + 4x$. $y' = -4x + 4$. $y'' = -4 < 0$ for all x, so
   the graph is concave down for all x, that is, on $(-\infty,\infty)$.

9. $y = 4x^3 + 12x^2 - 12x$. $y' = 12x^2 + 24x - 12$.
   $y'' = 24x + 24 = 24(x + 1)$. Possible inflection point

when x = -1. Concave down on $(-\infty, -1)$; concave up on $(-1, \infty)$; inflection point when x = -1.

11. $y = 4x^3 - 21x^2 + 5x$. $y' = 12x^2 - 42x + 5$.
$y'' = 24x - 42 = 24\left(x - \frac{7}{4}\right)$. Possible inflection point when x = 7/4. Concave down on $(-\infty, 7/4)$; concave up on $(7/4, \infty)$; inflection point when x = 7/4.

13. $y = x^4 - 6x^2 + 5x - 6$. $y' = 4x^3 - 12x + 5$.
$y'' = 12x^2 - 12 = 12(x^2 - 1) = 12(x + 1)(x - 1)$. Possible inflection points when x = ±1. Concave up on $(-\infty, -1)$ and on $(1, \infty)$; concave down on $(-1, 1)$; inflection points when x = ±1.

15. $y = x^{1/5}$. $y' = \frac{1}{5}x^{-4/5}$. $y'' = -\frac{4}{25}x^{-9/5} = -\frac{4}{25x^{9/5}}$.
y" is not defined when x = 0 and y is continuous there. Thus there is a possible inflection point when x = 0. Concave up on $(-\infty, 0)$; concave down on $(0, \infty)$; inflection point when x = 0.

17. $y = \frac{x^4}{2} + \frac{19x^3}{6} - \frac{7x^2}{2} + x + 5$. $y' = 2x^3 + \frac{19}{2}x^2 - 7x + 1$.
$y'' = 6x^2 + 19x - 7 = (3x - 1)(2x + 7)$. Possible inflection points when x = -7/2, 1/3. Concave up on $(-\infty, -7/2)$ and $(1/3, \infty)$; concave down on $(-7/2, 1/3)$; inflection points when x = -7/2, 1/3.

19. $y = \frac{1}{20}x^5 - \frac{1}{4}x^4 + \frac{1}{6}x^3 - \frac{1}{2}x - \frac{2}{3}$. $y' = \frac{1}{4}x^4 - x^3 - \frac{1}{2}x^2 - \frac{1}{2}$.
$y'' = x^3 - 3x^2 - x = x(x^2 - 3x - 1)$. y" is 0 when x = 0 or $x^2 - 3x - 1 = 0$. Using the quadratic formula to solve $x^2 - 3x - 1 = 0$ gives $x = \frac{3 \pm \sqrt{13}}{2}$. Thus possible inflection points occur when x = 0, $\frac{3 \pm \sqrt{13}}{2}$. Concave down on $\left(-\infty, \frac{3 - \sqrt{13}}{2}\right)$ and $\left(0, \frac{3 + \sqrt{13}}{2}\right)$; concave up on $\left(\frac{3 - \sqrt{13}}{2}, 0\right)$ and $\left(\frac{3 + \sqrt{13}}{2}, \infty\right)$; inflection points when

$$x = 0, \frac{3 \pm \sqrt{13}}{2}.$$

21. $y = \frac{1}{30}x^6 - \frac{7}{12}x^4 + 5x^2 + 2x - 1.$ $y' = \frac{1}{5}x^5 - \frac{7}{3}x^3 + 10x + 2.$

$y'' = x^4 - 7x^2 + 10 = (x^2 - 2)(x^2 - 5)$

$= (x + \sqrt{2})(x - \sqrt{2})(x + \sqrt{5})(x - \sqrt{5}).$

Possible inflection points when $x = \pm\sqrt{2}, \pm\sqrt{5}.$ Concave up on $(-\infty, -\sqrt{5}), (-\sqrt{2}, \sqrt{2}),$ and $(\sqrt{5}, \infty)$; concave down on $(-\sqrt{5}, -\sqrt{2})$ and $(\sqrt{2}, \sqrt{5})$; inflection points when $x = \pm\sqrt{5}, \pm\sqrt{2}.$

23. $y = (x + 1)/(x - 1).$ $y' = -2/(x - 1)^2.$ $y'' = 4/(x - 1)^3.$
No possible inflection point, but we consider $x = 1$ in the concavity analysis. Concave down on $(-\infty, 1)$; concave up on $(1, \infty).$

25. $y = \frac{x^2}{x^2 + 1}.$

$y' = \frac{(x^2 + 1)(2x) - x^2(2x)}{(x^2 + 1)^2} = = \frac{2x}{(x^2 + 1)^2}.$

$y'' = \frac{(x^2 + 1)^2(2) - 2x(2)(x^2 + 1)(2x)}{(x^2 + 1)^4} = \frac{(x^2 + 1)(2) - 8x^2}{(x^2 + 1)^3}$

$= \frac{2(1 - 3x^2)}{(x^2 + 1)^3} = \frac{2(1 + \sqrt{3}x)(1 - \sqrt{3}x)}{(x^2 + 1)^3}.$

Possible inflection points when $x = \pm 1/\sqrt{3}.$ Concave down on $(-\infty, -1/\sqrt{3})$ and $(1/\sqrt{3}, \infty)$; concave up on $(-1/\sqrt{3}, 1/\sqrt{3})$; inflection points when $x = \pm 1/\sqrt{3}.$

27. $y = \frac{21x + 40}{6(x + 3)^2}.$

$y' = \frac{1}{6} \cdot \frac{(x+3)^2(21) - (21x+40)[2(x+3)]}{(x+3)^4}$

$= \frac{1}{6} \cdot \frac{(x+3)(21) - (21x+40)[2]}{(x+3)^3} = \frac{1}{6} \cdot \frac{-21x-17}{(x+3)^3} = -\frac{1}{6} \cdot \frac{21x+17}{(x+3)^3}.$

$y'' = -\frac{1}{6} \cdot \frac{(x+3)^3(21) - (21x+17)[3(x+3)^2]}{(x+3)^6}.$ Simplifying,

$$y'' = -\frac{1}{6} \cdot \frac{(x+3)(21) - (21x+17)[3]}{(x+3)^4} = -\frac{1}{6} \cdot \frac{-42x+12}{(x+3)^4} = \frac{7x-2}{(x+3)^4}.$$

Possible inflection point when $x = 2/7$ ($x = -3$ must be considered in concavity analysis). Concave down on $(-\infty,-3)$ and $(-3,2/7)$; concave up on $(2/7,\infty)$; inflection point when $x = 2/7$.

29. $y = e^x$. $y' = e^x$. $y'' = e^x$. Thus $y'' > 0$ for all x. Concave up on $(-\infty,\infty)$.

31. $y = xe^x$. $y' = xe^x + e^x = e^x(x + 1)$.
    $y'' = e^x(1) + (x + 1)e^x = e^x(x + 2)$.
    $y'' = 0$ if $x = -2$. Concave down on $(-\infty,-2)$; concave up on $(-2,\infty)$; inflection point when $x = -2$.

33. $y = \frac{\ln x}{x}$. (Note: $x > 0$.) $y' = \frac{x \cdot \frac{1}{x} - (\ln x)(1)}{x^2} = \frac{1 - \ln x}{x^2}$.

    $$y'' = \frac{x^2\left(-\frac{1}{x}\right) - (1 - \ln x)(2x)}{x^4} = \frac{-x - (1 - \ln x)(2x)}{x^4}$$

    $$= \frac{-1 - (1 - \ln x)(2)}{x^3} = \frac{2\ln(x) - 3}{x^3}.$$

    y'' is 0 if $2\ln(x) - 3 = 0 \Rightarrow \ln x = 3/2 \Rightarrow x = e^{3/2}$. Concave down on $(0,e^{3/2})$; concave up on $(e^{3/2},\infty)$; inflection point when $x = e^{3/2}$.

35. $y = x^2 + 4x + 3 = (x + 3)(x + 1)$.
    Intercepts $(-3,0)$, $(-1,0)$, and $(0,3)$.
    $y' = 2x + 4 = 2(x + 2)$. CV: $x = -2$.
    Decreasing on $(-\infty,-2)$; increasing on $(-2,\infty)$; rel. min. at $(-2,-1)$. $y'' = 2$. No possible inflection point. Concave up on $(-\infty,\infty)$.

37. $y = 4x - x^2 = x(4 - x)$. Intercepts $(0,0)$ and $(4,0)$. $y' = 4 - 2x = 2(2 - x)$. CV: $x = 2$. Increasing on $(-\infty, 2)$; decreasing on $(2,\infty)$; rel. max. at $(2,4)$. $y'' = -2$. No possible inflection point. Concave down on $(-\infty,\infty)$.

39. $y = x^3 - 9x^2 + 24x - 19$. The x-intercept
is not convenient to find; the y-intercept
is $(0,-19)$.

$y' = 3x^2 - 18x + 24 = 3(x - 2)(x - 4)$.
CV: $x = 2$ and $x = 4$. Increasing on $(-\infty,2)$
and $(4,\infty)$; decreasing on $(2,4)$; rel. max.
at $(2,1)$; rel. min. at $(4,-3)$.
$y'' = 6x - 18 = 6(x - 3)$. Possible inflection point when
$x = 3$. Concave down on $(-\infty,3)$; concave up on $(3,\infty)$;
inflection point at $(3,-1)$.

41. $y = \frac{x^3}{3} - 4x = \frac{x^3-12x}{3} = \frac{1}{3}x(x+2\sqrt{3})(x-2\sqrt{3})$.

Intercepts $(0,0)$ and $(\pm2\sqrt{3},0)$.

$y' = x^2 - 4 = (x + 2)(x - 2)$. CV: $x = \pm2$.
Increasing on $(-\infty,-2)$ and $(2,\infty)$;
decreasing on $(-2, 2)$; rel. max. at
$(-2,16/3)$; rel. min. at $(2,-16/3)$. $y'' = 2x$. Possible
inflection point when $x = 0$. Concave down on $(-\infty,0)$;
concave up on $(0,\infty)$; inflection point at $(0,0)$. Symmetric
about the origin.

43. $y = x^3 - 3x^2 + 3x - 3$. Intercept $(0,-3)$.
$y' = 3x^2 - 6x + 3 = 3(x - 1)^2$. CV: $x = 1$.
Increasing on $(-\infty,1)$ and $(1,\infty)$; no rel. max.
or min.
$y'' = 6(x - 1)$. Possible inflection point
when $x = 1$. Concave down on $(-\infty,1)$;
concave up on $(1,\infty)$; inflection point at $(1,-2)$.

45. $y = 4x^3 - 3x^4 = x^3(4 - 3x)$.
Intercepts $(0,0)$, $(4/3,0)$.
$y' = 12x^2 - 12x^3 = 12x^2(1 - x)$.
CV: $x = 0$ and $x = 1$.
Increasing on $(-\infty,0)$ and $(0,1)$;
decreasing on $(1,\infty)$; rel. max. at $(1,1)$.

$y'' = 24x - 36x^2 = 12x(2 - 3x)$. Possible inflection points
at $x = 0$ and $x = 2/3$. Concave down on $(-\infty,0)$ and $(2/3,\infty)$;
concave up on $(0,2/3)$; inflection point at $(0,0)$ and
$(2/3, 16/27)$.

47. $y = -2 + 12x - x^3$. Intercept $(0,-2)$.

$y' = 12 - 3x^2 = 3(2 + x)(2 - x)$.
CV: $x = \pm 2$. Decreasing on $(-\infty,-2)$ and
$(2,\infty)$; increasing on $(-2,2)$; rel. min.
at $(-2,-18)$; rel. max. at $(2,14)$.
$y'' = -6x$. Possible inflection point
when $x = 0$. Concave up on $(-\infty,0)$; concave down on $(0,\infty)$;
inflection point at $(0,-2)$.

49. $y = x^3 - 6x^2 + 12x - 6$. Intercept $(0,-6)$.

$y' = 3x^2 - 12x + 12 = 3(x - 2)^2$.
CV: $x = 2$. Increasing on $(-\infty,2)$ and $(2,\infty)$.
$y'' = 6(x - 2)$. Possible inflection point
when $x = 2$. Concave down on $(-\infty,2)$;
concave up on $(2,\infty)$; inflection point at
$(2,2)$.

51. $y = 5x - x^5 = x(5 - x^4)$
$= x(\sqrt{5} + x^2)(\sqrt{5} - x^2)$
$= x(\sqrt{5} + x^2)(\sqrt[4]{5} + x)(\sqrt[4]{5} - x)$.
Intercepts $(0,0)$ and $(\pm\sqrt[4]{5},0)$.
Symmetric about the origin.
$y' = 5 - 5x^4 = 5(1 - x^4) = 5(1 - x^2)(1 + x^2) =$
$5(1 - x)(1 + x)(1 + x^2)$. CV: $x = \pm 1$. Decreasing on
$(-\infty,-1)$ and $(1,\infty)$; increasing on $(-1,1)$; rel. min. at
$(-1,-4)$; rel. max. at $(1,4)$.
$y'' = -20x^3$. Possible inflection point when $x = 0$.
Concave up on $(-\infty,0)$; concave down on $(0,\infty)$; inflection
point at $(0,0)$.

53. $y = 3x^4 - 4x^3 + 1$. Intercepts $(0,1)$
and $(1,0)$ [the latter is found by
inspection of the equation].
No symmetry.
$y' = 12x^3 - 12x^2 = 12x^2(x - 1)$.
CV: $x = 0$ and $x = 1$. Decreasing on $(-\infty,0)$ and

(0,1); increasing on (1,∞); relative minimum at (1,0).
$y'' = 36x^2 - 24x = 12x(3x - 2)$. Possible inflection points
at $x = 0$ and $x = 2/3$. Concave up on $(-∞,0)$ and $(2/3,∞)$;
concave down on $(0,2/3)$; inflection points at $(0,1)$ and
$(2/3,11/27)$.

55. $y = 4x^2 - x^4 = x^2(2 + x)(2 - x)$.
Intercepts $(0,0)$ and $(±2,0)$.
Symmetric about the y-axis.
$y' = 8x - 4x^3 = 4x(2 - x^2)$
$\quad = 4x(\sqrt{2} + x)(\sqrt{2} - x)$.

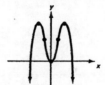

CV: $x = ±\sqrt{2}$.
Increasing on $(-∞,-\sqrt{2})$ and $(0,\sqrt{2})$;
decreasing on $(-\sqrt{2},0)$ and $(\sqrt{2},∞)$; rel. max.
at $(±\sqrt{2},4)$; rel. min. at $(0,0)$.
$y'' = 8 - 12x^2 = 12[(2/3) - x^2] = 12(\sqrt{2/3} - x)(\sqrt{2/3} + x)$.
Possible inflection points when $x = ±\sqrt{2/3}$. Concave down
on $(-∞,-\sqrt{2/3})$ and $(\sqrt{2/3},∞)$; concave up on
$(-\sqrt{2/3}, \sqrt{2/3})$; inflection points at $(±\sqrt{2/3}, 20/9)$.

57. $y = x^{1/3}(x - 8)$.
Intercepts $(0,0)$ and $(8,0)$.
Since $y = x^{4/3} - 8x^{1/3}$,
$y' = \frac{4}{3}x^{1/3} - \frac{8}{3}x^{-2/3}$
$\quad = \frac{4}{3}\left[x^{1/3} - \frac{2}{x^{2/3}}\right] = \frac{4(x - 2)}{3x^{2/3}}$.

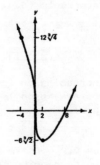

CV: $x = 0, 2$.
Decreasing on $(-∞,0)$ and $(0,2)$;
increasing on $(2,∞)$; rel. min.
at $(2,-6\sqrt[3]{2}) ≈ (2,-7.56)$.
$y'' = \frac{4}{9}x^{-2/3} + \frac{16}{9}x^{-5/3}$
$\quad = \frac{4}{9}\left[\frac{1}{x^{2/3}} + \frac{4}{x^{5/3}}\right] = \frac{4(x + 4)}{9x^{5/3}}$.
Possible inflection points when $x = -4, 0$. Concave up on
$(-∞,-4)$ and $(0,∞)$; concave down on $(-4,0)$; inflection

points at $(-4,12\sqrt[3]{4})$ and $(0,0)$. Observe that at the origin the tangent line exists but it is vertical.

59. $y = 4x^{1/3} + x^{4/3} = x^{1/3}(4 + x)$.
Intercepts $(0,0)$ and $(-4,0)$.

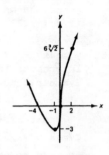

$$y' = \frac{4}{3}x^{-2/3} + \frac{4}{3}x^{1/3} = \frac{4}{3}\left[\frac{1}{x^{2/3}} + x^{1/3}\right]$$

$$= \frac{4(1 + x)}{3x^{2/3}}. \quad \text{CV: } x = 0, -1.$$

Decreasing on $(-\infty,-1)$; increasing on
$(-1,0)$ and $(0,\infty)$; rel. min. at $(-1,-3)$.

$$y'' = -\frac{8}{9}x^{-5/3} + \frac{4}{9}x^{-2/3} = \frac{4}{9}\left[\frac{1}{x^{2/3}} - \frac{2}{x^{5/3}}\right]$$

$$= \frac{4(x - 2)}{9x^{5/3}}.$$

Possible inflection points when $x = 0, 2$. Concave up on $(-\infty,0)$ and $(2,\infty)$; concave down on $(0,2)$; inflection point at $(0,0)$ and $(2,6\sqrt[3]{2})$. Observe that at the origin the tangent line exists but it is vertical.

61. $y = 2x + 3x^{2/3} = 2x^{2/3}\left(x^{1/3} + \frac{3}{2}\right)$.

Intercepts $(0,0)$ and $(-27/8, 0)$.

$$y' = 2 + 2x^{-1/3} = 2\left(1 + \frac{1}{\sqrt[3]{x}}\right)$$

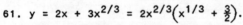

$$= 2\left(\frac{\sqrt[3]{x} + 1}{\sqrt[3]{x}}\right). \quad \text{CV: } x = 0, -1.$$

Increasing on $(-\infty,-1)$ and $(0,\infty)$;
decreasing on $(-1,0)$; rel. max. at $(-1,1)$; rel. min. at $(0,0)$.

$$y'' = -\frac{2}{3}x^{-4/3} = -\frac{2}{3x^{4/3}}.$$ Possible inflection point at $x = 0$. Concave down on $(-\infty,0)$ and $(0,\infty)$. Observe that at the origin the tangent line exists but it is vertical.

63.

65.

67. $p = 100/(q + 2)$. $dp/dq = -100/(q + 2)^2 < 0$ for $q > 0$, so p is decreasing. Since $d^2p/dq^2 = 200/(q + 2)^3 > 0$ for $q > 0$, the demand curve is concave up.

69. $S = f(A) = 12\sqrt[4]{A}$, $0 \leq A \leq 625$. For the given values of A we have $S' = 3A^{-3/4} > 0$ and $S" = -(9/4)A^{-7/4} < 0$. Thus y is increasing and concave down.

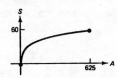

71. $y = 12.5 + 5.8(0.42)^X$. $y' = 5.8(0.42)^X \ln(0.42)$. Since $\ln(0.42) < 0$, we have $y' < 0$, so the function is decreasing. $y" = 5.8(0.42)^X \ln^2(0.42) > 0$, so the function is concave up.

73. $n = f(r) = 0.1 \ln(r) + \frac{7}{r} - 0.8$, $1 \leq r \leq 10$.

(a) $\frac{dn}{dr} = \frac{0.1}{r} - \frac{7}{r^2} = \frac{0.1r - 7}{r^2} = \frac{0.1(r - 70)}{r^2} < 0$ for $1 \leq r \leq 10$. Thus the graph of f is always falling. Also,

$$\frac{d^2n}{dr^2} = -\frac{0.1}{r^2} + \frac{14}{r^3} = \frac{14 - 0.1r}{r^3} = \frac{0.1(140 - r)}{r^3} > 0$$

for $1 \leq r \leq 10$ Thus the graph is concave up.

(b) [graph: n(r), 6.2, from 1 to 10]

(c) $\frac{dn}{dr}\Big|_{r=5} = -0.26$, so the rate of decrease is 0.26.

75. two inflection points

77. $y = x^3 - 2x^2 + x + 3$. $y' = 3x^2 - 4x + 1$. When x = 2,
    then y = 5 and y' = 5. Thus an equation of the tangent
    line at x = 2 is y - 5 = 5(x - 2), or y = 5x - 5.
    Graphing the curve and the tangent line indicates that the
    curve lies above the tangent line around x = 2. Thus the
    curve is concave up at x = 2.

EXERCISE 14.4

1. $y = x^2 - 5x + 6$. $y' = 2x - 5 = 0$. CV: x = 5/2. $y'' = 2$.
   $y''(5/2) = 2 > 0$. Thus there is a relative minimum when
   x = 5/2. Because there is only one relative extremum
   and f is continuous, the relative minimum is an absolute
   minimum.

3. $y = -4x^2 + 2x - 8$. $y' = -8x + 2 = 0$. CV: x = 1/4.
   $y'' = -8$. $y''(1/4) = -8 < 0$. Thus there is a relative
   maximum when x = 1/4. Because there is only one relative
   extremum and f is continuous, the relative maximum is an
   absolute maximum.

5. $y = x^3 - 27x + 1$.
   $y' = 3x^2 - 27 = 3(x^2 - 9) = 3(x + 3)(x - 3) = 0$.
   CV: x = ±3. $y'' = 6x$.
   $y''(-3) = -18 < 0 \Rightarrow$ relative maximum when x = -3
   $y''(3) = 18 > 0 \Rightarrow$ relative minimum when x = 3

7. $y = -x^3 + 3x^2 + 1$. $y' = -3x^2 + 6x = -3x(x - 2) = 0$.
   CV: x = 0, 2. $y'' = -6x + 6$.
   $y''(0) = 6 > 0 \Rightarrow$ rel. min. when x = 0
   $y''(2) = -6 < 0 \Rightarrow$ rel. max. when x = 2

9. $y = 2x^4 + 2$. $y' = 8x^3 = 0$. CV: x = 0. $y'' = 24x^2$.
   Since $y''(0) = 0$, the second-derivative test fails. Using
   the first-derivative test, we see that f decreases for
   x < 0 and f increases for x > 0, so there is a relative
   minimum when x = 0.

11. $y = 81x^5 - 5x$.

$y' = 81 \cdot 5x^4 - 5 = 5(81x^4 - 1) = 5(9x^2 - 1)(9x^2 + 1)$
$\quad = 5(3x + 1)(3x - 1)(9x^2 + 1) = 0$.   CV: $x = \pm 1/3$.

$y'' = 81 \cdot 5 \cdot 4x^3$.

$y''(-1/3) = -60 < 0 \Rightarrow$ rel. max. when $x = -1/3$

$y''(1/3) = 60 > 0 \Rightarrow$ rel. min. when $x = 1/3$

13. $y = (x^2 + 7x + 10)^2$.

$y' = 2(x^2 + 7x + 10)(2x + 7) = 2(x + 2)(x + 5)(2x + 7) = 0$.

CV: $x = -2, -5, -7/2$.

$y'' = 2[(x^2 + 7x + 10)(2) + (2x + 7)(2x + 7)]$

$y''(-5) = 18 > 0 \Rightarrow$ rel. min. when $x = -5$

$y''(-7/2) = -9 < 0 \Rightarrow$ rel. max. when $x = -7/2$

$y''(-2) = 18 > 0 \Rightarrow$ rel. min. when $x = -2$

EXERCISE 14.5

1. $y = f(x) = \frac{x}{x + 1}$. When $x = -1$ the denominator is zero but the numerator is not zero. Thus the line $x = -1$ is a vertical asymptote. Testing for horizontal asymptotes,

$$\lim_{x \to \infty} \frac{x}{x + 1} = \lim_{x \to \infty} \frac{x}{x} = \lim_{x \to \infty} 1 = 1.$$

Thus the line $y = 1$ is a horizontal asymptote. Similarly $\lim_{x \to -\infty} f(x) = 1$.

3. $f(x) = (x - 1)/(2x + 3)$. When $x = -3/2$ the denominator is zero but the numerator is not. Thus $x = -3/2$ is a vertical asymptote. $\lim_{x \to \infty} f(x) = \lim_{x \to \infty} \frac{x}{2x} = \lim_{x \to \infty} \frac{1}{2} = \frac{1}{2}$.

Similarly $\lim_{x \to -\infty} f(x) = 1/2$. Thus $y = 1/2$ is a horizontal asymptote.

5. $y = f(x) = \frac{4}{x}$. When $x = 0$ the denominator is zero but the numerator is not zero, so $x = 0$ is a vertical asymptote. Taking limits at infinity, $\lim_{x \to \infty} (4/x) = 0 = \lim_{x \to -\infty} (4/x)$, so the line $y = 0$ is a horizontal asymptote.

7. $y = f(x) = 1/(x^2 - 1) = 1/[(x - 1)(x + 1)]$. Vertical asymptotes are $x = 1$ and $x = -1$. $\lim\limits_{x \to \infty} \dfrac{1}{x^2 - 1} = 0$. Similarly, $\lim\limits_{x \to -\infty} f(x) = 0$. Thus $y = 0$ is a horizontal asymptote.

9. $y = f(x) = x^2 - 5x + 8$ is a polynomial function, so there are no horizontal or vertical asymptotes.

11. $f(x) = 2x^2/(x^2 + x - 6) = 2x^2/[(x + 3)(x - 2)]$. Vertical asymptotes are $x = -3$ and $x = 2$. $\lim\limits_{x \to \infty} f(x) = \lim\limits_{x \to \infty} 2x^2/x^2 = \lim\limits_{x \to \infty} 2 = 2$, and $\lim\limits_{x \to -\infty} f(x) = 2$. Thus $y = 2$ is a horizontal asymptote.

13. $y = f(x) = \dfrac{4 + 2x - 7x^2}{x^2 - 5} = \dfrac{4 + 2x - 7x^2}{(x - \sqrt{5})(x + \sqrt{5})}$.
When $x = -\sqrt{5}$ or $x = \sqrt{5}$, the denominator is zero but the numerator is not. Thus $x = -\sqrt{5}$ and $x = \sqrt{5}$ are vertical asymptotes. $\lim\limits_{x \to \infty} f(x) = \dfrac{-7x^2}{x^2} = \lim\limits_{x \to \infty} -7 = -7$. Similarly, $\lim\limits_{x \to -\infty} f(x) = -7$. Thus $y = -7$ is a horizontal asymptote.

15. $y = f(x) = \dfrac{4}{x - 6} + 4 = \dfrac{4x - 20}{x - 6}$. From the denominator we find that the line $x = 6$ is a vertical asymptote. Taking limits at infinity, $\lim\limits_{x \to \infty} f(x) = \lim\limits_{x \to \infty} 4x/x = \lim\limits_{x \to \infty} 4 = 4$, and $\lim\limits_{x \to -\infty} f(x) = 4$. Thus $y = 4$ is a horizontal asymptote.

17. $f(x) = (3 - x^4)/(x^3 + x^2) = (3 - x^4)/[x^2(x + 1)]$. Vertical asymptotes are $x = 0$ and $x = -1$. Because the degree of the numerator (4) is greater than the degree of the denominator (3), there is no horizontal asymptote.

19. $y = f(x) = \dfrac{x^2 - 3x - 4}{1 + 4x + 4x^2} = \dfrac{x^2 - 3x - 4}{(1 + 2x)^2}$.

From the denominator, $x = -1/2$ is a vertical asymptote.

Also, $\lim\limits_{x \to \infty} f(x) = \lim\limits_{x \to \infty} \dfrac{x^2}{4x^2} = \lim\limits_{x \to \infty} \dfrac{1}{4} = \dfrac{1}{4}$, and $\lim\limits_{x \to -\infty} f(x) = \dfrac{1}{4}$,

so $y = 1/4$ is a horizontal asymptote.

21. $y = f(x) = \dfrac{9x^2 - 16}{(3x + 4)^2} = \dfrac{(3x + 4)(3x - 4)}{(3x + 4)^2}$. When $x = -4/3$,

both the numerator and denominator are zero. Since

$\lim\limits_{x \to -4/3^+} f(x) = \lim\limits_{x \to -4/3^+} \dfrac{3x - 4}{3x + 4} = -\infty$, the line $x = -4/3$ is a

vertical asymptote. $\lim\limits_{x \to \infty} \dfrac{9x^2 - 16}{(3x + 4)^2} = \lim\limits_{x \to \infty} \dfrac{9x^2}{9x^2} = \lim\limits_{x \to \infty} 1 = 1$.

Similarly, $\lim\limits_{x \to -\infty} f(x) = 1$. Thus $y = 1$ is a vertical

asymptote.

23. $y = f(x) = 2e^{x+2} + 4$. We have $\lim\limits_{x \to \infty} f(x) = +\infty$ and

$\lim\limits_{x \to -\infty} f(x) = 2 \cdot \lim\limits_{x \to -\infty} e^x + \lim\limits_{x \to -\infty} 4 = 2(0) + 4 = 4$.

Thus $y = 4$ is a horizontal asymptote. There is no vertical asymptote because $f(x)$ neither increases nor decreases without bound around any fixed value of x.

25. $y = \dfrac{3}{x}$. Symmetric about the origin.

Vertical asymptote is $x = 0$.
$\lim\limits_{x \to \infty} 3/x = 0 = \lim\limits_{x \to -\infty} 3/x$, so $y = 0$
is a horizontal asymptote.

$$y' = -\dfrac{3}{x^2}.$$

CV: none, however $x = 0$ must be included in the inc.-dec. analysis. Decreasing on $(-\infty,0)$ and $(0,\infty)$.

$y'' = \dfrac{6}{x^3}$. No possible inflection point, but we include

$x = 0$ in the concavity analysis. Concave down on $(-\infty,0)$; concave up on $(0,\infty)$.

27. $y = \frac{x}{x + 1}$. Intercept (0, 0).

Vertical asymptote is x = -1.

$\lim\limits_{x\to\infty} y = 1 = \lim\limits_{x\to-\infty} y$, so y = 1 is

a horizontal asymptote.

$y' = \frac{(x+1)(1) - x(1)}{(x+1)^2} = \frac{1}{(x+1)^2}$.

CV: none, but x = 1 must be included in the inc.-dec.
analysis. Increasing on $(-\infty,-1)$ and $(-1,\infty)$.

$y'' = -2/(x + 1)^3$. No possible inflection point, but
x = -1 must be included in concavity analysis. Concave up
on $(-\infty,-1)$, concave down on $(-1,\infty)$.

29. $y = = x^2 + \frac{1}{x^2} = \frac{x^4 + 1}{x^2}$.

x ≠ 0, so there is no y-intercept.
Setting y = 0 ⟹ no x-intercept.
Replacing x by -x yields symmetry
about the y-axis.

Setting $x^2 = 0$ gives x = 0 as the
only vertical asymptote. Because the degree of the
numerator is greater than the degree of the denominator,
no horizontal asymptote exists.

$y = x^2 + x^{-2}$, $y' = 2x - 2x^{-3} = 2x - \frac{2}{x^3} = \frac{2x^4 - 2}{x^3} =$

$\frac{2(x^4 - 1)}{x^3}$. So $y' = \frac{2(x^2 + 1)(x + 1)(x - 1)}{x^3}$. CV: x = ±1,

but x = 0 must be included in the inc.-dec. analysis.
Decreasing on $(-\infty,-1)$ and (0,1); increasing on (-1,0) and
$(1,\infty)$; rel. min. at (-1,2) and (1,2),

$y'' = 2 + \frac{6}{x^4} > 0$ for all x ≠ 0. Concave up on $(-\infty,0)$ and

$(0,\infty)$.

31. $y = \dfrac{1}{x^2 - 1} = \dfrac{1}{(x + 1)(x - 1)}$.

Intercept $(0,-1)$. Symmetric about the y-axis. Vertical asymptotes are $x = -1$ and $x = 1$. More precisely,

$$\lim_{x \to -1^-} f(x) = \infty, \quad \lim_{x \to -1^+} f(x) = -\infty;$$

$$\lim_{x \to 1^-} f(x) = -\infty, \quad \lim_{x \to 1^+} f(x) = \infty.$$

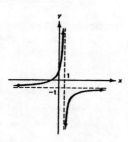

$\lim\limits_{x \to \infty} \dfrac{1}{x^2 - 1} = 0 = \lim\limits_{x \to -\infty} \dfrac{1}{x^2 - 1}$, so $y = 0$ is a horizontal asymptote.

$y' = \dfrac{-2x}{(x^2 - 1)^2}$. CV: $x = 0$, but $x = \pm 1$ must be included in the inc.-dec. analysis. Increasing on $(-\infty,-1)$ and $(-1,0)$; decreasing on $(0,1)$ and $(1,\infty)$; rel. max. at $(0,-1)$.

$y'' = -2 \cdot \dfrac{(x^2-1)^2(1) - x[4x(x^2-1)]}{(x^2-1)^4} = -2 \cdot \dfrac{(x^2-1)[(x^2-1) - 4x^2]}{(x^2-1)^4}$

$= \dfrac{2(3x^2+1)}{(x^2-1)^3} = \dfrac{2(3x^2+1)}{[(x+1)(x-1)]^3}$. No possible inflection point, but $x = \pm 1$ must be considered in the concavity analysis. Concave up on $(-\infty,-1)$ and $(1,\infty)$; concave down on $(-1,1)$.

33. $y = \dfrac{1 + x}{1 - x}$.

Intercepts: Setting $x = 0 \Rightarrow y = 1$; setting $y = 0 \Rightarrow x = -1$. Thus the only intercepts are $(0,1)$ and $(-1,0)$. Setting $1 - x = 0 \Rightarrow x = 1$ is the only vertical asymptote.

Since $\lim\limits_{x \to \infty} \dfrac{1 + x}{1 - x} = \lim\limits_{x \to \infty} \dfrac{x}{-x} =$

$\lim\limits_{x \to \infty} -1 = -1 = \lim\limits_{x \to -\infty} \dfrac{1 + x}{1 - x}$,

the only horizontal asymptote is $y = -1$.

$y' = \dfrac{(1 - x)(1) - (1 + x)(1)}{(1 - x)^2} = \dfrac{2}{(1 - x)^2}$.

No critical values, but $x = 1$ must be considered in the

inc.-dec. analysis. Increasing on $(-\infty,1)$ and $(1,\infty)$.
$y'' = \dfrac{4}{(1 - x)^3}$. No possible inflection point, but $x = 1$
must be included in the concavity analysis. Concave up on
$(-\infty,1)$; concave down on $(1,\infty)$.

35. $y = \dfrac{x^2}{7x + 4}$. Intercept: $(0,0)$.
Vertical asymptote is $x = -4/7$.
Because the degree of the
numerator is greater than the
degree of the denominator,
no horizontal asymptote exists.

$y' = \dfrac{(7x+4)(2x) - x^2(7)}{(7x+4)^2}$

$ = \dfrac{7x^2+8x}{(7x+4)^2} = \dfrac{x(7x+8)}{(7x+4)^2}$.

CV: $x = 0, -8/7$, but $x = -4/7$ must be included in the
inc.-dec. analysis. Increasing on $(-\infty,-8/7)$ and $(0,\infty)$;
decreasing on $(-8/7,-4/7)$ and $(-4/7,0)$; rel. max. at
$(-8/7, -16/49)$; rel. min. at $(0,0)$.

$y'' = \dfrac{(7x^2+4)^2(14x+8) - (7x^2+8x)[14(7x+4)]}{(7x+4)^4}$

$ = \dfrac{(7x+4)[(7x+4)(14x+8) - 14(7x^2+8x)]}{(7x+4)^4} = \dfrac{32}{(7x+4)^3}$.

No possible inflection point but $x = -4/7$ must be included
in concavity analysis. Concave down on $(-\infty,-4/7)$; concave
up on $(-4/7,\infty)$.

37. $y = \dfrac{9}{9x^2-6x-8} = \dfrac{9}{(3x+2)(3x-4)}$.
Intercept: $(0,-9/8)$.
Vert. asym.: $x = -\frac{2}{3}$, $x = \frac{4}{3}$

$\lim\limits_{x\to\infty} y = \lim\limits_{x\to\infty} \dfrac{9}{9x^2} = \lim\limits_{x\to\infty} \dfrac{1}{x^2} = 0 =$
$\lim\limits_{x\to-\infty} y$. Thus $y = 0$ is a
horizontal asymptote.
Since $y = 9(9x^2-6x-8)^{-1}$,

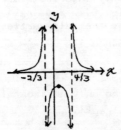

$$y' = 9(-1)(9x^2-6x-8)^{-2}(18x-6) = -\frac{54(3x-1)}{[(3x+2)(3x-4)]^2}.$$

CV: x = 1/3, but x = -2/3 and x = 4/3 must be included in inc.-dec. analysis. Increasing on $(-\infty,-2/3)$ and $(-2/3,1/3)$; decreasing on $(1/3,4/3)$ and $(4/3,\infty)$; rel. max. at $(1/3,-1)$.

$$y'' = -54\cdot\frac{(9x^2-6x-8)^2(3) - (3x-1)[2(9x^2-6x-8)(18x-6)]}{(9x^2-6x-8)^4}$$

$$= -54\cdot\frac{3(9x^2-6x-8)[(9x^2-6x-8) - 4(3x-1)(3x-1)]}{(9x^2-6x-8)^4}$$

$$= \frac{-162(-27x^2+18x-12)}{(9x^2-6x-8)^3} = \frac{486(9x^2-6x+4)}{[(3x+2)(3x-4)]^3}.$$

Since $9x^2-6x+4 = 0$ has no real roots, y" is never zero. No possible inflection point, but x = -2/3 and x = 4/3 must be included in concavity analysis. Concave up on $(-\infty,-2/3)$ and $(4/3,\infty)$; concave down on $(-2/3,4/3)$.

39. $y = \dfrac{2x - 3}{(2x - 9)^2}.$

Intercepts: (3/2,0), (0,-1/27).
Vertical asymptote is x = 9/2.

$$\lim_{x\to\infty} y = \lim_{x\to\infty} \frac{2x}{4x^2} = \lim_{x\to\infty} \frac{1}{2x} = 0 = \lim_{x\to-\infty} y.$$ Thus y = 0 is horiz. asym.

$$y' = \frac{(2x-9)^2(2) - (2x-3)[4(2x-9)]}{(2x-9)^4}$$

$$= \frac{2(2x-9)[(2x-9) - 2(2x-3)]}{(2x-9)^4} = \frac{2(-2x-3)}{(2x-9)^3} = -2\cdot\frac{2x+3}{(2x-9)^3}.$$

CV: x = -3/2, but x = 9/2 must be included in inc.-dec. analysis. Decreasing on $(-\infty,-3/2)$ and $(9/2,\infty)$; increasing on $(-3/2,9/2)$; rel. min. at $(-3/2,-1/24)$.

$$y'' = -2\cdot\frac{(2x-9)^3(2) - (2x+3)[6(2x-9)^2]}{(2x-9)^6}$$

$$= -2\cdot\frac{2(2x-9)^2[(2x-9) - 3(2x+3)]}{(2x-9)^6} = -4\cdot\frac{-4x-18}{(2x-9)^4} = 8\cdot\frac{2x+9}{(2x-9)^4}.$$

Possible inflection point when x = -9/2, but x = 9/2 must be included in concavity analysis. Concave down on $(-\infty,-9/2)$; concave up on $(-9/2,9/2)$ and $(9/2,\infty)$; inflection point at $(-9/2,-1/27)$.

41. $y = \dfrac{x^2 - 1}{x^3} = \dfrac{(x + 1)(x - 1)}{x^3}$.

Intercepts are $(-1, 0)$ and $(1, 0)$.
Symmetric about the origin.
Vertical asymptote $x = 0$.

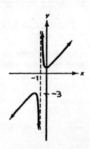

$\lim\limits_{x\to\infty} \dfrac{x^2 - 1}{x^3} = \lim\limits_{x\to\infty} \dfrac{x^2}{x^3} = \lim\limits_{x\to\infty} \dfrac{1}{x} = 0 =$

$\lim\limits_{x\to-\infty} \dfrac{1 - x}{x^2}$, so $y = 0$ is the only

horizontal asymptote. Since $y = x^{-1} - x^{-3}$, then

$y' = -x^{-2} + 3x^{-4} = x^{-4}(-x^2 + 3) = \dfrac{3 - x^2}{x^4}$. CV: $x = \pm\sqrt{3}$,

but $x = 0$ must be included in the inc.-dec. analysis.
Increasing on $(-\sqrt{3}, 0)$ and $(0, \sqrt{3})$; decreasing on
$(-\infty, -\sqrt{3})$ and $(\sqrt{3}, \infty)$; rel. max. at $(\sqrt{3}, 2\sqrt{3}/9)$; rel.
min. at $(-\sqrt{3}, -2\sqrt{3}/9)$.

$y'' = 2x^{-3} - 12x^{-5} = 2x^{-5}(x^2 - 6) = \dfrac{2(x^2 - 6)}{x^5}$. Possible

inflection points when $x = \pm\sqrt{6}$, but $x = 0$ must be
included in the concavity analysis. Concave down on
$(-\infty, -\sqrt{6})$ and $(0, \sqrt{6})$; concave up on $(-\sqrt{6}, 0)$ and $(\sqrt{6}, \infty)$;
inflection points at $(\sqrt{6}, 5\sqrt{6}/36)$ and $(-\sqrt{6}, -5\sqrt{6}/36)$.

43. $y = x + \dfrac{1}{x + 1} = \dfrac{x^2 + x + 1}{x + 1}$.
Intercepts: Setting $x = 0 \Rightarrow y = 1$;
setting $y = 0$ yields no real roots.
Thus the only intercept is $(0,1)$.
Setting $x + 1 = 0 \Rightarrow x = -1$ is the
only vertical asymptote. Because
the degree of the numerator is
greater than the degree of the
denominator, there is no horizontal
asymptote.

$y' = \dfrac{(x + 1)(2x + 1) - (x^2 + x + 1)}{(x + 1)^2} = \dfrac{x^2 + 2x}{(x + 1)^2} = \dfrac{x(x + 2)}{(x + 1)^2}$.

CV: 0 and -2, but $x = -1$ must be included in the inc.-dec.
analysis. Inc. on $(-\infty, -2)$ and $(0, \infty)$; dec. on $(-2, -1)$ and

$(-1,0)$; rel. max. at $(-2,-3)$; rel. min. at $(0,1)$.

$$y" = \frac{(x + 1)^2(2x + 2) - (x^2 + 2x)[2(x + 1)]}{(x + 1)^4}$$

$$= \frac{(x + 1)(2x + 2) - (x^2 + 2x)[2]}{(x + 1)^3} = \frac{2}{(x + 1)^3}.$$

No possible inflection point, but $x = -1$ must be included in the concavity analysis. Concave down on $(-\infty,-1)$; concave up on $(-1,\infty)$.

45. $y = \frac{-3x^2+2x-5}{3x^2-2x-1} = \frac{-3x^2+2x-5}{(3x+1)(x-1)}$.

Note that $-3x^2+2x-5$ is never zero. Intercept: $(0,5)$.
Vertical asymptotes are $x = -1/3$

and $x = 1$. $\lim\limits_{x\to\infty} y = \lim\limits_{x\to\infty} \frac{-3x^2}{3x^2} =$

$\lim\limits_{x\to\infty} -1 = -1 = \lim\limits_{x\to-\infty} y$. Thus
$y = -1$ is horizontal asymptote.

$$y' = \frac{(3x^2-2x-1)(-6x+2) - (-3x^2+2x-5)(6x-2)}{(3x^2-2x-1)^2}$$

$$= \frac{2(3x-1)[(3x^2-2x-1)(-1) - (-3x^2+2x-5)]}{(3x^2-2x-1)^2}$$

$$= \frac{12(3x-1)}{(3x^2-2x-1)^2} = \frac{12(3x-1)}{[(3x+1)(x-1)]^2}.$$

CV: $x = 1/3$, but $x = -1/3$ and $x = 1$ must be included in inc.-dec. analysis. Decreasing on $(-\infty,-1/3)$ and $(-1/3,1/3)$; increasing on $(1/3,1)$ and $(1,\infty)$; rel. min. at $(1/3,7/2)$.

$$y" = 12 \cdot \frac{(3x^2-2x-1)^2(3) - (3x-1)[2(3x^2-2x-1)(6x-2)]}{(3x^2-2x-1)^4}$$

$$= 12 \cdot \frac{(3x^2-2x-1)[3(3x^2-2x-1) - 2(3x-1)(6x-2)]}{(3x^2-2x-1)^4}$$

$$= 12 \cdot \frac{-27x^2+18x-7}{(3x^2-2x-1)^3} = \frac{-12(27x^2-18x+7)}{[(3x+1)(x-1)]^3}.$$

Since $27x^2-18x+7$ is never zero, there is no possible inflection point, but $x = -1/3$ and $x = 1$ must be included in concavity analysis. Concave down on $(-\infty,-1/3)$ and $(1,\infty)$; concave up on $(-1/3,1)$.

47. 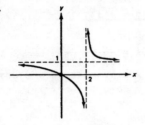 49.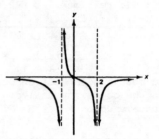

51. $\lim\limits_{x\to\infty} \dfrac{x}{a + bx} = \lim\limits_{x\to\infty} \dfrac{x}{bx} = \lim\limits_{x\to\infty} \dfrac{1}{b} = \dfrac{1}{b}$. Thus $y = \dfrac{1}{b}$ is a horizontal asymptote.

53. $\lim\limits_{t\to\infty} (150 - 76e^{-t}) = \lim\limits_{t\to\infty} \left(150 - \dfrac{76}{e^t}\right) = 150 - 0 = 150$. Thus $y = 150$ is a horizontal asymptote.

55. $x = \pm 2.45$, $x = 0.67$, $y = 2$.

57. From the graph, it appears that $\lim\limits_{x\to\infty} y = 0.48$. Thus a horizontal asymptote is $y = 0.48$. Algebraically, we have

$$\lim_{x\to\infty} \frac{0.34e^{0.7x}}{4.2 + 0.71e^{0.7x}} = \lim_{x\to\infty} \frac{\dfrac{0.34e^{0.7x}}{e^{0.7x}}}{\dfrac{4.2 + 0.71e^{0.7x}}{e^{0.7x}}}$$

$$= \lim_{x\to\infty} \frac{0.34}{\dfrac{4.2}{e^{0.7x}} + 0.71} = \frac{0.34}{0 + 0.71} \approx 0.48.$$

CHAPTER 14 - REVIEW PROBLEMS

1. $y = \dfrac{3x^2}{x^2 - 16} = \dfrac{3x^2}{(x + 4)(x - 4)}$. When $x = \pm 4$ the denominator is zero and the numerator is not zero. Thus $x = 4$ and $x = -4$ are vertical asymptotes.

$\lim\limits_{x\to\infty} \dfrac{3x^2}{x^2 - 16} = \lim\limits_{x\to\infty} \dfrac{3x^2}{x^2} = \lim\limits_{x\to\infty} 3 = 3$. Similarly, $\lim\limits_{x\to-\infty} y = 3$.

Thus $y = 3$ is the only horizontal asymptote.

3. $y = \dfrac{5x^2 - 3}{(3x + 2)^2} = \dfrac{5x^2 - 3}{9x^2 + 12x + 4}$. When $x = -2/3$, the

denominator is zero and the numerator is not zero. Thus $x = -2/3$ is a vertical asymptote.

$\lim\limits_{x\to\infty} y = \lim\limits_{x\to\infty} \dfrac{5x^2}{9x^2} = \lim\limits_{x\to\infty} \dfrac{5}{9} = \dfrac{5}{9}$. Similarly, $\lim\limits_{x\to-\infty} y = \dfrac{5}{9}$. Thus $y = 5/9$ is the only horizontal asymptote.

5. $f(x) = \dfrac{x^2}{2 - x}$;

$f'(x) = \dfrac{(2-x)(2x) - x^2(-1)}{(2-x)^2} = \dfrac{x[2(2-x) + x]}{(2-x)^2} = \dfrac{x(4-x)}{(2-x)^2}$.

Thus $x = 0$ and $x = 4$ are the critical values. Note: Although $f'(2)$ is not defined, 2 is not a critical value because 2 is not in the domain of f.

7. $f(x) = \dfrac{\sqrt[3]{x + 1}}{3 - 4x}$.

$f'(x) = \dfrac{(3-4x)\left[\frac{1}{3}(x+1)^{-2/3}\right] - (x+1)^{1/3}(-4)}{(3-4x)^2}$

$= \dfrac{\frac{1}{3}(x+1)^{-2/3}[(3-4x) + 12(x+1)]}{(3-4x)^2} = \dfrac{8x+15}{3(x+1)^{2/3}(3-4x)^2}$.

$f'(x)$ is zero when $x = -15/8$; $f'(x)$ is not defined when $x = -1$ or $x = 3/4$. However $3/4$ is not in the domain of f. Thus $x = -15/8$ and $x = -1$ are critical values.

9. $f(x) = -x^3 + 6x^2 - 9x$.

$f'(x) = -3x^2 + 12x - 9 = -3(x^2 - 4x + 3) = -3(x-1)(x-3)$.

CV: $x = 1$ and $x = 3$. Increasing on $(1,3)$; decreasing on $(-\infty,1)$ and $(3,\infty)$.

11. $f(x) = \dfrac{x^4}{x^2 - 3}$. $f'(x) = \dfrac{(x^2-3)(4x^3) - x^4(2x)}{(x^2-3)^3} =$

$\dfrac{2x^3[2(x^2-3) - x^2]}{(x^2-3)^2} = \dfrac{2x^3(x^2-6)}{(x^2-3)^2} = \dfrac{2x^3(x+\sqrt{6})(x-\sqrt{6})}{[(x+\sqrt{3})(x-\sqrt{3})]^2}$.

CV: $x = 0, \pm\sqrt{6}$, but $x = \pm\sqrt{3}$ is also considered in the inc.-dec. analysis. Decreasing on $(-\infty,-\sqrt{6})$, $(0,\sqrt{3})$, and

$(\sqrt{3},\sqrt{6})$; increasing on $(-\sqrt{6},-\sqrt{3})$, $(-\sqrt{3},0)$, and $(\sqrt{6},\infty)$.

13. $f(x) = x^4 - x^3 - 14$, $f'(x) = 4x^3 - 3x^2$, $f''(x) = 12x^2 - 6x = 6x(2x - 1)$. $f''(x) = 0$ when $x = 0$ or $x = 1/2$. Concave up on $(-\infty,0)$ and $(1/2,\infty)$; concave down on $(0,1/2)$.

15. $f(x) = \frac{1}{2x - 1} = (2x - 1)^{-1}$, $f'(x) = -2(2x - 1)^{-2}$,

$f''(x) = 8(2x - 1)^3 = \frac{8}{(2x - 1)^3}$. $f''(x)$ is not defined when $x = 1/2$. Concave down on $(-\infty,1/2)$; concave up on $(1/2,\infty)$.

17. $f(x) = (4x + 1)^3(4x + 9)$.

$f'(x) = (4x+1)^3(4) + (4x+9)[12(4x+1)^2]$
    $= 4(4x+1)^2[(4x+1) + 3(4x+9)] = 4(4x+1)^2(16x+28)$
    $= 16(4x+1)^2(4x+7)$.

$f''(x) = 16\{(4x+1)^2(4) + (4x+7)[8(4x+1)]\}$
    $= 64(4x+1)[(4x+1) + 2(4x+7)]$
    $= 64(4x+1)(12x+15) = 192(4x+1)(4x+5)$.

$f''(x) = 0$ when $x = -1/4$ or $x = -5/4$. Concave up on $(-\infty,-5/4)$ and $(-1/4,\infty)$; concave down on $(-5/4,-1/4)$.

19. $f(x) = 2x^3 - 9x^2 + 12x + 7$. $f'(x) = 6x^2 - 18x + 12 = 6(x^2 - 3x + 2) = 6(x - 1)(x - 2)$. CV: $x = 1$ and $x = 2$. Increasing on $(-\infty,1)$ and $(2,\infty)$; decreasing on $(1,2)$. Rel. max. when $x = 1$; rel min. when $x = 2$.

21. $f(x) = \frac{x^6}{6} + \frac{x^3}{3}$. $f'(x) = x^5 + x^2 = x^2(x^3 + 1)$. CV: $x = 0$ and $x = -1$. Decreasing on $(-\infty,-1)$; increasing on $(-1,0)$ and $(0,\infty)$; rel. min. when $x = -1$.

23. $f(x) = x^{2/3}(x + 1) = x^{5/3} + x^{2/3}$.
$f'(x) = \frac{5}{3}x^{2/3} + \frac{2}{3}x^{-1/3} = \frac{1}{3}x^{-1/3}(5x + 2) = \frac{5x + 2}{3x^{1/3}}$.
CV: $x = 0$ and $x = -2/5$. Increasing on $(-\infty,-2/5)$ and $(0,\infty)$; decreasing on $(-2/5,0)$. Rel. max. when $x = -2/5$; rel. min. when $x = 0$.

25. $y = x^5 - 5x^4 + 3x$, $y' = 5x^4 - 20x^3 + 3$, $y'' = 20x^3 - 60x^2 = 20x^2(x - 3)$. Possible inflection points occur when $x = 0$ or $x = 3$. Concave down on $(-\infty,0)$ and $(0,3)$; concave up on $(3,\infty)$. Concavity changes around $x = 3$, so there is an inflection point at $x = 3$.

27. $y = (3x - 5)(x^4 + 2) = 3x^5 - 5x^4 + 6x - 10$, $y' = 15x^4 - 20x^3 + 6$, $y'' = 60x^3 - 60x^2 = 60x^2(x - 1)$. Possible inflection points occur when $x = 0$ or $x = 1$. Concave down on $(-\infty,0)$ and $(0,1)$; concave up on $(1,\infty)$. Inflection point when $x = 1$.

29. $y = \dfrac{x^2}{e^x} = x^2 e^{-x}$, $y' = x^2(-e^{-x}) + e^{-x}(2x) = -e^{-x}(x^2 - 2x)$,

$y'' = -[e^{-x}(2x - 2) + (x^2 - 2x)(-e^{-x})]$
$\quad = e^{-x}[-(2x - 2) + (x^2 - 2x)] = e^{-x}(x^2 - 4x + 2)$.
y'' is defined for all x and y'' is zero only when $x^2 - 4x + 2 = 0$. By the quadradic formula the only possible points of inflection occur when $x = 2 \pm \sqrt{2}$. Concave up on $(-\infty, 2-\sqrt{2})$ and $(2+\sqrt{2}, \infty)$; concave down on $(2-\sqrt{2}, 2+\sqrt{2})$. Inflection points when $x = 2 \pm \sqrt{2}$.

31. $f(x) = 3x^4 - 4x^3$ and f is continuous on $[0,2]$.
$f'(x) = 12x^3 - 12x^2 = 12x^2(x - 1)$. The only critical value on $(0,2)$ is $x = 1$. Evaluating f at this value and at the endpoints gives $f(0) = 0$, $f(1) = -1$, and $f(2) = 16$. Ab. maximum: $f(2) = 16$; ab. minimum: $f(1) = -1$.

33. $f(x) = \dfrac{x}{(5x - 6)^2}$ and f is continuous on $[-2, 0]$.

$f'(x) = \dfrac{(5x-6)^2(1) - x[10(5x-6)]}{(5x-6)^4} = \dfrac{(5x-6)[(5x-6) - 10x]}{(5x-6)^4} =$
$\dfrac{-5x-6}{(5x-6)^3} = -\dfrac{5x+6}{(5x-6)^3}$. The only critical value on $(-2,0)$ is $x = -6/5$. Evaluating f at this value and at the endpoints gives $f(-2) = -1/128$, $f(-6/5) = -1/120$, and $f(0) = 0$. Ab. maximum: $f(0) = 0$; ab. minimum: $f(-6/5) = -1/120$.

35. $f(x) = (x^2 + 1)e^{-x}$.

(a) $f'(x) = (x^2+1)(-e^{-x}) + e^{-x}(2x)$

$= -e^{-x}[(x^2+1) - 2x] = -e^{-x}(x^2-2x+1) = -e^{-x}(x-1)^2$.

CV: $x = 1$. Decreasing on $(-\infty,1)$ and $(1,\infty)$. No relative extrema.

(b) $f''(x) = -\{e^{-x}[2(x-1)] + (x-1)^2(-e^{-x})\}$

$= e^{-x}(x-1)[-2 + (x-1)] = e^{-x}(x-1)(x-3)$.

Possible inflection points when $x = 1, 3$. Concave up on $(-\infty,1)$ and $(3,\infty)$; concave down on $(1,3)$.

Inflection points at $(1, f(1)) = (1, 2e^{-1})$ and $(3, f(3)) = (3, 10e^{-3})$.

37. $y = x^2 - 2x - 24 = (x + 4)(x - 6)$.
Intercepts: $(-4,0)$, $(6,0)$, $(0,-24)$.
No symmetry. No asymptotes.
$y' = 2(x - 1)$. CV: $x = 1$.
Inc. on $(1,\infty)$; dec. on $(-\infty,1)$;
rel. min. at $(1,-25)$.
$y'' = 2$. No possible inflection
point. Concave up on $(-\infty,\infty)$.

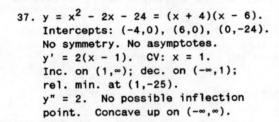

39. $y = x^3 - 12x + 20$.
Intercept: $(0,20)$.
No symmetry. No asymptotes.
$y' = 3x^2 - 12$

$= 3(x^2 - 4) = 3(x + 2)(x - 2)$.
CV: $x = \pm 2$. Increasing on $(-\infty,-2)$
and $(2,\infty)$; decreasing on $(-2,2)$; rel. max. at $(-2,36)$;
rel. min. at $(2,4)$.
$y'' = 6x$. Possible inflection point when $x = 0$. Concave up on $(0,\infty)$; concave down on $(-\infty,0)$; inflection point at $(0,20)$.

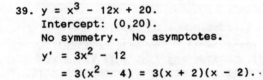

41. $y = x^3 + x = x(x^2 + 1)$.
    Intercept (0,0).
    Symmetric about the origin.
    No asymptotes.
    $y' = 3x^2 + 1$.
    CV: none. Increasing on $(-\infty,\infty)$.
    $y'' = 6x$. Possible inflection
    point when $x = 0$. Concave down
    on $(-\infty,0)$; concave up on $(0,\infty)$; inflection point at (0,0).

43. $y = f(x) = \dfrac{100(x + 5)}{x^2}$.
    Intercept: (-5,0).
    No symmetry.
    $x = 0$ is the only vertical asymptote.
    $\lim\limits_{x\to\infty} y = 100 \lim\limits_{x\to\infty} \dfrac{x}{x^2} = 100 \lim\limits_{x\to\infty} \dfrac{1}{x} = 0$,
    and $\lim\limits_{x\to-\infty} y = 0$, so $y = 0$ is the only horizontal asymptote.

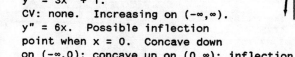

$$y = 100[x^{-1} + 5x^{-2}],$$
$$y' = 100[-x^{-2} - 10x^{-3}] = -100\left[\dfrac{1}{x^2} + \dfrac{10}{x^3}\right] = \dfrac{-100(x + 10)}{x^3}.$$

CV: $x = -10$, but $x = 0$ must be included in analysis
for inc.-dec. Increasing on (-10,0); decreasing on
$(-\infty,-10)$ and $(0,\infty)$; rel. min. at (-10,-5).

$$y'' = 100[2x^{-3} + 30x^{-4}] = 200\left[\dfrac{1}{x^3} + \dfrac{15}{x^4}\right] = \dfrac{200(x + 15)}{x^4}.$$

Possible inflection point when $x = -15$, but $x = 0$ must
also be considered in concavity analysis. Concave up on
(-15,0) and $(0,\infty)$; concave down on $(-\infty,-15)$; inflection
point at (-15,-40/9).

45. $y = \dfrac{x}{(2x - 1)^3}$.   Intercept: $(0,0)$.

No symmetry. Vertical asymptote

is $x = 1/2$. $\lim\limits_{x\to\infty} y = \lim\limits_{x\to\infty} \dfrac{x}{8x^3} =$

$\lim\limits_{x\to\infty} \dfrac{1}{8x^2} = 0 = \lim\limits_{x\to-\infty} y$, so $y = 0$

is a horizontal asymptote.

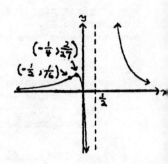

$y' = \dfrac{(2x-1)^3(1) - x[6(2x-1)^2]}{(2x-1)^6}$

$= \dfrac{(2x-1)^2[(2x-1) - 6x]}{(2x-1)^6}$

$= - \dfrac{4x+1}{(2x-1)^4}$.

CV: $x = -1/4$, but $x = 1/2$ must be considered in inc.-dec.
analysis. Increasing on $(-\infty,-1/4)$; decreasing on
$(-1/4,1/2)$ and $(1/2,\infty)$; rel. max. at $(-1/4,2/27)$.

$y'' = - \dfrac{(2x-1)^4(4) - (4x+1)[8(2x-1)^3]}{(2x-1)^8}$

$= - \dfrac{4(2x-1)^3[(2x-1) - 2(4x+1)]}{(2x-1)^8} = - \dfrac{4(-6x-3)}{(2x-1)^5} = \dfrac{12(2x+1)}{(2x-1)^5}$.

Possible inflection point when $x = -1/2$, but $x = 1/2$ must
be considered in concavity analysis. Concave up on
$(-\infty,-1/2)$ and $(1/2,\infty)$; concave down on $(-1/2,1/2)$;
inflection point at $(-1/2,1/16)$.

47. $f(x) = \dfrac{e^x + e^{-x}}{2}$.   Intercept $(0,1)$.

Symmetric about the y-axis.
No asymptotes.

$f'(x) = \dfrac{e^x - e^{-x}}{2}$.

Setting $f'(x) = 0 \Rightarrow e^x = e^{-x} \Rightarrow x = -x \Rightarrow x = 0$.
CV: $x = 0$. Increasing on $(0,\infty)$; decreasing on $(-\infty,0)$;
rel. min. at $(0,1)$

$f''(x) = \dfrac{e^x + e^{-x}}{2}$. Note: $f''(x) > 0$ for all x. No possible

inflection point. Concave up on $(-\infty,\infty)$.

49. (a) False. $f'(x_0) = 0$ only indicates the *possibility* of a
       relative extremum at x_0. For example, if $f(x) = x^3$,
       then $f'(x) = 3x^2$ and $f'(0) = 0$. However there is no
       relative extremum at $x = 0$.
    (b) False. For example, let $x_1 = -1$ and $x_2 = 1$. Then
       $x_1 < x_2$ and $f(x_1) = -1 < f(x_2) = 1$.

    (c) True.
       The absolute maximum
       is $f(1) = 1$.
       The absolute minimum
       is $f(0) = 0$.

    (d) False. If concavity does not change around x_0, then
       $(x_0, f(x_0))$ is not an inflection point. For example,
       consider $f(x) = x^4$. If $x_0 = 0$, then $f''(x_0) = 0$, but
       $(x_0, f(x_0))$ is not an inflection point. See graph in
       part (c).
    (e) False. Consider the
       function f whose graph
       is shown. On (-2,2) it
       has exactly one relative
       maximum [at the point (0,1)]
       but no absolute maximum.

51. $c = q^3 - 6q^2 + 12q + 18$.
    Marginal cost = $dc/dq = 3q^2 - 12q + 12$. Marginal cost
    is increasing when it's derivative, which is d^2c/dq^2,
    is positive.
    $$d^2c/dq^2 = 6q - 12 = 6(q - 2).$$
    $d^2c/dq^2 > 0$ for $q > 2$. Thus marginal cost is increasing
    for $q > 2$.

53. $p = 150 - \frac{\sqrt{q}}{10}$, $q > 0$. The revenue function r is given

by $r = pq = \left(150 - \frac{\sqrt{q}}{10}\right)q = 150q - \frac{q^{3/2}}{10}$.

$r' = 150 - \frac{3}{20}q^{1/2}$. $r'' = -\frac{3}{40}q^{-1/2} = -\frac{3}{40\sqrt{q}}$.

Since $r'' < 0$ for $q > 0$, the graph of the revenue function is concave down for $q > 0$.

55. $f(t) = At^3 + Bt^2 + Ct + D$. $f'(t) = 3At^2 + 2Bt + C$.
$f''(t) = 6At + 2B$, which gives an inflection point when
$6At + 2B = 0$, that is for $t_0 = -B/(3A)$. This value of t_0
must be such that $f'(t_0) = 0$.

$$3A\left(-\frac{B}{3A}\right)^2 + 2B\left(-\frac{B}{3A}\right) + C = 0$$

$$\frac{1}{3}\left(\frac{B^2}{A}\right) - \frac{2}{3}\left(\frac{B^2}{A}\right) + C = 0$$

$$C = \frac{1}{3}\left(\frac{B^2}{A}\right)$$

$$3AC = B^2,$$

which was to be shown.

57. Rel. max. (-1.32, 12.28); rel. min. (0.44, 1.29).

59. The x-value of the inflection point of f corresponds to the x-intercept of f''. Thus the x-value of the inflection point is $x = -0.60$.

15

Applications of Differentiation

1. Let the numbers be x and 40 - x. Then if P = x(40 - x) = 40x - x^2, we have P' = 40 - 2x. Setting P' = 0 $\Rightarrow$ x = 20. Since P"(20) = -2 < 0, there is a maximum when x = 20. Because 40 - x = 20, the required numbers are 20 and 20.

3. We are given that 5x + 3(2y) = 3000, or y = (3000 - 5x)/6. We want to maximize area A, where A = xy.

$$A = xy = x\left(\frac{3000 - 5x}{6}\right) = \frac{1}{6}(3000 - 5x^2).$$

$$A' = \frac{1}{6}(3000 - 10x).$$

Setting A' = 0 $\Rightarrow$ x = 300. Since A"(300) = $\frac{1}{6}$(-10) < 0, we have a maximum at x = 300. Thus y = [3000 - 5(300)]/6 = 250. The dimensions are 300 ft by 250 ft.

5. $c = 0.05q^2 + 5q + 500$.

     Avg. Cost per unit $= \bar{c} = \frac{c}{q} = 0.05q + 5 + \frac{500}{q}$.

$\bar{c}' = 0.05 - \frac{500}{q^2}$. Setting $\bar{c}' = 0$ yields $0.05 = \frac{500}{q^2}$,

$q^2 = 10,000$, $q = \pm100$. We exclude $q = -100$ because q
represents number of units. Since $\bar{c}'' = \frac{1000}{q^3} > 0$ for $q > 0$,

$\bar{c}$ is an absolute minimum when $q = 100$ units.

7. $p = -5q + 30$. Since total revenue = (price)(quantity),
            $r = pq$
            $r = (-5q + 30)q = -5q^2 + 30q$.
Setting $r' = -10q + 30 = 0 \Longrightarrow q = 3$. Since $r'' = -10 < 0$,
r is maximum at $q = 3$ units, for which the corresponding
price is $p = -5(3) + 30 = \$15$.

9. $f(p) = 160 - p - \frac{900}{p + 10}$, where $0 \leq p \leq 100$.

     Setting $f'(p) = 0$ gives $-1 + \frac{900}{(p + 10)^2} = 0$, $\frac{900}{(p + 10)^2} = 1$,

$(p + 10)^2 = 900$, $p + 10 = \pm30$, from which $p = 20$.

    (a) Since $f''(p) = \frac{-1800}{(p + 10)^3} < 0$ for $p = 20$, we have an
        absolute maximum of $f(20) = 110$ grams.

    (b) $f(0) = 70$ and $f(100) = 51\frac{9}{11}$, so we have an absolute

        minimum of $f(100) = 51\frac{9}{11}$ grams.

11. $p = 72 - 0.04q$, $c = 500 + 30q$.
      Profit = Total Revenue - Total Cost
          $P = pq - c = (72 - 0.04q)q - (500 + 30q)$
            $= -(0.04q^2 - 42q + 500)$.

Setting $P' = -(0.08q - 42) = 0$ yields $q = 525$. Since P
is increasing on $[0,525)$ and decreasing on $(525,\infty)$, P is
maximum when $q = 525$ units. This corresponds to a price
of $p = 72 - 0.04(525) = \$51$ and a profit of $P = \$10,525$.

13. $p = 42 - 4q$, $\bar{c} = 2 + \frac{80}{q}$. Total Cost = $c = \bar{c}q = 2q + 80$.

   Profit = Total Revenue - Total Cost
   $$P = pq - c = (42 - 4q)q - (2q + 80)$$
   $$= -(4q^2 - 40q + 80).$$
   $$P' = -(8q - 40).$$

   Setting $P' = -(8q - 40) = 0$ gives $q = 5$. We find that $P" = -8 < 0$, so P has a maximum value when $q = 5$. The corresponding value of price p is $42 - 4(5) = \$22$.

15. $p = q^2 - 100q + 3200$ on $[0, 120]$. $\bar{c} = \frac{2}{3}q^2 - 40q + \frac{10,000}{q}$.

   Profit = Total Revenue - Total Cost

   Since total revenue $r = pq$ and total cost = $c = \bar{c}q$,

   $$P = pq - \bar{c}q$$
   $$= q^3 - 100q^2 + 3200q - \left(\frac{2}{3}q^3 - 40q^2 + 10,000\right)$$
   $$= \frac{1}{3}q^3 - 60q^2 + 3200q - 10,000.$$

   $$P' = q^2 - 120q + 3200 = (q - 40)(q - 80).$$

   Setting $P' = 0$ gives $q = 40$ or 80. Evaluating profit at $q = 0, 40, 80,$ and 120 gives

   $$P(0) = -10,000, \qquad P(40) = \frac{130,000}{3} = 43,333\tfrac{1}{3}$$
   $$P(80) = \frac{98,000}{3} = 32,666\tfrac{2}{3} \qquad P(120) = 86,000$$

   Thus the profit maximizing output is $q = 120$ units, and the corresponding maximum profit is $\$86,000$.

17. Total fixed costs = $\$1200$, material-labor costs/unit = $\$2$, and the demand equation is $p = 100/\sqrt{q}$.

   Profit = Total Revenue - Total Cost
   $$P = pq - c = \frac{100}{\sqrt{q}} \cdot q - (2q + 1200)$$
   $$= 100\sqrt{q} - 2q - 1200 = 2(50\sqrt{q} - q - 600).$$

   Setting $P' = 2\left(\frac{25}{\sqrt{q}} - 1\right) = 0$ yields $q = 625$. We see that

   $P" = -25q^{-3/2} < 0$ for $q > 0$, so P is maximum when $q = 625$. When MR equals MC, then $\frac{50}{\sqrt{q}} = 2$ or, equivalently, $q = 625$. When $q = 625$, then $p = \$4$.

19. If x = number of $0.10 decreases, where $0 \leq x \leq 50$, then the monthly fee for each subscriber is 5 - 0.10x, and the total number of subscribers is 1000 + 100x. Let r be the total (monthly) revenue.

revenue = (monthly rate)(number of subscribers).

$$r = (5 - 0.10x)(1000 + 100x)$$
$$r' = (5 - 0.10x)(100) + (1000 + 100x)(-0.10)$$
$$= 400 - 20x = 20(20 - x).$$

Evaluating r when x = 0, 20, and 50, we find that r is a maximum when x = 20. This corresponds to a monthly fee of 5 - 0.10(20) = $3 and a monthly revenue r of $9,000.

21. See Fig. 15.6 in the text. Given that $x^2y = 32$, we want to minimize $S = 4(xy) + x^2$. Since $y = 32/x^2$, where x > 0, we have $S = 4x(32/x^2) + x^2 = (128/x) + x^2$, from which $S' = (-128/x^2) + 2x$. Setting S' = 0 gives $2x^3 = 128$, $x^3 = 64$, x = 4. Since $S'' = (256/x^3) + 2$, we get S''(4) > 0, so x = 4 gives a minimum. If x = 4, then y = 32/16 = 2. The dimensions are 4 ft × 4 ft × 2 ft.

23.

$$V = (12 - 2x)^2$$
$$= 144x - 48x^2 + 4x^3$$
$$= 4(36x - 12x + x^3),$$

where 0 < x < 6.

$$V' = 4(36 - 24x + 3x^2)$$
$$= 12(x^2 - 8x + 12)$$
$$= 12(x - 6)(x - 2).$$

For 0 < x < 6, setting V' = 0 gives x = 2. Since V' > 0 on (0, 2) and V' < 0 on (2, 6), V is maximum when x = 2. Thus the length of the side of the square must be 2 in., which results in a volume of $(12 - 4)^2(2) = 128$ in.^{3}.

25. See Fig. 15.9 in text.

$$V = K = \pi r^2 h \qquad (1)$$
$$S = 2\pi rh + \pi r^2 \qquad (2)$$

From Eq. (1), $h = K/(\pi r^2)$. Thus Eq. (2) becomes

$$S = \frac{2K}{r} + \pi r^2.$$

$$\frac{dS}{dr} = -\frac{2K}{r^2} + 2\pi r$$

$$= \frac{2(\pi r^3 - K)}{r^2}.$$

If $S' = 0$, then $\pi r^3 - K = 0$, $\pi r^3 = K$, $r = \sqrt[3]{K/\pi}$. Thus

$$h = \frac{K}{\pi \left(\frac{K}{\pi}\right)^{2/3}} = \left(\frac{K}{\pi}\right)^{1/3} = \sqrt[3]{\frac{K}{\pi}}.$$

Note that since $S'' = 2\pi + (4K/r^3) > 0$ for $r > 0$, we have a minimum.

27. $p = 600 - 2q$, $c = 0.2q^2 + 28q + 200$.

Profit = Total Revenue - Total Cost

$$P = pq - c$$

$$P = (600 - 2q)q - (0.2q^2 + 28q + 200)$$

$$= -(2.2q^2 - 572q + 200).$$

$$P' = -(4.4q - 572).$$

Setting $P' = 0$ yields $q = 130$. Since $P'' = -4.4 < 0$, P is maximum when $q = 130$ units. The corresponding price is $p = 600 - 2(130) = \$340$, and the profit is $P = \$36,980$.

If a tax of \$22/unit is imposed on the manufacturer, then the cost equation is

$$c_1 = (0.2q^2 + 28q + 200) + 22q$$

$$= 0.2q^2 + 50q + 200.$$

The demand equation remains the same. Thus

$$P_1 = pq - c_1$$

$$= (600 - 2q)q - (0.2q^2 + 50q + 200)$$

$$= -(2.2q^2 - 550q + 200).$$

$$P_1' = -(4.4q - 550).$$

Setting $P_1' = 0$ yields $q = 125$. Since $P_1'' = -4.4 < 0$, P_1 is maximum when $q = 125$ units. The corresponding price is $p = \$350$ and the profit is $P_1 = \$34,175$.

29. Let q = number of units in a production run. Since inventory is depleted at a uniform rate, assume that the average inventory is q/2. The value of average inventory is 10(q/2), and carrying costs are 0.128[10(q/2)]. The number of production runs per year is 1000/q, and total set-up costs are 40(1000/q). We want to minimize the sum C of carrying costs and set-up costs.

$$C = 0.128[10(q/2)] + 40(1000/q)$$

$$= 0.64q + \frac{40,000}{q}.$$

$$C' = 0.64 - \frac{40,000}{q^2}.$$

Setting C' = 0 yields $q^2 = \frac{40,000}{0.64} = 62,500$, q = 250 (since q > 0). Since $C'' = \frac{80,000}{q^3} > 0$, C is minimum when q = 250. Thus the economic lot size is 250/lot (4 lots).

31. Let x = number of people over the 30. Note: $0 \le x \le 10$. Revenue = r = (number attending)(charge/person).

$$= (30 + x)(50 - 1.25x)$$

$$= 1500 + 12.5x - 1.25x^2.$$

$$r' = 12.5 - 2.5x.$$

Setting r' = 0 yields x = 5. Since r" = -2.5 < 0, r is maximum when x = 5, that is, when 35 attend.

33. The cost per mile of operating the truck = 0.11 + (s/300). Driver's salary is $12/hr. The number of hours for 700 mi trip is 700/s. Driver's salary for trip is 12(700/s), or 8400/s. The cost of operating the truck for the trip is 700[0.11 + (s/300)].

Total cost for trip = $C = \frac{8400}{s} + 700\left(0.11 + \frac{s}{300}\right)$.

Setting $C' = -\frac{8400}{s^2} + \frac{7}{3} = 0$ yields $s^2 = 3600$, or s = 60 (since s > 0). Since $C'' = \frac{16,800}{s^3} > 0$ for s > 0, C is a minimum when s = 60 mi/hr.

35. Profit P is given by
    P = Total revenue - Total cost
      = Total revenue - (salaries + fixed coct)
      = 50q - (750m + 2500)
      = $50(m^3 - 12m^2 + 60m) - 750m - 2500$
      = $50(m^3 - 12m^2 + 45m - 50)$, where $0 \le m \le 7$.
    P' = $50(3m^2 - 24m + 45)$
       = $150(m^2 - 8m + 15) = 150(m - 3)(m - 5)$.
    Setting P' = 0 gives the critical values 3 and 5. We
    now evaluate P at these critical values and also at the
    endpoints 0 and 7.
        P(0) = -2500,   P(30 = 200,   P(5) = 0,   P(7) = 1000.
    Thus Ms. Jones should hire 7 salespeople to obtain a
    maximum weekly profit of $1000.

37. x = tons of chemical A (x $\le$ 4), y = (24 - 6x)/(5 - x) =
    tons of chemical B, profit on A = $2000/ton, and profit
    on B = $1000/ton.
    Total Profit = P_T = $2000x + 1000\left(\dfrac{24 - 6x}{5 - x}\right)$

    $$= 2000\left[x + \dfrac{12 - 3x}{5 - x}\right].$$

    $$P_T' = 2000\left[1 + \dfrac{(5 - x)(-3) - (12 - 3x)(-1)}{(5 - x)^2}\right]$$

    $$= 2000\left[1 - \dfrac{3}{(5 - x)^2}\right]$$

    $$= 2000\left[\dfrac{x^2 - 10x + 22}{(5 - x)^2}\right].$$

    Setting P_T' = 0 yields (by the quadratic formula)

    $$x = \dfrac{10 \pm 2\sqrt{3}}{2} = 5 \pm \sqrt{3}.$$

    Because x $\le$ 4, choose x = 5 - $\sqrt{3}$. Since P_T is increasing
    on [0,5 - $\sqrt{3}$) and decreasing on (5 - $\sqrt{3}$,4], P_T is a
    maximum for x = 5 - $\sqrt{3}$ tons. If profit on A is P/ton and
    profit on B is (P/2)/ton, then

$$P_T = Px + \frac{P}{2}\left(\frac{24 - 6x}{5 - x}\right) = P\left[x + \frac{12 - 6x}{5 - x}\right].$$

$$P_T' = P\left[\frac{x^2 - 10x + 22}{(5 - x)^2}\right].$$

Setting $P_T' = 0$ and using an argument similar to that above, we find that P_T is a maximum when $x = 5 - \sqrt{3}$ tons.

39. $P(j) = Aj\frac{L^4}{V} + B\frac{V^3 L^2}{1+j}$. $\frac{dP}{dj} = \frac{AL^4}{V} - \frac{BV^3 L^2}{(1+j)^2} = 0$. Solving

for $(1 + j)^2$ gives $(1 + j)^2 = \frac{BV^4}{AL^2}$.

41. $\bar{c} = \frac{c}{q} = 3q + 50 - 18 \ln(q) + \frac{120}{q}$, $q > 0$.

$$\frac{d\bar{c}}{dq} = 3 - \frac{18}{q} - \frac{120}{q^2} = \frac{3q^2 - 18q - 120}{q^2} = \frac{3(q^2 - 6q - 40)}{q^2}$$

$$= \frac{3(q - 10)(q + 4)}{q^2}.$$ Critical value $q = 10$.

Since $\frac{d\bar{c}}{dq} < 0$ for $0 < q < 10$, and $\frac{d\bar{c}}{dq} > 0$ for $q > 10$, we have a minimum when $q = 10$ cases. This minimum average cost is $3(10) + 50 - 18 \ln 10 + 12 \approx \$50.55$.

EXERCISE 15.2

1. $y = 3x - 4$. $dy = \frac{d}{dx}(3x - 4) \, dx = 3 \, dx$

3. $d[f(x)] = f'(x) \, dx = \frac{1}{2}(x^4 + 2)^{-1/2}(4x^3) \, dx = \frac{2x^3}{\sqrt{x^4 + 2}} \, dx$

5. $u = x^{-2}$. $du = \frac{d}{dx}(x^{-2}) \, dx = -2x^{-3} \, dx = -\frac{2}{x^3} \, dx$

7. $dp = \frac{d}{dx}[\ln(x^2 + 7)] = \frac{1}{x^2 + 7}(2x) \, dx = \frac{2x}{x^2 + 7} \, dx$

9. $dy = y' \, dx = [(4x + 3)e^{2x^2+3}(4x) + e^{2x^2+3}(4)] \, dx$

$= 4e^{2x^2+3}[(4x + 3)x + 1] \, dx$

$= 4e^{2x^2+3}(4x^2 + 3x + 1) \, dx$

11. $\Delta y = [4 - 7(3.02)] - [4 - 7(3)] = -0.14$

$dy = -7 \, dx = -7(0.02) = -0.14$

13. $\Delta y = [4(-0.75)^2 - 3(-0.75) + 10] - [4(-1)^2 - 3(-1) + 10]$

$= -2.5$

$dy = (8x - 3) \, dx = [8(-1) - 3](0.25) = -2.75$

15. $\Delta y = \sqrt{25 - (2.9)^2} - \sqrt{25 - 3^2}$

$= \sqrt{16.59} - \sqrt{16} \approx 4.073 - 4 = 0.073$

$dy = \dfrac{-x}{\sqrt{25 - x^2}} \, dx = \dfrac{-3}{\sqrt{16}}(-0.1) = 0.075$

17. (a) $f(x) = \dfrac{x + 5}{x + 1}$

$f'(x) = \dfrac{(x + 1)(1) - (x + 5)(1)}{(x + 1)^2} = \dfrac{-4}{(x + 1)^2}.$

$f'(1) = \dfrac{-4}{4} = -1.$

(b) We use $f(x + dx) \approx f(x) + dy$ with $x = 1$, $dx = 0.1$.

$f(1.1) = f(1 + 0.1) \approx f(1) + f'(1) \, dx$

$= \dfrac{6}{2} + (-1)(0.1) = 2.9$

19. $\sqrt{101}$. Let $y = f(x) = \sqrt{x}$.

$f(x + dx) \approx f(x) + dy = \sqrt{x} + \dfrac{1}{2\sqrt{x}} \, dx.$

If $x = 100$ and $dx = 1$, then

$\sqrt{101} = f(100 + 1) \approx \sqrt{100} + \dfrac{1}{2\sqrt{100}}(1) = 10.05.$

21. $\sqrt[3]{63}$. Let $y = f(x) = \sqrt[3]{x}$.

$$f(x + dx) \approx f(x) + dy = \sqrt[3]{x} + \frac{1}{3x^{2/3}} \, dx.$$

If $x = 64$ and $dx = -1$, then

$$\sqrt[3]{63} = f[64 + (-1)] \approx \sqrt[3]{64} + \frac{1}{3(\sqrt[3]{64})^2}(-1)$$

$$= 4 + \frac{-1}{3 \cdot 4^2} = 3\frac{47}{48}.$$

23. $\ln(0.97)$. Let $y = f(x) = \ln x$.

$$f(x + dx) \approx f(x) + dy = \ln(x) + \frac{1}{x} \, dx.$$

If $x = 1$ and $dx = -0.03$, then

$$\ln(0.97) = f[1 + (-0.03)] \approx \ln(1) + \frac{1}{1}(-0.03) = -0.03.$$

25. $e^{0.01}$. Let $y = f(x) = e^x$.

$$f(x + dx) \approx f(x) + dy = e^x + e^x \, dx.$$

If $x = 0$ and $dx = 0.01$, then

$$e^{0.01} = f(0 + 0.01) \approx e^0 + e^0(0.01) = 1.01.$$

27. $\frac{dy}{dx} = 2$, so $\frac{dx}{dy} = \frac{1}{dy/dx} = \frac{1}{2}$

29. $\frac{dq}{dp} = 6p(p^2 + 5)^2$, so $\frac{dp}{dq} = \frac{1}{6p(p^2 + 5)^2}$

31. $q = p^{-1}$, $\frac{dq}{dp} = -1p^{-2} = \frac{-1}{p^2}$, so $\frac{dp}{dq} = -p^2$

33. $\frac{dx}{dy} = \frac{1}{dy/dx} = \frac{1}{10x + 3}$. If $x = 1$, then $\frac{dx}{dy} = \frac{1}{13}$

35. $p = \frac{500}{q + 2}$, $\frac{dp}{dq} = \frac{-500}{(q + 2)^2}$, $\frac{dq}{dp} = -\frac{(q + 2)^2}{500}$.

$$\frac{dq}{dp}\bigg|_{q=18} = -\frac{(q + 2)^2}{500}\bigg|_{q=18} = -\frac{4}{5}.$$

37. $P = 396q - 2.2q^2 - 400$, q changes from 80 to 81.

$$\Delta P \approx dP = P' \, dq = (396 - 4.4q) \, dq.$$

Choosing q = 80 and dq = 1,

$$\Delta P \approx [396 - 4.4(80)](1) = [396 - 352](1) = 44.$$

True change is $P(81) - P(80) = 17,241.80 - 17,200 = 41.80$.

True change is $r(41) - r(40) = 16,974 - 18,000 = -1026$.

39. $p = \dfrac{10}{\sqrt{q}}$. We approximate p when q = 24.

$$p(q + dq) \approx p + dp = \frac{10}{\sqrt{q}} - \frac{5}{\sqrt{q^3}} \, dq.$$

If q = 25 and dq = -1, then

$$\frac{10}{\sqrt{24}} = p[25 + (-1)] \approx \frac{10}{\sqrt{25}} - \frac{5}{\sqrt{(25)^3}}(-1)$$

$$= 2 + \frac{1}{25} = \frac{51}{25} = 2.04.$$

41. $c = (q^4/2) + 3q + 400$. If q = 10 and dq = 2,

$$\frac{dc}{c} = \frac{(2q^3 + 3) \, dq}{(q^4/2) + 3q + 400} = \frac{(2003)(2)}{5430} \approx 0.7.$$

43. $V = \frac{4}{3}\pi r^3$, $dV = 4\pi r^2 \, dr$. Now, the change in r is

$$dr = (6.6 \times 10^{-4}) - (6.5 \times 10^{-4})$$

$$= 0.1 \times 10^{-4} = 10^{-5}.$$

Thus $dV = 4\pi(6.5 \times 10^{-4})^2(10^{-5}) \approx (1.69 \times 10^{-11})\pi \; cm^3$.

45. $2 + \dfrac{q^2}{200} = \dfrac{4000}{p^2}$

(a) We substitute q = 40 and p = 20.

$$2 + \frac{40^2}{200} = \frac{4000}{20^2}, \quad 2 + 8 = 10, \quad 10 = 10.$$

(b) We differentiate implicitly with respect to p.

$$0 + \frac{1}{200}\left(2q \, \frac{dq}{dp}\right) = -\frac{8000}{p^3}$$

From part (a) q = 40 when p = 20. Substituting gives

$$\frac{1}{200}\left(2 \cdot 40 \, \frac{dq}{dp}\right) = -\frac{8000}{20^3}$$

$$\frac{dq}{dp} = -2.5$$

Exercise 15.2 -358-

(c) $q(p + dp) \approx q(p) + dq = q(p) + q'(p)\, dp$
$q(19.20) = q[20 + (-0.8)] \approx q(20) + q'(20)\, dp$
$\qquad = 40 + (-2.5)(-0.8) = 42$ units

EXERCISE 15.3

1. $\eta = \dfrac{p/q}{dp/dq} = \dfrac{p/q}{-2}$. When $q = 5$ then $p = 40 - 2(5) = 30$, so
$\eta = \dfrac{30/5}{-2} = -3$. Because $|\eta| > 1$, demand is elastic.

3. $p = \dfrac{1000}{q} = 1000q^{-1}$, $\dfrac{dp}{dq} = -1000q^{-2} = -\dfrac{1000}{q^2}$.
$\qquad \eta = \dfrac{p/q}{dp/dq} = \dfrac{p/q}{-1000/q^2} = \dfrac{(1000/q)/q}{-1000/q^2} = -1$.
Because $|\eta| = 1$, demand has unit elasticity.

5. $\eta = \dfrac{p/q}{dp/dq} = \dfrac{p/q}{-500/(q + 2)^2} = \dfrac{[500/(q + 2)]/q}{-500/(q + 2)^2} = -\dfrac{q + 2}{q}$. If
$q = 100$, $\eta = -\dfrac{102}{100} = -1.02$. Since $|\eta| > 1$, elastic demand.

7. $\eta = \dfrac{p/q}{dp/dq} = \dfrac{p/q}{-e^{q/100}/100}$. When $q = 100$, then $p = 150 - e$
and $\eta = \dfrac{(150 - e)/100}{-e/100} = -\left(\dfrac{150}{e} - 1\right)$. Because $|\eta| > 1$,
demand is elastic.

9. $q = 600 - 100p$. $\eta = \dfrac{p/q}{dp/dq} = \dfrac{p}{q} \cdot \dfrac{dq}{dp} = \dfrac{p}{q}(-100)$. If $p = 3$,
then $q = 600 - 100(3) = 300$, so $\eta = \dfrac{3}{300}(-100) = -1$. Since
$|\eta| = 1$, demand has unit elasticity.

11. $q = \cdot\sqrt{2500 - p}$. $\eta = \dfrac{p/q}{dp/dq} = \dfrac{p}{q} \cdot \dfrac{dq}{dp}$.
$\qquad \dfrac{dq}{dp} = \dfrac{1}{2}(2500 - p)^{-1/2}(-1) = \dfrac{-1}{2\sqrt{2500 - p}} = \dfrac{-1}{2q}$.
$\eta = \dfrac{p}{q}\left(\dfrac{-1}{2q}\right) = \dfrac{-p}{2q^2}$. If $p = 900$, then $q = \sqrt{2500 - 900} = 40$,
so $\eta\big|_{p=900} = \dfrac{-900}{3200} = \dfrac{-9}{32}$. $|\eta| < 1$, so demand is inelastic.

13. $q = \frac{(p - 100)^2}{2}$. $\eta = \frac{p/q}{dp/dq} = \frac{p}{q} \cdot \frac{dq}{dp}$.

$\frac{dq}{dp} = \frac{1}{2}(2)(p - 100)(1) = p - 100$, so $\eta = \frac{p}{q}(p - 100)$.

If $p = 20$, then $q = \frac{(20 - 100)^2}{2} = 3200$. Thus we have

$\eta\big|_{p=20} = \frac{20}{3200}(20 - 100) = -\frac{1}{2}$. Demand is inelastic.

15. $p = 13 - 0.05q$. $\eta = \frac{p/q}{dp/dq} = -\frac{p}{0.05q}$.

p	q	η	demand
10	60	-10/3	elastic
3	200	-3/10	inelastic
6.50	130	-1	unit elasticity

17. $q = 500 - 40p + p^2$. $\eta = \frac{p/q}{dp/dq} = \frac{p}{q} \cdot \frac{dq}{dp}$.

$\frac{dq}{dp} = -40 + 2p$, so $\eta = \frac{p}{q}(2p - 40)$.

When $p = 15$, then $q = 500 - 40(15) + 15^2 = 125$, so

$\eta\big|_{p=15} = \frac{15}{125}(30 - 40) = -\frac{6}{5} = -1.2$.

Now, (% change in price)(η) = % change in demand. Thus
if the price of 15 increases $\frac{1}{2}$%, then the change in demand
is approximately $(1/2\%)(-1.2) = -0.6\%$. Thus demand
decreases approximately 0.6%.

19. $p = 500 - 2q$. $\eta = \frac{p/q}{dp/dq} = \frac{(500 - 2q)/q}{-2} = \frac{q - 250}{q}$.

If demand is elastic, then $\eta = \frac{q - 250}{q} < -1$. For $q > 0$,
we have $q - 250 < -q$, $2q < 250$, so $q < 125$. Conversely,
if $0 < q < 125$, then $\eta > -1$ and demand is elastic. If
demand is inelastic, then $\eta = \frac{q - 250}{q} > -1$. For $q > 0$,
the inequality implies $q > 125$. Thus if $125 < q < 250$,
then demand is inelastic.

Since Total Revenue = $r = pq = 500q - 2q^2$, then
$$r' = 500 - 4q = 4(125 - q).$$
If $0 < q < 125$, then $r' > 0$, so r is increasing.
If $125 < q < 250$, then $r' < 0$, so r is decreasing.

21. $p = \dfrac{1000}{q^2}$. $r = pq = \dfrac{1000}{q}$. $\dfrac{dr}{dq} = -1000q^{-2} = -\dfrac{1000}{q^2}$.

$$\eta = \dfrac{p/q}{dp/dq} = \dfrac{1000/q^3}{-2000/q^3} = -\dfrac{1}{2}.$$

$$p\left(1 + \dfrac{1}{\eta}\right) = \dfrac{1000}{q^2}(1 - 2) = -\dfrac{1000}{q^2} = \dfrac{dr}{dq}.$$

23. (a) $p = \dfrac{200}{\sqrt{6000 + 10q^2}}$. Substituting $p = 2$ and $q = 20$:

$$2 = \dfrac{200}{\sqrt{6000 + 10(20)^2}}, \quad 2 = \dfrac{200}{100}, \quad 2 = 2.$$

(b) $\dfrac{dp}{dq} = 200\left(-\dfrac{1}{2}\right)(6000 + 10q^2)^{-3/2}(20q)$

$$= \dfrac{-2000q}{(6000 + 10q^2)^{3/2}} = \dfrac{200}{\sqrt{6000 + 10q^2}} \cdot \dfrac{-10q}{6000 + 10q^2}$$

$$= \dfrac{-10pq}{6000 + 10q^2}.$$

Thus $\eta = \dfrac{p/q}{dp/dq} = \dfrac{\frac{p}{q}}{\dfrac{-10pq}{6000 + 10q^2}} = \dfrac{-(6000 + 10q^2)}{10q^2}$.

When $p = 2$ we have $q = 20$, so

$$\eta\big|_{p=2} = \dfrac{-[6000 + 10(20)^2]}{10(20)^2} = -2.5.$$

Thus demand is elastic.

(c) If the price (when $p = 2$) is decreased by 2%, then the percentage change in demand is $(-2)(-2.5) = 5\%$. Since $q = 20$ when $p = 2$, this means that q will increase approximately by 5% of 20, which is 1. That is, demand increases by 1 unit (approx).

(d) Total revenue will increase when the price is lowered because demand is elastic (see the discussion of elasticity and revenue in the text).

25. (a) $q = \dfrac{60}{p} + \ln(65 - p^3)$.

$$\eta = \dfrac{p/q}{dp/dq} = \dfrac{p}{q}\dfrac{dq}{dp} = \dfrac{p}{q}\left[-\dfrac{60}{p^2} - \dfrac{3p^2}{65 - p^3}\right].$$

If $p = 4$, then $q = \dfrac{60}{4} + \ln 1 = 15$, so

$$\eta = \frac{4}{15}\left[-\frac{60}{16} - \frac{3(16)}{65 - 64}\right] = -\frac{207}{15} \approx -13.8, \quad \text{elastic}$$

(b) The percentage change in q is $(-2)(-13.8) = 27.6\%$, so q increases by approximately 27.6%

(c) Lowering the price increases revenue because demand is elastic.

27. The percentage change in price is $\frac{-5}{80}\cdot 100 = -\frac{25}{4}\%$ and the percentage change in quantity is $\frac{50}{500}\cdot 100 = 10\%$. Thus

(elasticity)(% change in price) $\approx$ % change in quantity

$$\text{(elasticity)}\left(-\frac{25}{4}\right) \approx 10$$

$$\text{elasticity} \approx -\frac{40}{25} = -\frac{8}{5} = -1.6.$$

To estimate dr/dq when p = 80, we have

$$\frac{dr}{dq} = p\left(1 + \frac{1}{\eta}\right) = 80\left(1 + \frac{1}{-8/5}\right) = \$30.$$

29. $p = \frac{200}{q + 5}$, $5 \leq q \leq 95$. $\frac{dp}{dq} = 200(-1)(q + 5)^{-2} = \frac{-200}{(q + 5)^2}$.

Thus $\eta = \frac{p/q}{dp/dq} = \frac{200/[q(q + 5)]}{-200/(q + 5)^2} = -\frac{q + 5}{q}$.

For $5 \leq q \leq 95$, $|\eta| = \frac{q + 5}{q} = 1 + \frac{5}{q}$ and $|\eta|' = -\frac{5}{q^2}$.

Since $|\eta|' < 0$, $|\eta|$ is decreasing on [5, 95], and thus $|\eta|$ is maximum at q = 5 and minimum at q = 95.

EXERCISE 15.4

1. We want a root of $f(x) = x^3 - 4x + 1 = 0$. We see that $f(0) = 1$ and $f(1) = -2$ have opposite signs, so there must be a root between 0 and 1. Moreover, $f(0)$ is closer to 0 than is $f(1)$, so we select $x_1 = 0$ as our initial estimate.

Since $f'(x) = 3x^2 - 4$, the recursion formula is

$$x_{n+1} = x_n - \frac{f(x_n)}{f'(x_n)} = x_n - \frac{x_n^3 - 4x_n + 1}{3x_n^2 - 4}.$$

Simplifying gives $x_{n+1} = \dfrac{2x_n^3 - 1}{3x_n^2 - 4}$.

n	x_n	x_{n+1}
1	0.00000	0.25000
2	0.25000	0.25410
3	0.25410	0.25410

Because $|x_4 - x_3| < 0.0001$, the root is approximately $x_4 = 0.25410$.

3. Let $f(x) = x^3 - x - 1$. We have $f(1) = -1$ and $f(2) = 5$ (note the sign change). Since $f(1)$ is closer to 0 than is $f(2)$, we choose $x_1 = 1$. We have $f'(x) = 3x^2 - 1$, so the recursion formula is

$$x_{n+1} = x_n - \frac{f(x_n)}{f'(x_n)} = x_n - \frac{x_n^3 - x_n - 1}{3x_n^2 - 1} = \frac{2x_n^3 + 1}{3x_n^2 - 1}.$$

n	x_n	x_{n+1}
1	1.00000	1.50000
2	1.50000	1.34783
3	1.34783	1.32520
4	1.32520	1.32472
5	1.32472	1.32472

Because $|x_6 - x_5| < 0.0001$, the root is approximately $x_6 = 1.32472$.

5. Let $f(x) = x^3 + x + 16$. We have $f(-3) = -14$ and $f(-2) = 6$ (note the sign change). Since $f(-2)$ is closer to 0 than is $f(-3)$, we choose $x_1 = -2$. Since $f'(x) = 3x^2 + 1$, the recursion formula is

$$x_{n+1} = x_n - \frac{f(x_n)}{f'(x_n)} = x_n - \frac{x_n^3 + x_n + 16}{3x_n^2 + 1} = \frac{2x_n^3 - 16}{3x_n^2 + 1}.$$

n	x_n	x_{n+1}
1	-2.00000	-2.46154
2	-2.46154	-2.38977
3	-2.38977	-2.38769
4	-2.38769	-2.38769

Because $|x_5 - x_4| < 0.0001$, the root is approximately $x_5 = -2.38769$.

7. $x^4 = 3x - 1$, or $f(x) = x^4 - 3x + 1 = 0$. Since $f(0) = 1$ and $f(1) = -1$ (note the sign change), $f(0)$ and $f(1)$ are equally close to 0. We shall choose $x_1 = 0$. Since $f'(x) = 4x^3 - 3$, the recursion formula is

$$x_{n+1} = x_n - \frac{f(x_n)}{f'(x_n)} = x_n - \frac{x_n^4 - 3x_n + 1}{4x_n^3 - 3}$$

$$= \frac{3x_n^4 - 1}{4x_n^3 - 3}.$$

n	x_n	x_{n+1}
1	0.00000	0.33333
2	0.33333	0.33766
3	0.33766	0.33767

Since $|x_4 - x_3| < 0.0001$, root is approx. $x_4 = 0.33767$.

9. Let $f(x) = x^4 - 2x^3 + x^2 - 3$. $f(1) = -3$ and $f(2) = 1$ (note the sign change), so $f(2)$ is closer to 0 than is $f(1)$. We choose $x_1 = 2$. Since $f'(x) = 4x^3 - 6x^2 + 2x$, the recursion formula is

$$x_{n+1} = x_n - \frac{f(x_n)}{f'(x_n)} = x_n - \frac{x_n^4 - 2x_n^3 + x_n^2 - 3}{4x_n^3 - 6x_n^2 + 2x_n}.$$

n	x_n	x_{n+1}
1	2.00000	1.91667
2	1.91667	1.90794
3	1.90794	1.90785

Since $|x_4 - x_3| < 0.0001$, the is approx. $x_4 = 1.90785$.

11. The desired number is x, where $x^3 = 70$, or $x^3 - 70 = 0$. Thus we want to find a root of $f(x) = x^3 - 70 = 0$. Since $4^3 = 64$, the solution should be close to 4, so we choose $x_1 = 4$ as our initial estimate. We have $f'(x) = 3x^2$, so the recursion formula is

$$x_{n+1} = x_n - \frac{f(x_n)}{f'(x_n)} = x_n - \frac{x_n^3 - 70}{3x_n^2} = \frac{2x_n^3 + 70}{3x_n^2}.$$

n	x_n	x_{n+1}
1	4	4.125
2	4.125	4.121
3	4.121	4.121

Thus to three decimal places, $\sqrt[3]{70} = 4.121$

13. We want real solutions of $e^x = x + 5$. Thus we want to find roots of $f(x) = e^x - x - 5 = 0$. A rough sketch of the exponential function $y = e^x$ and the line $y = x + 5$ shows that there are two intersection points: one when x is near -5, and the other when x is near 3. Thus we must find two roots. Since $f'(x) = e^x - 1$, the recursion formula is

$$x_{n+1} = x_n - \frac{f(x_n)}{f'(x_n)} = x_n - \frac{e^{x_n} - x_n - 5}{e^{x_n} - 1}$$

If $x_1 = -5$, we obtain

n	x_n	x_{n+1}
1	-5	-4.99
2	-4.99	-4.99

If $x_1 = 3$, we obtain:

n	x_n	x_{n+1}
1	3	2.37
2	2.37	2.03
3	2.03	1.94
4	1.94	1.94

Thus the solutions are -4.99 and 1.94.

15. The break-even quantity is the value of q when total
    revenue and total cost are equal: r = c, or r - c = 0.
    Thus we must find a root of $3q - (250 + 2q - 0.1q^3) = 0$,
    or $f(q) = q - 250 + 0.1q^3 = 0$, where $f'(q) = 1 + 0.3q^2$.
    The recursion formula is
    $$q_{n+1} = q_n - \frac{f(q_n)}{f'(q_n)} = q_n - \frac{q - 250 + 0.1q^3}{1 + 0.3q^2}.$$
    We choose $q_1 = 13$, as suggested.

n	q_n	q_{n+1}
1	13	13.33
2	13.33	13.33

Thus $q \approx 13.33$.

17. The equilibrium quantity is the value of q for which
    supply and demand are equal, that is, it is a root of
    $2q + 5 = \frac{100}{q^2 + 1}$, or of $f(q) = 2q + 5 - \frac{100}{q^2 + 1} = 0$. Since
    $f'(q) = 2 + \frac{200q}{(q^2 + 1)^2}$, the recursion formula is
    $$q_{n+1} = q_n - \frac{f(q_n)}{f'(q_n)} = q_n - \frac{2q + 5 - \frac{100}{q^2 + 1}}{2 + \frac{200q}{(q^2 + 1)^2}}$$
    A rough sketch shows that the graph of the supply equation
    intersects the graph of the demand equation when q is near

3. Thus we select $q_1 = 3$.

n	x_n	x_{n+1}
1	3	2.875
2	2.875	2.880
3	2.880	2.880

Thus $q \approx 2.880$

19. For a critical value of $f(x) = \frac{x^3}{3} - x^2 - 5x + 1$, we want a root of $f'(x) = x^2 - 2x - 5 = 0$. Since $\frac{d}{dx}[f'(x)] = 2x - 2$, the recursion formula is

$$x_{n+1} = x_n - \frac{x^2 - 2x - 5}{2x - 2}.$$

For the given interval $[3, 4]$, note that $f'(3) = -2$ and $f'(4) = 3$ have opposite signs. Thus there is a root x between 3 and 4. Since 3 is closer to 0, we shall select $x_1 = 3$.

n	x_n	x_{n+1}
1	3.	3.5
2	3.5	3.45
3	3.45	3.45

Thus $x \approx 3.45$.

CHAPTER 15 - REVIEW PROBLEMS

1. $q = 80m^2 - 0.1m^4$. $\frac{dq}{dm} = 160m - 0.4m^3 = 0.4m(400 - m^2) = 0.4m(20 + m)(20 - m)$. Setting $\frac{dq}{dm} = 0$ yields $m = 0$ or $m = 20$ (for $m \geq 0$). We find that q is increasing on $(0, 20)$ and decreasing on $(20, \infty)$, so q is maximum at $m = 20$.

3. $p = \sqrt{600 - q}$, where $100 \leq q \leq 300$.

Total Revenue $= r = pq = q\sqrt{600 - q}$.

$r' = q\left(\frac{1}{2}\right)(600 - q)^{-1/2}(-1) + \sqrt{600 - q}\,(1)$

$\quad = \frac{1}{2}(600 - q)^{-1/2}[-q + 2(600 - q)] = \frac{1200 - 3q}{2\sqrt{600 - q}}$

$\quad = \frac{3(400 - q)}{2\sqrt{600 - q}}$. No critical values over $[100, 300]$.

Since $r' > 0$ on $[100, 300]$, r is increasing on $[100, 300]$, so r must have a maximum at $q = 300$.

5. $p = 400 - 2q$. $\bar{c} = q + 160 + \frac{2000}{q}$.

Total Cost $= c = \bar{c}q = q^2 + 160q + 2000$.

Profit = Total Revenue - Total Cost

$\qquad P = pq - c = (400 - 2q)q - (q^2 + 160q + 2000)$

$\qquad\quad = -(3q^2 - 240q + 2000)$.

$\qquad P' = -(6q - 240) = -6(q - 40)$.

Setting $P' = 0$ yields $q = 40$. Since $P'' = -6 < 0$, P is maximum when $q = 40$, and the corresponding profit is

$P = -[3\cdot40^2 - 240(40) + 2000] = \$2800$.

7. $2x + 4y = 800$; thus $x = 400 - 2y$

Area $= A = xy = (400 - 2y)y$

$\qquad = 400y - 2y^2$.

$DA/dy = 400 - 4y = 4(100 - y)$.

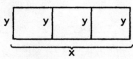

Setting $dA/dy = 0$ gives $y = 100$. Since $d^2A/dy^2 = -4 < 0$, A is maximum when $y = 100$. When $y = 100$, then $x = 200$. The dimensions are 200 ft by 100 ft.

9. (a) $c = 2q^3 - 9q^2 + 12q + 20$, where $\frac{3}{4} \leq q \leq 6$.

$\qquad \frac{dc}{dq} = 6q^2 - 18q + 12 = 6(q^2 - 3q + 2)$

$\qquad\quad = 6(q - 1)(q - 2)$.

Setting $dc/dq = 0$ gives $q = 1$ or 2. Evaluating c at these critical values and the endpoints:

$c\left(\frac{3}{4}\right) = \frac{793}{32} \approx 24.78$, $c(1) = 25$, $c(2) = 24$, $c(6) = 200$.

Thus a minimum occurs at $q = 2$, which corresponds to

200 desks and a total cost of $24,000. This gives an
average cost per desk of 24,000/200 = $120.

(b) There are no critical values for $3 \leq q \leq 6$, so we only
evaluate c at the endpoints: c(3) = 29, c(6) = 200.
Thus a minimum occurs at q = 3, which corresponds to
300 desks.

11. $d\left[x^2 \ln(x + 5)\right] = \frac{d}{dx}\left[x^2 \ln(x + 5)\right] dx$

$= \left[\frac{x^2}{x + 5} + 2x \ln(x + 5)\right] dx.$

13. $F = \frac{9}{5}C + 32.$ $dF = F' \, dC = \frac{9}{5} \, dC = \frac{9}{5}\left(\frac{1}{2}\right)^\circ = \left(\frac{9}{10}\right)^\circ.$

15. $e^{-0.01}.$ Let $y = f(x) = e^x.$

$f(x + dx) \approx f(x) + dy = e^x + e^x \, dx.$

If x = 0 and dx = -0.01, then

$e^{-0.01} = f[0 + (-0.01)] \approx e^0 + e^0(-0.01)$

$= 1 - 0.01 = 0.99.$

17. $x = 4y^2 + 7y - 3.$ $\frac{dy}{dx} = \frac{1}{dx/dy} = \frac{1}{8y + 7}.$

19. $p = 900 - q^2.$ $\eta = \frac{p/q}{dp/dq} = \frac{(900 - q^2)/q}{-2q} = -\frac{900 - q^2}{2q^2}.$

When q = 10, then $\eta = -4.$ Since $|\eta| > 1$, demand is elastic

21. $p = 30 - \sqrt{q}.$

$\eta = \frac{p/q}{dp/dq} = \frac{p/q}{-1/(2\sqrt{q})} = \frac{-2p}{\sqrt{q}} = \frac{-2p}{30 - p} = \frac{2p}{p - 30}.$

(a) When p = 10, then $\eta = \frac{2(10)}{10 - 30} = -1.$

(b) $\eta = \frac{2p}{p - 30}.$ If $0 < p < 10$, then $-1 < \eta < 0$, so
$|\eta| < 1$ and demand is inelastic.

23. (a) $\eta = \dfrac{p/q}{dp/dq} = \dfrac{p}{q} \cdot \dfrac{dq}{dp}$. $q = \sqrt{100 - p}$, where $0 < p < 100$.

$\dfrac{dq}{dp} = \dfrac{-1}{2\sqrt{100 - p}}$. Thus

$\eta = \dfrac{p}{\sqrt{100 - p}} \cdot \dfrac{-1}{2\sqrt{100 - p}} = \dfrac{-p}{2(100 - p)} = \dfrac{p}{2p - 200}$.

For elastic demand we want $\dfrac{p}{2p - 200} < -1$. Noting that
the denominator is negative for $0 < p < 100$, we
multiply both sides of the inequality by $2p - 200$ and
reverse the direction of the inequality.

$p > -2p + 200 \implies 3p > 200 \implies p > \dfrac{200}{3}$.

Thus $\dfrac{200}{3} < p < 100$ for elastic demand.

(b) $|\eta|_{p=40} = \dfrac{40}{80 - 200} = -\dfrac{1}{3}$.

% change in q $\approx$ (% change in price)(η)

$= 5\left(-\dfrac{1}{3}\right) = -\dfrac{5}{3}\% = -1.67\%$.

Thus demand decreases by approximately 1.67%.

25. We want real solutions of $e^x = 3x$. Thus we want to
find roots of $f(x) = e^x - 3x = 0$. A rough sketch of
the exponential function $y = e^x$ and the line $y = 3x$
shows that there are two intersection points: one when x
is near 0.5, and the other when x is near 1.5. Thus we
must find two roots. Since $f'(x) = e^x - 3$, the recursion
formula is

$$x_{n+1} = x_n - \dfrac{f(x_n)}{f'(x_n)} = x_n - \dfrac{e^{x_n} - 3x_n}{e^{x_n} - 3}$$

If $x_1 = 0.5$, we obtain

n	x_n	x_{n+1}
1	0.5	0.610
2	0.610	0.619
3	0.619	0.619

If $x_1 = 1.5$, we obtain:

n	x_n	x_{n+1}
1	1.5	1.512
2	1.512	1.512

Thus the solutions are 0.619 and 1.512.

16

Integration

1. $\int 5 \ dx = 5x + C$

3. $\int x^8 \ dx = \dfrac{x^{8+1}}{8+1} + C = \dfrac{x^9}{9} + C$

5. $\int 5x^{-7} \ dx = 5\int x^{-7} \ dx = 5\cdot\dfrac{x^{-7+1}}{-7+1} + C = 5\cdot\dfrac{x^{-6}}{-6} + C = -\dfrac{5}{6x^6} + C$

7. $\int \dfrac{1}{x^{10}} \ dx = \int x^{-10} \ dx = \dfrac{x^{-10+1}}{-10+1} + C = \dfrac{x^{-9}}{-9} + C = -\dfrac{1}{9x^9} + C$

9. $\int \dfrac{1}{y^{11/5}} \ dy = \int y^{-11/5} \ dy = \dfrac{y^{(-11/5)+1}}{(-11/5)+1} + C$

$$= \dfrac{y^{-6/5}}{-6/5} + C = -\dfrac{5}{6y^{6/5}} + C$$

11. $\int (8 + u)\ du = \int 8\ du + \int u\ du = 8u + \frac{u^{1+1}}{1+1} + C = 8u + \frac{u^2}{2} + C$

13. $\int (y^5 - 5y)\ dy = \int y^5\ dy - \int y\ dy = \frac{y^{5+1}}{5+1} - 5 \cdot \frac{y^{1+1}}{1+1} + C$

$$= \frac{y^6}{6} - 5 \cdot \frac{y^2}{2} + C = \frac{y^6}{6} - \frac{5y^2}{2} + C$$

15. $\int (3t^2 - 4t + 5)\ dt = 3 \int t^2\ dt - 4 \int t\ dt + \int 5\ dt$

$$= 3 \cdot \frac{t^3}{3} - 4 \cdot \frac{t^2}{2} + 5t + C = t^3 - 2t^2 + 5t + C$$

17. Since $7 + e$ is a constant, $\int (7 + e)\ dx = (7 + e)x + C.$

19. $\int \left(\frac{x}{7} - \frac{3}{4}x^4\right)\ dx = \frac{1}{7} \int x\ dx - \frac{3}{4} \int x^4\ dx = \frac{1}{7} \cdot \frac{x^2}{2} - \frac{3}{4} \cdot \frac{x^5}{5} + C$

$$= \frac{x^2}{14} - \frac{3x^5}{20} + C$$

21. $\int 3e^x\ dx = 3 \int e^x\ dx = 3e^x + C$

23. $\int (x^{8.3} - 9x^6 + 3x^{-4} + x^{-3})\ dx$

$$= \frac{x^{9.3}}{9.3} - 9 \cdot \frac{x^7}{7} + 3 \cdot \frac{x^{-3}}{-3} + \frac{x^{-2}}{-2} + C = \frac{x^{9.3}}{9.3} - \frac{9x^7}{7} - \frac{1}{x^3} - \frac{1}{2x^2} + C$$

25. $\int \frac{-2\sqrt{x}}{3}\ dx = -\frac{2}{3} \int x^{1/2}\ dx = -\frac{2}{3} \cdot \frac{x^{(1/2)+1}}{(1/2)+1} + C$

$$= -\frac{2}{3} \cdot \frac{x^{3/2}}{3/2} + C = -\frac{4x^{3/2}}{9} + C$$

27. $\int \frac{1}{4\sqrt[8]{x^7}}\ dx = \frac{1}{4} \int x^{-7/8}\ dx = \frac{1}{4} \cdot \frac{x^{(-7/8)+1}}{(-7/8)+1} + C$

$$= \frac{1}{4} \cdot \frac{x^{1/8}}{1/8} + C = 2\sqrt[8]{x} + C$$

29. $\int\left(\frac{x^3}{3} - \frac{3}{x^3}\right) dx = \frac{1}{3}\int x^3 dx - 3\int x^{-3} dx$

$= \frac{1}{3}\cdot\frac{x^{3+1}}{3+1} - 3\cdot\frac{x^{-3+1}}{-3+1} + C = \frac{1}{3}\cdot\frac{x^4}{4} - 3\cdot\frac{x^{-2}}{-2} + C = \frac{x^4}{12} + \frac{3}{2x^2} + C$

31. $\int\left(\frac{3w^2}{2} - \frac{2}{3w^2}\right) dw = \frac{3}{2}\int w^2 dw - \frac{2}{3}\int w^{-2} dw$

$= \frac{3}{2}\cdot\frac{w^3}{3} - \frac{2}{3}\cdot\frac{w^{-1}}{-1} + C = \frac{w^3}{2} + \frac{2}{3w} + C$

33. $\int\frac{2z - 5}{7} dz = \frac{1}{7}\int(2z - 5) dz = \frac{1}{7}\left(2\int z dz - \int 5 dz\right)$

$= \frac{1}{7}\left(2\cdot\frac{z^{1+1}}{1 + 1} - 5z\right) + C = \frac{1}{7}\left(2\cdot\frac{z^2}{2} - 5z\right) + C = \frac{1}{7}(z^2 - 5z) + C$

35. $\int(x^e + e^x) dx = \int x^e dx + \int e^x dx = \frac{x^{e+1}}{e + 1} + e^x + C$

37. $\int(2\sqrt{x} - 3\sqrt[4]{x}) dx = \int(2x^{1/2} - 3x^{1/4}) dx$

$= 2\int x^{1/2} dx - 3\int x^{1/4} dx = 2\cdot\frac{x^{(1/2)+1}}{(1/2) + 1} - 3\cdot\frac{x^{(1/4)+1}}{(1/4) + 1} + C$

$= 2\cdot\frac{x^{3/2}}{3/2} - 3\cdot\frac{x^{5/4}}{5/4} + C = \frac{4x^{3/2}}{3} - \frac{12x^{5/4}}{5} + C$

39. $\int\left(- \frac{\sqrt[3]{x^2}}{5} - \frac{7}{2\sqrt{x}} + 6x\right) dx = \int\left(- \frac{x^{2/3}}{5} - \frac{7x^{-1/2}}{2} + 6x\right) dx$

$= - \frac{1}{5}\int x^{2/3} dx - \frac{7}{2}\int x^{-1/2} dx + 6\int x dx$

$= - \frac{1}{5}\cdot\frac{x^{5/3}}{5/3} - \frac{7}{2}\cdot\frac{x^{1/2}}{1/2} + 6\cdot\frac{x^2}{2} + C = - \frac{3x^{5/3}}{25} - 7x^{1/2} + 3x^2 + C$

41. $\int(x^2 + 5)(x - 3) dx = \int(x^3 - 3x^2 + 5x - 15) dx$

$= \frac{x^4}{4} - 3\cdot\frac{x^3}{3} + 5\cdot\frac{x^2}{2} - 15x + C = \frac{x^4}{4} - x^3 + \frac{5x^2}{2} - 15x + C$

43. $\int \sqrt{x}(x + 3)\ dx = \int(x^{3/2} + 3x^{1/2})\ dx = \frac{x^{5/2}}{5/2} + 3 \cdot \frac{x^{3/2}}{3/2} + C$

$$= \frac{2x^{5/2}}{5} + 2x^{3/2} + C$$

45. $\int(2u + 1)^2\ du = \int(4u^2 + 4u + 1)\ du$

$$= 4 \cdot \frac{u^3}{3} + 4 \cdot \frac{u^2}{2} + u + C = \frac{4u^3}{3} + 2u^2 + u + C$$

47. $\int v^{-2}(2v^4 + 3v^2 - 2v^{-3})\ dv = \int(2v^2 + 3 - 2v^{-5})\ dv$

$$= 2 \cdot \frac{v^3}{3} + 3v - 2 \cdot \frac{v^{-4}}{-4} + C = \frac{2v^3}{3} + 3v + \frac{1}{2v^4} + C$$

49. $\int \frac{z^4 + 10z^3}{5z^2}\ dz = \frac{1}{5}\int\left(\frac{z^4}{z^2} + \frac{10z^3}{z^2}\right)\ dz = \frac{1}{5}\int(z^2 + 10z)\ dz$

$$= \frac{1}{5}\left(\frac{z^3}{3} + 10 \cdot \frac{z^2}{2}\right) + C = \frac{z^3}{15} + z^2 + C$$

51. $\int \frac{e^x + e^{2x}}{e^x}\ dx = \int\left(\frac{e^x}{e^x} + \frac{e^{2x}}{e^x}\right)\ dx = \int(1 + e^x)\ dx = x + e^x + C$

53. No, $F(x) - G(x)$ must be a constant.

55. Because an antiderivative of the derivative of a function is the function itself, we have

$$\int \frac{d}{dx}\left[\frac{1}{\sqrt{x^2 + 1}}\right] dx = \frac{1}{\sqrt{x^2 + 1}} + C.$$

EXERCISE 16.2

1. $\frac{dy}{dx} = 3x - 4$. $y = \int(3x - 4)\ dx = \frac{3x^2}{2} - 4x + C$. Using

$y(-1) = \frac{13}{2}$ gives $\frac{13}{2} = \frac{3(-1)^2}{2} - 4(-1) + C$, $\frac{13}{2} = \frac{11}{2} + C$.

Thus $C = 1$, so $y = \frac{3x^2}{2} - 4x + 1$.

3. $y' = \frac{4}{\sqrt{x}}$. $y = \int \frac{4}{\sqrt{x}} dx = \int 4x^{-1/2} dx = 4 \cdot \frac{x^{1/2}}{1/2} + C$. Thus

$y = 8x^{1/2} + C$. Now, $y(4) = 10$ implies $10 = 8 \cdot (4)^{1/2} + C$,

$10 = 16 + C$, $C = -6$. Thus $y = 8\sqrt{x} - 6$. Evaluating at

$x = 9$ gives $y(9) = 8 \cdot 3 - 6 = 18$.

5. $y'' = -x^2 - 2x$. $y' = \int (-x^2 - 2x) dx = -\frac{x^3}{3} - x^2 + C_1$.

$y'(1) = 0$ implies $0 = -\frac{1}{3} - 1 + C_1$, so $C_1 = \frac{4}{3}$. Thus

$y' = \frac{x^3}{3} - x^2 + \frac{4}{3}$. Now we integrate again to find y.

$y = \int \left(\frac{x^3}{3} - x^2 + \frac{4}{3} \right) dx = -\frac{x^4}{12} - \frac{x^3}{3} + \frac{4x}{3} + C_2$. $y(1) = 1$, so

$1 = -\frac{1}{12} - \frac{1}{3} + \frac{4}{3} + C_2$, from which $C_2 = \frac{1}{12}$. Thus we have

$y = -\frac{x^4}{12} - \frac{x^3}{3} + \frac{4x}{3} + \frac{1}{12}$.

7. $y''' = 2x$. $y'' = \int 2x \, dx = x^2 + C_1$. $y''(-1) = 3$ implies that

$3 = 1 + C_1$, so $C_1 = 2$. $y' = \int (x^2 + 2) dx = \frac{x^3}{3} + 2x + C_2$.

$y'(3) = 10$ implies $10 = 9 + 6 + C_2$, so $C_2 = -5$. Thus

$y = \int \left(\frac{x^3}{3} + 2x - 5 \right) dx = \frac{x^4}{12} + x^2 - 5x + C_3$. $y(0) = 2$

implies that $2 = 0 + 0 - 0 + C_3$, so $C_3 = 2$. Therefore

$y = \frac{x^4}{12} + x^2 - 5x + 2$.

9. $\frac{dr}{dq} = 0.7$. $r = \int 0.7 \, dq = 0.7q + C$. If $q = 0$, r must be 0,

so $0 = 0 + C$, or $C = 0$. Thus $r = 0.7q$. Since $r = pq$, we

have $p = \frac{r}{q} = \frac{0.7q}{q} = 0.7$. The demand function is $p = 0.7$.

11. $\frac{dr}{dq} = 275 - q - 0.3q^2$. Thus $r = \int (275 - q - 0.3q^2)\ dq = 275q - 0.5q^2 - 0.1q^3 + C$. When $q = 0$, r must be 0. Thus $0 = 0 - 0 - 0 + C$, so $C = 0$ and $r = 275q - 0.5q^2 - 0.1q^3$. Since $r = pq$, then $p = r/q = 275 - 0.5q - 0.1q^2$. Thus the demand function is $p = 275 - 0.5q - 0.1q^2$.

13. $dc/dq = 1.35$. $c = \int 1.35\ dq = 1.35q + C$. When $q = 0$, then $c = 200$, so $200 = 0 + C$, or $C = 200$. Thus $c = 1.35q + 200$.

15. $\frac{dc}{dq} = 0.09q^2 - 1.2q + 4.5$. $c = \int (0.09q^2 - 1.2q + 4.5)\ dq = 0.03q^3 - 0.6q^2 + 4.5q + C$. If $q = 0$, then $c = 7700$, from which $C = 0$. Hence $c = 0.03q^3 - 0.6q^2 + 4.5q + 7700$. If $q = 10$, substituting gives $c = 7715$.

17. $G = \int \left[-\frac{P}{25} + 2 \right]\ dP = -\frac{P^2}{50} + 2P + C$. When $P = 10$, then $G = 38$, so $38 = -2 + 20 + C$, from which $C = 20$. Thus $G = -\frac{1}{50}P^2 + 2P + 20$

19. $v = \int -\frac{(P_1 - P_2)r}{2L\eta}\ dr = -\frac{(P_1 - P_2)r^2}{4L\eta} + C$. Since $v = 0$ when $r = R$, then $0 = -\frac{(P_1-P_2)R^2}{4L\eta} + C$, so $C = \frac{(P_1 - P_2)R^2}{4L\eta}$. Thus

$$v = -\frac{(P_1 - P_2)r^2}{4L\eta} + \frac{(P_1 - P_2)R^2}{4L\eta} = \frac{(P_1 - P_2)(R^2 - r^2)}{4L\eta}.$$

21. $dc/dq = 0.003q^2 - 0.4q + 40$. $c = \int (0.003q^2 - 0.4q + 40)\ dq = 0.003 \cdot \frac{q^3}{3} - 0.04 \cdot \frac{q^2}{2} + 40q + C = 0.001q^3 - 0.2q^2 + 40q + C$. When $q = 0$, then $c = 5000$, so $5000 = 0 - 0 + 0 + C$, or

$C = 5000$. Thus $c = 0.001q^3 - 0.2q^2 + 40q + 5000$. When $q = 100$, then $c = 8000$. Since

$$\text{Avg. Cost} = \bar{c} = \frac{\text{Total Cost}}{\text{Quantity}} = \frac{c}{q},$$

when $q = 100$ we have $\bar{c} = 8000/100 = \$80$. (Observe that knowing $dc/dq = 27.50$ when $q = 50$ is not relevant to the problem.)

EXERCISE 16.3

1. Let $u = x + 5 \implies du = 1\ dx = dx$.

$$\int (x + 5)^7 [dx] = \int u^7\ du = \frac{u^8}{8} + C = \frac{(x + 5)^8}{8} + C$$

3. Let $u = x^2 + 3 \implies du = 2x\ dx$. $\qquad \int 2x(x^2 + 3)^5\ dx =$

$$\int (x^2 + 3)^5 [2x\ dx] = \int u^5\ du = \frac{u^6}{6} + C = \frac{(x^2 + 3)^6}{6} + C$$

5. Let $u = y^3 + 3y^2 + 1 \implies du = (3y^2 + 6y)\ dy$.

$$\int (3y^2 + 6y)(y^3 + 3y^2 + 1)^{2/3}\ dy$$

$$= \int (y^3 + 3y^2 + 1)^{2/3} [(3y^2 + 6y)\ dy] = \int u^{2/3}\ du = \frac{u^{5/3}}{5/3} + C$$

$$= \tfrac{3}{5}(y^3 + 3y^2 + 1)^{5/3} + C$$

7. Let $u = 3x - 1 \implies du = 3\ dx$.

$$\int \frac{3}{(3x - 1)^3}\ dx = \int \frac{1}{(3x - 1)^3} [3\ dx] = \int \frac{1}{u^3}\ du = \int u^{-3}\ du$$

$$= \frac{u^{-2}}{-2} + C = -\frac{(3x - 1)^{-2}}{2} + C$$

9. Let $u = 2x - 1 \implies du = 2\ dx$.

$$\int \sqrt{2x - 1}\ dx = \int (2x - 1)^{1/2}\ dx = \tfrac{1}{2} \int (2x - 1)^{1/2} [2\ dx]$$

$$= \tfrac{1}{2} \int u^{1/2}\ du = \tfrac{1}{2} \cdot \frac{u^{3/2}}{3/2} + C = \tfrac{1}{3}(2x - 1)^{3/2} + C$$

11. Let $u = 7x - 6 \implies du = 7\,dx$.

$$\int (7x - 6)^4\,dx = \frac{1}{7}\int (7x - 6)^4[7\,dx] = \frac{1}{7}\int u^4\,du = \frac{1}{7}\cdot\frac{u^5}{5} + C$$

$$= \frac{(7x - 6)^5}{35} + C$$

13. Let $u = x^2 + 3 \implies du = 2x\,dx$.

$$\int x(x^2 + 3)^{12}\,dx = \frac{1}{2}\int (x^2 + 3)^{12}[2x\,dx] = \frac{1}{2}\int u^{12}\,du$$

$$= \frac{1}{2}\cdot\frac{u^{13}}{13} + C = \frac{(x^2 + 3)^{13}}{26} + C$$

15. Let $u = 27 + x^5 \implies du = 5x^4\,dx$.

$$\int x^4(27 + x^5)^{1/3}\,dx = \frac{1}{5}\int (27 + x^5)^{1/3}[5x^4\,dx] = \frac{1}{5}\int u^{1/3}\,du$$

$$= \frac{1}{5}\cdot\frac{u^{4/3}}{4/3} + C = \frac{3}{20}(27 + x^5)^{4/3} + C$$

17. Let $u = 3x \implies du = 3\,dx$.

$$\int 3e^{3x}\,dx = \int e^{3x}[3\,dx] = \int e^u\,du = e^u + C = e^{3x} + C$$

19. Let $u = t^2 + t \implies du = (2t + 1)\,dt$.

$$\int (2t + 1)e^{t^2+t}\,dt = \int e^{t^2+t}[(2t + t)\,dt]$$

$$= \int e^u\,du = e^u + C = e^{t^2+t} + C$$

21. $\displaystyle\int xe^{5x^2}\,dx = \frac{1}{10}\int e^{5x^2}[10x\,dx] = \frac{1}{10}e^{5x^2} + C$

23. $\displaystyle\int 6e^{-2x}\,dx = 6\left(-\frac{1}{2}\right)\int e^{-2x}[-2\,dx] = -3e^{-2x} + C$

25. Let $u = x + 5 \implies du = dx$

$$\int \frac{1}{x + 5}[dx] = \int \frac{1}{u}\,du = \ln|u| + C = \ln|x + 5| + C$$

27. $\displaystyle\int \frac{3x^2 + 4x^3}{x^3 + x^4}\ dx = \int \frac{1}{x^3 + x^4}[(3x^2 + 4x^3)\ dx] = \ln|x^3 + x^4| + C$

29. $\displaystyle\int \frac{6z}{(z^2 - 6)^5}\ dz = 6 \cdot \frac{1}{2}\int (z^2 - 6)^{-5}[2z\ dz] = 3 \cdot \frac{(z^2 - 6)^{-4}}{-4} + C$

$$= -\frac{3}{4}(z^2 - 6)^{-4} + C$$

31. $\displaystyle\int \frac{4}{x}\ dx = 4\int \frac{1}{x}\ dx = 4\ \ln|x| + C$

33. $\displaystyle\int \frac{s^2}{s^3 + 5}\ ds = \frac{1}{3}\int \frac{1}{s^3 + 5}[3s^2\ ds] = \frac{1}{3}\ \ln|s^3 + 5| + C$

35. $\displaystyle\int \frac{7}{5 - 3x}\ dx = 7\left(-\frac{1}{3}\right)\int \frac{1}{5 - 3x}[-3\ dx] = -\frac{7}{3}\ \ln|5 - 3x| + C$

37. $\displaystyle\int \sqrt{5x}\ dx = \frac{1}{5}\int (5x)^{1/2}[5\ dx] = \frac{1}{5} \cdot \frac{(5x)^{3/2}}{3/2} + C$

$$= \frac{2}{15}(5x)^{3/2} + C = \frac{2}{15}(5x)(5x)^{1/2} + C$$

$$= \frac{2}{3}x\sqrt{5x} + C = \frac{2\sqrt{5}}{3}x^{3/2} + C$$

39. $\displaystyle\int \frac{x}{\sqrt{x^2 - 4}}\ dx = \frac{1}{2}\int (x^2 - 4)^{-1/2}[2x\ dx] = \frac{1}{2} \cdot \frac{(x^2 - 4)^{1/2}}{1/2} + C$

$$= \sqrt{x^2 - 4} + C$$

41. $\displaystyle\int 2y^3 e^{y^4+1}\ dy = 2\int y^3 e^{y^4+1}\ dy = 2 \cdot \frac{1}{4}\int e^{y^4+1}[4y^3\ dy]$

$$= 2 \cdot \frac{1}{4} \cdot e^{y^4+1} + C = \frac{1}{2}e^{y^4+1} + C$$

43. $\displaystyle\int v^2 e^{-2v^3+1}\ dv = -\frac{1}{6}\int e^{-2v^3+1}[-6v^2\ dv] = -\frac{1}{6}e^{-2v^3+1} + C$

45. $\int (e^{-5x} + 2e^x)\, dx = \int e^{-5x}\, dx + 2\int e^x\, dx$

$= -\frac{1}{5}\int e^{-5x}[-5\, dx] + 2\int e^x\, dx = -\frac{1}{5}e^{-5x} + 2e^x + C$

47. $\int (x + 1)(3 - 3x^2 - 6x)^3\, dx$

$= -\frac{1}{6}\int (3 - 3x^2 - 6x)^3[(-6x - 6)\, dx]$

$= -\frac{1}{6}\cdot\frac{(3 - 3x^2 - 6x)^4}{4} + C = -\frac{1}{24}(3 - 3x^2 - 6x)^4 + C$

49. $\int \frac{x^2 + 2}{x^3 + 6x}\, dx = \frac{1}{3}\int \frac{1}{x^3 + 6x}[(3x^2 + 6)\, dx] = \frac{1}{3}\ln|x^3 + 6x| + C$

51. $\int \frac{16s - 4}{3 - 2s + 4s^2}\, ds = 2\int \frac{1}{3 - 2s + 4s^2}[(8s - 2)\, ds]$

$= 2\ln|3 - 2s + 4s^2| + C$

53. $\int x(2x^2 + 1)^{-1}\, dx = \int \frac{x}{2x^2 + 1}\, dx = \frac{1}{4}\int \frac{1}{2x^2 + 1}[4x\, dx]$

$= \frac{1}{4}\ln(2x^2 + 1) + C$

55. $\int -(x^2 - 2x^5)(x^3 - x^6)^{-10}\, dx$

$= -\frac{1}{3}\int (x^3 - x^6)^{-10}[3(x^2 - 2x^5)\, dx]$

$= -\frac{1}{3}\cdot\frac{(x^3 - x^6)^{-9}}{-9} + C = \frac{1}{27}(x^3 - x^6)^{-9} + C$

57. $\int (2x^3 + x)(x^4 + x^2)\, dx = \frac{1}{2}\int (x^4 + x^2)^1[2(2x^3 + x)\, dx]$

$= \frac{1}{2}\cdot\frac{(x^4 + x^2)^2}{2} + C = \frac{1}{4}(x^4 + x^2)^2 + C$

59. $\displaystyle\int \frac{18 + 12x}{(4 - 9x - 3x^2)^5} \, dx = -2\int (4 - 9x - 3x^2)^{-5}[(-9 - 6x) \, dx]$

$= -2 \cdot \frac{(4 - 9x - 3x^2)^{-4}}{-4} + C = \frac{1}{2}(4 - 9x - 3x^2)^{-4} + C$.

61. $u = 4x^3 + 3x^2 - 4 \implies du = (12x^2 + 6x) \, dx = 6x(2x + 1) \, dx$

$\displaystyle\int x(2x + 1)e^{4x^3+3x^2-4} \, dx = \frac{1}{6}\int e^{4x^3+3x^2-4}[6x(2x + 1) \, dx]$

$= \frac{1}{6}\int e^u \, du = \frac{1}{6}e^u + C = \frac{1}{6}e^{4x^3+3x^2-4} + C$

63. $\displaystyle\int x\sqrt{(7 - 5x^2)^3} \, dx = -\frac{1}{10}\int (7 - 5x^2)^{3/2}[-10x \, dx]$

$= -\frac{1}{10} \cdot \frac{(7 - 5x^2)^{5/2}}{5/2} + C = -\frac{1}{25}(7 - 5x^2)^{5/2} + C$

65. $\displaystyle\int \left(\sqrt{2x} - \frac{1}{\sqrt{2x}}\right) \, dx = \int \sqrt{2x} \, dx - \int \frac{1}{\sqrt{2x}} \, dx$

$= \frac{1}{2}\int (2x)^{1/2}[2 \, dx] - \frac{1}{2}\int (2x)^{-1/2}[2 \, dx]$

$= \frac{1}{2} \cdot \frac{(2x)^{3/2}}{3/2} - \frac{1}{2} \cdot \frac{(2x)^{1/2}}{1/2} + C = \frac{(2x)^{3/2}}{3} - \sqrt{2x} + C$

$= \frac{2\sqrt{2}}{3}x^{3/2} - \sqrt{2}x^{1/2} + C$

67. $\displaystyle\int (x^2 + 1)^2 \, dx = \int (x^4 + 2x^2 + 1) \, dx = \frac{x^5}{5} + \frac{2x^3}{3} + x + C$

69. $\displaystyle\int \left[\frac{x}{x^2 + 1} + \frac{x^5}{(x^6 + 1)^2}\right] \, dx = \int \frac{x}{x^2 + 1} \, dx + \int \frac{x^5}{(x^6 + 1)^2} \, dx$

$= \frac{1}{2}\int \frac{1}{x^2 + 1}[2x \, dx] + \frac{1}{6}\int (x^6 + 1)^{-2}[6x^5 \, dx]$

$= \frac{1}{2} \ln(x^2 + 1) + \frac{1}{6} \cdot \frac{(x^6 + 1)^{-1}}{-1} + C$

$= \frac{1}{2} \ln(x^2 + 1) - \frac{1}{6(x^6 + 1)} + C$

71. $\int \left[\frac{1}{3x - 5} - (x^2 - 2x^5)(x^3 - x^6)^{-10} \right] dx$

$= \frac{1}{3} \int \frac{1}{3x - 5} [3\ dx] - \frac{1}{3} \int (x^3 - x^6)^{-10} [(3x^2 - 6x^5)\ dx]$

$= \frac{1}{3}\ \ln|3x - 5| - \frac{1}{3} \cdot \frac{(x^3 - x^6)^{-9}}{-9} + C$

$= \frac{1}{3}\ \ln|3x - 5| + \frac{1}{27}(x^3 - x^6)^{-9} + C$

73. $\int \left[\sqrt{3x + 1} - \frac{x}{x^2 + 3} \right]\ dx = \int (3x + 1)^{1/2}\ dx - \int \frac{x}{x^2 + 3}\ dx$

$= \frac{1}{3} \int (3x + 1)^{1/2} [3\ dx] - \frac{1}{2} \int \frac{1}{x^2 + 3} [2x\ dx]$

$= \frac{1}{3} \cdot \frac{(3x + 1)^{3/2}}{3/2} - \frac{1}{2}\ \ln(x^2 + 3) + C$

$= \frac{2}{9}(3x + 1)^{3/2} - \ln\sqrt{x^2 + 3} + C$

75. Let $u = \sqrt{x} \implies du = \frac{1}{2}x^{-1/2}\ dx = \frac{1}{2\sqrt{x}}\ dx$.

$\int \frac{e^{\sqrt{x}}}{\sqrt{x}}\ dx = 2 \int e^{\sqrt{x}} \left[\frac{1}{2\sqrt{x}}\ dx \right] = 2 \int e^u\ du = 2e^u + C = 2e^{\sqrt{x}} + C$

77. $\int \frac{1 + e^{2x}}{e^x}\ dx = \int \left(\frac{1}{e^x} + \frac{e^{2x}}{e^x} \right)\ dx = \int (e^{-x} + e^x)\ dx$

$= - \int e^{-x} [-1\ dx] + \int e^x\ dx = -e^{-x} + e^x + C$

79. Let $u = \ln(x^2 + 2x) \implies du = \frac{1}{x^2 + 2x}(2x + 2)\ dx$.

$\int \frac{x + 1}{x^2 + 2x}\ \ln(x^2 + 2x)\ dx = \frac{1}{2} \int \ln(x^2 + 2x) \left[\frac{2x + 2}{x^2 + 2x}\ dx \right]$

$= \frac{1}{2} \int u\ du = \frac{1}{2} \cdot \frac{u^2}{2} + C = \frac{1}{4}\ \ln^2(x^2 + 2x) + C$

81. $y = \int (3 - 2x)^2\ dx = - \frac{1}{2} \int (3 - 2x)^2 [-2\ dx] = - \frac{1}{2} \cdot \frac{(3-2x)^3}{3} + C$

$= - \frac{1}{6}(3 - 2x)^3 + C.\quad y(0) = 1$ implies $1 = - \frac{1}{6}(27) + C$, so

$C = \frac{11}{2}.$ Thus $y = - \frac{1}{6}(3 - 2x)^3 + \frac{11}{2}$

83. $y'' = \dfrac{1}{x^2}$. $y' = \displaystyle\int x^{-2}\, dx = -x^{-1} + C_1$. $y'(-1) = 1$ implies

$1 = 1 + C_1$, so $C_1 = 0$. Thus $y' = -x^{-1}$. $y = \displaystyle\int (-x^{-1})\, dx =$

$-\displaystyle\int \dfrac{1}{x}\, dx = -\ln|x| + C_2$. $y(1) = 0$ implies that $0 = 0 + C_2$,

so $C_2 = 0$. Thus $y = -\ln|x| = \ln\left|\dfrac{1}{x}\right|$.

85. Note that $r > 0$. $C = \displaystyle\int \left[\dfrac{Rr}{2K} + \dfrac{B_1}{r}\right] dr = \int \dfrac{Rr}{2K}\, dr + \int \dfrac{B_1}{r}\, dr =$

$\dfrac{R}{2K}\displaystyle\int r\, dr + B_1\int \dfrac{1}{r}\, dr = \dfrac{R}{2K}\cdot\dfrac{r^2}{2} + B_1\ln(r) + B_2$.   Thus we obtain

$C = \dfrac{Rr^2}{4K} + B_1\ln(r) + B_2$.

EXERCISE 16.4

1. $\displaystyle\int \dfrac{2x^4 + 3x^3 - x^2}{x^3}\, dx = \int \left(\dfrac{2x^4}{x^3} + \dfrac{3x^3}{x^3} - \dfrac{x^2}{x^3}\right) dx = \int \left(2x + 3 - \dfrac{1}{x}\right) dx$

    $= 2\cdot\dfrac{x^2}{2} + 3x - \ln|x| + C = x^2 + 3x - \ln|x| + C$

3. $\displaystyle\int (3x^2 + 2)\sqrt{2x^3 + 4x + 1}\, dx = \dfrac{1}{2}\int (2x^3 + 4x + 1)^{1/2}[2(3x^2 + 2)\, dx]$

    $= \dfrac{1}{2}\cdot\dfrac{(2x^3 + 4x + 1)^{3/2}}{3/2} + C = \dfrac{1}{3}(2x^3 + 4x + 1)^{3/2} + C$

5. $\displaystyle\int \dfrac{3}{\sqrt{4 - 5x}}\, dx = 3\int (4 - 5x)^{-1/2}\, dx = 3\cdot\left(-\dfrac{1}{5}\right)\int (4 - 5x)^{-1/2}[-5\, dx]$

                $= -\dfrac{3}{5}\cdot\dfrac{(4 - 5x)^{1/2}}{1/2} + C = -\dfrac{6}{5}\sqrt{4 - 5x} + C$

7. $\int 4^{7x} \, dx = \int (e^{\ln 4})^{7x} \, dx = \int e^{(\ln 4)(7x)} \, dx$

$= \frac{1}{7 \ln 4} \int e^{(\ln 4)(7x)} [7 \ln 4 \, dx] = \frac{1}{7 \ln 4} \cdot e^{(\ln 4)(7x)} + C$

$= \frac{1}{7 \ln 4} (e^{\ln 4})^{(7x)} + C = \frac{4^{7x}}{7 \ln 4} + C$

9. $\int 2x(7 - e^{x^2/4}) \, dx = \int (14x - 2xe^{x^2/4}) \, dx$

$= 14 \int x \, dx - 2 \int xe^{x^2/4} \, dx = 14 \int x \, dx - 2 \cdot 2 \int e^{x^2/4} [\frac{1}{2}x \, dx]$

$= 14 \cdot \frac{x^2}{2} - 4 \cdot e^{x^2/4} + C = 7x^2 - 4e^{x^2/4} + C$

11. By long division, $\frac{6x^2 - 11x + 5}{3x - 1} = 2x - 3 + \frac{2}{3x-1}$. Thus

$\int \frac{6x^2 - 11x + 5}{3x - 1} \, dx = \int \left(2x - 3 + \frac{2}{3x-1}\right) dx =$

$2 \int x \, dx - \int 3 \, dx + 2 \cdot \frac{1}{3} \int \frac{1}{3x-1} [3 \, dx] = x^2 - 3x + \frac{2}{3} \ln|3x-1| + C$

9

13. $\int \frac{3e^{2x}}{e^{2x}+1} \, dx = \frac{3}{2} \int \frac{1}{e^{2x}+1} [2e^{2x} \, dx] = \frac{3}{2} \ln(e^{2x}+1) + C$

15. $\int \frac{e^{7/x}}{x^2} \, dx = \int e^{7/x} \cdot \frac{1}{x^2} \, dx = -\frac{1}{7} \int e^{7/x} \left[- \frac{7}{x^2} \, dx\right] = -\frac{1}{7}e^{7/x} + C$

17. By using long division on the integrand,

$\int \frac{2x^3}{x^2 - 4} \, dx = \int \left(2x + \frac{8x}{x^2-4}\right) dx = \int 2x \, dx + 4 \int \frac{1}{x^2-4} [2x \, dx]$

$= x^2 + 4 \ln|x^2-4| + C.$

19. $\int \frac{(\sqrt{x}+2)^2}{3\sqrt{x}} \, dx = \frac{2}{3} \int (\sqrt{x}+2)^2 \left[\frac{1}{2\sqrt{x}} \, dx\right] = \frac{2}{3} \cdot \frac{(\sqrt{x}+2)^3}{3} + C$

$= \frac{2}{9}(\sqrt{x} + 2)^3 + C$

21. $\displaystyle\int \frac{(x^{1/3} + 2)^4}{\sqrt[3]{x^2}}\, dx = 3\int (x^{1/3} + 2)^4\left[\frac{1}{3}x^{-2/3}\, dx\right]$

$$= \frac{3}{5}(x^{1/3} + 2)^5 + c$$

23. $\displaystyle\int \frac{\ln x}{x}\, dx = \int (\ln x)\left[\frac{1}{x}\, dx\right] = \frac{(\ln x)^2}{2} + c = \frac{1}{2}(\ln^2 x) + c$

25. $\displaystyle\int \frac{\ln^2(r+1)}{r+1}\, dr = \int [\ln(r+1)]^2\left(\frac{1}{r+1}\, dr\right) = \frac{1}{3}\ln^3(r+1) + c$

27. $\displaystyle\int \frac{3^{\ln x}}{x}\, dx = \int \frac{(e^{\ln 3})^{\ln x}}{x}\, dx = \frac{1}{\ln 3}\int e^{(\ln 3)\ln x}\left[\frac{\ln 3}{x}\, dx\right]$

$$= \frac{1}{\ln 3}\cdot e^{(\ln 3)\ln x} + c = \frac{1}{\ln 3}(e^{\ln 3})^{\ln x} + c = \frac{3^{\ln x}}{\ln 3} + c$$

29. $\displaystyle\int x\sqrt{e^{x^2+3}}\, dx = \int x(e^{x^2+3})^{1/2}\, dx = \int e^{(x^2+3)/2}[x\, dx]$

$$= e^{(x^2+3)/2} + c$$

31. $\displaystyle\int \frac{dx}{(x+3)\,\ln(x+3)} = \int \frac{1}{\ln(x+3)}\left[\frac{1}{x+3}\, dx\right] = \ln|\ln(x+3)| + c$

33. By using long division on the integrand,

$$\int \frac{x^3+x^2-x-3}{x^2-3}\, dx = \int \left(x + 1 + \frac{2x}{x^2-3}\right) dx$$

$$= \int (x + 1)\, dx + \int \frac{1}{x^2-3}[2x\, dx] = \frac{x^2}{2} + x + \ln|x^2-3| + c$$

35. $\displaystyle\int \frac{3x\sqrt{\ln(x^2+1)^2}}{x^2+1}\, dx = \frac{3}{4}\int [2\,\ln(x^2+1)]^{1/2}\left[\frac{4x}{x^2+1}\, dx\right]$

$$= \frac{3}{4}\cdot\frac{[2\,\ln(x^2+1)]^{3/2}}{3/2} + c = \frac{1}{2}\ln^{3/2}[(x^2+1)^2] + c$$

37. $\int \left[\frac{x^3}{\sqrt{x^4-1}} - \ln 4 \right] dx = \frac{1}{4}\int (x^4-1)^{-1/2}[4x^3 \ dx] - \int (\ln 4) \ dx$

$= \frac{1}{4}\cdot\frac{(x^4-1)^{1/2}}{1/2} - (\ln 4)x + C = \frac{1}{2}\sqrt{x^4-1} - (\ln 4)x + C$

39. $\int \frac{2x^4 - 8x^3 - 6x^2+4}{x^3} \ dx = \int \left(2x - 8 - \frac{6}{x} + \frac{4}{x^3} \right) dx =$

$2\int x \ dx - \int 8 \ dx - 6\int \frac{1}{x} \ dx + 4\int x^{-3} \ dx =$

$2\cdot\frac{x^2}{2} - 8x - 6 \ \ln|x| + 4\cdot\frac{x^{-2}}{-2} + C = x^2 - 8x - 6 \ \ln|x| - \frac{2}{x^2} +$

41. By using long division on the integrand,

$$\int \frac{x}{x-1} \ dx = \int \left(1 + \frac{1}{x-1} \right) dx = x + \ln|x-1| + C$$

43. $\int \frac{xe^{x^2}}{\sqrt{e^{x^2}+2}} \ dx = \frac{1}{2}\int (e^{x^2}+2)^{-1/2}[2xe^{x^2} \ dx] = \frac{1}{2}\cdot\frac{(e^{x^2}+2)^{1/2}}{1/2} + C$

$$= \sqrt{e^{x^2} + 2} + C$$

45. $\int \frac{(e^{-x} + 6)^2}{e^x} \ dx = -\int (e^{-x} + 6)^2[-e^{-x} \ dx] = -\frac{(e^{-x} + 6)^3}{3} + C$

47. $\int (x^3 + ex)\sqrt{x^2 + e} \ dx = \int x(x^2 + e)(x^2 + e)^{1/2} \ dx$

$= \frac{1}{2}\int (x^2+1)^{3/2}[2x \ dx] = \frac{1}{2}\cdot\frac{(x^2+e)^{5/2}}{5/2} + C = \frac{1}{5}(x^2+e)^{5/2} + C$

49. $\int \sqrt{x}\sqrt{(8x)^{3/2} + 3} \ dx = \int (8^{3/2}x^{3/2} + 3)^{1/2}\cdot x^{1/2} \ dx$

$= \frac{2}{3\cdot 8^{3/2}}\int (8^{3/2}x^{3/2} + 3)^{1/2}[8^{3/2}\cdot\frac{3}{2}\cdot x^{1/2} \ dx]$

$= \frac{2}{3\cdot 16\sqrt{2}}\cdot\frac{(8^{3/2}x^{3/2}+3)^{3/2}}{3/2} + C = \frac{1}{36\sqrt{2}}[(8x)^{3/2} + 3]^{3/2} + C$

51. $\int \dfrac{\sqrt{s}}{e^{\sqrt{s^3}}}\, ds = -\dfrac{2}{3}\int e^{-s^{3/2}}\left[-\dfrac{3}{2}s^{1/2}\, ds\right] = -\dfrac{2}{3}e^{-\sqrt{s^3}} + C$

53. $e^{\ln(x+2)}$ is simply $x + 2$. Thus

$$\int e^{\ln(x+2)}\, dx = \int (x + 2)\, dx = \dfrac{x^2}{2} + 2x + C$$

55. $\displaystyle\int \dfrac{\ln(xe^x)}{x}\, dx = \int \dfrac{\ln x + \ln e^x}{x}\, dx = \int \dfrac{\ln x + x}{x}\, dx$

$= \displaystyle\int \left(\dfrac{\ln x}{x} + 1\right) dx = \int (\ln x)\left[\dfrac{1}{x}\, dx\right] + \int 1\, dx = \dfrac{\ln^2 x}{2} + x + C$

57. $\dfrac{dr}{dq} = \dfrac{200}{(q+2)^2}$. $r = \displaystyle\int 200(q + 2)^{-2}\, dq = 200 \cdot \dfrac{(q + 2)^{-1}}{-1} + C =$

$-\dfrac{200}{q + 2} + C$. When $q = 0$, then $r = 0$, so $0 = -100 + C$, or

$C = 100$. Hence $r = -\dfrac{200}{q + 2} + 100 = \dfrac{100q}{q + 2}$. Since $r = pq$,

then $p = r/q = 100/(q + 2)$.

59. $\dfrac{dc}{dq} = \dfrac{20}{q+5}$. $c = \displaystyle\int \dfrac{20}{q+5}\, dq = 20\int \dfrac{1}{q+5}\, dq = 20\ln|q+5| + C$.

When $q = 0$, then $c = 2000$, so $2000 = 20\ln(5) + C$, or
$C = 2000 - 20\ln 5$. Hence $c = 20\ln|q+5| + 2000 - 20\ln 5 =$
$20(\ln|q+5| - \ln 5) + 2000 = 20\ln|(q+5)/5| + 2000$.

61. $\dfrac{dC}{dI} = \dfrac{1}{\sqrt{I}}$. $C = \displaystyle\int I^{-1/2}\, dI = \dfrac{I^{1/2}}{1/2} + C_1 = 2\sqrt{I} + C_1$.

$C(9) = 8$ implies that $8 = 2\cdot 3 + C_1$, or $C_1 = 2$.   Thus
$C = 2\sqrt{I} + 2 = 2(\sqrt{I} + 1)$.

63. $\dfrac{dC}{dI} = \dfrac{3}{4} - \dfrac{1}{6\sqrt{I}}$. $C = \displaystyle\int \left[\dfrac{3}{4} - \dfrac{I^{-1/2}}{6}\right] dI =$

$\displaystyle\int \dfrac{3}{4}\, dI - \dfrac{1}{6}\int I^{-1/2}\, dI = \dfrac{3}{4}I - \dfrac{1}{6}\cdot\dfrac{I^{1/2}}{1/2} + C_1 = \dfrac{3}{4}I - \dfrac{\sqrt{I}}{3} + C_1$. Thus

$C = \frac{3}{4}I - \frac{4\sqrt{I}}{3} + C_1$. $C(25) = 23$ implies that

$23 = \frac{3}{4}\cdot25 - \frac{5}{3} + C_1$, so $C_1 = \frac{71}{12}$. Thus $C = \frac{3}{4}I - \frac{1}{3}\sqrt{I} + \frac{71}{12}$

65. $\frac{dc}{dq} = \frac{100q^2 - 4998q + 50}{q^2 - 50q + 1}$.

(a) $\frac{dc}{dq}\Big|_{q=50} = \frac{100(50)^2 - 4998(50) + 50}{(50)^2 - 50(50) + 1} = \$150$ per unit

(b) To find c, we integrate dc/dq by using long division:

$c = \int\frac{100q^2-4998q+50}{q^2-50q+1} dq = \int\left(100 + \frac{2q-50}{q^2-50q+1}\right) dq =$

$\int 100 dq + \int\frac{1}{q^2-50q+1}[(2q-50) dq]$. Thus

$c = 100q + \ln|q^2-50q+1| + C$. When q = 0, then
c = 10,000, so 10,000 = 0 + ln(1) + C, or C = 10,000.
Hence $c = 100q + \ln|q^2-50q+1| + 10,000$. When q = 50,
then c = 5000 + ln(1) + 10,000 = \$15,000.

(c) If c = f(q), then $f(q+dq) \approx f(q) + dc = f(q) + \frac{dc}{dq}dq$.

Letting q = 50 and dq = 2, we have

$f(52) = f(50+2) \approx f(50) + \frac{dc}{dq}\Big|_{q=50}\cdot(2)$

$= 15,000 + 150(2) = \$15,300.$

67. $\frac{dV}{dt} = \frac{8t^3}{\sqrt{0.2t^4 + 8000}}$.

$V = \int\frac{8t^3}{\sqrt{0.2t^4+8000}} dt = \int(0.2t^4+8000)^{-1/2}\cdot8t^3 dt =$

$10\int(0.2t^4+8000)^{-1/2}[0.8t^3 dt] = 10\frac{(0.2t^4+8000)^{1/2}}{1/2} + C.$

Thus $V = 20\sqrt{0.2t^4+8000} + C$. If t = 0, then V = 500, so
$500 = 20\sqrt{8000} + C$, $500 = 20\sqrt{1600\cdot5} + C$, $500 = 800\sqrt{5} + C$,
or $C = 500 - 800\sqrt{5}$. Hence
$V = 20\sqrt{0.2t^4+8000} + 500 - 800\sqrt{5}$. When t = 10, then

$$V = 20\sqrt{10,000} + 500 - 800\sqrt{5}$$
$$\approx 20(100) + 500 - 800(2.236) \approx \$711 \text{ per acre.}$$

69. $S = \int \frac{dS}{dI} \, dI = \int \frac{5}{(I+2)^2} \, dI = 5\int (I+2)^{-2} \, dI = 5 \cdot \frac{(I+2)^{-1}}{-1} + C_1$.

Thus $S = -\frac{5}{I+2} + C_1$. If C is the total national consumption (in billions of dollars), then $C + S = I$, or $C = I - S$. Hence $C = I + \frac{5}{I+2} - C_1$. When $I = 8$, then $C = 7.5$, so $7.5 = 8 + \frac{1}{2} - C_1$, or $C_1 = 1$. Thus $S = 1 - \frac{5}{I+2}$. If $S = 0$, then $0 = 1 - \frac{5}{I+2} \Rightarrow \frac{5}{I+2} = 1 \Rightarrow 5 = I+2 \Rightarrow I = 3$.

EXERCISE 16.5

1. $5+6+7+8+9 = 35$

3. $(-1)+1+(-1)+1+(-1)+1+(-1)+1+(-1)+1 = 0$

5. $5+20 = 25$ 7. $\left(-\frac{1}{2}\right) + \frac{5}{16} = -\frac{3}{16}$

9. $0 + \frac{3}{2} + \left(-\frac{8}{3}\right) = -\frac{7}{6}$

11. $\sum\limits_{k=1}^{15} k$ 13. $\sum\limits_{k=1}^{4} (2k-1)$ 15. $\sum\limits_{k=1}^{12} k^2$

17. $\sum\limits_{k=1}^{450} k = \frac{450(451)}{2} = 101,475$

19. $\sum\limits_{j=1}^{6} 4j = 4 \sum\limits_{j=i}^{6} j = 4 \cdot \frac{6(7)}{2} = 84$

21. $\displaystyle\sum_{i=1}^{6} 3i^2 = 3\sum_{i=1}^{6} i^2 = 3\cdot\frac{6(7)(13)}{6} = 273$

23. After n years,

   Total maint. cost = 100 + 200 + ... + 100n
   = 100(1 + 2 + ... + n)
   = $100\cdot\frac{n(n+1)}{2}$ = 50n(n+1).

   Thus total cost after n years = 3200 + 50n(n+1).

   Avg. annual total cost = C = $\dfrac{3200 + 50n(n+1)}{n} = \dfrac{3200}{n} + 50(n+1)$.

   $\dfrac{dC}{dn} = -\dfrac{3200}{n^2} + 50 = \dfrac{50n^2 - 3200}{n^2} = \dfrac{50(n-8)(n+8)}{n^2}$.

   We want n > 0, so the critical point to consider is n = 8.
   Since dC/dn < 0 for 0 < n < 8 and dC/dn > 0 for n > 8,
   then C is minimum when n = 8. At this value, C = $850.

EXERCISE 16.6

1. f(x) = x, y = 0, x = 1. S_3, $\Delta x = 1/3$.

   $S_3 = \frac{1}{3}f\left(\frac{1}{3}\right) + \frac{1}{3}f\left(\frac{2}{3}\right) + \frac{1}{3}f\left(\frac{3}{3}\right) = \frac{1}{3}\left[f\left(\frac{1}{3}\right) + f\left(\frac{2}{3}\right) + f\left(\frac{3}{3}\right)\right]$

   $= \frac{1}{3}\left[\frac{1}{3} + \frac{2}{3} + \frac{3}{3}\right] = \frac{1}{3}\cdot\frac{6}{3} = \frac{2}{3}$. <u>Ans.</u> 2/3 sq unit

3. f(x) = x^2, y = 0, x = 1. S_3, $\Delta x = 1/3$.

   $S_3 = \frac{1}{3}f\left(\frac{1}{3}\right) + \frac{1}{3}f\left(\frac{2}{3}\right) + \frac{1}{3}f\left(\frac{3}{3}\right) = \frac{1}{3}\left[\frac{1}{9} + \frac{4}{9} + \frac{9}{9}\right] = \frac{1}{3}\cdot\frac{14}{9} = \frac{14}{27}$.
   <u>Ans.</u> 14/27 sq unit

5. f(x) = 4x; [0,1]. $\Delta x = 1/n$.

   $S_n = \frac{1}{n}f\left(\frac{1}{n}\right) + ... + \frac{1}{n}f\left(n\cdot\frac{1}{n}\right) = \frac{1}{n}\left[f\left(\frac{1}{n}\right) + ... + f\left(n\cdot\frac{1}{n}\right)\right]$

   $= \frac{1}{n}\left[4\cdot\frac{1}{n} + ... + 4\cdot\frac{n}{n}\right] = \frac{4}{n^2}[1 + ... + n] = \frac{4}{n^2}\cdot\frac{n(n+1)}{2}$

   $= \frac{2(n+1)}{n}$

7. (a) $S_n = \frac{1}{n}\left[\left(\frac{1}{n} + 1\right) + \left(\frac{2}{n} + 1\right) + \ldots + \left(\frac{n}{n} + 1\right)\right]$

$= \frac{1}{n}\left[\frac{1}{n}(1 + 2 + \ldots + n) + n\right] = \frac{1}{n}\left[\frac{1}{n}\cdot\frac{n(n+1)}{2} + n\right]$

$= \frac{n+1}{2n} + 1$

(b) $\lim_{n\to\infty} S_n = \lim_{n\to\infty}\left[\frac{n+1}{2n} + 1\right] = \lim_{n\to\infty}\left[\frac{1}{2} + \frac{1}{2n} + 1\right]$

$= \frac{1}{2} + 0 + 1 = \frac{3}{2}$

9. $f(x) = x$, $y = 0$, $x = 1$. $\Delta x = 1/n$.

$S_n = \frac{1}{n}f\left(\frac{1}{n}\right) + \ldots + \frac{1}{n}f\left(n\cdot\frac{1}{n}\right) = \frac{1}{n}\left[f\left(\frac{1}{n}\right) + \ldots + f\left(n\cdot\frac{1}{n}\right)\right]$

$= \frac{1}{n}\left[\frac{1}{n} + \ldots + \frac{n}{n}\right] = \frac{1}{n^2}[1 + \ldots + n] = \frac{1}{n^2}\cdot\frac{n(n+1)}{2}$

$= \frac{1}{2}\cdot\frac{n+1}{n} = \frac{1}{2}\left[1 + \frac{1}{n}\right]$. $\lim_{n\to\infty} S_n = \frac{1}{2}$. __Ans.__ 1/2 sq unit

11. $f(x) = x^2$, $y = 0$, $x = 1$. $\Delta x = 1/n$.

$S_n = \frac{1}{n}f\left(\frac{1}{n}\right) + \ldots + \frac{1}{n}f\left(n\cdot\frac{1}{n}\right) = \frac{1}{n}\left[\left(\frac{1}{n}\right)^2 + \ldots + \left(n\cdot\frac{1}{n}\right)^2\right]$

$= \frac{1}{n^3}[1^2 + \ldots + n^2]$

$= \frac{1}{n^3}\cdot\frac{n(n+1)(2n+1)}{6} = \frac{1}{6}\cdot\frac{2n^2+3n+1}{n^2}$

$= \frac{1}{6}\left[2 + \frac{3}{n} + \frac{1}{n^2}\right]$. $\lim_{n\to\infty} S_n = \frac{1}{3}$. __Ans.__ 1/3 sq unit

13. $f(x) = 2x^2$, $y = 0$, $x = 2$. $\Delta x = 2/n$.

$S_n = \frac{2}{n}f\left(\frac{2}{n}\right) + \ldots + \frac{2}{n}f\left(n\cdot\frac{2}{n}\right) = \frac{2}{n}\left[f\left(\frac{2}{n}\right) + \ldots + f\left(n\cdot\frac{2}{n}\right)\right]$

$= \frac{2}{n}\left[2\left(\frac{2}{n}\right)^2 + \ldots + 2\left(n\cdot\frac{2}{n}\right)^2\right]$

$= \frac{16}{n^3}[1^2 + \ldots + n^2]$

$= \frac{16}{n^3}\cdot\frac{n(n+1)(2n+1)}{6} = \frac{8}{3}\cdot\frac{2n^2+3n+1}{n^2} = \frac{8}{3}\left[2 + \frac{3}{n} + \frac{1}{n^2}\right]$.

$\lim_{n\to\infty} S_n = \frac{16}{3}$. __Ans.__ 16/3 sq units

15. $\int_0^2 3x\ dx$. Let $f(x) = 3x$. $\Delta x = 2/n$.

$S_n = \frac{2}{n}f\left(\frac{2}{n}\right) + \ldots + \frac{2}{n}f\left(n\cdot\frac{2}{n}\right) = \frac{2}{n}\left[3\left(\frac{2}{n}\right) + \ldots + 3\left(n\cdot\frac{2}{n}\right)\right]$

$\qquad = \frac{12}{n^2}\left[1 + \ldots + n\right] = \frac{12}{n^2}\cdot\frac{n(n+1)}{2} = 6\cdot\frac{n+1}{n} = 6\left[1 + \frac{1}{n}\right]$.

$\int_0^2 3x\ dx = \lim_{n\to\infty} S_n = 6$.

17. $\int_0^3 -4x\ dx$. Let $f(x) = -4x$. $\Delta x = 3/n$.

$S_n = \frac{3}{n}f\left(\frac{3}{n}\right) + \ldots + \frac{3}{n}f\left(n\cdot\frac{3}{n}\right) = \frac{3}{n}\left[f\left(\frac{3}{n}\right) + \ldots + f\left(n\cdot\frac{3}{n}\right)\right]$

$\qquad = \frac{3}{n}\left[-4\left(\frac{3}{n}\right) - \ldots - 4\left(n\cdot\frac{3}{n}\right)\right] = -\frac{36}{n^2}[1 + \ldots + n]$

$\qquad = -\frac{36}{n^2}\cdot\frac{n(n+1)}{2} = -18\cdot\frac{n+1}{n} = -18\left[1 + \frac{1}{n}\right]$.

$\int_0^3 -4x\ dx = \lim_{n\to\infty} S_n = -18$.

19. $\int_0^1 (x^2+x)\ dx$. Let $f(x) = x^2+x$. $\Delta x = 1/n$.

$S_n = \frac{1}{n}f\left(\frac{1}{n}\right) + \ldots + \frac{1}{n}f\left(n\cdot\frac{1}{n}\right)$

$\qquad = \frac{1}{n}\left\{\left[\left(\frac{1}{n}\right)^2 + \frac{1}{n}\right] + \ldots + \left[\left(n\cdot\frac{1}{n}\right)^2 + n\cdot\frac{1}{n}\right]\right\}$

$\qquad = \frac{1}{n}\left\{\left(\frac{1}{n}\right)^2[1^2 + \ldots + n^2] + \frac{1}{n}[1 + \ldots + n]\right\}$

$\qquad = \frac{1}{n^3}\cdot\frac{n(n+1)(2n+1)}{6} + \frac{1}{n^2}\cdot\frac{n(n+1)}{2} = \frac{1}{6}\cdot\frac{2n^2+3n+1}{n^2} + \frac{1}{2}\cdot\frac{n+1}{n}$

$\qquad = \frac{1}{6}\left[2 + \frac{3}{n} + \frac{1}{n^2}\right] + \frac{1}{2}\left[1 + \frac{1}{n}\right]$.

$\int_0^1 (x^2+x)\ dx = \lim_{n\to\infty} S_n = \frac{1}{3} + \frac{1}{2} = \frac{5}{6}$.

21. $\int_2^3 \sqrt{x^2 + 1}\ dx$ is simply a real number. Thus

$D_X\left[\int_2^3 \sqrt{x^2 + 1}\ dx\right] = D_X(\text{real number}) = 0.$

23. $f(x) = \begin{cases} 1, & \text{if } x \leq 1, \\ 2-x, & \text{if } 1 \leq x \leq 2, \\ -1+\frac{x}{2}, & \text{if } x > 2. \end{cases}$

f is continuous and $f(x) \geq 0$
on $[-1,3]$. Thus $\int_{-1}^3 f(x)\ dx$ gives the area A bounded by
$y = f(x)$, $y = 0$, $x = -1$, and $x = 3$. From the diagram,
this area is composed of three subareas, A_1, A_2, and A_3,
and $A = A_1 + A_2 + A_3$.

A_1 = area of rectangle = $(2)(1) = 2$ sq units,

A_2 = area of triangle = $\frac{1}{2}(1)(1) = \frac{1}{2}$ sq unit,

A_3 = area of triangle = $\frac{1}{2}(1)\left(\frac{1}{2}\right) = \frac{1}{4}$ sq unit.

Since $A = A_1 + A_2 + A_3 = 2 + \frac{1}{2} + \frac{1}{4} = \frac{11}{4}$ sq units, we have

$\int_{-1}^3 f(x)\ dx = \frac{11}{4}.$

EXERCISE 16.7

1. $\int_0^2 5\ dx = 5x\Big|_0^2 = 5(2) - 5(0) = 10 - 0 = 10$

3. $\int_1^2 7x\ dx = 7 \cdot \frac{x^2}{2}\Big|_1^2 = 14 - \frac{7}{2} = \frac{21}{2}$

5. $\int_{-3}^1 (2x - 3)\ dx = (x^2 - 3x)\Big|_{-3}^1 = -2 - (18) = -20$

7. $\int_{2}^{3} (y^2 - 2y + 1)\ dy = \int_{2}^{3} (y - 1)^2\ dy = \frac{(y-1)^3}{3}\Big|_{2}^{3} = \frac{8}{3} - \frac{1}{3} = \frac{7}{3}$

9. $\int_{-2}^{-1} (3w^2 - w - 1)\ dw = \left(w^3 - \frac{w^2}{2} - w\right)\Big|_{-2}^{-1} = -\frac{1}{2} - (-8) = \frac{15}{2}$

11. $\int_{1}^{2} (-4t^{-4})\ dt = \frac{4}{3}t^{-3}\Big|_{1}^{2} = \frac{4}{3t^3}\Big|_{1}^{2} = \frac{1}{6} - \frac{8}{6} = -\frac{7}{6}$

13. $\int_{-1}^{1} \sqrt[3]{x^5}\ dx = \int_{-1}^{1} x^{5/3}\ dx = \frac{3x^{8/3}}{8}\Big|_{-1}^{1} = \frac{3}{8} - \frac{3}{8} = 0$

15. $\int_{1/2}^{3} \frac{1}{x^2}\ dx = -\frac{1}{x}\Big|_{1/2}^{3} = -\frac{1}{3} - (-2) = \frac{5}{3}$

17. $\int_{-1}^{1} (z + 1)^5\ dz = \frac{(z + 1)^6}{6}\Big|_{-1}^{1} = \frac{32}{3} - 0 = \frac{32}{3}$

19. $\int_{0}^{1} 2x^2(x^3 - 1)^3\ dx = \frac{2}{3}\int_{0}^{1} (x^3 - 1)^3[3x^2\ dx] = \frac{1}{6}(x^3 - 1)^4\Big|_{0}^{1}$

$$= 0 - \frac{1}{6} = -\frac{1}{6}$$

21. $\int_{1}^{8} \frac{4}{y}\ dy = 4\ \ln|y|\ \Big|_{1}^{8} = 4(\ln 8 - \ln 1) = 4(\ln 8 - 0) = 4\ \ln 8$

23. $\int_{0}^{1} e^5\ dx = e^5 x\Big|_{0}^{1} = e^5 - 0 = e^5.$

25. $\int_{0}^{2} x^2 e^{x^3}\ dx = \frac{1}{3}\int_{0}^{2} e^{x^3}[3x^2\ dx] = \frac{1}{3}e^{x^3}\Big|_{0}^{2} = \frac{1}{3}(e^8 - e^0)$

$$= \frac{1}{3}(e^8 - 1)$$

27. $\displaystyle\int_4^5 \frac{2}{(x-3)^3}\,dx = 2\int_4^5 (x-3)^{-3}\,dx = 2\cdot\frac{(x-3)^{-2}}{-2}\Big|_4^5$

$\displaystyle = -\frac{1}{(x-3)^2}\Big|_4^5 = -\frac{1}{4} - (-1) = \frac{3}{4}$

29. $\displaystyle\int_{1/3}^2 \sqrt{10-3p}\,dp = -\frac{1}{3}\int_{1/3}^2 (10-3p)^{1/2}[-3\,dp]$

$\displaystyle = -\frac{2}{9}(10-3p)^{3/2}\Big|_{1/3}^2 = -\frac{2}{9}(8-27) = \frac{38}{9}$

31. $\displaystyle\int_0^1 x^2\sqrt[3]{7x^3+1}\,dx = \frac{1}{21}\int_0^1 (7x^3+1)^{1/3}[21x^2\,dx]$

$\displaystyle = \frac{1}{21}\cdot\frac{(7x^3+1)^{4/3}}{4/3}\Big|_0^1 = \frac{(7x^3+1)^{4/3}}{28}\Big|_0^1 = \frac{16}{28} - \frac{1}{28} = \frac{15}{28}$

33. $\displaystyle\int_0^1 \frac{2x^3+x}{x^2+x^4+1}\,dx = \frac{1}{2}\int_0^1 \frac{1}{x^4+x^2+1}[2(2x^3+x)\,dx]$

$\displaystyle = \frac{1}{2}\ln(x^4+x^2+1)\Big|_0^1 = \frac{1}{2}[\ln 3 - \ln 1] = \frac{1}{2}\ln 3$

35. $\displaystyle\int_0^1 (e^x - e^{-2x})\,dx = \int_0^1 e^x\,dx - \left(-\frac{1}{2}\right)\int_0^1 e^{-2x}[-2\,dx]$

$\displaystyle = \left(e^x + \frac{e^{-2x}}{2}\right)\Big|_0^1 = \left(e + \frac{e^{-2}}{2}\right) - \left(1 + \frac{1}{2}\right) = e + \frac{1}{2e^2} - \frac{3}{2}$

37. $\displaystyle\int_1^e (x^{-1} + x^{-2} - x^{-3})\,dx = \left(\ln|x| + \frac{x^{-1}}{-1} - \frac{x^{-2}}{-2}\right)\Big|_1^e$

$\displaystyle = \left(1 - \frac{1}{e} + \frac{1}{2e^2}\right) - \left(0 - 1 + \frac{1}{2}\right) = \frac{3}{2} - \frac{1}{e} + \frac{1}{2e^2}$

39. $\displaystyle\int_1^3 (x + 1)e^{x^2+2x}\, dx = \frac{1}{2}\int_1^3 e^{x^2+2x}[2(x + 1)\, dx] = \frac{1}{2}e^{x^2+2x}\Big|_1^3$

$$= \frac{1}{2}\left(e^{15} - e^3\right) = \frac{e^3}{2}(e^{12} - 1)$$

41. Using long division on the integrand, we obtain

$$\int_0^2 \left[x^3 + x + \frac{3x^3 + 5}{x^3 + 5x + 1}\right] dx = \left[\frac{x^4}{4} + \frac{x^2}{2} + \ln|x^3 + 5x + 1|\right]\Big|_0^2$$

$$= (6 + \ln 19) - 0 = 6 + \ln 19$$

43. $\displaystyle\int_0^2 f(x)\, dx = \int_0^{1/2} 4x^2\, dx + \int_{1/2}^2 2x\, dx = \frac{4x^3}{3}\Big|_0^{1/2} + x^2\Big|_{1/2}^2$

$$= \left(\frac{1}{6} - 0\right) + \left(4 - \frac{1}{4}\right) = \frac{47}{12}$$

45. $\displaystyle f(x) = \int_1^x \frac{1}{t^2}\, dt = -\frac{1}{t}\Big|_1^x = -\frac{1}{x} + 1 = 1 - \frac{1}{x}.$

$$\int_e^1 f(x)\, dx = \int_e^1 \left(1 - \frac{1}{x}\right) dx = (x - \ln|x|)\Big|_e^1$$

$$= (1 - 0) - (e - 1) = 2 - e$$

47. $\displaystyle\int_1^3 f(x)\, dx = \int_1^2 f(x)\, dx + \int_2^3 f(x)\, dx,$ so

$$\int_1^2 f(x)\, dx = \int_1^3 f(x)\, dx - \int_2^3 f(x)\, dx = \int_1^3 f(x)\, dx + \int_3^2 f(x)\, dx$$

$$= 4 + 3 = 7.$$

49. $\displaystyle\int_1^2 e^{x^2}\, dx$ is a constant, so $\dfrac{d}{dx}\displaystyle\int_1^2 e^{x^2}\, dx$ is 0. Thus

$$\int_1^2 \left[\frac{d}{dx}\int_1^2 e^{x^2}\, dx\right] dx = \int_1^2 0\, dx = C\Big|_1^2 = C - C = 0.$$

51. $\int_0^T \alpha^{5/2} \, dt = \alpha^{5/2}t \Big|_0^T = \alpha^{5/2}T - 0 = \alpha^{5/2}T$

53. The total number receiving between a and b dollars equals the number $N(a)$ receiving a or more dollars minus the number $N(b)$ receiving b or more dollars. Thus

$$N(a) - N(b) = \int_b^a -Ax^{-B} \, dx.$$

55.
$$\int_0^5 2000e^{-0.06t} \, dt = 2000 \cdot \frac{1}{-0.06} \int_0^5 e^{-0.06t}[-0.06 \, dt]$$
$$= -\frac{2000}{0.06}e^{-0.06t} \Big|_0^5 = -\frac{2000}{0.06}(e^{-0.3} - 1)$$
$$\approx -\frac{2000}{0.06}(0.74082 - 1) \approx \$8639$$

57.
$$\int_{36}^{64} 10,000\sqrt{100-t} \, dt = (-1)(10,000)\int_{36}^{64}(100-t)^{1/2}[(-1) \, dt]$$
$$= -\frac{2}{3}(10,000)(100 - t)^{3/2} \Big|_{36}^{64} = -\frac{2}{3}(10,000)[216 - 512]$$
$$\approx 1,973,333$$

59. $\int_{60}^{70}(0.2q + 3) \, dq = (0.1q^2 + 3q) \Big|_{60}^{70} = 700 - 540 = \$160$

61.
$$\int_{400}^{900} \frac{1000}{\sqrt{100q}} \, dq = \int_{400}^{900} \frac{1000}{10\sqrt{q}} \, dq = 100\int_{400}^{900} q^{-1/2} \, dq$$
$$= 100 \cdot \frac{q^{1/2}}{1/2} \Big|_{400}^{900} = 200\sqrt{q} \Big|_{400}^{900} = 200(30 - 20) = \$2000$$

63. $\int_0^{12} (8t + 10)\, dt = (4t^2 + 10t)\Big|_0^{12} = 696 - 0 = 696;$

$\int_6^{12} (8t + 10)\, dt = (4t^2 + 10t)\Big|_6^{12} = 696 - 204 = 492.$

65. $G = \int_{-R}^{R} i\, dx = ix\Big|_{-R}^{R} = iR - (-iR) = 2Ri.$

67. $A = \dfrac{\displaystyle\int_0^R (m+x)[1-(m+x)]\, dx}{\displaystyle\int_0^R [1-(m+x)]\, dx} = \dfrac{\displaystyle\int_0^R (m+x-m^2-2mx-x^2)\, dx}{\displaystyle\int_0^R (1-m-x)\, dx}$

$= \dfrac{\left[mx + \dfrac{x^2}{2} - m^2x - mx^2 - \dfrac{x^3}{3}\right]\Big|_0^R}{\left[x - mx - \dfrac{x^2}{2}\right]\Big|_0^R}$

$= \dfrac{\left[mR + \dfrac{R^2}{2} - m^2R - mR^2 - \dfrac{R^3}{3}\right] - 0}{\left[R - mR - \dfrac{R^2}{2}\right] - 0}$

$= \dfrac{R\left[m + \dfrac{R}{2} - m^2 - mR - \dfrac{R^2}{3}\right]}{R\left[1 - m - \dfrac{R}{2}\right]} = \dfrac{m + \dfrac{R}{2} - m^2 - mR - \dfrac{R^2}{3}}{1 - m - \dfrac{R}{2}}$

EXERCISE 16.8

In Problems 1-33, answers are assumed to be expressed in square units.

1. $y = 4x$, $x = 2$.

Area $= \int_0^2 4x\, dx$

$= 2x^2\Big|_0^2$

$= 8 - 0 = 8.$

3. $y = 3x + 2$, $x = 2$, $x = 3$.

Area $= \int_{2}^{3}(3x + 2)\ dx$

$= \left(\dfrac{3x^2}{2} + 2x\right)\Big|_{2}^{3}$

$= \dfrac{39}{2} - 10 = \dfrac{19}{2}.$

5. $y = x - 1$, $x = 5$.

Area $= \int_{1}^{5}(x - 1)\ dx$

$= \left(\dfrac{x^2}{2} - x\right)\Big|_{1}^{5}$

$= \dfrac{15}{2} - \left(-\dfrac{1}{2}\right) = \dfrac{16}{2} = 8.$

7. $y = x^2$, $x = 2$, $x = 3$.

Area $= \int_{2}^{3}x^2\ dx = \dfrac{x^3}{3}\Big|_{2}^{3}$

$= 9 - \dfrac{8}{3} = \dfrac{19}{3}.$

9. $y = x^2 + 2$, $x = -1$, $x = 2$.

Area $= \int_{-1}^{2}(x^2 + 2)\ dx$

$= \left(\dfrac{x^3}{3} + 2x\right)\Big|_{-1}^{2}$

$= \dfrac{20}{3} - \left(-\dfrac{7}{3}\right) = \dfrac{27}{3} = 9.$

11. $y = x^2 - 2x$, $x = -3$, $x = -1$.

Area $= \int_{-3}^{-1}(x^2 - 2x)\ dx$

$= \left(\dfrac{x^3}{3} - x^2\right)\Big|_{-3}^{-1}$

$= -\dfrac{4}{3} - (-18) = \dfrac{50}{3}.$

13. $y = 9 - x^2$.

Area $= \int_{-3}^{3} (9 - x^2)\, dx$

$= \left(9x - \dfrac{x^3}{3}\right)\Big|_{-3}^{3}$

$= 18 - (-18) = 36.$

15. $y = 1 - x - x^3$, $x = -2$, $x = 0$.

Area $= \int_{-2}^{0} (1 - x - x^3)\, dx$

$= \left(x - \dfrac{x^2}{2} - \dfrac{x^4}{4}\right)\Big|_{-2}^{0}$

$= 0 - (-8) = 8.$

17. $A = 3 + 2x - x^2$.

Area $= \int_{-1}^{3} (3 + 2x - x^2)\, dx$

$= \left(3x + x^2 - \dfrac{x^3}{3}\right)\Big|_{-1}^{3}$

$= 9 - \left(-\dfrac{5}{3}\right) = \dfrac{32}{3}.$

19. $y = \dfrac{1}{x}$, $x = 1$, $x = e$.

Area $= \int_{1}^{e} \dfrac{1}{x}\, dx$

$= \ln|x|\ \Big|_{1}^{e}$

$= \ln e - \ln 1 = 1 - 0 = 1.$

21. $y = \sqrt{x + 9}$, $x = -9$, $x = 0$.

Area $= \int_{-9}^{0} \sqrt{x + 9}\, dx = \int_{-9}^{0} (x + 9)^{1/2}\, dx$

$= \dfrac{(x + 9)^{3/2}}{3/2}\Big|_{-9}^{0} = \dfrac{2(x + 9)^{3/2}}{3}\Big|_{-9}^{0}$

$= 18 - 0 = 18.$

23. $y = \sqrt{2x - 1}$, $x = 1$, $x = 5$.

    Area $= \int_1^5 \sqrt{2x - 1} \; dx$

         $= \frac{1}{2}\int_1^5 (2x - 1)^{1/2} [2 \; dx]$

         $= \left.\frac{(2x-1)^{3/2}}{3}\right|_1^5 = 9 - \frac{1}{3} = \frac{26}{3}.$

25. $y = \sqrt[3]{x}$, $x = 2$.

    Area $= \int_0^2 \sqrt[3]{x} \; dx = \int_0^2 x^{1/3} \; dx$

         $= \left.\frac{3x^{4/3}}{4}\right|_0^2 = \frac{3(2)^{4/3}}{4} - 0$

         $= \frac{3(2\sqrt[3]{2})}{4} = \frac{3}{2}\sqrt[3]{2}.$

27. $y = e^x$, $x = 0$, $x = 2$.

    Area $= \int_0^2 e^x \; dx$

         $= \left. e^x \right|_0^2 = e^2 - 1.$

29. $y = x + \frac{2}{x}$, $x = 1$, $x = 2$.

    Area $= \int_1^2 \left(x + \frac{2}{x}\right) dx$

         $= \left.\left(\frac{x^2}{2} + 2 \ln|x|\right)\right|_1^2$

         $= (2 + 2 \ln 2) - \frac{1}{2}$

         $= \frac{3}{2} + 2 \ln 2 = \frac{3}{2} + \ln 4.$

31. $y = x^3$, $x = -2$, $x = 4$.

Area $= \int_{-2}^{0} -x^3 \, dx + \int_{0}^{4} x^3 \, dx$

$= -\left.\frac{x^4}{4}\right|_{-2}^{0} + \left.\frac{x^4}{4}\right|_{0}^{4}$

$= [0 - (-4)] + [64 - 0] = 68.$

33. $y = 2x - x^2$, $x = 1$, $x = 3$.

Area $= \int_{1}^{2} (2x - x^2) \, dx + \int_{2}^{3} -(2x - x^2) \, dx$

$= \left.\left(x^2 - \frac{x^3}{3}\right)\right|_{1}^{2} - \left.\left(x^2 - \frac{x^3}{3}\right)\right|_{2}^{3}$

$= \left[\frac{4}{3} - \frac{2}{3}\right] - \left[0 - \frac{4}{3}\right] = \frac{6}{3} = 2.$

35. $f(x) = \begin{cases} 3x^2, & \text{if } 0 \leq x \leq 2 \\ 16 - 2x, & \text{if } x \geq 2. \end{cases}$

Area $= \int_{0}^{3} f(x) \, dx$

$= \int_{0}^{2} 3x^2 \, dx + \int_{2}^{3} (16 - 2x) \, dx$

$= \left.x^3\right|_{0}^{2} + \left.(16x - x^2)\right|_{2}^{3}$

$= [8 - 0] + [39 - 28] = 19 \text{ sq units.}$

37. (a) $P(0 \leq x \leq 1) = \int_{0}^{1} \frac{1}{8}x \, dx = \left.\frac{x^2}{16}\right|_{0}^{1} = \frac{1}{16} - 0 = \frac{1}{16}.$

(b) $P(2 \leq x \leq 4) = \int_{2}^{4} \frac{1}{8}x \, dx = \left.\frac{x^2}{16}\right|_{2}^{4} = 1 - \frac{1}{4} = \frac{3}{4}.$

(c) $P(x \geq 3) = \int_{3}^{4} \frac{1}{8}x \, dx = \left.\frac{x^2}{16}\right|_{3}^{4} = 1 - \frac{9}{16} = \frac{7}{16}.$

39. (a) $P(3 \leq x \leq 5) = \int_3^5 \frac{1}{x} dx = \ln|x| \Big|_3^5 = \ln 5 - \ln 3 = \ln \frac{5}{3}.$

(b) $P(x \leq 4) = \int_e^4 \frac{1}{x} dx = \ln|x| \Big|_e^4 = \ln(4) - \ln e = \ln(4) - 1.$

(c) $P(x \geq 3) = \int_3^{e^2} \frac{1}{x} dx = \ln|x| \Big|_3^{e^2} = \ln e^2 - \ln 3$

$= 2 - \ln 3.$

(d) $P(e \leq x \leq e^2) = \int_e^{e^2} \frac{1}{x} dx = \ln|x| \Big|_e^{e^2} = \ln e^2 - \ln e$

$= 2 - 1 = 1.$

EXERCISE 16.9

1. Area $= \int_a^b (y_{UPPER} - y_{LOWER}) dx = \int_{-2}^3 [(x + 6) - x^2] dx.$

3. Intersection points:

$x^2 - x = 2x, \quad x^2 - 3x = 0, \quad x(x - 3) = 0 \implies x = 0 \text{ or } x = 3.$

Area $= \int_0^3 (y_{UPPER} - y_{LOWER}) dx + \int_3^4 (y_{UPPER} - y_{LOWER}) dx$

$= \int_0^3 [2x - (x^2 - x)] dx + \int_3^4 [(x^2 - x) - 2x] dx.$

5. The graphs of $y = 1 - x^2$ and $y = x - 1$ intersect when
$1 - x^2 = x - 1$, $0 = x^2 + x - 2$, $0 = (x - 1)(x + 2) \implies$
$x = 1$ or $x = -2$. When $x = 1$, then $y = 0$. We use
horizontal elements, where y ranges from 0 to 1. Solving
$y = x - 1$ for x gives $x = y + 1$, and solving $y = 1 - x^2$
for x gives $x^2 = 1 - y$, $x = \pm\sqrt{1 - y}$. We must choose
$x = \sqrt{1 - y}$ because x is not negative over the given
region.

Area $= \int_0^1 (x_{RIGHT} - x_{LEFT}) dy = \int_0^1 [(y + 1) - \sqrt{1 - y}] dy.$

7. The graphs of $y = x^2 - 4$ and
$y = 11 - 2x^2$ intersect when
$x^2 - 4 = 11 - 2x^2$, $3x^2 = 15$,
$x^2 = 5$, so $x = \pm\sqrt{5}$. We use
vertical elements.

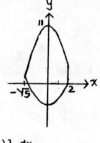

$$\text{Area} = \int_{-\sqrt{5}}^{2} (y_{UPPER} - y_{LOWER})\ dx$$

$$= \int_{-\sqrt{5}}^{2} [(11 - 2x^2) - (x^2 - 4)]\ dx.$$

In Problems 9-33, the answers are assumed to be expressed in square units.

9. $y = x^2$, $y = 2x$. Region appears below. Intersection:
$x^2 = 2x$, $x^2 - 2x = 0$, $x(x - 2) = 0$, so $x = 0$ or 2.

$$\text{Area} = \int_{0}^{2} (2x - x^2)\ dx = \left(x^2 - \frac{x^3}{3}\right)\Big|_{0}^{2} = \left(4 - \frac{8}{3}\right) - 0 = \frac{4}{3}.$$

9.

11.

11. $y = x^2$, $x = 0$, $y = 4$ ($x \geq 0$). Region appears above.
Intersection: $x^2 = 4$, so $x = \pm 2$.

$$\text{Area} = \int_{0}^{2} (4 - x^2)\ dx = \left(4x - \frac{x^3}{3}\right)\Big|_{0}^{2} = \left(8 - \frac{8}{3}\right) - 0 = \frac{16}{3}.$$

13. $y = x^2 + 3$, $y = 9$. Region appears below.

Intersection: $x^2 + 3 = 9$, $x^2 = 6$, so $x = \pm\sqrt{6}$.

$$\text{Area} = \int_{-\sqrt{6}}^{\sqrt{6}} [9 - (x^2 + 3)]\, dx = \int_{-\sqrt{6}}^{\sqrt{6}} (6 - x^2)\, dx \cdot$$

$$= \left(6x - \frac{x^3}{3}\right)\Bigg|_{-\sqrt{6}}^{\sqrt{6}}$$

$$= \left(6\sqrt{6} - \frac{6\sqrt{6}}{3}\right) - \left(-6\sqrt{6} + \frac{6\sqrt{6}}{3}\right) = 8\sqrt{6}.$$

13.

15.

15. $x = 8 + 2y$, $x = 0$, $y = -1$, $y = 3$. Region appears above.

$$\text{Area} = \int_{-1}^{3} (8 + 2y)\, dy = (8y + y^2)\Bigg|_{-1}^{3}$$

$$= (24 + 9) - (-8 + 1) = 40.$$

17. $y = 4 - x^2$, $y = -3x$. Region appears below. Intersection:
$-3x = 4 - x^2$, $x^2 - 3x - 4 = 0$, $(x + 1)(x - 4) = 0$, so
$x = -1$ or 4.

$$\text{Area} = \int_{-1}^{4} [(4 - x^2) - (-3x)]\, dx = \left(4x - \frac{x^3}{3} + \frac{3x^2}{2}\right)\Bigg|_{-1}^{4}$$

$$= \left(16 - \frac{64}{3} + 24\right) - \left(-4 + \frac{1}{3} + \frac{3}{2}\right) = \frac{125}{6}.$$

17.

19.

19. $y^2 = x$, $y = x - 2$. Region appears above.
    Intersection:
    $$y^2 = y + 2, \quad y^2 - y - 2 = 0, \quad (y + 1)(y - 2) = 0,$$
    so $y = -1$ or 2.
    $$\text{Area} = \int_{-1}^{2} [(y + 2) - y^2] \, dy = \left(\frac{y^2}{2} + 2y - \frac{y^3}{3}\right)\Big|_{-1}^{2}$$
    $$= \left(2 + 4 - \frac{8}{3}\right) - \left(\frac{1}{2} - 2 + \frac{1}{3}\right) = \frac{9}{2}.$$

21. $2y = 4x - x^2$, $2y = x - 4$. Region appears below.
    Intersection: $x - 4 = 4x - x^2$, $x^2 - 3x - 4 = 0$,
    $(x + 1)(x - 4) = 0$, so $x = -1$ or 4. Note that the
    y-values of the curves are given by $y = \frac{4x - x^2}{2}$ and
    $y = \frac{x - 4}{2}$.
    $$\text{Area} = \int_{-1}^{4} \left[\left(\frac{4x - x^2}{2}\right) - \left(\frac{x - 4}{2}\right)\right] dx$$
    $$= \int_{-1}^{4} \left(\frac{3}{2}x - \frac{x^2}{2} + 2\right) dx$$
    $$= \left(\frac{3x^2}{4} - \frac{x^3}{6} + 2x\right)\Big|_{-1}^{4} = \left(12 - \frac{64}{6} + 8\right) - \left(\frac{3}{4} + \frac{1}{6} - 2\right)$$
    $$= \frac{125}{12}$$

21.

23.

23. $y^2 = x$, $3x - 2y = 1$ $\left(\text{or } x = \frac{1 + 2y}{3}\right)$ Region appears above.
    Intersection: $y^2 = \frac{1 + 2y}{3}$, $3y^2 = 1 + 2y$, $3y^2 - 2y - 1 = 0$,
    $(3y + 1)(y - 1) = 0$, so $y = -1/3$ or 1.
    Thus the area is given by

$$\text{Area} = \int_{-1/3}^{1} \left[\left(\frac{1 + 2y}{3}\right) - y^2\right] dy = \int_{-1/3}^{1} \left(\frac{1}{3} + \frac{2y}{3} - y^2\right) dy$$

$$= \left(\frac{y}{3} + \frac{y^2}{3} - \frac{y^3}{3}\right)\Big|_{-1/3}^{1}$$

$$= \left(\frac{1}{3} + \frac{1}{3} - \frac{1}{3}\right) - \left(-\frac{1}{9} + \frac{1}{27} + \frac{1}{81}\right) = \frac{32}{81}.$$

25. $y = 8 - x^2$, $y = x^2$, $x = -1$, $x = 1$. Region appears below.

Intersection: $x^2 = 8 - x^2$, $2x^2 = 8$, $x^2 = 4$, so $x = \pm 2$.

$$\text{Area} = \int_{-1}^{1} [(8 - x^2) - x^2] dx = \int_{-1}^{1} (8 - 2x^2) dx$$

$$= \left(8x - \frac{2x^3}{3}\right)\Big|_{-1}^{1} = \left(8 - \frac{2}{3}\right) - \left(-8 + \frac{2}{3}\right) = \frac{44}{3}.$$

25.   27.

27. $y = x^2$, $y = 2$, $y = 5$. Region appears above.

$$\text{Area} = \int_{2}^{5} [\sqrt{y} - (-\sqrt{y})] dy = \int_{2}^{5} 2\sqrt{y}\, dy = 2 \cdot \frac{y^{3/2}}{3/2}\Big|_{2}^{5}$$

$$= \frac{4y^{3/2}}{3}\Big|_{2}^{5} = \frac{4 \cdot 5\sqrt{5}}{3} - \frac{4 \cdot 2\sqrt{2}}{3} = \frac{4}{3}(5\sqrt{5} - 2\sqrt{2}).$$

29. $y = x^3$, $y = x$. Region appears below. Intersection:
$x^3 = x$, $x^3 - x = 0$, $x(x^2 - 1) = 0$, $x(x + 1)(x - 1) = 0$, so
$x = 0$ or $x = \pm 1$.

$$\text{Area} = \int_{-1}^{0} (x^3 - x)\, dx + \int_{0}^{1} (x - x^3)\, dx$$

$$= \left(\frac{x^4}{4} - \frac{x^2}{2}\right)\Big|_{-1}^{0} + \left(\frac{x^2}{2} - \frac{x^4}{4}\right)\Big|_{0}^{1}$$

$$= \left[0 - \left(\frac{1}{4} - \frac{1}{2}\right)\right] + \left[\left(\frac{1}{2} - \frac{1}{4}\right) - 0\right] = \frac{1}{2}.$$

29.

31.

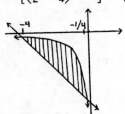

31. $y = \frac{-17 - 4x}{4}$, $y = \frac{1}{x}$. Region appears above. Intersection:

$\frac{-17 - 4x}{4} = \frac{1}{x}$, $-17x - 4x^2 = 4$, $4x^2 + 17x + 4 = 0$,

$(4x + 1)(x + 4) = 0$, so $x = -1/4$ and 4.

$$\text{Area} = \int_{-4}^{-1/4} \left[\frac{1}{x} - \left(\frac{-17 - 4x}{4}\right)\right] dx = \left(\ln|x| + \frac{17}{4}x + \frac{x^2}{2}\right)\Big|_{-4}^{-1/4}$$

$$= \left(\ln \frac{1}{4} - \frac{17}{16} + \frac{1}{32}\right) - \left(\ln 4 - 17 + 8\right) = \frac{255}{32} - 4 \ln 2.$$

33. $y = x - 1$, $y = 5 - 2x$. Region appears below.
Intersection: $x - 1 = 5 - 2x$, $3x = 6$, so $x = 2$.

$$\text{Area} = \int_0^2 [(5 - 2x) - (x - 1)]dx + \int_2^4 [(x - 1) - (5 - 2x)]dx$$

$$= \int_0^2 (6 - 3x) \, dx + \int_2^4 (3x - 6) \, dx$$

$$= -\frac{1}{3}\int_0^2 (6 - 3x)[-3 \, dx] + \frac{1}{3}\int_2^4 (3x - 6)[3 \, dx]$$

$$= -\frac{(6 - 3x)^2}{6}\Big|_0^2 + \frac{(3x - 6)^2}{6}\Big|_2^4$$

$$= -[0 - 6] + [6 - 0] = 6 + 6 = 12.$$

35. $\dfrac{\text{Area between curve and diag.}}{\text{Area under diagonal}} = \dfrac{\int_0^1 \left[x - \left(\frac{20}{21}x^2 + \frac{1}{21}x\right)\right]\, dx}{\int_0^1 x\, dx}$.

Numerator $= \displaystyle\int_0^1 \left[\frac{20}{21}x - \frac{20}{21}x^2\right]\, dx = \frac{20}{21}\int_0^1 (x - x^2)\, dx$

$= \dfrac{20}{21}\left(\dfrac{x^2}{2} - \dfrac{x^3}{3}\right)\Big|_0^1 = \dfrac{20}{21}\left[\left(\dfrac{1}{2} - \dfrac{1}{3}\right) - 0\right]$

$= \dfrac{20}{21}\cdot\dfrac{1}{6} = \dfrac{10}{63}$.

Denominator $= \displaystyle\int_0^1 x\, dx = \dfrac{x^2}{2}\Big|_0^1 = \dfrac{1}{2}$.

Coefficient of inequality $= \dfrac{10/63}{1/2} = \dfrac{20}{63}$.

37. $y^2 = 4x$, $y = mx$. Intersection:

$(mx)^2 = 4x$ $m^2 x^2 = 4x$, $m^2 x^2 - 4x = 0$,

$x(m^2 x - 4) = 0$, $x = 0$ or $x = \dfrac{4}{m^2}$.

If $x = 0$, then $y = 0$; if $x = \dfrac{4}{m^2}$,

then $y = \dfrac{4}{m}$.   With horizontal elements,

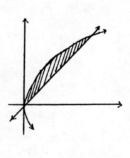

area $= \displaystyle\int_0^{4/m}\left(\frac{y}{m} - \frac{y^2}{4}\right) dy = \left(\frac{y^2}{2m} - \frac{y^3}{12}\right)\Big|_0^{4/m}$

$= \dfrac{8}{m^3} - \dfrac{16}{3m^3} = \dfrac{8}{3m^3}$ square units.

Note: With vertical elements,

area $= \displaystyle\int_0^{4/m^2} (2\sqrt{x} - mx)\, dx$.

39. $y = x^2$ and $y = k$ intersect when
$x^2 = k$, $x = \pm\sqrt{k}$. Equating areas
gives

$$\int_{-\sqrt{k}}^{\sqrt{k}} (k - x^2) \, dx = \frac{1}{2}\int_{-2}^{2} (4 - x^2) \, dx$$

$$\left(kx - \frac{x^3}{3}\right)\Bigg|_{-\sqrt{k}}^{\sqrt{k}} = \frac{1}{2}\left(4x - \frac{x^3}{3}\right)\Bigg|_{-2}^{2}$$

$$\frac{4}{3}k^{3/2} = \frac{16}{3},$$

$$k^{3/2} = 4 \implies k = 4^{2/3} = (2^2)^{2/3} = 2^{4/3}.$$

EXERCISE 16.10

1. D: $p = 20 - 0.8q$
   S: $p = 4 + 1.2q$ $\quad$ Equilibrium pt. $= (q_0, p_0) = (8, 13.6)$.

$$CS = \int_0^{q_0} [f(q) - p_0] \, dq$$

$$= \int_0^8 [(20 - 0.8q) - 13.6] \, dq = \int_0^8 (6.4 - 0.8q) \, dq$$

$$= (6.4q - 0.4q^2)\Bigg|_0^8 = (51.2 - 25.6) - 0 = 25.6.$$

$$PS = \int_0^{q_0} [p_0 - g(q)] \, dq$$

$$= \int_0^8 [13.6 - (4 + 1.2q)] \, dq = \int_0^8 (9.6 - 1.2q) \, dq$$

$$= (9.6q - 0.6q^2)\Bigg|_0^8 = (76.8 - 38.4) - 0 = 38.4.$$

3. D: $p = 50/(q + 5)$
   S: $p = (q/10) + 4.5$ } Equilibrium pt. $= (q_0, p_0) = (5, 5)$.

$$CS = \int_0^{q_0} [f(q) - p_0]\, dq$$

$$= \int_0^5 \left[\frac{50}{q + 5} - 5\right]\, dq = (50\ \ln|q + 5| - 5q)\Big|_0^5$$

$$= [50\ \ln(10) - 25] - [50\ \ln(5)]$$

$$= 50[\ln(10) - \ln(5)] - 25 = 50\ \ln(2) - 25.$$

$$PS = \int_0^{q_0} [p_0 - g(q)]\, dq$$

$$= \int_0^5 \left[5 - \left(\frac{q}{10} + 4.5\right)\right]\, dq = \int_0^5 \left(0.5 - \frac{q}{10}\right)\, dq$$

$$= \left(0.5q - \frac{q^2}{20}\right)\Big|_0^5 = (2.5 - 1.25) - 0 = 1.25$$

5. D: $q = 100(10 - p)$
   S: $q = 80(p - 10$ } Equilibrium pt. $= (q_0, p_0) = (400, 6)$.
   We use horizontal strips and integrate with respect to p.

$$CS = \int_6^{10} 100(10 - p)\, dp$$

$$= 100\left(10p - \frac{p^2}{2}\right)\Big|_6^{10}$$

$$= 100[(100 - 50) - (60 - 18)]$$

$$= 800.$$

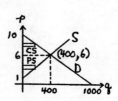

$$PS = \int_1^6 80(p - 1)\, dp$$

$$= 80\left(\frac{p^2}{2} - p\right)\Big|_1^6$$

$$= 80\left[(18 - 6) - \left(\frac{1}{2} - 1\right)\right] = 1000.$$

7. We integrate with respect to p. $p = 84 \Rightarrow q = 10\sqrt{100} = 100$.

$$CS = \int_{84}^{100} 10\sqrt{100 - p}\ dp = \int_{84}^{100} -10(100 - p)^{1/2}\ [-dp]$$

$$= -\frac{20}{3}(100 - p)^{3/2}\Big|_{84}^{100} = -\frac{20}{3}[0 - (16)^{3/2}] = -\frac{20}{3}(-64)$$

$$= 426\frac{2}{3} \approx \$426.67$$

9. At equilibrium, $2^{11-q} = 2^{q+1} \Rightarrow 11 - q = q + 1 \Rightarrow q = 5$, and so $p = 2^{11-5} = 64$.

$$CS = \int_0^5 (2^{11-q} - 64)\ dq = \left(-\frac{2^{11-q}}{\ln 2} - 64q\right)\Big|_0^5$$

$$= -\frac{64}{\ln 2} - 320 - \left(-\frac{2^{11}}{\ln 2} - 0\right)$$

$$\approx 2542.307 \text{ hundred} \approx \$254,000$$

11. (a) $\frac{d}{dx}\Big[(q + 20)\ \ln(q + 20) - q + C\Big]$

$$= (q + 20)\frac{1}{q + 20} + [\ln(q + 200](1) - 1 + 0 = \ln(q + 20$$

(b) Setting $q = 80$ in the demand equation gives
$$p = 60 - \frac{4000}{\sqrt{10,000}} = 20,$$
and setting $q = 80$ in the supply equation gives
$$p = 10\ \ln(100) - 26 = 10(4.6) - 26 = 20.$$
Thus both expressions for p become 20 when $q = 80$.

(c) $$CS = \int_0^{80} \left[\left(60 - \frac{50q}{\sqrt{q^2 + 3600}}\right) - 20\right]\ dq$$

$$= \left[40q - 50(q^2 + 3600)^{1/2}\right]\Big|_0^{80}$$

$$= 3200 - 50(100) - [0 - 50(60)] = 1200.$$

(d) $$PS = \int_0^{80} \Big\{20 - [10\ \ln(q + 20) - 26]\Big\}\ dq$$

$$= \{46q - 10[(q + 20)\ \ln(q + 20) - q]\}\Big|_0^{80}$$

$$= 3680 - 1000\ \ln(100) + 800 - (0 - 200\ \ln 20)$$

$$= 3680 - 4600 + 800 + 600 = 480$$

CHAPTER 16 - REVIEW PROBLEMS

1. $\int (x^3 + 2x -) \, dx = \frac{x^4}{4} + 2 \cdot \frac{x^2}{2} - 7x + C = \frac{x^4}{4} + x^2 - 7x + C$

3. $\int_0^9 (\sqrt{x} + x) \, dx = \int_0^9 (x^{1/2} + x) \, dx = \left(\frac{2x^{3/2}}{3} + \frac{x^2}{2} \right) \Big|_0^9$

$$= \left(18 + \frac{81}{2} \right) - 0 = \frac{117}{2}$$

5. $\int \frac{2}{(x + 5)^3} \, dx = 2\int (x + 5)^{-3} \, dx = \frac{2(x + 5)^{-2}}{-2} + C$

$$= -(x + 5)^{-2} + C$$

7. $\int \frac{6x^2 - 12}{x^3 - 6x + 1} \, dx = 2\int \frac{1}{x^3 - 6x + 1} [(3x^2 - 6) \, dx]$

$$= 2 \ln|x^3 - 6x + 1| + C$$

9. $\int_0^1 \sqrt[3]{3t + 8} \, dt = \frac{1}{3}\int_0^1 (3t + 8)^{1/3} [3 \, dt] = \frac{1}{3} \cdot \frac{(3t + 8)^{4/3}}{4/3} \Big|_0^1$

$$= \frac{(3t + 8)^{4/3}}{4} \Big|_0^1 = \frac{11\sqrt[3]{11}}{4} - 4$$

11. $\int y(y + 1)^2 \, dy = \int (y^3 + 2y^2 + y) \, dy = \frac{y^4}{4} + \frac{2y^3}{3} + \frac{y^2}{2} + C$

13. $\int \frac{\sqrt[4]{z} - \sqrt[3]{z}}{\sqrt{z}} \, dz = \int \left(\frac{z^{1/4}}{z^{1/2}} - \frac{z^{1/3}}{z^{1/2}} \right) dz = \int \left(z^{-1/4} - z^{-1/6} \right) dz$

$$= \frac{z^{3/4}}{3/4} - \frac{z^{5/6}}{5/6} + C = \frac{4z^{3/4}}{3} - \frac{6z^{5/6}}{5} + C$$

15. $\int_1^2 \frac{t^2}{2 + t^3} \, dt = \frac{1}{3}\int_1^2 \frac{1}{2 + t^3} [3t^2 \, dt] = \frac{1}{3} \ln|2 + t^3| \Big|_1^2$

$$= \frac{1}{3}(\ln 10 - \ln 3) = \frac{1}{3} \ln \frac{10}{3}$$

17. $\int x^2\sqrt{3x^3 + 2}\ dx = \frac{1}{9}\int (3x^3 + 2)^{1/2}[9x^2\ dx]$

$= \frac{1}{9}\cdot\frac{(3x^3 + 2)^{3/2}}{3/2} + C = \frac{2}{27}(3x^3 + 2)^{3/2} + C$

19. $\int (e^{2y} - e^{-2y})\ dy = \frac{1}{2}\int e^{2y}[2\ dy] - \left(-\frac{1}{2}\right)\int e^{-2y}[-2\ dy]$

$= \frac{1}{2}e^{2y} + \frac{1}{2}e^{-2y} + C = \frac{1}{2}(e^{2y} + e^{-2y}) + C$

21. $\int \left(\frac{1}{x} + \frac{2}{x^2}\right)\ dx = \int \frac{1}{x}\ dx + 2\int x^{-2}\ dx = \ln|x| + 2\cdot\frac{x^{-1}}{-1} + C$

$= \ln|x| - \frac{2}{x} + C$

23. $\int_{-2}^{1} (y^4 - y + 1)\ dy = \left(\frac{y^5}{5} - \frac{y^2}{2} + y\right)\Big|_{-2}^{1}$

$= \left(\frac{1}{5} - \frac{1}{2} + 1\right) - \left(-\frac{32}{5} - 2 - 2\right) = \frac{111}{10} = 11.1$

25. $\int_{\sqrt{3}}^{2} 7x\sqrt{4 - x^2}\ dx = 7\int_{\sqrt{3}}^{2} x(4 - x^2)^{1/2}\ dx$

$= 7\cdot\left(-\frac{1}{2}\right)\int_{\sqrt{3}}^{2} (4 - x^2)^{1/2}[-2x\ dx] = -\frac{7}{2}\cdot\frac{(4 - x^2)^{3/2}}{3/2}\Big|_{\sqrt{3}}^{2}$

$= -\frac{7}{3}\cdot(4 - x^2)^{3/2}\Big|_{\sqrt{3}}^{2} = -\frac{7}{3}(0 - 1) = \frac{7}{3}$

27. $\int_{0}^{1}\left[2x - \frac{1}{(x + 1)^{2/3}}\right]\ dx = 2\int_{0}^{1} x\ dx - \int_{0}^{1}(x + 1)^{-2/3}[dx]$

$= \left[2\cdot\frac{x^2}{2} - \frac{(x + 1)^{1/3}}{1/3}\right]\Big|_{0}^{1} = [x^2 - 3(x + 1)^{1/3}]\Big|_{0}^{1}$

$= \left[1 - 3\sqrt[3]{2}\right] - \left[0 - 3\right] = 4 - 3\sqrt[3]{2}$

29. $\displaystyle\int \frac{\sqrt{t} - 3}{t^2}\, dt = \int \left[\frac{t^{1/2}}{t^2} - \frac{3}{t^2}\right] dt = \int (t^{-3/2} - 3t^{-2})\, dt$

$\displaystyle = \frac{t^{-1/2}}{-1/2} - 3\cdot\frac{t^{-1}}{-1} + C = -2t^{-1/2} + 3t^{-1} + C = \frac{3}{t} - \frac{2}{\sqrt{t}} + C$

31. $\displaystyle\int_{-1}^{0} \frac{x^2 + 4x - 1}{x + 2}\, dx = \int_{-1}^{0} \left(x + 2 - \frac{5}{x + 2}\right) dx$

$\displaystyle = \left(\frac{x^2}{2} + 2x - 5\ln|x + 2|\right)\Big|_{-1}^{0} = (-5\ln 2) - \left(\frac{1}{2} - 2 - 0\right)$

$\displaystyle = \frac{3}{2} - 5\ln 2$

33. $\displaystyle\int \sqrt{x}\sqrt{x^{3/2} + 1}\, dx = \frac{2}{3}\int (x^{3/2} + 1)^{1/2}\left[\frac{3}{2}x^{1/2}\, dx\right]$

$\displaystyle = \frac{2}{3}\cdot\frac{(x^{3/2} + 1)^{3/2}}{3/2} + C = \frac{4}{9}(x^{3/2} + 1)^{3/2} + C$

35. $\displaystyle\int_{1}^{e} \frac{e^{\ln x}}{x^2}\, dx = \int_{1}^{e} \frac{x}{x^2}\, dx = \int_{1}^{e} \frac{1}{x}\, dx = \ln|x|\Big|_{1}^{e} = \ln e - \ln 1$

$= 1 - 0 = 1$

37. $\displaystyle\int \frac{(1 + e^{3x})^2}{e^{-3x}}\, dx = \frac{1}{3}\int (1 + e^{3x})^2[3e^{3x}\, dx] = \frac{(1 + e^{3x})^3}{9} + C$

39. $\displaystyle\int \sqrt{10^{3x}}\, dx = \int 10^{3x/2}\, dx = \frac{2}{3}\int 10^{3x/2}\left[\frac{3}{2}\, dx\right]$

$\displaystyle = \frac{2}{3}\cdot\frac{10^{3x/2}}{\ln 10} + C = \frac{2\sqrt{10^{3x}}}{3\,\ln 10} + C$

41. $\displaystyle y = \int (e^{2x} + 3)\, dx = \int e^{2x}\, dx + \int 3\, dx = \frac{1}{2}\int e^{2x}[2\, dx] + \int 3\, dx$

$\displaystyle = \frac{1}{2}e^{2x} + 3x + C.$

$y(0) = -\frac{1}{2}$ implies that $-\frac{1}{2} = \frac{1}{2} + 0 + C$, so $C = -1$. Thus

$y = \frac{1}{2}e^{2x} + 3x - 1$

In Problems 43-57, answers are assumed to be expressed in square units.

43. $y = x^2 - 1$, $x = 2$, $y \geq 0$. Region appears below.

Area $= \displaystyle\int_1^2 (x^2 - 1) \, dx = \left(\frac{x^3}{3} - x\right)\Big|_1^2 = \left(\frac{8}{3} - 2\right) - \left(\frac{1}{3} - 1\right) = \frac{4}{3}$

43.   45.

45. $y = \sqrt{x + 4}$, $x = 0$.   Region appears above.

Area $= \displaystyle\int_{-4}^0 \sqrt{x + 4} \, dx = \int_{-4}^0 (x + 4)^{1/2} [dx] = \frac{(x + 4)^{3/2}}{3/2}\Big|_{-4}^0$

$= \dfrac{2(x + 4)^{3/2}}{3}\Big|_{-4}^0 = \dfrac{16}{3} - 0 = \dfrac{16}{3}$

47. $y = 5x - x^2$. Region appears below.

Area $= \displaystyle\int_0^5 (5x - x^2) \, dx = \left(\frac{5x^2}{2} - \frac{x^3}{3}\right)\Big|_0^5 = \left(\frac{125}{2} - \frac{125}{3}\right) - 0$

$= \dfrac{125}{6}$

47.   49.

49. $y = \frac{1}{x} + 3$, $x = 1$, $x = 3$.   Region appears above.

Area $= \displaystyle\int_1^3 \left(\frac{1}{x} + 3\right) dx = (\ln|x| + 3x)\Big|_1^3 = [\ln(3) + 9] - 3$

$= 6 + \ln 3$

51. $y^2 = 4x$, $x = 0$, $y = 2$. Region appears below.

$$\text{Area} = \int_0^2 \frac{y^2}{4} \, dy = \frac{y^3}{12}\Big|_0^2 = \frac{8}{12} - 0 = \frac{2}{3}$$

51.

53.

53. $y = x^2 + 4x - 5$, $y = 0$. Region appears above.

$x^2 + 4x - 5 = 0$, $(x + 5)(x - 1) = 0$, so $x = -5, 1$.

$$\text{Area} = \int_{-5}^1 -(x^2 + 4x - 5) \, dx = -\left(\frac{x^3}{3} + 2x^2 - 5x\right)\Big|_{-5}^1$$

$$= -\left(\frac{1}{3} + 2 - 5\right) + \left(-\frac{125}{3} + 50 + 25\right) = 36$$

55. $y = x^2 - 2x$, $y = 12 - x^2$. Region appears below.

$x^2 - 2x = 12 - x^2$, $2x^2 - 2x - 12 = 0$, $2(x + 2)(x - 3) = 0$,
so $x = -2, 3$.

$$\text{Area} = \int_{-2}^3 [(12 - x^2) - (x^2 - 2x)] \, dx = 2\int_{-2}^3 (6 + x - x^2) \, dx$$

$$= 2\left(6x + \frac{x^2}{2} - \frac{x^3}{3}\right)\Big|_{-2}^3$$

$$= 2\left[\left(18 + \frac{9}{2} - 9\right) - \left(-12 + 2 + \frac{8}{3}\right)\right] = \frac{125}{3}$$

55.

57.

57. $y = \ln x$, $x = 0$, $y = 0$, $y = 1$. Region appears above.
$y = \ln x \Rightarrow x = e^y$. Thus

$$\text{Area} = \int_0^1 e^y \, dy = e^y \Big|_0^1 = e - 1$$

59. $r = \int \left(100 - \frac{3}{2}\sqrt{2q}\right) dq = \int 100 \, dq - \frac{3}{2}\sqrt{2}\int q^{1/2} \, dq$

$= 100q - \frac{3}{2}\sqrt{2} \cdot \frac{q^{3/2}}{3/2} + C = 100q - \sqrt{2}q^{3/2} + C.$

When $q = 0$, then $r = 0$. Thus $0 = 0 - 0 + C$, so $C = 0$.
Hence $r = 100q - \sqrt{2}q^{3/2}$. Since $r = pq$, then $p = r/q =$
$100 - \sqrt{2}q^{1/2} = 100 - \sqrt{2q}$. Thus $p = 100 - \sqrt{2q}$.

61. $\int_{10}^{20} (275 - q - 0.3q^2) \, dq = \left(275q - \frac{q^2}{2} - \frac{0.3q^3}{3}\right)\Big|_{10}^{20}$

$= (5500 - 200 - 800) - (2750 - 50 - 100) = \$1900$

63. $\int_0^{100} 0.008e^{-0.008t} \, dt = -\int_0^{100} e^{0.008t}[-0.008 \, dt]$

$= -e^{-0.008t}\Big|_0^{100} = -e^{-0.8} + 1 \approx -0.44933 + 1 \approx 0.5507.$

65. $y = 9 - 2x$, $y = x$; from $x = 0$ to $x = 4$. Region appears below. Intersection: $x = 9 - 2x$, $3x = 9$, so $x = 3$.

$$\text{Area} = \int_0^3 [(9 - 2x) - x] \, dx + \int_3^4 [x - (9 - 2x)] \, dx$$

$$= \int_0^3 (9 - 3x) \, dx + \int_3^4 (3x - 9) \, dx$$

$$= \left(9x - \frac{3x^2}{2}\right)\Big|_0^3 + \left(\frac{3x^2}{2} - 9x\right)\Big|_3^4$$

$$= \left[\left(27 - \frac{27}{2}\right) - 0\right] + \left[(24 - 36) - \left(\frac{27}{2} - 27\right)\right]$$

$$= 15 \text{ square units}$$

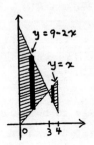

67. D: $p = 0.01q^2 - 1.1q + 30$
    S: $p = 0.01q^2 + 8$ } Equil. pt. $= (q_0, p_0) = (20, 12)$

$$CS = \int_0^{q_0} [f(q) - p_0]\, dq = \int_0^{20} [(0.01q^2 - 1.1q + 30) - 12]\, dq$$

$$= \int_0^{20} (0.01q^2 - 1.1q + 18)\, dq$$

$$= \left(\frac{0.01q^3}{3} - \frac{1.1q^2}{2} + 18q\right)\bigg|_0^{20}$$

$$= \left(\frac{80}{3} - 220 + 360\right) - 0 = 166\frac{2}{3}.$$

$$PS = \int_0^{q_0} [p_0 - g(q)]\, dq = \int_0^{20} [12 - (0.01q^2 + 8)]\, dq$$

$$= \int_0^{20} (4 - 0.01q^2)\, dq = \left(4q - \frac{0.01q^3}{3}\right)\bigg|_0^{20}$$

$$= \left(80 - \frac{80}{3}\right) - 0 = 53\frac{1}{3}.$$

69. $\displaystyle\int_{q_0}^{q_n} \frac{dq}{q - \hat{q}} = -(u + v)\int_0^n dt, \qquad \ln|q - \hat{q}| \Big|_{q_0}^{q_n} = -(u + v)t \Big|_0^n,$

$\ln|q_n - \hat{q}| - \ln|q_0 - \hat{q}| = -(u + v)n,$

$\ln|q_0 - \hat{q}| - \ln|q_n - \hat{q}| = (u + v)n, \qquad \ln\left|\dfrac{q_0 - \hat{q}}{q_n - \hat{q}}\right| = (u + v)n,$

$n = \dfrac{1}{u + v} \ln\left|\dfrac{q_0 - \hat{q}}{q_n - \hat{q}}\right|,$ as was to be shown.

71. <u>Case 1.</u> $r \neq -1.$

$g(x) = \dfrac{1}{k}\displaystyle\int_1^{1/x} ku^r\, du = \int_1^{1/x} u^r\, du = \dfrac{u^{r+1}}{r + 1}\Big|_1^{1/x}$

$= \dfrac{1}{r + 1}(x^{-r-1} - 1).$

$g'(x) = \dfrac{1}{r + 1}[-(r + 1)x^{-r-2}] = -\dfrac{1}{x^{r+2}}.$

<u>Case 2.</u> $r = -1.$

$g(x) = \dfrac{1}{k}\displaystyle\int_1^{1/x} ku^{-1}\, du = \int_1^{1/x} \dfrac{1}{u}\, du$

$= \ln|u| \Big|_1^{1/x} = \ln\left(\dfrac{1}{x}\right) - 0 = -\ln x.$

$g'(x) = -\dfrac{1}{x} = -\dfrac{1}{x^{r+2}}.$

CHAPTER 16 — MATHEMATICAL SNAPSHOT

1. $\displaystyle\int_0^5 f(t)\, dt = \int_0^5 (50 - 2t)\, dt = (50t - t^2)\Big|_0^5$

$= (250 - 25) - 0 = 225.$

$\displaystyle\int_{10}^{15} f(t)\, dt = \int_{10}^{15} (50 - 2t)\, dt = (50t - t^2)\Big|_{10}^{15}$

$= (750 - 225) - (500 - 100) = 125$

3. (a)   Total revenue

$$= \int_0^R (m + st)f(t)\ dt = \int_0^{30} (100 + t)(60 - 2t)\ dt$$

$$= 2\int_0^{30} (100 + t)(30 - t)\ dt = 2\int_0^{30} (3000 - 70t - t^2)\ dt$$

$$= 2\left(3000t - 35t^2 - \frac{t^3}{3}\right)\Big|_0^{30}$$

$$= 2[(90,000 - 31,500 - 9000) - 0] = 2[49,500] = \$99,000$$

(b)   Total number of units sold

$$= \int_0^R f(t)\ dt = \int_0^{30} (60-2t)\ dt = 2\int_0^{30} (30-t)\ dt$$

$$= 2\left(30t - \frac{t^2}{2}\right)\Big|_0^{30} = 2[(900 - 450) - 0] = 900$$

(c) Average delivered price $= \dfrac{\text{total revenue}}{\text{total number of units sold}}$

$$= \frac{99,000}{900} = \$110$$

17

Methods and Applications of Integration

1. $\int f(x)\ dx = uv - \int v\ du = x \cdot \frac{2}{3}(x+5)^{3/2} - \int \frac{2}{3}(x+5)^{3/2}\ dx =$

   $\frac{2}{3}x(x+5)^{3/2} - \frac{2}{3} \cdot \frac{2}{5}(x+5)^{5/2} + C = \frac{2}{3}x(x+5)^{3/2} - \frac{4}{15}(x+5)^{5/2} + C.$

3. $\int xe^{-x}\ dx$. Letting $u = x$, $dv = e^{-x}\ dx$, then $du = dx$,

   $v = -e^{-x}$. $\int xe^{-x}\ dx = -xe^{-x} - \int -e^{-x}\ dx =$

   $-xe^{-x} - \int e^{-x}[-dx] = -xe^{-x} - e^{-x} + C = -e^{-x}(x+1) + C.$

5. $\int y^3 \ln y\ dy$. Letting $u = \ln y$, $dv = y^3\ dy$, then $du =$

   $(1/y)\ dy$, $v = y^4/4$. $\int y^3 \ln y\ dy = \frac{y^4 \ln y}{4} - \int \frac{y^4}{4}\left(\frac{1}{y}\ dy\right) =$

   $\frac{y^4 \ln y}{4} - \int \frac{y^3}{4}\ dy = \frac{y^4 \ln y}{4} - \frac{y^4}{16} + C = \frac{y^4}{4}\left[\ln(y) - \frac{1}{4}\right] + C.$

7. $\int \ln(4x)\, dx$. Letting $u = \ln(4x)$, $dv = dx$, then $du =$ $(1/x)\, dx$, $v = x$. $\int \ln(4x)\, dx = x\ln(4x) - \int x\left(\frac{1}{x}\, dx\right) =$ $x\ln(4x) - \int dx = x\ln(4x) - x + C = x[\ln(4x)-1] + C$.

9. $\int x\sqrt{x+1}\, dx$. Letting $u = x$, $dv = \sqrt{x+1}\, dx$, then $du = dx$, $v = (2/3)(x+1)^{3/2}$. $\int x\sqrt{x+1}\, dx = \frac{2x(x+1)^{3/2}}{3} -$ $\int \frac{2(x+1)^{3/2}}{3}\, dx = \frac{2x(x+1)^{3/2}}{3} - \frac{4}{15}(x+1)^{5/2} + C =$ $\frac{2}{15}(x+1)^{3/2}[5x-2(x+1)] + C = \frac{2}{15}(x+1)^{3/2}(3x-2) + C$.

11. $\int \frac{x}{(2x+1)^2}\, dx$. Letting $u = x$, $dv = (2x+1)^{-2}\, dx$, then $du = dx$, $v = \frac{(2x+1)^{-1}}{-2} = -\frac{1}{2(2x+1)}$. $\int \frac{x}{(2x+1)^2}\, dx = -\frac{x}{2(2x+1)} - \int -\frac{1}{2(2x+1)}\, dx =$ $-\frac{x}{2(2x+1)} + \frac{1}{2}\cdot\frac{1}{2}\int \frac{1}{2x+1}[2\, dx] = -\frac{x}{2(2x+1)} + \frac{1}{4}\ln|2x+1| + C$.

13. $\int \frac{\ln x}{x^2}\, dx$. Letting $u = \ln x$, $dv = x^{-2}\, dx$, then $du = \frac{1}{x}\, dx$, $v = -x^{-1}$. $\int \frac{\ln x}{x^2}\, dx = -\frac{\ln x}{x} - \int -x^{-1}\left(\frac{1}{x}\, dx\right) =$ $-\frac{\ln x}{x} + \int x^{-2}\, dx = -\frac{\ln x}{x} - \frac{1}{x} + C = -\frac{1}{x}(1 + \ln x) + C$.

15. $\int_1^2 xe^{2x}\, dx$. Letting $u = x$, $dv = e^{2x}\, dx$, then $du = dx$, $v = \frac{1}{2}e^{2x}$. $\int_1^2 xe^{2x}\, dx = \left[\frac{xe^{2x}}{2} - \int \frac{1}{2}e^{2x}\, dx\right]\Big|_1^2 =$ $\left[\frac{xe^{2x}}{2} - \frac{e^{2x}}{4}\right]\Big|_1^2 = \frac{e^{2x}}{4}(2x-1)\Big|_1^2 = \frac{e^4}{4}(3) - \frac{e^2}{4}(1) = \frac{e^2}{4}(3e^2-1)$.

17. $\int_0^1 xe^{-x^2}\ dx = -\frac{1}{2}\int_0^1 e^{-x^2}(-2x\ dx)$ (Form: $\int e^u\ du$)

$$= -\frac{1}{2}e^{-x^2}\Big|_0^1 = -\frac{1}{2}(e^{-1} - 1) = \frac{1}{2}(1 - e^{-1})$$

19. $\int_1^2 (x/\sqrt{4-x})\ dx$. Letting $u = x$, $dv = (4-x)^{-1/2}\ dx$, then

$du = dx$, $v = -2(4-x)^{1/2}$.

$\int_1^2 (x/\sqrt{4-x})\ dx = \left[-2x(4-x)^{1/2} - \int -2(4-x)^{1/2}\ dx\right]\Big|_1^2 =$

$\left[-2x(4-x)^{1/2} - \frac{4(4-x)^{3/2}}{3}\right]\Big|_1^2 = \left\{-\frac{2\sqrt{4-x}}{3}[3x+2(4-x)]\right\}\Big|_1^2 =$

$\left\{-\frac{2\sqrt{4-x}}{3}(x+8)\right\}\Big|_1^2 = -\frac{2}{3}(10\sqrt{2} - 9\sqrt{3}) = \frac{2}{3}(9\sqrt{3} - 10\sqrt{2})$.

21. $\int (2x-1)\ \ln(x-1)\ dx$. Letting $u = \ln(x-1)$, $dv = (2x-1)\ dx$,

then $du = \frac{1}{x-1}\ dx$, $v = x^2-x = x(x-1)$. $\int (2x-1)\ \ln(x-1)\ dx =$

$x(x-1)\ \ln(x-1) - \int x(x-1)\frac{1}{x-1}\ dx = x(x-1)\ \ln(x-1) - \int x\ dx =$

$x(x-1)\ \ln(x-1) - \frac{x^2}{2} + C$.

23. $\int x^2 e^x\ dx$. Letting $u = x^2$, $dv = e^x\ dx$, then $du = 2x\ dx$

and $v = e^x$.

     $\int x^2 e^x\ dx = x^2 e^x - \int e^x(2x\ dx) = x^2 e^x - 2\int xe^x\ dx$.

For $\int xe^x\ dx$, let $u = x$, $dv = e^x\ dx$. Then $du = dx$, $v = e^x$

and

     $\int xe^x\ dx = xe^x - \int e^x\ dx = xe^x - e^x + C_1 = e^x(x-1) + C_1$.

Thus $\int x^2 e^x\ dx = x^2 e^x - 2[e^x(x-1)] + C = e^x(x^2-2x+2) + C$.

25. $\int (x-e^{-x})^2\ dx = \int (x^2-2xe^{-x}+e^{-2x})\ dx =$

$\frac{x^3}{3} - \frac{e^{-2x}}{2} - 2\int xe^{-x}\ dx$. Using Problem 3 for $\int xe^{-x}\ dx$,

$$\int (x - e^{-x})^2 \, dx = \frac{x^3}{3} - \frac{e^{-2x}}{2} + 2e^{-x}(x+1) + C.$$

27. $\int x^3 e^{x^2} \, dx.$ Letting $u = x^2$, $dv = xe^{x^2} \, dx$, then $du =$

$2x \, dx$, $v = (1/2)e^{x^2}$. $\int x^3 e^{x^2} \, dx = \frac{x^2 e^{x^2}}{2} - \int \frac{e^{x^2}}{2}(2x \, dx) =$

$\frac{x^2 e^{x^2}}{2} - \frac{e^{x^2}}{2} + C = \frac{e^{x^2}}{2}(x^2 - 1) + C.$

29. $\int (2^x + x)^2 \, dx = \int (2^{2x} + 2x2^x + x^2) \, dx =$

$\int 2^{2x} \, dx + \int x2^{x+1} \, dx + \int x^2 \, dx.$ For $\int x2^{x+1} \, dx$, let $u = x$,

$dv = 2^{x+1} \, dx.$ Then $du = dx$, $v = \frac{1}{\ln 2} \cdot 2^{x+1}$ and

$$\int x2^{x+1} \, dx = \frac{x}{\ln 2} \cdot 2^{x+1} - \frac{1}{\ln 2} \int 2^{x+1} \, dx$$

$$= \frac{x}{\ln 2} \cdot 2^{x+1} - \frac{1}{\ln^2 2} \cdot 2^{x+1} + C_1.$$

Thus $\int (2^x + x)^2 \, dx = \int 2^{2x} \, dx + \int x2^{x+1} \, dx + \int x^2 \, dx =$

$\frac{1}{2}\int 2^{2x}[2 \, dx] + \int x2^{x+1} \, dx + \int x^2 \, dx =$

$\frac{1}{2 \ln 2} \cdot 2^{2x} + \frac{x}{\ln 2} \cdot 2^{x+1} - \frac{1}{\ln^2 2} \cdot 2^{x+1} + \frac{x^3}{3} + C =$

$\frac{1}{\ln 2} \cdot 2^{2x-1} + \frac{x}{\ln 2} \cdot 2^{x+1} - \frac{1}{\ln^2 2} \cdot 2^{x+1} + \frac{x^3}{3} + C.$

31. area $= \int_1^{e^3} (\ln x) \, dx.$ Letting $u = \ln x$, $dv = dx$, then

$du = (1/x) \, dx$, $v = x.$

area $= \int_1^{e^3} (\ln x) \, dx = \left[(x \ln x) - \int x \cdot \frac{1}{x} \, dx \right] \Big|_1^{e^3} =$

$\left[(x \ln x) - \int dx \right] \Big|_1^{e^3} = [x \ln(x) - x] \Big|_1^{e^3} =$

$[e^3 \cdot 3 - e^3] - [1 \cdot 0 - 1] = 2e^3 + 1$ <u>Ans.</u> $2e^3 + 1$ sq units

Exercise 17.1 -426-

33. area = $\int_0^4 x\sqrt{2x+1}\, dx$. Letting $u = x$, $dv = (2x+1)^{1/2}\, dx$,

then $du = dx$, $v = \frac{1}{3}(2x+1)^{3/2}$.

area = $\int_0^4 x\sqrt{2x+1}\, dx = \left[\frac{x}{3}(2x+1)^{3/2} - \frac{1}{3}\int(2x+1)^{3/2}\, dx\right]\Big|_0^4 =$

$\left[\frac{x}{3}(2x+1)^{3/2} - \frac{1}{3}\cdot\frac{1}{2}\int(2x+1)^{3/2}(2\, dx)\right]\Big|_0^4 =$

$\left[\frac{x}{3}(2x+1)^{3/2} - \frac{1}{15}(2x+1)^{5/2}\right]\Big|_0^4 =$

$\left[\frac{4}{3}(27) - \frac{1}{15}(243)\right] - \left[0 - \frac{1}{15}\right] = \frac{298}{15}$. **Ans.** $\frac{298}{15}$ sq units

35. (a) Consider $\int p\, dq$. Letting $u = p$, $dv = dq$, then

$du = \frac{dp}{dq}dq$, $v = q$. Thus $\int p\, dq = pq - \int q\frac{dp}{dq}\, dq =$

$r - \int q\frac{dp}{dq}\, dq$ (since $r = pq$).

(b) From (a), $r = \int p\, dq + \int q\frac{dp}{dq}\, dq$. Combining the

integrals gives $r = \int\left(p + q\frac{dp}{dq}\right)dq$.

(c) From (b), $\frac{dr}{dq} = p + q\frac{dp}{dq}$. Thus $\int_0^{q_0}\left(p + q\frac{dp}{dq}\right)dq =$

$\int_0^{q_0}\frac{dr}{dq}\, dq = r(q_0) - r(0) = r(q_0)$ [since $r(0) = 0$].

EXERCISE 17.2

1. $\frac{10x}{x^2+7x+6} = \frac{10x}{(x+6)(x+1)} = \frac{A}{x+6} + \frac{B}{x+1}$. $10x = A(x+1) + B(x+6)$.

If $x = -1$, then $-10 = 5B$, or $B = -2$. If $x = -6$, then

$-60 = -5A$, or $A = 12$. **Ans.** $\frac{12}{x+6} - \frac{2}{x+1}$

3. $\dfrac{x^2}{x^2+6x+8} = 1 + \dfrac{-6x-8}{x^2+6x+8}$ (by long division).

$\dfrac{-6x-8}{x^2+6x+8} = \dfrac{-6x-8}{(x+2)(x+4)} = \dfrac{A}{x+2} + \dfrac{B}{x+4}$. $-6x-8 = A(x+4) + B(x+2)$.
If $x = -4$, then $16 = -2B$, or $B = -8$. If $x = -2$, then
$4 = 2A$, or $A = 2$. **Ans.** $1 + \dfrac{2}{x+2} - \dfrac{8}{x+4}$

5. $\dfrac{4x-5}{x^2+2x+1} = \dfrac{4x-5}{(x+1)^2} = \dfrac{A}{x+1} + \dfrac{B}{(x+1)^2}$. $4x-5 = A(x+1) + B$.
If $x = -1$, then $-9 = B$. If $x = 0$, then $-5 = A + B$,
$-5 = A - 9$, or $A = 4$. **Ans.** $\dfrac{4}{x+1} - \dfrac{9}{(x+1)^2}$

7. $\dfrac{x^2+3}{x^3+x} = \dfrac{x^2+3}{x(x^2+1)} = \dfrac{A}{x} + \dfrac{Bx+C}{x^2+1}$. $x^2+3 = A(x^2+1) + (Bx+C)x$.

$x^2+3 = (A+B)x^2 + Cx + A$. Thus $A+B = 1$, $C = 0$, $A = 3$.
This gives $A = 3$, $B = -2$, $C = 0$.
Ans. $\dfrac{3}{x} - \dfrac{2x}{x^2+1}$

9. $\dfrac{5x-2}{x^2-x} = \dfrac{5x-2}{x(x-1)} = \dfrac{A}{x} + \dfrac{B}{x-1}$. $5x-2 = A(x-1) + Bx$.
If $x = 1$, then $3 = B$. If $x = 0$, then $-2 = -A$, or $A = 2$.
$\displaystyle\int \dfrac{5x-2}{x^2-x}\,dx = \int \left(\dfrac{2}{x} + \dfrac{3}{x-1}\right)\,dx$.
Ans. $2\ln|x| + 3\ln|x-1| + C = \ln|x^2(x-1)^3| + C$

11. $\dfrac{x+10}{x^2-x-2} = \dfrac{x+10}{(x+1)(x-2)} = \dfrac{A}{x+1} + \dfrac{B}{x-2}$. $x+10 = A(x-2) + B(x+1)$.
If $x = 2$, then $12 = 3B$, or $B = 4$. If $x = -1$, then
$9 = -3A$, or $A = -3$. $\displaystyle\int \dfrac{x+10}{x^2-x-2}\,dx = \int \left(\dfrac{-3}{x+1} + \dfrac{4}{x-2}\right)\,dx$.
Ans. $-3\ln|x+1| + 4\ln|x-2| + C = \ln|(x-2)^4/(x+1)^3| + C$

13. $\dfrac{3x^3-3x+4}{4x^2-4} = \dfrac{1}{4}\cdot\dfrac{3x^3-3x+4}{x^2-1} = \dfrac{1}{4}\left(3x + \dfrac{4}{x^2-1}\right)$.

$\dfrac{4}{x^2-1} = \dfrac{4}{(x-1)(x+1)} = \dfrac{A}{x-1} + \dfrac{B}{x+1}$. $4 = A(x+1) + B(x-1)$.

If x = -1, then 4 = -2B, or B = -2. If x = 1, then
4 = 2A, or A = 2.

$$\int \frac{3x^3-3x+4}{4x^2-4} \, dx = \frac{1}{4}\int \left(3x + \frac{2}{x-1} + \frac{-2}{x+1}\right) dx.$$

Ans. $(1/4)[(3x^2/2) + 2 \ln|x-1| - 2 \ln|x+1|] + C =$

$$(1/4)[(3x^2/2) + \ln\{(x-1)/(x+1)\}^2] + C$$

15. $\frac{17x-12}{x^3-x^2-12x} = \frac{7x-12}{x(x-4)(x+3)} = \frac{A}{x} + \frac{B}{x-4} + \frac{C}{x+3}.$

17x-12 = A(x-4)(x+3) +Bx(x+3) + Cx(x-4). If x = 0, then
-12 = -12A, or A = 1. If x = 4, then 56 = 28B, or B = 2.
If x = -3, then -63 = 21C, or C = -3.

$$\int \frac{17x-12}{x^3-x^2-12x} \, dx = \int \left(\frac{1}{x} + \frac{2}{x-4} + \frac{-3}{x+3}\right) dx.$$

Ans. $\ln|x| + 2 \ln|x-4| - 3 \ln|x+3| + C =$

$$\ln|x(x-4)^2/(x+3)^3| + C$$

17. $\int \frac{3x^5+4x^3-x}{x^6+2x^4-x^2-2} \, dx = \frac{1}{2}\int \frac{1}{x^6+2x^4-x^2-2}[2(3x^5+4x^3-x) \, dx].$

(Form: $\int(1/u) \, du$.) (Partial fractions not required.)

Ans. $(1/2) \ln|x^6+2x^4-x^2-2| + C$

19. $\frac{2x^2-5x-2}{(x-2)^2(x-1)} = \frac{A}{x-1} + \frac{B}{x-2} + \frac{C}{(x-2)^2}.$

$2x^2-5x-2 = A(x-2)^2 + B(x-1)(x-2) + C(x-1)$. If x = 1, then
-5 = A. If x = 2, the -4 = C. If x = 0, then
-2 = 4A + 2B - C, -2 = -20 + 2B + 4, or B = 7.

$$\int \frac{2x^2-5x-2}{(x-2)^2(x-1)} \, dx = \int \left[\frac{-5}{x-1} + \frac{7}{x-2} + \frac{-4}{(x-2)^2}\right] dx.$$

Ans. $[4/(x-2)] - 5 \ln|x-1| + 7 \ln|x-2| + C =$

$$[4/(x-2)] + \ln|(x-2)^7/(x-1)^5| + C$$

21. $\dfrac{x^2+8}{x^3+4x} = \dfrac{x^2+8}{x(x^2+4)} = \dfrac{A}{x} + \dfrac{Bx+C}{x^2+4}$. $x^2+8 = A(x^2+4) + (Bx+C)x$.

$x^2+8 = (A+B)x^2 + Cx + 4A$.   Thus $A+B = 1$, $C = 0$, $4A = 8$.
This gives $A = 2$, $B = -1$, $C = 0$.

$$\int\dfrac{x^2+8}{x^3+4x}\,dx = \int\left(\dfrac{2}{x} + \dfrac{-x}{x^2+4}\right)dx = 2\int\dfrac{1}{x}\,dx - \dfrac{1}{2}\int\dfrac{1}{x^2+4}[2\,dx].$$

Ans. $2\ln|x| - \dfrac{1}{2}\ln(x^2+4) + C = \dfrac{1}{2}\ln[x^4/(x^2+4)] + C$

23. $\dfrac{-x^3+8x^2-9x+2}{(x^2+1)(x-3)^2} = \dfrac{Ax+B}{x^2+1} + \dfrac{C}{x-3} + \dfrac{D}{(x-3)^2}$.

$-x^3+8x^2-9x+2 = (Ax+B)(x-3)^2 + C(x-3)(x^2+1) + D(x^2+1) =$
$(Ax+B)(x^2-6x+9) + C(x^3-3x^2+x-3) + D(x^2+1) =$
$(A+C)x^3 + (B-6A-3C+D)x^2 + (9A-6B+C)x + (9B-3C+D)$. Thus
$A+C = -1$, $B-6A-3C+D = 8$, $9A-6B+C = -9$, $9B-3C+D = 2$. This
gives $A = -1$, $B = 0$, $C = 0$, $D = 2$.

$$\int\dfrac{-x^3+8x^2-9x+2}{(x^2+1)(x-3)^2}\,dx = \int\left(\dfrac{-x}{x^2+1} + \dfrac{0}{x-3} + \dfrac{2}{(x-3)^2}\right)dx.$$

Ans. $(-1/2)\ln(x^2+1) - [2/(x-3)] + C$

25. $\dfrac{14x^3+24x}{(x^2+1)(x^2+2)} = \dfrac{Ax+B}{x^2+1} + \dfrac{Cx+D}{x^2+2}$.

$14x^3+24x = (x^2+2)(Ax+B) + (x^2+1)(Cx+D)$
$\qquad\qquad = (A+C)x^3 + (B+D)x^2 + (2A+C)x + (2B+D)$.
Thus $A+C = 14$, $B+D = 0$, $2A+C = 24$, $2B+D = 0$.
This gives $A = 10$, $B = 0$, $C = 4$, $D = 0$.

$$\int\dfrac{14x^3+24x}{(x^2+1)(x^2+2)}\,dx = \int\left(\dfrac{10x}{x^2+1} + \dfrac{4x}{x^2+2}\right)dx$$

$$= 5\int\dfrac{1}{x^2+1}[2\,dx] + 2\int\dfrac{1}{x^2+2}[2\,dx].$$

Ans. $5\ln(x^2+1) + 2\ln(x^2+2) + C = \ln[(x^2+1)^5(x^2+2)^2] + C$

27. $\dfrac{3x^3+x}{(x^2+1)^2} = \dfrac{Ax+B}{x^2+1} + \dfrac{Cx+D}{(x^2+1)^2}.$ $3x^3+x = (Ax+B)(x^2+1) + (Cx+D) =$

$Ax^3 + Bx^2 + (A+C)x + (B+D).$ Thus A = 3, B = 0, A+C = 1,
B+D = 0. This gives A = 3, B = 0, C = -2, D = 0.

$$\int \dfrac{3x^3+x}{(x^2+1)^2}\, dx = \int \left[\dfrac{3x}{x^2+1} + \dfrac{-2x}{(x^2+1)^2} \right]\, dx.$$

Ans. $\dfrac{3}{2}\ln(x^2+1) + \dfrac{1}{x^2+1} + C$

29. $\dfrac{2-2x}{x^2+7x+12} = \dfrac{2-2x}{(x+3)(x+4)} = \dfrac{A}{x+3} + \dfrac{B}{x+4}.$ $2-2x = A(x+4)+B(x+3).$
If x = -4, then 10 = -B, or B = -10. If x = -3, then
8 = A.

$$\int_0^1 \dfrac{2-2x}{x^2+7x+12}\, dx = \int_0^1 \left(\dfrac{8}{x+3} + \dfrac{-10}{x+4} \right)\, dx =$$

$$\left[8\ln|x+3| - 10\ln|x+4| \right]\Big|_0^1.$$

Ans. $18\ln(4) - 10\ln(5) - 8\ln(3)$

31. Note that $(x^2+1)/(x+2)^2 \geqq 0$ on [0,1]. Area = $\displaystyle\int_0^1 \dfrac{x^2+1}{(x+2)^2}\, dx.$

$\dfrac{x^2+1}{(x+2)^2} = 1 + \dfrac{-4x-3}{(x+2)^2}$ (by long division).

$\dfrac{-4x-3}{(x+2)^2} = \dfrac{A}{x+2} + \dfrac{B}{(x+2)^2}.$ $-4x-3 = A(x+2) + B.$ If x = -2,
then 5 = B. If x = 0, then -3 = 2A + B, -3 = 2A + 5, or

A = -4. Area = $\displaystyle\int_0^1 \dfrac{x^2+1}{(x+2)^2}\, dx = \int_0^1 \left[1 + \dfrac{-4}{x+2} + \dfrac{5}{(x+2)^2} \right]\, dx =$

$\left[x - 4\ln|x+2| - \dfrac{5}{x+2} \right]\Big|_0^1.$ Ans. $\dfrac{11}{6} + 4\ln\dfrac{2}{3}$ sq units

EXERCISE 17.3

1. Let $u = x$, $a^2 = 9$. Then $du = dx$.   <u>Ans.</u> $\dfrac{x}{9\sqrt{9-x^2}} + C$

3. Let $u = 4x$, $a^2 = 3$. Then $du = 4\,dx$.

$$\int \frac{dx}{x^2\sqrt{16x^2+3}} = 4 \int \frac{(4\,dx)}{(4x)^2\sqrt{(4x)^2+3}} = 4\left[-\frac{\sqrt{(4x)^2+3}}{3(4x)}\right] + C.$$

<u>Ans.</u> $-\dfrac{\sqrt{16x^2+3}}{3x} + C$

5. Form. 5 with $u = x$, $a = 6$, $b = 7$. Then $du = dx$.
<u>Ans.</u> $\frac{1}{6}\ln\left|\frac{x}{6+7x}\right| + C$

7. Form. 28 with $u = x$, $a = 3$. Then $du = dx$.

<u>Ans.</u> $\frac{1}{3}\ln\left|\frac{\sqrt{x^2+9}-3}{x}\right| + C$

9. Form. 12 with $u = x$, $a = 2$, $b = 3$, $c = 4$, $k = 5$. Then $du = dx$.   <u>Ans.</u> $\frac{1}{2}\left[\frac{4}{5}\ln|4+5x| - \frac{2}{3}\ln|2+3x|\right] + C$

11. Form. 45 with $u = x$, $a = 4$, $b = 3$, $c = 2$. Then $du = dx$.
<u>Ans.</u> $\frac{1}{8}(2x - \ln[4 + 3e^{2x}]) + C$

13. Form. 9 with $u = x$, $a = 1$, $b = 1$. Then $du = dx$.

$$\int \frac{2\,dx}{x(1+x)^2} = 2\int \frac{dx}{x(1+x)^2} = 2\left[\frac{1}{1+x} + \ln\left|\frac{x}{1+x}\right|\right] + C$$

15. Form. 3 with $u = x$, $a = 2$, $b = 1$. Then $du = dx$.
$(x - 2\ln|2+x|)\Big|_0^1 = 1 - 2\ln 3 + 2\ln 2 =$
$1 - \ln 9 + \ln 4$.   <u>Ans.</u> $1 + \ln(4/9)$

17. Form. 23 with $u = x$, $a^2 = 3$. Then $du = dx$.
<u>Ans.</u> $\frac{1}{2}\left(x\sqrt{x^2-3} - 3\ln\left|x+\sqrt{x^2-3}\right|\right) + C$

19. Form. 38 with u = x, a = 12. Then du = dx.

$$\frac{e^{12x}}{144}(12x-1)\Big|_0^{1/12} = \frac{1}{144}[e(0) - 1(-1)].$$ <u>Ans.</u> $\frac{1}{144}$

21. Form. 39 with u = x, n = 2, a = 1. Then du = dx.

$\int x^2 e^x \, dx = x^2 e^x - 2\int xe^x \, dx.$ Applying Form. 38 on $\int xe^x \, dx$
with u = x, a = 1 (so du = dx) gives $\int xe^x \, dx =$
$e^x(x-1) + C_1.$ Thus $\int x^2 e^x \, dx = x^2 e^x - 2[e^x(x-1)] + C =$
$e^x[x^2-2(x-1)] + C = e^x(x^2-2x+2) + C.$
<u>Ans.</u> $e^x(x^2-2x+2) + C$

23. Form. 26 with u = 2x, a^2 = 1. Then du = 2 dx.

$$\int \frac{\sqrt{4x^2+1}}{x^2} \, dx = 2\int \frac{\sqrt{(2x)^2+1}}{(2x)^2}[2 \, dx].$$

<u>Ans.</u> $2\left(- \frac{\sqrt{4x^2+1}}{2x} + \ln|2x + \sqrt{4x^2+1}|\right) + C$

25. Form. 7 with u = x, a = 1, b = 3. Then du = dx.
<u>Ans.</u> $\frac{1}{9}\left(\ln|1+3x| + \frac{1}{1+3x}\right) + C$

27. Form. 34 with u = $\sqrt{5}$x, a = $\sqrt{7}$. Then du = $\sqrt{5}$ dx.

$$\int \frac{dx}{7-5x^2} = \frac{1}{\sqrt{5}}\int \frac{1}{(\sqrt{7})^2 - (\sqrt{5}x)^2}[\sqrt{5} \, dx].$$

<u>Ans.</u> $\frac{1}{\sqrt{5}}\left(\frac{1}{2\sqrt{7}} \ln\left|\frac{\sqrt{7} + \sqrt{5}x}{\sqrt{7} - \sqrt{5}x}\right|\right) + C$

29. Form. 42 with u = 3x, n = 5. Then du = 3 dx.

$$\int x^5 \ln(3x) \, dx = \frac{1}{3^6} \int (3x)^5 \ln(3x) [3 \, dx].$$

<u>Ans.</u> $\frac{1}{3^6}\left[\frac{(3x)^6}{6} \ln(3x) - \frac{(3x)^6}{36}\right] + C = \frac{x^6}{36}[6 \ln(3x) - 1] + C$

31. Form. 13 with u = x, a = 1, b = 3. Then du = dx.

$\int 2x\sqrt{1+3x}\ dx = 2\int x\sqrt{1+3x}\ dx.$

<u>Ans.</u> $2\left[\dfrac{2(9x-2)(1+3x)^{3/2}}{15\cdot 9}\right] + C = \dfrac{4(9x-2)(1+3x)^{3/2}}{135} + C$

33. Form. 27 with u = 2x, a^2 = 13. Then du = 2 dx.

$\int \dfrac{dx}{\sqrt{4x^2-13}} = \dfrac{1}{2}\int \dfrac{1}{\sqrt{(2x)^2-13}}[2\ dx].$

<u>Ans.</u> $\dfrac{1}{2}\ \ln\left|2x + \sqrt{4x^2-13}\right| + C$

35. Form. 21 with u = 2x, a^2 = 9. Then du = 2 dx.

$\int \dfrac{dx}{x^2\sqrt{9-4x^2}} = 2\int \dfrac{[2\ dx]}{(2x)^2\sqrt{9-(2x)^2}} = 2\left[-\dfrac{\sqrt{9-4x^2}}{9(2x)}\right] + C.$

<u>Ans.</u> $-\dfrac{\sqrt{9-4x^2}}{9x} + C$

37. Form. 45 with u = $\sqrt{x}$, a = π, b = 7, c = 4. Then
du = $\dfrac{1}{2\sqrt{x}}$ dx.

$\int \dfrac{dx}{\sqrt{x}(\pi+7e^{4\sqrt{x}})} = 2\int \dfrac{1}{\pi+7e^{4\sqrt{x}}}\left[\dfrac{1}{2\sqrt{x}}\ dx\right] =$

$2\left[\dfrac{1}{4\pi}\left(4\sqrt{x} - \ln|\pi+7e^{4\sqrt{x}}|\right)\right] + C.$

<u>Ans.</u> $\dfrac{1}{2\pi}\left(4\sqrt{x} - \ln|\pi+7e^{4\sqrt{x}}|\right) + C$

39. Can be put in the form $\int(1/u)\ du.$

$\int \dfrac{x\ dx}{x^2+1} = \dfrac{1}{2}\int \dfrac{1}{x^2+1}[2x\ dx] = \dfrac{1}{2}\ \ln(x^2+1) + C$

41. Can be put in the form $\int u^n\ du.$

$\int x\sqrt{2x^2+1}\ dx = \dfrac{1}{4}\int(2x^2+1)^{1/2}[4x\ dx] = \dfrac{1}{4}\cdot\dfrac{(2x^2+1)^{3/2}}{\frac{3}{2}} + C$

<u>Ans.</u> $\dfrac{1}{6}(2x^2+1)^{3/2} + C$

43. Form. 11 or partial fractions. For partial fractions,

$$\int \frac{x^2}{x^2-5x+6} \, dx = \int \left[\frac{1}{x-3} + \frac{-1}{x-2} \right] \, dx.$$ **Ans.** $\ln \left| \frac{x-3}{x-2} \right| + C$

45. Integration by parts or formula 42. For formula 42, let u = x, n = 3. Then du = dx. **Ans.** $\frac{x^4}{4} \left[\ln(x) - \frac{1}{4} \right] + C$

47. Integration by parts or formula 38. For formula 38, let u = x, a = 2. Then du = dx. **Ans.** $\frac{e^{2x}}{4}(2x-1) + C$

49. Integration by parts (applied twice) or formula 43 and then formula 41. For formula 43, let u = x, n = 0, m = 2. Then du = dx.

$\int \ln^2 x \, dx = x \ln^2 x - 2 \int \ln x \, dx.$ Now we apply formula 41 to the last integral with u = x (so du = dx).
Ans. $x(\ln x)^2 - 2x(\ln x) + 2x + C$

51. Integration by parts or formula 15. For formula 15, let u = x, a = 4, b = -1. Then du = dx.

$$\int_1^2 \frac{x \, dx}{\sqrt{4-x}} = \left. \frac{2(-x-8)\sqrt{4-x}}{3} \right|_1^2.$$ **Ans.** $\frac{2}{3}(9\sqrt{3} - 10\sqrt{2})$

53. Can be put in the form $\int u^n \, du.$

$$\int_0^1 \frac{2x \, dx}{\sqrt{8-x^2}} = -\int_0^1 (8-x^2)^{-1/2}[-2x \, dx] = -\left. \frac{(8-x^2)^{1/2}}{1/2} \right|_0^1 =$$

$-2(8-x^2)^{1/2} \Big|_0^1 = -2(\sqrt{7} - \sqrt{8}) = -2(\sqrt{7} - 2\sqrt{2}) =$

$2(2\sqrt{2} - \sqrt{7}).$ **Ans.** $2(2\sqrt{2} - \sqrt{7})$

55. Integration by parts or formula 42. For formula 42, let u = 2x, n = 1. Then du = 2 dx.

$$\int_1^2 x \ln(2x) \, dx = \frac{1}{4} \int_1^2 (2x) \ln(2x) \, [2 \, dx] =$$

$$\frac{1}{4}\left[\frac{(2x)^2}{2}\ln(2x) - \frac{(2x)^2}{4}\right]\Big|_1^2 = 2\ln(4) - \frac{1}{2}\ln(2) - \frac{3}{4} =$$

$$2\ln(2^2) - \frac{1}{2}\ln(2) - \frac{3}{4} = 4\ln(2) - \frac{1}{2}\ln(2) - \frac{3}{4}.$$

<u>Ans.</u> $\frac{7}{2}(\ln 2) - \frac{3}{4}$

57. Partial fractions or formula 5. For formula 5, let u = q, a = 1, b = -1. Then du = dq.

$$\int_{q_0}^{q_n} \frac{dq}{q(1-q)} = \ln\left|\frac{q}{1-q}\right|\ \Big|_{q_0}^{q_n} = \ln\left|\frac{q_n}{1-q_n}\right| - \ln\left|\frac{q_0}{1-q_0}\right|.$$

<u>Ans.</u> $\ln\left|\frac{q_n(1-q_0)}{q_0(1-q_n)}\right|$

59. (a) For $\int_0^{10} 5000e^{-0.06t}\ dt$, the form $\int e^u\ du$ can be applied.

$$\int_0^{10} 5000e^{-0.06t}\ dt = -\frac{5000}{0.06}\int_0^{10} e^{-0.06t}(-0.06\ dt) =$$

$$-\frac{5000}{0.06}e^{-0.06t}\Big|_0^{10} = -\frac{5000}{0.06}(e^{-0.6} - 1) \approx$$

$$-\frac{5000}{0.06}(0.54881 - 1). \quad \underline{Ans.} \quad \$37,599$$

(b) $\int_0^8 200te^{-0.05t}\ dt = 200\int_0^8 te^{-0.05t}\ dt =$

$$= 200\left[\frac{e^{-0.05t}}{0.0025}(-0.05t - 1)\right]\Big|_0^8 \quad \text{(Form. 38: u = t,}$$
$$\qquad\qquad\qquad\qquad\qquad\qquad\qquad\quad a = -0.05;\ du = dt)$$

$$= \frac{200}{0.0025}[e^{-0.4}(-1.4) + 1]$$

$$\approx \frac{200}{0.0025}[(0.67032)(-1.4) + 1]. \quad \underline{Ans.} \quad \$4924$$

<u>Ans.</u> (a) \$37,599; (b) \$4924

61. (a) $\int_0^{10} 400e^{0.06(10-t)}\ dt = 400\int_0^{10} e^{0.6-0.06t}\ dt$

$$= 400\int_0^{10} e^{0.6}e^{-0.06t}\ dt = 400e^{0.6}\int_0^{10} e^{-0.06t}\ dt$$

$$= 400e^{0.6}\left(\frac{1}{-0.06}\right)\int_0^{10} e^{-0.06t}[-0.06\ dt] \qquad \left[\text{Form } \int e^u\ du\right]$$

$$= \frac{400e^{0.6}}{-0.06}e^{-0.06t}\Big|_0^{10} = \frac{400e^{0.6}}{-0.06}[e^{-0.6} - 1]$$

$$\approx \frac{400(1.8221)}{-0.06}[0.54881 - 1] \approx 5481. \quad \underline{Ans.} \quad \$5481$$

(b) $\int_0^5 40te^{0.04(5-t)} \, dt = 40\int_0^5 te^{0.2}e^{-0.04t} \, dt$

$$= 40e^{0.2}\int_0^5 te^{-0.04t} \, dt$$

$$= 40e^{0.2}\left[\frac{e^{-0.04t}}{0.0016}(-0.04t-1)\right]\Big|_0^5 \quad \begin{array}{l}\text{(Form. 38: } u = t, \\ a = -0.04; \; du = dt)\end{array}$$

$$= \frac{40e^{0.2}}{0.0016}[e^{-0.2}(-0.2 - 1) - 1(-1)]$$

$$\approx \frac{40(1.2214)}{0.0016}[0.017524] \approx 535. \quad \underline{Ans.} \quad \$535$$

<u>Ans.</u> (a) \$5481; (b) \$535

<u>EXERCISE 17.4</u>

1. $\bar{f} = \frac{1}{4-0}\int_0^4 x^2 \, dx = \frac{1}{4}\cdot\frac{x^3}{3}\Big|_0^4 = \frac{16}{3} - 0 = \frac{16}{3}$

3. $\bar{f} = \frac{1}{2-(-1)}\int_{-1}^2 (2-3x^2) \, dx = \frac{1}{3}(2x-x^3)\Big|_{-1}^2 = -1$

5. $\bar{f} = \frac{1}{2-(-2)}\int_{-2}^2 4t^3 \, dt = \frac{1}{4}t^4\Big|_{-2}^2 = 4 - 4 = 0$

7. $\bar{f} = \frac{1}{9-1}\int_1^9 \sqrt{x} \, dx = \frac{1}{8}\left(\frac{2x^{3/2}}{3}\right)\Big|_1^9 = \frac{13}{6}$

9. $\bar{P} = \frac{1}{100-0}\int_0^{100} (396q-2.1q^2-400) \, dq =$

$\frac{1}{100}(198q^2 - 0.7q^3 - 400q)\Big|_0^{100} =$

$\frac{1}{100}(1,980,000 - 700,000 - 40,000) - 0 = 12,400.$
<u>Ans.</u> \$12,400

11. $\frac{1}{2-0}\int_0^2 3000e^{0.1t} \, dt = \frac{3000}{2} \cdot \frac{1}{0.1}\int_0^2 e^{0.1t}[0.1 \, dt] =$

$15,000e^{0.1t}\Big|_0^2 = 15,000(e^{0.2} - 1) \approx 15,000(1.2214 - 1) =$

$3321. <u>Ans.</u> $3321

13. Avg. value $= \frac{1}{q_0-0}\int \frac{dr}{dq} \, dq = \frac{1}{q_0}[r(q_0) - r(0)]$. But $r(0) = 0$,

so avg. value $= \frac{r(q_0)}{q_0}$. Since $r(q_0) = \begin{bmatrix} \text{price per unit} \\ \text{when } q_0 \text{ units} \\ \text{are sold} \end{bmatrix} \cdot q_0$,

we have

avg. value $= \dfrac{\begin{bmatrix} \text{price per unit} \\ \text{when } q_0 \text{ units} \\ \text{are sold} \end{bmatrix} \cdot q_0}{q_0} = \begin{matrix} \text{price per unit when} \\ q_0 \text{ units are sold.} \end{matrix}$

EXERCISE 17.5

1. $f(x) = 170/(1+x^2)$, $n = 6$, $a = -2$, $b = 4$. Trapezoidal.

$h = \frac{b-a}{n} = \frac{4-(-2)}{6} = \frac{6}{6} = 1$.

$$
\begin{aligned}
f(-2) &= 34 &&= 34 \\
2f(-1) &= 2(85) &&= 170 \\
2f(0) &= 2(170) &&= 340 \\
2f(1) &= 2(85) &&= 170 \\
2f(2) &= 2(34) &&= 68 \\
2f(3) &= 2(17) &&= 34 \\
f(4) &= 10 &&= \underline{10} \\
&&& 826
\end{aligned}
$$

$\displaystyle\int_{-2}^4 \frac{170}{1+x^2} \, dx \approx \frac{1}{2}(826) = 413.$ <u>Ans.</u> 413

3. $f(x) = x^2$, n = 5, a = 0, b = 1.   Trapezoidal.

    $h = \frac{b-a}{n} = \frac{1-0}{5} = \frac{1}{5} = 0.2$.

$$
\begin{aligned}
f(0) &= 0.0000 \\
2f(0.2) &= 0.0800 \\
2f(0.4) &= 0.3200 \\
2f(0.6) &= 0.7200 \\
2f(0.8) &= 1.2800 \\
f(1) &= \underline{1.0000} \\
& 3.4000
\end{aligned}
$$

$\int_0^1 x^2 \, dx \approx \frac{0.2}{2}(3.4000) = 0.340$.

Actual value: $\int_0^1 x^2 \, dx = \frac{x^3}{3}\Big|_0^1 = \frac{1}{3} \approx 0.333$.

<u>Ans.</u>   0.340; 0.333

5. $f(x) = \frac{1}{x}$, n = 6, a = 1, b = 4.   Simpson.

    $h = \frac{b-a}{n} = \frac{4-1}{6} = 0.5$.

$$
\begin{aligned}
f(1) &= 1.0000 \\
4f(1.5) &= 2.6667 \\
2f(2) &= 1.0000 \\
4f(2.5) &= 1.6000 \\
2f(3) &= 0.6667 \\
4f(3.5) &= 1.1429 \\
f(4) &= \underline{0.2500} \\
& 8.3263
\end{aligned}
$$

$\int_1^4 \frac{1}{x} \, dx \approx \frac{0.5}{3}(8.3263) \approx 1.388$.

Actual value: $\int_1^4 \frac{1}{x} \, dx = \ln|x|\,\Big|_1^4 = \ln 4 - \ln 1 =$

$\ln 4 - 0 = \ln 4 \approx 1.386$.   <u>Ans.</u>   1.388; 1.386

7. $f(x) = \frac{x}{x+1}$, n = 4, a = 0, b = 2.   Trapezoidal.

    $h = \frac{b-a}{n} = \frac{2}{4} = 0.5$.

$$f(0) = 0.0000$$
$$2f(0.5) = 0.6667$$
$$2f(1) = 1.0000$$
$$2f(1.5) = 1.2000$$
$$f(2) = \underline{0.6667}$$
$$3.5334$$

$$\int_0^2 \frac{x}{x+1}\, dx \approx \frac{0.5}{2}(3.5334) \approx 0.883. \qquad \underline{Ans.} \quad 0.883$$

9. $\int_{15}^{40} l(t)\, dt$, males, n = 5, a = 15, b = 40. h = $\frac{40-15}{5}$ = 5.

$$l(15) = 96,672$$
$$2l(20) = 191,922$$
$$2l(25) = 190,000$$
$$2l(30) = 188,194$$
$$2l(35) = 186,134$$
$$l(40) = \underline{91,628}$$
$$944,550$$

$$\int_{15}^{40} l(t)\, dt \approx \frac{5}{2}(944,550) = 2,361,375. \qquad \underline{Ans.} \quad 2,361,375$$

1. a = 1, b = 5, h = 1.
$$f(1) = 0.4 \qquad = 0.4$$
$$4f(2) = 4(0.6) = 2.4$$
$$2f(3) = 2(1.2) = 2.4$$
$$4f(4) = 4(0.8) = 3.2$$
$$f(5) = 0.5 \qquad = \underline{0.5}$$
$$8.9$$

$$\int_1^5 f(x)\, dx \approx \frac{1}{3}(8.9) \approx 3.0. \qquad \underline{Ans.} \quad 3.0 \text{ sq units}$$

3. $\int_1^3 f(x)\, dx$, n = 4, a = 1, b = 3. h = $\frac{3-1}{4}$ = 0.5.

$$f(1) = 1 \qquad\qquad = 1$$
$$4f(1.5) = 4(2) \qquad = 8$$
$$2f(2) = 2(2) \qquad = 4$$
$$4f(2.5) = 4(0.5) = 2$$
$$f(3) = 1 \qquad\qquad = \underline{1}$$
$$16$$

$$\int_1^3 f(x)\ dx \approx \frac{0.5}{3}(16) = \frac{8}{3}.\qquad \underline{\text{Ans.}}\quad \frac{8}{3}$$

15. $f(x) = \sqrt{1-x^2}$, $a = 0$, $b = 1$, $n = 4$, $h = \frac{1-0}{4} = 0.25$.
    Simpson.

$$
\begin{aligned}
f(0) &= 1.0000 \\
4f(0.25) &= 3.8730 \\
2f(0.50) &= 1.7321 \\
4f(0.75) &= 2.6458 \\
f(1) &= \underline{0.0000} \\
&\ \ \ 9.2509
\end{aligned}
$$

$$\int_0^1 \sqrt{1-x^2}\ dx \approx \frac{0.25}{3}(9.2509) \approx 0.771.\qquad \underline{\text{Ans.}}\quad 0.771$$

17. Let $f(x)$ = distance from near to far shore at point x on highway. Then area $\approx \int_0^4 f(x)\ dx$. Using Simpson's rule with $h = 0.5$:

$$
\begin{aligned}
f(0) &= 0.5-0.5 &&= 0 &&= 0 \\
4f(0.5) &= 4(2.3-0.3) &&= 4(2) &&= 8 \\
2f(1) &= 2(2.2-0.7) &&= 2(1.5) &&= 3 \\
4f(1.5) &= 4(3-1) &&= 4(2) &&= 8 \\
2f(2) &= 2(2.5-0.5) &&= 2(2) &&= 4 \\
4f(2.5) &= 4(2.2-0.2) &&= 4(2) &&= 8 \\
2f(3) &= 2(1.5-0.5) &&= 2(1) &&= 2 \\
4f(3.5) &= 4(1.3-0.8) &&= 4(0.5) &&= 2 \\
f(4) &= 1-1 &&= 0 &&= \underline{0} \\
&&&&&\ \ 35
\end{aligned}
$$

Area $\approx \int_0^4 f(x)\ dx \approx \frac{0.5}{3}(35) = \frac{35}{6}$ km^2 $\underline{\text{Ans.}}\quad \frac{35}{6}$ km^2

19. (a) MC $= \frac{dc}{dq}$. $\int_0^{120} \frac{dc}{dq}\ dq = c(120) - c(0) =$
    (total cost of 120 units) - (fixed costs) =
    total variable costs of 120 units. Using the
    trapezoidal rule with $h = 20$ and $f(q) = dc/dq$ to
    estimate the integral:

$$f(0) = 260 \quad = \quad 260$$
$$2f(20) = 2(255) = 510$$
$$2f(40) = 2(240) = 480$$
$$2f(60) = 2(240) = 480$$
$$2f(80) = 2(245) = 490$$
$$2f(100) = 2(250) = 500$$
$$f(120) = 255 \quad = \underline{\ 255}$$
$$2975$$

$$\int_0^{120} \frac{dc}{dq}\, dq \approx \frac{20}{2}(2975) = 29{,}750. \quad \underline{\text{Ans.}}\ \ \$29{,}750$$

(b) MR = dr/dq.

$$\int_0^{120} \frac{dr}{dq}\, dq = r(120) - r(0) = r(120) \quad [\text{since } r(0) = 0]$$
$$= \text{total revenue from sale of 120 units.}$$

Using Simpson's rule with h = 20 and g(q) = dr/dq to estimate the integral:

$$g(0) = 415 \quad = \quad 415$$
$$4g(20) = 4(360) = 1440$$
$$2g(40) = 2(320) = \quad 640$$
$$4g(60) = 4(290) = 1160$$
$$2g(80) = 2(270) = \quad 540$$
$$4g(100) = 4(260) = 1040$$
$$g(120) = 255 \quad = \underline{\ 255}$$
$$5490$$

$$\int_0^{120} \frac{dr}{dq}\, dq \approx \frac{20}{3}(5490) = 36{,}600. \quad \underline{\text{Ans.}}\ \ \$36{,}600$$

(c) At q = 120:

total revenue $\approx$ 36,600,

total cost = (tot. var. costs) + (fixed costs)

$\approx$ 29,750 + 1500 = 31,250.

Thus maximum profit = (total rev.) - (total costs)

$\approx$ 36,600 - 31,250 = $5350.

<u>Ans.</u> $5350

EXERCISE 17.6

1. $y' = 2xy^2$. $\frac{dy}{dx} = 2xy^2$. $\frac{dy}{y^2} = 2x\ dx$. $\int y^{-2}\ dy = \int 2x\ dx$.

   $-\frac{1}{y} = x^2 + C$. $y = -\frac{1}{x^2 + C}$.

3. $\frac{dy}{dx} - x\sqrt{x^2 + 1} = 0$. $dy = x(x^2 + 1)^{1/2}\ dx$.

   $\int dy = \int x(x^2 + 1)^{1/2}\ dx$. $\int dy = \frac{1}{2}\int (x^2 + 1)^{1/2}[2x\ dx]$.

   $y = \frac{1}{2}\cdot\frac{(x^2 + 1)^{3/2}}{3/2} + C$, $y = \frac{1}{3}(x^2 + 1)^{3/2} + C$.

5. $\frac{dy}{dx} = y$, where $y > 0$. $\frac{dy}{y} = dx$. $\int\frac{dy}{y} = \int dx$. $\ln y = x + C_1$.

   $y = e^{x+C_1} = e^{C_1}e^x = Ce^x$, where $C = e^{C_1}$. Thus $y = Ce^x$,

   where $C > 0$.

7. $y' = \frac{y}{x}$, where $x, y > 0$. $\frac{dy}{dx} = \frac{y}{x}$, $\frac{dy}{y} = \frac{dx}{x}$. $\int\frac{dy}{y} = \int\frac{dx}{x}$.

   $\ln y = \ln x + C_1$, $\ln y = \ln x + \ln C$, where $C > 0$.

   $\ln y = \ln(Cx) \Rightarrow y = Cx$, where $C > 0$.

9. $y' = \frac{1}{y}$, where $y > 0$ and $y(2) = 2$. $\frac{dy}{dx} = \frac{1}{y}$. $y\ dy = dx$.

   $\int y\ dy = \int dx$. $\frac{y^2}{2} = x + C$. Given $y(2) = 2$, we obtain

   $\frac{2^2}{2} = 2 + C$, so $C = 0$. Thus $y^2 = 2x$, $y = \pm\sqrt{2x}$. But given

   that $y > 0$, we must choose $y = \sqrt{2x}$.

11. $e^y y' - x^2 = 0$, where $y = 0$ when $x = 0$. $e^y \frac{dy}{dx} = x^2$,

    $e^y\ dy = x^2\ dx$. $\int e^y\ dy = \int x^2\ dx$. $e^y = \frac{x^3}{3} + C$. Given that

    $y(0) = 0$, we have $e^0 = 0 + C$, so $1 = C \Rightarrow e^y = \frac{x^3}{3} + 1$,

$e^y = \frac{x^3 + 3}{3}$, so $y = \ln \frac{x^3 + 3}{3}$.

13. $(4x^2 + 3)^2 \frac{dy}{dx} - 4xy^2 = 0$, $(4x^2 + 3)^2 \frac{dy}{dx} = 4xy^2$.

$\frac{dy}{y^2} = \frac{4x}{(4x^2 + 3)^2} dx$. $\int \frac{dy}{y^2} = \int \frac{4x}{(4x^2 + 3)^2} dx$,

$\int y^{-2} dy = 4 \cdot \frac{1}{8} \int (4x^2 + 3)^{-2} [8x \, dx]$. $-\frac{1}{y} = -\frac{1}{2(4x^2 + 3)} + C$.

Now, $y(0) = \frac{3}{2}$ implies $-\frac{1}{3/2} = -\frac{1}{2(3)} + C$, so $C = -\frac{1}{2}$. Thus

$$-\frac{1}{y} = -\frac{1}{2(4x^2 + 3)} - \frac{1}{2} = -\frac{1 + (4x^2 + 3)}{2(4x^2 + 3)}$$

$$= -\frac{4x^2 + 4}{2(4x^2 + 3)} = -\frac{2(x^2 + 1)}{4x^2 + 3}.$$

$$y = \frac{4x^2 + 3}{2(x^2 + 1)}.$$

15. $\frac{dy}{dx} = \frac{3x\sqrt{1 + y^2}}{y}$, where $y > 0$ and $y(1) = \sqrt{8}$.

$\frac{y \, dy}{\sqrt{1 + y^2}} = 3x \, dx$. $\frac{1}{2} \int (1 + y^2)^{-1/2} [2y \, dy] = 3 \int x \, dx$,

$(1 + y^2)^{1/2} = \frac{3x^2}{2} + C$. $y(1) = \sqrt{8} \Rightarrow (1 + 8)^{1/2} = \frac{3}{2} + C$,

$C = \frac{3}{2}$. Thus $(1 + y^2)^{1/2} = \frac{3x^2}{2} + \frac{3}{2}$, $1 + y^2 = \left[\frac{3x^2}{2} + \frac{3}{2}\right]^2$,

$y^2 = \left[\frac{3x^2}{2} + \frac{3}{2}\right]^2 - 1$. Since $y > 0$, $y = \sqrt{\left[\frac{3x^2}{2} + \frac{3}{2}\right]^2 - 1}$.

17. $2\frac{dy}{dx} = \frac{xe^{-y}}{\sqrt{x^2 + 3}}$, where $y(1) = 0$. $\frac{dy}{e^y} = \frac{1}{2}x(x^2 + 3)^{-1/2} dy$

$\int e^y \, dy = \frac{1}{2} \cdot \frac{1}{2} \int (x^2 + 3)^{-1/2} [2x \, dx]$,

$e^y = \frac{1}{2}(x^2 + 3)^{1/2} + C$. Now, $y(1) = 0 \Rightarrow e^0 = \frac{1}{2}(2) + C$,

so $C = 0$. Thus $e^y = \frac{1}{2}(x^2 + 3)^{1/2} \Rightarrow y = \ln\left(\frac{1}{2}\sqrt{x^2 + 3}\right)$.

19. $(q + 1)^2 \frac{dc}{dq} = cq$, $\int \frac{1}{c} \, dc = \int \frac{q}{(q + 1)^2} \, dq$. Using partial

fractions or formula 7 for $\int \frac{q}{(q + 1)^2} \, dq$, we obtain

$\ln c = \ln(q + 1) + \frac{1}{q + 1} + C$. Now, fixed cost is given to
be e, which means that c = e when q = 0. This implies
$1 = 0 + 1 + C$, so C = 0. Thus $\ln c = \ln(q + 1) + \frac{1}{q + 1} \Rightarrow$
$c = e^{\ln(q+1)+1/(q+1)}$, $c = e^{\ln(q+1)}e^{1/(q+1)}$, or
$c = (q + 1)e^{1/(q+1)}$.

21. $\frac{dy}{dt} = -0.05y$. $\int \frac{1}{y} \, dy = -0.05 \int dt$, $\ln|y| = -0.05t + C$. Given
that y = 900 when t = 0, we have $\ln 900 = -0 + C = C$.
Thus $\ln|y| = -0.05t + \ln 900$. To find t when money is 90%
new, we note that y would be 10%(900) = 90. Solving
$\ln 90 = -0.05t + \ln 900$ gives
$$t = \frac{\ln 900 - \ln 90}{0.05} \approx 46 \text{ weeks.}$$

23. Let N be the population at time t, where t = 0 corresponds
to 1980. Since N follows exponential growth, $N = N_0 e^{kt}$.
Now, N = 20,000 when t = 0, so N_0 = 20,000. Therefore
$N = 20,000e^{kt}$. Since N = 24,000 when t = 10, we have
$24,000 = 20,000e^{10k}$, $1.2 = e^{10k}$, $\ln 1.2 = 10k$, $k = \frac{\ln 1.2}{10}$.
Thus
$$N = 20,000e^{(\ln 1.2)(t/10)} \qquad (*)$$
$$N = 20,000e^{0.18(t/10)}$$
$$N = 20,000e^{0.018t}. \qquad \text{(First form)}$$
From (*), we have $N = 20,000\left[e^{\ln 1.2}\right]^{t/10}$, so
$$N = 20,000(1.2)^{t/10}. \qquad \text{(Second form)}$$
At year 2000, t = 20 and so
$$N = 20,000(1.2)^{20/10} = 20,000(1.2)^2 = 28,800.$$

25. Let N be the population (in billions) at time t, where t is the number of years past 1930. N follows exponential growth, so $N = N_0 e^{kt}$. When t = 0, then N = 2, so $N_0 = 2$. Thus $N = 2e^{kt}$. Since N = 3 when t = 30, then $3 = 2e^{30k}$, $3/2 = e^{30k}$, 30k = ln(3/2), or k = ln(3/2)/30. Thus $N = 2e^{t \ln(3/2)/30} \approx 2e^{t(0.40547/30)} \approx 2e^{(0.013516)t}$. In 2000, t = 70 and so $N \approx 2e^{(0.013516)70} \approx 2e^{0.946}$ billion.

27. Let N be amount of sample that remains after t seconds. Then $N = N_0 e^{-\lambda t}$, where N_0 is the initial amount present. When t = 100, then $N = 0.3N_0$. Thus $0.3N_0 = N_0 e^{-100\lambda}$, $0.3 = e^{-100\lambda}$, $-100\lambda = \ln 0.3 = \ln 3 - \ln 10$, $-100\lambda \approx$ 1.09861 - 2.30259 = -1.20398. Thus $\lambda \approx 0.01204$. The half-life is $0.69315/\lambda \approx 57.57$ s.

29. Let N be the amount of ^{14}C present in the scroll t years after it was made. Then $N = N_0 e^{-\lambda t}$, where N_0 is amount of ^{14}C present when t = 0. We must find t when $N = 0.7N_0$. $0.7N_0 = N_0 e^{-\lambda t}$, $0.7 = e^{-\lambda t}$, $-\lambda t = \ln 0.7 = \ln(7/10)$, $-\lambda t = \ln 7 - \ln 10 \approx 1.94591 - 2.30259 = -0.35668$, so $t \approx 0.35668/\lambda$. By Eq. 15 in text, $\lambda \approx 0.69315/5600$, so t $\approx 0.35668(5600/0.69315) \approx 2900$ years.

31. $\frac{dN}{dt} = kN$, $N = Ae^{kt}$, $N_0 = Ae^{kt_0}$, $A = \frac{N_0}{e^{kt_0}}$. Thus $N = \frac{N_0}{e^{kt_0}}(e^{kt}) = N_0 e^{kt-kt_0}$, or $N = N_0 e^{k(t-t_0)}$, where $t \geq t_0$.

33. $N = N_0 e^{-\lambda t}$. When t = 2, then N = 10. Thus $10 = N_0 e^{-2\lambda}$, $N_0 = 10e^{2\lambda}$. By Eq. 15, $6 = (\ln 2)/\lambda$, $\lambda = (\ln 2)/6$. Thus $N_0 = 10e^{2(\ln 2)/6} \approx 10e^{0.69315/3} \approx 10e^{0.23} \approx 12.6$ units.

35. $\frac{dA}{dt} = 200 - 0.50A,$ $\int \frac{dA}{200 - 0.50A} = \int dt.$

$$- \frac{1}{0.50} \ln(200 - 0.50A) = t + C_1,$$
$$\ln(200 - 0.50A) = -0.50t - 0.50C_1$$
$$= -0.50t + C_2.$$

Thus

$$200 - 0.50A = e^{-0.50t + C_2}$$
$$= e^{-0.50t} e^{C_2}.$$
$$200 - \frac{A}{2} = Ce^{-0.50t}.$$

Given that $A = 0$ when $t = 0$, we have $C = 200$, so
$200 - (A/2) = 200e^{-t/2},$ $200 - 200e^{-t/2} = A/2,$
$200(1 - e^{-t/2}) = A/2.$ Thus $A = 400(1 - e^{-t/2}).$
If $t = 1$, $A = 400(1 - e^{-1/2}) \approx 400(1 - 0.60653) \approx 157$ gm.

37. (a) $\frac{dV}{dt} = kV,$ $\int \frac{1}{V} dV = \int k \, dt,$ $\ln V = kt + C_1,$ $V = e^{kt} e^{C_1},$ or

$V = Ce^{kt}.$ Now $t = 0$ corresponds to July 1, 1989 where
$V = 7000$, so $7000 = C(1)$. Thus $V = 7000e^{kt}$. Also,
$V = 6300$ for $t = \frac{1}{2}$, so $6300 = 7000e^{k/2},$ $0.9 = e^{k/2},$
$\frac{k}{2} = \ln 0.9,$ $k = 2 \ln 0.9$. Thus $V = 7000e^{(2 \ln 0.9)t}.$

(b) $3500 = 7000e^{(2 \ln 0.9)t},$ $\frac{1}{2} = e^{(2 \ln 0.9)t},$

$\ln \frac{1}{2} = (2 \ln 0.9)t,$ $t = \frac{\ln \frac{1}{2}}{2 \ln 0.9} \approx 3.289$ yr.
This corresponds to approx. 39.47 months beyond July
1, 1989 $\Rightarrow$ Oct. 1992.

EXERCISE 17.7

1. $N = M/(1 + be^{-ct})$. $M = 40,000$. Since $N = 20,000$ at $t = 0$ (1987), we have $20,000 = \frac{40,000}{1 + b}$, so $1 + b = \frac{40,000}{20,000} = 2$, or $b = 1$. Hence $N = \frac{40,000}{1 + e^{-ct}}$. If $t = 5$, then $N = 25,000$, so

   $25,000 = \frac{40,000}{1 + e^{-5c}}$, $1 + e^{-5c} = \frac{40,000}{25,000} = \frac{8}{5}$, $e^{-5c} = \frac{8}{5} - 1 = \frac{3}{5}$,

   $e^{-c} = \left(\frac{3}{5}\right)^{1/5}$. Hence $N = \frac{40,000}{1 + (3/5)^{t/5}}$. In 1997, $t = 10$ so

   $N = \frac{40,000}{1 + (3/5)^2} \approx 29,400$.

3. $N = \frac{M}{1 + be^{-ct}}$. $M = 3,000,000$, and $N = 50$ when $t = 0$, so

   $50 = \frac{3,000,000}{1 + b}$, $1 + b = \frac{3,000,000}{50} = 60,000$, $b = 59,999$.

   Hence $N = \frac{3,000,000}{1 + 59,999e^{-ct}}$. Since $t = 1$ when $N = 5000$,

   $5000 = \frac{3,000,000}{1 + 59,999e^{-c}}$, $1 + 59,999e^{-c} = \frac{3,000,000}{5000} = 600$,

   $e^{-c} = \frac{599}{59,999}$. Hence $N = \frac{3,000,000}{1 + 59,999\left(\frac{599}{59,999}\right)^t}$. If $t = 2$,

   then $N = \frac{3,000,000}{1 + 59,999\left(\frac{599}{59,999}\right)^2} \approx 430,000$.

5. $N = \frac{M}{1 + be^{-ct}}$. $M = 100,000$, and since $N = 500$ when

   $t = 0$, we have $500 = \frac{100,000}{1 + b}$, $1 + b = \frac{100,000}{500} = 200$,

   $b = 199$. Hence $N = \frac{100,000}{1 + 199e^{-ct}}$. If $t = 1$, then $N = 1000$.

   Thus $1000 = \frac{100,000}{1 + 199e^{-c}}$, $1 + 199e^{-c} = \frac{100,000}{1000} = 100$,

   $199e^{-c} = 99$, $e^{-c} = \frac{99}{199}$. Hence $N = \frac{100,000}{1 + 199\left(\frac{99}{199}\right)^t}$.

   If $t = 2$, then $N = \frac{100,000}{1 + 199\left(\frac{99}{199}\right)^2} \approx 1990$.

7. (a) $N = \dfrac{375}{1 + e^{5.2-2.3t}} = \dfrac{375}{1 + e^{5.2}e^{-2.3t}} = \dfrac{375}{1 + 181.27e^{-2.3t}}.$

  (b) $\lim\limits_{t\to\infty} N = \dfrac{375}{1 + 181.27(0)} = 375.$

9. $\dfrac{dT}{dt} = k(T - a)$ where $a = 10.$ $\dfrac{dT}{T - a} = k\,dt,$ $\displaystyle\int \dfrac{dT}{T - a} = \int k\,dt.$

Thus $\ln(T - 10) = kt + C.$ At $t = 0,$ we have $T = 32,$ so
$\ln(32 - 10) = 0 + C,$ $C = \ln 22,$ so $\ln(T - 10) = kt + \ln 22,$
$\ln(T - 10) - \ln 22 = kt.$ Hence $\ln\left(\dfrac{T - 10}{22}\right) = kt.$ If $t = 1,$

then $T = 30.$ Thus $\ln\left(\dfrac{30 - 10}{22}\right) = k\cdot1,$ so $k = \ln(20/22) =$
$\ln(10/11) = \ln(1/1.1) = -\ln 1.1 \approx -0.09531.$ Hence
$\ln\left(\dfrac{T-10}{22}\right) \approx -0.09531t.$ If $T = 37,$ then $\ln\left(\dfrac{27}{22}\right) \approx -0.09531t,$

$\ln\left(\dfrac{2.7}{2.2}\right) \approx -0.09531t,$ $t \approx \dfrac{\ln 2.7 - \ln 2.2}{-0.09531}$

                   $\approx \dfrac{0.99325 - 0.78846}{-0.09531}$

                   $\approx \dfrac{0.20479}{-0.09531} \approx -2.15$ hr,

which corresponds to 2 hr 9 min. Time of murder:
     3:15 A.M. - 2 hr 9 min = 1:06 A.M.

11. $\dfrac{dx}{dt} = k(70,000 - x).$ $\displaystyle\int \dfrac{dx}{70,000 - x} = \int k\,dt,$

$-\ln(70,000 - x) = kt + C,$ $\ln(70,000 - x) = -kt - C,$
$70,000 - x = e^{-kt-C} = e^{-C}e^{-kt} = Ae^{-kt},$ where $A = e^{-C}.$
Thus $x = 70,000 - Ae^{-kt}.$ If $t = 0,$ then $x = 10,000,$ so
$10,000 = 70,000 - A \implies A = 60,000.$ Thus
         $x = 70,000 - 60,000e^{-kt}.$

If $t = 1,$ then $x = 40,000,$ so $40,000 = 70,000 - 60,000e^{-k},$
$60,000e^{-k} = 30,000,$ $e^{-k} = \dfrac{30,000}{60,000} = \dfrac{1}{2}.$ Thus

         $x = 70,000 - 60,000\left(\dfrac{1}{2}\right)^t.$

If $t = 3,$ then $x = 70,000 - 60,000\left(\dfrac{1}{8}\right) = \$62,500.$

13. $\frac{dN}{dt} = k(M - N)$, $\int \frac{dN}{M - N} = \int k \, dt$, $-\ln(M - N) = kt + C$.

If $t = 0$, then $N = N_0$, so $-\ln(M - N_0) = C$. Thus we have

$-\ln(M - N) = kt - \ln(M - N_0)$, $\ln(M - N_0) - \ln(M - N) = kt$,

$\ln \frac{M - N_0}{M - N} = kt$, $\ln \frac{M - N}{M - N_0} = -kt$, $\frac{M - N}{M - N_0} = e^{-kt}$,

$M - N = (M - N_0)e^{-kt}$, $N = M - (M - N_0)e^{-kt}$.

EXERCISE 17.8

1. $\displaystyle\int_3^\infty \frac{1}{x^2} \, dx = \lim_{r\to\infty} \int_3^r x^{-2} \, dx = \lim_{r\to\infty} \frac{x^{-1}}{-1}\bigg|_3^r = \lim_{r\to\infty} \left(-\frac{1}{x}\right)\bigg|_3^r$

$\qquad = \lim_{r\to\infty} \left(-\frac{1}{r} + \frac{1}{3}\right) = 0 + \frac{1}{3} = \frac{1}{3}$.

3. $\displaystyle\int_1^\infty \frac{1}{x} \, dx = \lim_{r\to\infty} \int_1^r \frac{1}{x} \, dx = \lim_{r\to\infty} \ln|x|\bigg|_1^r = \lim_{r\to\infty} (\ln|r| - 0)$

$\qquad = \lim_{r\to\infty} \ln|r| = \infty \implies$ diverges.

5. $\displaystyle\int_1^\infty e^{-x} \, dx = \lim_{r\to\infty} -\int_1^r e^{-x}[-dx] = \lim_{r\to\infty} -e^{-x}\bigg|_1^r$

$\qquad = \lim_{r\to\infty} (-e^{-r} + e^{-1}) = \lim_{r\to\infty} \left(-\frac{1}{e^r} + \frac{1}{e}\right) = 0 + \frac{1}{e} = \frac{1}{e}$.

7. $\displaystyle\int_1^\infty \frac{1}{\sqrt{x}} \, dx = \lim_{r\to\infty} \int_1^r x^{-1/2} \, dx = \lim_{r\to\infty} 2x^{1/2}\bigg|_1^r$

$\qquad = \lim_{r\to\infty} (2\sqrt{r} - 2) = \infty \implies$ diverges.

9. $\displaystyle\int_{-\infty}^{-2} \frac{1}{(x + 1)^3} \, dx = \lim_{r\to-\infty} \int_r^{-2} (x + 1)^{-3} \, dx$

$\qquad = \lim_{r\to-\infty} \frac{(x + 1)^{-2}}{-2}\bigg|_r^{-2} = \lim_{r\to-\infty} -\frac{1}{2(x + 1)^2}\bigg|_r^{-2} =$

$$\lim_{r \to -\infty} \left[-\frac{1}{2} + \frac{1}{2(r+1)^2} \right] = -\frac{1}{2} + 0 = -\frac{1}{2}.$$

11. $\displaystyle\int_{-\infty}^{\infty} xe^{-x^2} \, dx = \int_{-\infty}^{0} xe^{-x^2} \, dx + \int_{0}^{\infty} xe^{-x^2} \, dx.$

$$\int_{-\infty}^{0} xe^{-x^2} \, dx = \lim_{r \to -\infty} -\frac{1}{2}\int_{r}^{0} e^{-x^2}[-2x \, dx] = \lim_{r \to -\infty} -\frac{1}{2}e^{-x^2}\Big|_{r}^{0}$$

$$= \lim_{r \to -\infty} \left[-\frac{1}{2} + \frac{1}{2e^{r^2}} \right] = -\frac{1}{2} + 0 = -\frac{1}{2}.$$

$$\int_{0}^{\infty} xe^{-x^2} \, dx = \lim_{r \to \infty} -\frac{1}{2}\int_{0}^{r} e^{-x^2}[-2x \, dx] = \lim_{r \to \infty} -\frac{1}{2}e^{-x^2}\Big|_{0}^{r}$$

$$= \lim_{r \to \infty} \left[-\frac{1}{2e^{r^2}} + \frac{1}{2} \right] = 0 + \frac{1}{2} = \frac{1}{2}.$$

Thus $\displaystyle\int_{-\infty}^{\infty} xe^{-x^2} \, dx = -\frac{1}{2} + \frac{1}{2} = 0.$

13. (a) $\displaystyle\int_{800}^{\infty} \frac{k}{x^2} \, dx = 1, \ \lim_{r \to \infty} k\int_{800}^{r} x^{-2} \, dx = 1, \ \lim_{r \to \infty} -\frac{k}{x}\Big|_{800}^{r} = 1,$

$$\lim_{r \to \infty} \left(-\frac{k}{r} + \frac{k}{800} \right) = 1, \ 0 + \frac{k}{800} = 1, \ k = 800.$$

(b) $\displaystyle\int_{1200}^{\infty} \frac{800}{x^2} \, dx = \lim_{r \to \infty} 800\int_{1200}^{r} x^{-2} \, dx = \lim_{r \to \infty} -\frac{800}{x}\Big|_{1200}^{r}$

$$= \lim_{r \to \infty} \left(-\frac{800}{r} + \frac{800}{1200} \right) = 0 + \frac{2}{3} = \frac{2}{3}.$$

15. $\displaystyle\int_{0}^{\infty} 240{,}000e^{-0.06t} \, dt$

$$= \lim_{r \to \infty} \frac{240{,}000}{-0.06}\int_{0}^{r} e^{-0.06t}[-0.06 \, dt] = \lim_{r \to \infty} -\frac{240{,}000}{0.06}e^{-0.06t}\Big|_{0}^{r}$$

$$= \lim_{r \to \infty} -\frac{240{,}000}{0.06}\left(\frac{1}{e^{0.06r}} - 1 \right) = -\frac{240{,}000}{0.06}(-1) = 4{,}000{,}000.$$

17. area = $\int_0^\infty e^{-2x} \, dx = \lim_{r\to\infty} -\frac{1}{2}\int_0^r e^{-2x}[-2 \, dx] = \lim_{r\to\infty} -\left.\frac{e^{-2x}}{2}\right|_0^r$

     $= \lim_{r\to\infty} \left[-\frac{1}{2e^{2r}} + \frac{1}{2}\right] = 0 + \frac{1}{2} = \frac{1}{2}$ square unit.

19. $\int_0^\infty \frac{10,000}{(t+2)^2} \, dt = \lim_{r\to\infty} \left.\frac{-10,000}{t+2}\right|_0^r = \lim_{r\to\infty}\left[-\frac{10,000}{r+2} + \frac{10,000}{2}\right]$

         $= 0 + \frac{10,000}{2} = 5000$ increase.

CHAPTER 17 - REVIEW PROBLEMS

1. Integration by parts or formula 42. For integration by
   parts, let u = ln x, dv = x dx. Then du = $\frac{1}{x}$ dx and

   $v = \frac{x^2}{2}$. Thus

   $\int x \ln x \, dx = \ln x \left(\frac{x^2}{2}\right) - \int \frac{x^2}{2} \cdot \frac{1}{x} \, dx = \frac{x^2}{2} \ln(x) - \int \frac{x}{2} \, dx$

           $= \frac{x^2}{2} \ln(x) - \frac{x^2}{4} + C = \frac{x^2}{4}[2 \ln(x) - 1] + C.$

   [Or, with formula 42, let u = x (so du = dx) and n = 1.]

3. Formula 23 with u = 2x, a^2 = 9. Then du = 2 dx.

   $\int_0^2 \sqrt{4x^2 + 9} \, dx = \frac{1}{2}\int_0^2 \sqrt{(2x)^2 + 9} \, [2 \, dx]$

   $= \frac{1}{2}\left[\frac{1}{2}\left((2x)\sqrt{4x^2 + 9} + 9 \ln|2x + \sqrt{4x^2 + 9}|\right)\right]\Big|_0^2$

   $= \left(5 + \frac{9}{4} \ln 9\right) - \left(0 + \frac{9}{4} \ln 3\right) = 5 + \frac{9}{4} \ln 3^2 - \frac{9}{4} \ln 3$

   $= 5 + \frac{18}{4} \ln 3 - \frac{9}{4} \ln 3 = 5 + \frac{9}{4} \ln 3.$

5. By partial fractions,

   $\int \frac{x \, dx}{(2+3x)(3+x)} = \int \left[\frac{-2/7}{2+3x} + \frac{3/7}{3+x}\right] dx$

           $= -\frac{2}{21} \ln|2 + 3x| + \frac{3}{7} \ln|3 + x| + C =$

$$\frac{1}{21}(9 \ln|3 + x| - 2 \ln|2 + 3x|) + C.$$

Alternatively, by formula 12 with u = x (so du = dx), a = 2, b = 3, c = 3, and k = 1,

$$\int \frac{x \, dx}{(2 + 3x)(3 + x)} = \frac{1}{7}\left[3 \ln|3 + x| - \frac{2}{3} \ln|2 + 3x|\right] + C.$$

$$= \frac{1}{21}(9 \ln|3 + x| - 2 \ln|2 + 3x|) + C.$$

7. Partial fractions or formula 9. For formula 9, let u = x (so du = dx), a = 2, and b = 1. For partial fractions,

$$\int \frac{dx}{x(x + 2)^2} = \int \left[\frac{1/4}{x} + \frac{-1/4}{x + 2} + \frac{-1/2}{(x + 2)^2}\right] dx$$

$$= \frac{1}{4}\ln|x| - \frac{1}{4}\ln|x + 2| + \frac{1}{2(x + 2)} + C$$

$$= \frac{1}{2(x + 2)} + \frac{1}{4} \ln\left|\frac{x}{x + 2}\right| + C.$$

9. Formula 21 with u = 4x (so du = 4 dx) and $a^2 = 9$.

$$\int \frac{dx}{x^2\sqrt{9 - 16x^2}} = 4\int \frac{[4 \, dx]}{(4x)^2\sqrt{9 - (4x)^2}} = 4\left[-\frac{\sqrt{9 - 16x^2}}{9(4x)}\right] + C$$

$$= -\frac{\sqrt{9 - 16x^2}}{9x} + C.$$

11. Partial fractions or formula 35. For partial fractions

$$\int \frac{9 \, dx}{x^2 - 9} = \int \left[\frac{3/2}{x - 3} + \frac{-3/2}{x + 3}\right] dx = \frac{3}{2}\ln|x-3| - \frac{3}{2}\ln|x+3| + C$$

$$= \frac{3}{2} \ln\left|\frac{x - 3}{x + 3}\right| + C.$$

To apply formula 35, first write the integral as $9\int \frac{dx}{x^2 - 9}$ and then set u = x (so du = dx) and a = 3.

13. Integration by parts or formula 38. With parts, let u = x and $dv = e^{7x} dx$. Then du = dx and $v = \frac{1}{7}e^{7x}$. Thus

$$\int xe^{7x} \, dx = x \cdot \frac{1}{7}e^{7x} - \int \frac{1}{7}e^{7x} \, dx = \frac{x}{7}e^{7x} - \frac{1}{7}\cdot\frac{1}{7}\int e^{7x}[7 \, dx]$$

$$= \frac{x}{7}e^{7x} - \frac{1}{49}e^{7x} + C = \frac{e^{7x}}{49}(7x - 1) + C.$$

With formula 38, let u = x (so du = dx) and a = 7.

15. The integral has the form $\int \frac{1}{u} du$.

$$\int \frac{dx}{2x \ln 2x} = \frac{1}{2} \int \frac{1}{\ln 2x} \left[\frac{2}{2x} dx \right] = \frac{1}{2} \ln|\ln 2x| + C.$$

17. Long division or formula 3. For long division,

$$\int \frac{2x}{3 + 2x} dx = \int \left[1 - \frac{3}{3 + 2x} \right] dx = x - 3 \cdot \frac{1}{2} \int \frac{1}{3 + 2x} [2 dx]$$

$$= x - \frac{3}{2} \ln|3 + 2x| + C.$$

To apply formula 3, write the integral as $2 \int \frac{x}{3 + 2x} dx$ and let u = x (so du = dx), a = 3 and b = 2.

19. Partial Fractions.

$$\int \frac{5x^2 + 2}{x^3 + x} dx = \int \left[\frac{2}{x} + \frac{3x}{x^2 + 1} \right] dx = 2 \ln|x| + \frac{3}{2} \ln(x^2 + 1) + C.$$

21. Integration by parts. u = ln(x + 1), dv = $(x + 1)^{-1/2}$ dx. Then du = $\frac{1}{x + 1}$ dx and v = $2(x + 1)^{1/2}$.

$$\int \frac{\ln(x + 1)}{\sqrt{x + 1}} dx = 2(x + 1)^{1/2} \ln(x + 1) - 2 \int (x + 1)^{-1/2} dx$$

$$= 2(x + 1)^{1/2} \ln(x + 1) - 4(x + 1)^{1/2} + C$$

$$= 2\sqrt{x + 1}[\ln(x + 1) - 2] + C.$$

23. $\bar{F} = \frac{1}{4 - 2} \int_2^4 (3x^2 + 2x) dx = \frac{1}{2}(x^3 + x^2) \Big|_2^4$

$$= \frac{1}{2}[(64 + 16) - (8 + 4)] = 34.$$

25. $f(x) = \frac{1}{x + 1}$, $n = 6$, $a = 0$, $b = 3$. $h = \frac{b-a}{n} = \frac{3-0}{6} = 0.5$.

(a) Trapezoidal

$f(0) = 1.0000$
$2f(0.5) = 1.3333$
$2f(1) = 1.0000$
$2f(1.5) = 0.8000$
$2f(2) = 0.6667$
$2f(2.5) = 0.5714$
$f(3) = \underline{0.2500}$

5.6214

$\frac{0.5}{2}(5.6214) \approx 1.405$

(b) Simpson

$f(0) = 1.0000$
$4f(0.5) = 2.6667$
$2f(1) = 1.0000$
$4f(1.5) = 1.6000$
$2f(2) = 0.6667$
$4f(2.5) = 1.1429$
$f(3) = \underline{0.2500}$

8.3263

$\frac{0.5}{3}(8.3263) \approx 1.388$

27. $y' = 3x^2 y + 2xy$, $y > 0$. $\frac{dy}{y} = (3x^2 + 2x)\, dx$.

$\int \frac{dy}{y} = \int (3x^2 + 2x)\, dx$. $\ln y = x^3 + x^2 + C_1$, from which

$y = e^{x^3 + x^2 + C_1}$, $y = Ce^{x^3 + x^2}$, where $C > 0$.

29. $\displaystyle\int_3^\infty \frac{1}{x^3}\, dx = \lim_{r \to \infty} \int_3^r x^{-3}\, dx = \lim_{r \to \infty} \frac{x^{-2}}{-2}\Big|_3^r = \lim_{r \to \infty} -\frac{1}{2x^2}\Big|_3^r$

$\qquad = \lim_{r \to \infty} \left[-\frac{1}{2r^2} + \frac{1}{18} \right] = 0 + \frac{1}{18} = \frac{1}{18}$.

31. $\displaystyle\int_1^\infty \frac{1}{2x}\, dx = \lim_{r \to \infty} \int_1^r \frac{1}{2x}\, dx = \lim_{r \to \infty} \frac{1}{2} \ln|x|\, \Big|_1^r$

$\qquad = \lim_{r \to \infty} \left[\frac{1}{2} \ln|r| - 0 \right] = \infty \implies$ diverges.

33. $N = N_0 e^{kt}$. Since $N = 100{,}000$ when $t = 0$, $N_0 = 100{,}000$.

Thus $N = 100{,}000 e^{kt}$. Since $N = 120{,}000$ when $t = 15$, then
$120{,}000 = 100{,}000 e^{15k}$, $1.2 = e^{15k}$, $\ln 1.2 = 15k$, or $k = (\ln 1.2)/15$. Thus

$\qquad N = 100{,}000 e^{t \ln(1.2)/15} = 100{,}000 (e^{\ln 1.2})^{t/15}$

$\qquad = 100{,}000 (1.2)^{t/15}$.

For the year 2000 we have $t = 30$ and

$\qquad N = 100{,}000 (1.2)^{30/15} = 100{,}000 (1.2)^2 = 144{,}000$.

35. $N = N_0 e^{-\lambda t}$, where N_0 is the original amount present. When

$t = 100$, then $N = 0.95N_0$, so we have $0.95N_0 = N_0 e^{-100\lambda}$,

$0.95 = e^{-100\lambda}$, $-100\lambda = \ln 0.95$, $\lambda = -\frac{1}{100}(\ln 9.5 - \ln 10)$,

$\lambda = 0.0005$ (decay constant).

After 200 years, $N = N_0 e^{-200\lambda}$. Thus

$\frac{N}{N_0} = e^{-200\lambda} = e^{-200(0.0005)} = e^{-0.1} \approx 0.90 = 90\%$.

37. $N = \frac{450}{1 + be^{-ct}}$. If $t = 0$, then $N = 2$. Thus $2 = \frac{4500}{1 + b}$,

$1 + b = \frac{450}{2} = 225$, $b = 224$, so $N = \frac{450}{1 + 224e^{-ct}}$.

If $t = 6$, then $N = 300 \Rightarrow 300 = \frac{450}{1 + 224e^{-6c}}$, $1 + 224e^{-6c} =$

$\frac{450}{300} = \frac{3}{2}$, $224e^{-6c} = \frac{3}{2} - 1 = \frac{1}{2}$, $e^{-6c} = \frac{1}{448}$, $e^{6c} = 448$,

$6c = \ln 448 \approx 6.10479$, $c \approx 1.02$. Thus $N = \frac{450}{1 + 224e^{-1.02t}}$.

39. $\frac{dT}{dt} = k(T - 25)$, $\frac{dT}{T - 25} = k\, dt$, $\int \frac{dT}{T - 25} = \int k\, dt$,

$\ln(T - 25) = kt + C$. If $t = 0$, then $T = 35$. Thus $\ln 10 =$

C, so $\ln\left(\frac{T-25}{10}\right) = kt$. If $t = 1$, then $T = 34$ and $\ln(9/10) =$

k, $k \approx -0.10537$. Thus $\ln\left(\frac{T-25}{10}\right) \approx -0.10537t$. If $T = 37$,

$\ln \frac{12}{10} \approx -0.10537t$, $\ln 1.2 \approx -0.10537t$, $t \approx -\frac{0.18232}{0.10537} \approx$

-1.73. Note that 1.73 hr corresponds approximately to 1
hr 44 min. Thus 6:00 P.M. - 1 hr 44 min = 4:16 P.M.

41. $\int_0^\infty f(x)\, dx = \lim_{r \to \infty} \int_0^r (0.008e^{-0.01x} + 0.00004e^{-0.0002x})\, dx$

$= \lim_{r \to \infty} (-0.8e^{-0.01x} - 0.2e^{-0.0002x})\Big|_0^r$

$= \lim_{r \to \infty} \left(-\frac{0.8}{e^{0.01r}} - \frac{0.2}{e^{0.0002r}} + 0.8 + 0.2\right)$

$= 0 - 0 + 0.8 + 0.2 = 1$.

43. (a) Total revenue = $r(12) - r(0) = \int_0^{12} \frac{dr}{dq}\, dq$. $f(q) = \frac{dr}{dq}$.

n = 4, a = 0, b = 12. $h = \frac{b-a}{n} = \frac{12-0}{4} = 3$

Trapezoidal   Simpson

f(0) =	25
2f(3) =	44
2f(6) =	36
2f(9) =	26
f(12) =	7
	138

f(0) =	25
4f(3) =	88
2f(6) =	36
4f(12) =	52
f(12) =	7
	208

TR ≈ $\frac{3}{2}(138) = 207$ TR ≈ $\frac{3}{3}(208) = 208$

(b) Total *variable* cost = $c(12) - c(0) = \int_0^{12} \frac{dc}{dq}\, dq$.

$f(q) = \frac{dc}{dq}$. a = 0, b = 12. Using as *little* data as possible, we choose n = 1 for Trapezoidal and n = 2 for Simpson (n must be even).

Trapezoidal (n = 1)   Simpson (n = 2)

$h = \frac{b-a}{n} = \frac{12-0}{1} = 12$. $h = \frac{b-a}{n} = \frac{12-0}{2} = 6$.

f(0) =	15
f(12) =	7
	22

f(0) =	15
4f(6) =	48
f(12) =	7
	70

VC ≈ $\frac{12}{2}(22) = 132$ VC ≈ $\frac{6}{3}(70) = 140$

To each of our results we must add on the fixed cost of 25 to obtain total cost. Thus for trapezoidal we get TC ≈ 132 + 25 = 157, and for Simpson we have TC ≈ 140 + 25 = 165.

(c) We use the relation $P(12) = \int_0^{12} \left[\frac{dr}{dq} - \frac{dc}{dq}\right] dq - 25$.

First we determine variable cost for each rule with

$$n = 4 \text{ and } h = \frac{b - a}{n} = \frac{12 - 0}{4} = 3.$$

Trapezoidal		Simpson	
f(0) =	15	f(0) =	15
2f(3) =	28	4f(3) =	56
2f(6) =	24	2f(6) =	24
2f(9) =	20	4f(12) =	40
f(12) =	7	f(12) =	7
	94		142

$$VC \approx \frac{3}{2}(94) = 141 \qquad VC \approx \frac{3}{3}(142) = 142$$

Using these results and those of part (a), we have

Trapezoidal	Simpson
P(12) ≈ 207 - 141 - 25	P(12) ≈ 208 - 142 - 25
= 41	= 41

CHAPTER 17 — MATHEMATICAL SNAPSHOT

1. $C = 2000$, $w_0 = 200$. $w_{eq} = \frac{C}{17.5} = \frac{2000}{17.5} \approx 114$.

$$w(t) = \frac{C}{17.5} + \left(w_0 - \frac{C}{17.5}\right)e^{-0.005t}$$

$$= \frac{2000}{17.5} + \left(200 - \frac{2000}{17.5}\right)e^{-0.005t}.$$

Letting $w(t) = 175$ and solving for t gives

$$175 = \frac{2000}{17.5} + \left(200 - \frac{2000}{17.5}\right)e^{-0.005t}$$

$$175 - \frac{2000}{17.5} = \left(200 - \frac{2000}{17.5}\right)e^{-0.005t}$$

$$\frac{175 - \frac{2000}{17.5}}{200 - \frac{2000}{17.5}} = e^{-0.005t}$$

$$-0.005t = \ln\left[\frac{175 - \frac{2000}{17.5}}{200 - \frac{2000}{17.5}}\right]$$

$$t = \frac{\ln\left[\frac{175 - \frac{2000}{17.5}}{200 - \frac{2000}{17.5}}\right]}{-0.005} \approx 69$$

Thus $w_{eq} = 114$ and $t = 69$.

3. $w(t) = \frac{C}{17.5} + \left(w_0 - \frac{C}{17.5}\right)e^{-0.005t}$. Since $\frac{C}{17.5} = w_{eq}$, we

   have $w(t) = w_{eq} + (w_0 - w_{eq})e^{-0.005t}$. Simplifying the

   equation $w(t + d) = w(t) - \frac{1}{2}[w(t) - w_{eq}]$ gives

   $w(t+d) = \frac{1}{2}[w(t) + w_{eq}]$. Thus

   $$w_{eq}+(w_0-w_{eq})e^{-0.005(t+d)} = \frac{1}{2}\left[w_{eq}+(w_0-w_{eq})e^{-0.005t}+w_{eq}\right],$$

   or

   $$w_{eq}+(w_0-w_{eq})e^{-0.005(t+d)} = w_{eq} + \frac{1}{2}(w_0-w_{eq})e^{-0.005t}.$$

   Solving for d gives

   $$e^{-0.005(t+d)} = \frac{1}{2}e^{-0.005t},$$

   $$e^{-0.005t}e^{-0.005d} = \frac{1}{2}e^{-0.005t},$$

   $$e^{-0.005d} = \frac{1}{2},$$

   $$-0.005d = \ln\frac{1}{2} = -\ln 2,$$

   $$d = \frac{\ln 2}{0.005}, \quad \text{as was to be shown.}$$

18

Continuous Random Variables

1. a. $P(1 < X < 2) = \int_1^2 \frac{1}{6}(x + 1) \, dx = \frac{(x + 1)^2}{12} \Big|_1^2 = \frac{9}{12} - \frac{4}{12} = \frac{5}{12}.$

   b. $P(X < 2.5) = \int_1^{2.5} \frac{1}{6}(x + 1) \, dx = \frac{(x + 1)^2}{12} \Big|_1^{2.5}$

   $$= \frac{49}{48} - \frac{4}{12} = \frac{11}{16} = 0.6875.$$

   c. $P\left(X \geq \frac{3}{2}\right) = \int_{3/2}^3 \frac{1}{6}(x + 1) \, dx = \frac{(x + 1)^2}{12} \Big|_{3/2}^3$

   $$= \frac{16}{12} - \frac{25}{48} = \frac{13}{16} = 0.8125.$$

   d. $\int_1^c \frac{1}{6}(x + 1) \, dx = \frac{1}{2}, \quad \frac{(x + 1)^2}{12} \Big|_1^c = \frac{1}{2}, \quad \frac{(c + 1)^2}{12} - \frac{1}{3} = \frac{1}{2},$

   $(c + 1)^2 - 4 = 6, \quad (c + 1)^2 = 10, \quad c + 1 = \pm\sqrt{10},$

   $c = -1 \pm \sqrt{10}.$ We choose $c = -1 + \sqrt{10}$ since $1 < c < 3.$

3. a. $f(x) = \begin{cases} 1/3, & \text{if } 1 \le x \le 4, \\ 0, & \text{otherwise.} \end{cases}$

b. $\frac{3-2}{4-1} = \frac{1}{3}$.

c. $P(0 < X < 1) = \int_0^1 0 \, dx = 0$.

d. $P(X \le 3.5) = P(1 \le X \le 3.5) = \frac{3.5 - 1}{4 - 1} = \frac{2.5}{3} = \frac{5}{6}$.

e. $P(X > 2) = P(2 < X \le 4) = \frac{4-2}{4-1} = \frac{2}{3}$.

f. 0.

g. $P(X < 5) = P(1 \le X \le 4) = \frac{4-1}{4-1} = 1$.

h. $\mu = \int_1^4 x\left(\frac{1}{3}\right) dx = \frac{x^2}{6}\Big|_1^4 = \frac{5}{2}$.

i. $\sigma^2 = \int_1^4 x^2\left(\frac{1}{3}\right) dx - \mu^2 = \frac{x^3}{9}\Big|_1^4 - \left(\frac{5}{2}\right)^2 = \left[\frac{64}{9} - \frac{1}{9}\right] - \frac{25}{4}$

$= 7 - \frac{25}{4} = \frac{3}{4}$. Thus $\sigma = \frac{\sqrt{3}}{2}$.

j. If $1 \le x \le 4$, $F(x) = \int_1^x \frac{1}{3} \, dt = \frac{t}{3}\Big|_1^x = \frac{x-1}{3}$.

Thus $F(x) = \begin{cases} 0, & \text{if } x < 1, \\ \frac{x-1}{3}, & \text{if } 1 \le x \le 4, \\ 1, & \text{if } x > 4. \end{cases}$

$P(X < 2) = F(2) = \frac{2-1}{3} = \frac{1}{3}$.

$P(1 < X < 3) = F(3) - F(1) = (2/3) - 0 = 2/3$.

5. a. $f(x) = \begin{cases} 1/(b - a), & \text{if } a \leq x \leq b, \\ 0, & \text{otherwise.} \end{cases}$

   b. $\mu = \int_a^b x\left(\dfrac{1}{b - a}\right) dx = \dfrac{x^2}{2(b - a)}\bigg|_a^b = \dfrac{b^2 - a^2}{2(b - a)} = \dfrac{a + b}{2}.$

   c. $\sigma^2 = \int_a^b x^2\left(\dfrac{1}{b - a}\right) dx - \mu^2 = \dfrac{x^3}{3(b - a)}\bigg|_a^b - \left(\dfrac{a + b}{2}\right)^2$

   $= \dfrac{b^3 - a^3}{3(b - a)} - \dfrac{(a + b)^2}{4} = \dfrac{b^2 + ab + a^2}{3} - \dfrac{a^2 + 2ab + b^2}{4}$

   $= \dfrac{b^2 - 2ab + a^2}{12} = \dfrac{(b - a)^2}{12}.$ Thus $\sigma = \dfrac{b - a}{\sqrt{12}}.$

7. $f(x) = \begin{cases} 2e^{-2x}, & \text{if } x \geq 0, \\ 0, & \text{if } x < 0. \end{cases}$

   a. $P(1 < X < 3) = \int_1^3 2e^{-2x}\, dx = -e^{-2x}\bigg|_1^3 = -e^{-6} + e^{-2}$

   $= -0.00248 + 0.13534 = 0.133.$

   b. $P(X < 2) = \int_0^2 2e^{-2x}\, dx = -e^{-2x}\bigg|_0^2 = -e^{-4} + 1$

   $= -0.01832 + 1 = 0.982.$

   c. $P(X > 2.5) = 1 - P(X \leq 2.5) = 1 - \int_0^{2.5} 2e^{-2x}\, dx$

   $= 1 - (-e^{-2x})\bigg|_0^{2.5} = 1 - (-e^{-5} + 1)$

   $= 1 - (-0.00674 + 1) = 0.007.$

   d. From the text, $\mu = \sigma = 1/k = 1/2.$

   $P(\mu-\sigma < X < \mu+\sigma) = P(0 < X < 1) = \int_0^1 2e^{-2x}\, dx = -e^{-2x}\bigg|_0^1$

   $= -e^{-2} + 1 = -0.13534 + 1 = 0.865.$

   e. $\int_0^{\infty} 2e^{-2x}\, dx = \lim_{r \to \infty} \int_0^r 2e^{-2x}\, dx = \lim_{r \to \infty} -e^{-2x}\bigg|_0^r$

   $= \lim_{r \to \infty} (-e^{-2r} + 1) = 0 + 1 = 1.$

9. a. $\int_0^4 kx\, dx = 1$, $\frac{kx^2}{2}\Big|_0^4 = 1$, $8k = 1$, $k = \frac{1}{8}$.

   b. $P(2 < X < 3) = \int_2^3 \frac{x}{8}\, dx = \frac{x^2}{16}\Big|_2^3 = \frac{9}{16} - \frac{4}{16} = \frac{5}{16}$.

   c. $P(X > 2.5) = \int_{2.5}^4 \frac{x}{8}\, dx = \frac{x^2}{16}\Big|_{2.5}^4 = 1 - \frac{25}{64} = \frac{39}{64} = 0.609$.

   d. $P(X > 0) = P(0 \leq X \leq 4) = 1$.

   e. $\mu = \int_0^4 x\left(\frac{x}{8}\right) dx = \frac{x^3}{24}\Big|_0^4 = \frac{64}{24} - 0 = \frac{8}{3}$.

   f. $\sigma^2 = \int_0^4 x^2\left(\frac{x}{8}\right) dx - \mu^2 = \frac{x^4}{32}\Big|_0^4 - \left(\frac{8}{3}\right)^2$

   $= 8 - \frac{64}{9} = \frac{8}{9}$. Thus $\sigma = \frac{\sqrt{8}}{3} = \frac{2\sqrt{2}}{3}$.

   g. $P(X < c) = \frac{1}{2}$, $\int_0^c \frac{x}{8}\, dx = \frac{1}{2}$, $\frac{x^2}{16}\Big|_0^c = \frac{1}{2}$, $\frac{c^2}{16} = \frac{1}{2}$,

   $c^2 = 8$, $c = \pm 2\sqrt{2}$. We choose $c = 2\sqrt{2}$ since $0 < c < 4$.

   h. $P(3 < X < 5) = P(3 < X < 4) = \int_3^4 \frac{x}{8}\, dx = \frac{x^2}{16}\Big|_3^4 = \frac{16}{16} - \frac{9}{16} = \frac{7}{16}$.

11. $P(X \leq 7) = \int_0^7 \frac{1}{10}\, dx = \frac{x}{10}\Big|_0^7 = \frac{7}{10}$.

   $E(X) = \int_0^{10} x\left(\frac{1}{10}\right) dx = \frac{x^2}{20}\Big|_0^{10} = 5$ min.

13. $P(X > 1) = 1 - P(X \leq 1) = 1 - \int_0^1 3e^{-3x}\, dx = 1 - (-e^{-3x})\Big|_0^1$

   $= 1 - (-e^{-3} + 1) = 1 - (-0.04979 + 1) = 0.050$.

EXERCISE 18.2

1. a. $P(0<Z<1.8) = A(1.8) = 0.4641$
   b. $P(0.45<Z<2.81) = A(2.81) - A(0.45)$
      $$= 0.4975 - 0.1736 = 0.3239$$
   c. $P(Z>-1.22) = 0.5 + A(1.22) = 0.5 + 0.3888 = 0.8888$
   d. $P(Z\leq2.93) = 0.5 + A(2.93) = 0.5 + 0.4983 = 0.9983$
   e. $P(-2.61<Z\leq1.4) = A(2.61) + A(1.4)$
      $$= 0.4955 + 0.4192 = 0.9147$$
   f. $P(Z>0.07) = 0.5 - A(0.07) = 0.5 - 0.0279 = 0.4721$

3. $P(Z<z_0) = 0.5517$, $0.5 + A(z_0) = 0.5517$, $A(z_0) = 0.0517$,
   $z_0 = 0.13$.   <u>Ans.</u>   0.13

5. $P(Z>z_0) = 0.8599$, $0.5 + A(-z_0) = 0.8599$, $A(-z_0) = 0.3599$,
   $-z_0 = 1.08$, $z_0 = -1.08$.   <u>Ans.</u>   -1.08

7. $P(-z_0<Z<z_0) = 0.2662$, $2A(z_0) = 0.2662$, $A(z_0) = 0.1331$,
   $z_0 = 0.34$.   <u>Ans.</u>   0.34

9. a. $P(X<22) = P\left(Z<\frac{22-16}{4}\right) = P(Z<1.5) = 0.5 + A(1.5)$
      $$= 0.5 + 0.4332 = 0.9332$$
   b. $P(X<10) = P\left(Z<\frac{10-16}{4}\right) = P(Z<-1.5) = 0.5 - A(1.5)$
      $$= 0.5 - 0.4332 = 0.0668$$
   c. $P(10.8<X<12.4) = P\left(\frac{10.8-16}{4}<Z<\frac{12.4-16}{4}\right)$
      $$= P(-1.3<Z<-0.9) = A(1.3) - A(0.9)$$
      $$= 0.4032 - 0.3159 = 0.0873$$

11. $P(X>-2) = P\left(Z>\frac{-2-(-3)}{2}\right) = P(Z>1/2) = 0.5 - A(1/2)$
    $$= 0.5 - 0.1915 = 0.3085$$

13. Since $\sigma^2 = 9$, $\sigma = 3$. Thus $P(19<X\leq28) = P\left(\frac{19-25}{3}<Z\leq\frac{28-25}{3}\right) =$
    $P(-2<Z\leq1) = A(2) + A(1) = 0.4772 + 0.3413 = 0.8185$.

15. $P(X>54) = P(Z>z_0) = 0.0401$, $0.5 - A(z_0) = 0.0401$,
    $A(z_0) = 0.4599$, $z_0 = 1.75$. Thus $(54-40)/\sigma = 1.75$,
    $14/\sigma = 1.75$, so $\sigma = 14/1.75 = 8$.   <u>Ans.</u>   8

17. Let X be score on test. $P(X>630) = P\left(Z>\frac{630-500}{100}\right) =$
$P(Z>1.3) = 0.5 - A(1.3) = 0.5 - 0.4032 = 0.0968.$
Ans. 9.68%

19. Let X be height of an adult. $P(X<72) = P\left(Z<\frac{72-68}{3}\right) =$
$P(Z<1.33) = 0.5 + A(1.33) = 0.5 + 0.4082 = 0.9082.$
Ans. 90.82%

21. Let X be I.Q. of a child in population.
(a) $P(X>125) = P\left(Z>\frac{125-100.4}{11.6}\right) = P(Z>2.12) = 0.5 - A(2.12) =$
$0.5 - 0.4830 = 0.0170.$ Thus 1.7% of the children have
I.Q.'s greater than 125.
(b) If x_0 is the value, then $P(X>x_0) = 0.90.$ Thus
$$P\left(Z > \frac{x_0-100.4}{11.6}\right) = 0.90 \text{ or } 0.5 + A\left(-\frac{x_0-100.4}{11.6}\right) = 0.90.$$
Hence $A\left(-\frac{x_0-100.4}{11.6}\right) = 0.4,$ so $-\frac{x_0-100.4}{11.6} = 1.28,$ or
$x_0 = 85.552 \approx 85.6.$
Ans. (a) 1.7%; (b) 85.6

EXERCISE 18.3

1. $n = 150,$ $p = 0.4,$ $q = 0.6,$ $\mu = np = 150(0.4) = 60,$
$\sigma = \sqrt{npq} = \sqrt{150(0.4)(0.6)} = \sqrt{36} = 6.$
$P(X\leq52) = P(X\leq52.5).$ $z = \frac{52.5-60}{6} = -1.25.$
$P(X\leq52) = P(X\leq52.5) \approx P(Z\leq-1.25) = 0.5 - A(1.25)$
$= 0.5 - 0.3944 = 0.1056;$
$P(X\geq74) = P(X\geq73.5) \approx P\left(Z\geq\frac{73.5-60}{6}\right) = P(Z\geq2.25)$
$= 0.5 - A(2.25) = 0.5 - 0.4878 = 0.0122.$
Ans. 0.1056; 0.0122

3. $n = 200$, $p = 0.6$, $q = 0.4$, $\mu = np = 120$,

   $\sigma = \sqrt{npq} = \sqrt{48} = 6.93$.

   $P(X=125) = P(124.5 \leq X \leq 125.5) \approx P\left(\frac{124.5-120}{6.93} \leq Z \leq \frac{125.5-120}{6.93}\right)$

   $= P(0.65 \leq Z \leq 0.79) = A(0.79) - A(0.65)$

   $= 0.2852 - 0.2422 = 0.0430;$

   $P(110 \leq X \leq 135) = P(109.5 \leq X \leq 135.5) \approx P\left(\frac{109.5-120}{6.93} \leq Z \leq \frac{135.5-120}{6.93}\right)$

   $= P(-1.52 \leq Z \leq 2.24) = A(1.52) + A(2.24)$

   $= 0.4357 + 0.4875 = 0.9232.$

   <u>Ans.</u> 0.0430; .0.9232

5. Let X = no. of times 5 occurs. Then X is binomial with

   $n = 300$, $p = 1/6$, $q = 5/6$, $\mu = np = 50$, $\sigma = \sqrt{npq} =$

   $\sqrt{41.666} = 6.45$.

   $P(45 \leq X \leq 60) = P(44.5 \leq X \leq 60.5) \approx P\left(\frac{44.5-50}{6.45} \leq Z \leq \frac{60.5-50}{6.45}\right)$

   $= P(-0.85 \leq Z \leq 1.63) = A(0.85) + A(1.63)$

   $= 0.3023 + 0.4484 = 0.7507.$ <u>Ans.</u> 0.7507

7. Let X = no. of trucks out of service. Then X can be
   considered binomial with $n = 60$, $p = 0.1$, $q = 0.9$,

   $\mu = np = 6$, $\sigma = \sqrt{npq} = \sqrt{5.4} = 2.32$.

   $P(X \geq 7) = P(X \geq 6.5) \approx P\left(Z \geq \frac{6.5-6}{2.32}\right) = P(Z \geq 0.22) = 0.5 - A(0.22)$

   $= 0.5 - 0.0871 = 0.4129.$ <u>Ans.</u> 0.4129

9. Let X = no. of correct answers. Then X is binomial and
   $p = 0.5$, $q = 0.5$.

   If $n = 20$, then $\mu = np = 20(0.5) = 10$, $\sigma = \sqrt{npq} =$

   $\sqrt{20(0.5)(0.5)} = \sqrt{5} = 2.24$, and

   $P(X \geq 12) = P(X \geq 11.5) \approx P\left(Z \geq \frac{11.5-10}{2.24}\right) = P(Z \geq 0.67)$

   $= 0.5 - A(0.67) = 0.5 - 0.2486 = 0.2514.$

   If $n = 100$, then $\mu = np = 100(0.5) = 50$, $\sigma = \sqrt{npq} =$

   $\sqrt{100(0.5)(0.5)} = \sqrt{25} = 5$, and

   $P(X \geq 60) = P(X \geq 59.5) \approx P\left(Z \geq \frac{59.5-50}{5}\right) = P(Z \geq 1.9)$

   $= 0.5 - A(1.9) = 0.5 - 0.4713 = 0.0287.$

   <u>Ans.</u> 0.2514; 0.0287

11. Let X = no. of deals consisting of three cards of one suit
    and two cards of another suit. Then X is binomial with
    n = 100, p = 0.1, q = 0.9, μ = np = 10,

    $\sigma = \sqrt{npq} = \sqrt{9} = 3$.

    $P(X \geq 16) = P(X \geq 15.5) \approx P\left(Z \geq \frac{15.5-10}{3}\right) = P(Z \geq 1.83)$

    $\qquad = 0.5 - A(1.83) = 0.5 - 0.4664 = 0.0336$.

    <u>Ans.</u> 0.0336

CHAPTER 18 - REVIEW PROBLEMS

1. a. $P(0 \leq X \leq 1) = 1$, $\int_0^1 \left(\frac{1}{3} + kx^2\right) dx = 1$, $\left(\frac{x}{3} + \frac{kx^3}{3}\right)\Big|_0^1 = 1$,

   $\frac{1}{3} + \frac{k}{3} = 1$, $\frac{k}{3} = \frac{2}{3}$, k = 2.   <u>Ans.</u> 2

   b. $P(\frac{1}{2} < X < \frac{3}{4}) = \int_{1/2}^{3/4} \left(\frac{1}{3} + 2x^2\right) dx = \left(\frac{x}{3} + \frac{2x^3}{3}\right)\Big|_{1/2}^{3/4} =$

   $\frac{1}{3}(x + 2x^3)\Big|_{1/2}^{3/4} = \frac{1}{3}\left[\left(\frac{3}{4} + \frac{27}{32}\right) - \left(\frac{1}{2} + \frac{1}{4}\right)\right] = \frac{9}{32}$.

   c. $P(X \geq \frac{1}{2}) = \int_{1/2}^1 \left(\frac{1}{3} + 2x^2\right) dx = \left(\frac{x}{3} + \frac{2x^3}{3}\right)\Big|_{1/2}^1 =$

   $\frac{1}{3}(x + 2x^3)\Big|_{1/2}^1 = \frac{1}{3}\left[(1 + 2) - \left(\frac{1}{2} + \frac{1}{4}\right)\right] = \frac{3}{4}$.

   d. If $0 \leq x \leq 1$, $F(x) = \int_0^x \left(\frac{1}{3} + 2t^2\right) dt =$

   $\left(\frac{t}{3} + \frac{2t^3}{3}\right)\Big|_0^x = \frac{x}{3} + \frac{2x^3}{3}$.

   <u>Ans.</u> $F(x) = \begin{cases} 0, & \text{if } x < 0 \\ \frac{x}{3} + \frac{2x^3}{3}, & \text{if } 0 \leq x \leq 1 \\ 1, & \text{if } x > 1 \end{cases}$

3. a. $\mu = \int_0^3 x\left(\frac{2}{9}x\right) dx = \frac{2x^3}{27}\Big|_0^3 = 2$.

   b. $\sigma^2 = \int_0^3 x^2\left(\frac{2}{9}x\right) dx - \mu^2 = \frac{x^4}{18}\Big|_0^3 - 2^2 = \frac{9}{2} - 4 = \frac{1}{2}$, so

   $\sigma = \frac{1}{\sqrt{2}} = 0.71$.

5. $P(X>22) = P\left(Z>\frac{22-20}{4}\right) = P(Z>0.5) = 0.5 - A(0.5)$
$$= 0.5 - 0.1915 = 0.3085$$

7. $P(12<X<18) = P\left(\frac{12-20}{4}<Z<\frac{18-20}{4}\right) = P(-2<Z<-0.5)$
$$= A(2) - A(0.5) = 0.4772 - 0.1915 = 0.2857$$

9. $P(X<16) = P\left(Z<\frac{16-20}{4}\right) = P(Z<-1) = 0.5 - A(1)$
$$= 0.5 - 0.3413 = 0.1587$$

1. $n = 100$, $p = 0.35$, $q = 0.65$, $\mu = np = 35$, $\sigma = \sqrt{npq} = \sqrt{22.75} = 4.77$.
$$P(25 \leq X \leq 47) = P(24.5 \leq X \leq 47.5) \approx P\left(\frac{24.5-35}{4.77} \leq Z \leq \frac{47.5-35}{4.77}\right)$$
$$= P(-2.20 \leq Z \leq 2.62) = A(2.20) + A(2.62)$$
$$= 0.4956 + 0.4861 = 0.9817 \quad \underline{\text{Ans.}} \quad 0.9817$$

3. Let X = height of an individual. X is normally distributed with $\mu = 68$ and $\sigma = 2$.
$$P(X>72) = P\left(Z>\frac{72-68}{2}\right) = P(Z>2) = 0.5 - A(2)$$
$$= 0.5 - 0.4772 = 0.0228. \quad \underline{\text{Ans.}} \quad 0.0228$$

19

Multivariable Calculus

1. $f(2,1) = 4(2) - (1)^2 + 3 = 8 - 1 + 3 = 10$

3. $g(0,-1,2) = e^0(2[-1] + 3[2]) = 1(-2 + 6) = 4$

5. $h(-3,3,5,4) = \dfrac{-3(3)}{5^2 - 4^2} = \dfrac{-9}{25 - 16} = \dfrac{-9}{9} = -1$

7. $g(4,8) = 2(4)(4^2 - 5) = 2(4)(11) = 88$

9. $F(2,0,-1) = 3$

11. $f(x,y) = 2x-5y+4.$ $f(x_0+h,y_0) = 2(x_0+h)-5y_0+4.$
    Ans. $2x_0+2h-5y_0+4$

13. $f(400, 400, 80) = 400(400)/80 = 2000$

15. A plane parallel to the x,z-plane has the form
    y = constant. Because $(0,-4,0)$ lies on the plane, the
    equation is $y = -4$.   Ans. $y = -4$

17. A plane parallel to the x,y-plane has the form
z = constant. Because (2,7,6) lies on the plane, the
equation is z = 6. <u>Ans.</u> z = 6

19. x+y+z = 1 can be put in the form
Ax+By+Cz+D = 0, so the graph is
a plane. The intercepts are
(1,0,0), (0,1,0), and (0,0,1).

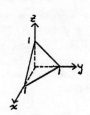

21. 3x+6y+2z = 12 can be put in the
form Ax+By+Cz+D = 0, so the
graph is a plane. The intercepts
are (4,0,0), (0,2,0), and (0,0,6).

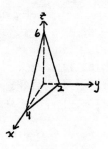

23. x+2y = 2 can be put in the form
Ax+By+Cz+D = 0, so the graph is a
plane. There are only two
intercepts: (2,0,0), and (0,1,0).
The x,y-trace is x+2y = 2, which
is a line. For any fixed value
of z, we obtain the line x+2y = 2.

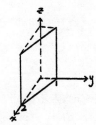

25. $z = 4-x^2$. The x,z-trace is
$z = 4-x^2$, which is a parabola.
For any fixed value of y, we
obtain the parabola $z = 4-x^2$.

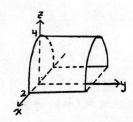

27. $x^2+y^2+z^2 = 1$. The x,y-trace is
$x^2+y^2 = 1$, which is a circle.
The x,z-trace is $x^2+z^2 = 1$, which
is a circle. The y,z-trace is
$y^2+z^2 = 1$, which is a circle.
The surface is a sphere with only
the top hemisphere shown in the
diagram.

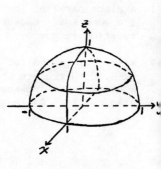

EXERCISE 19.2

1. $f_x(x,y) = 4(2x) + 0 + 0 = 8x$;
$f_y(x,y) = 0 + 3(2y) + 0 = 6y$

3. $f_x(x,y) = 0 + 0 = 0$; $f_y(x,y) = 2(1) + 0 = 2$

5. $g_x(x,y) = (3x^2)y^2 + 2(2x)y - 3(1)y + 0 = 3x^2y^2 + 4xy - 3y$;
$g_y(x,y) = x^3(2y) + 2x^2(1) - 3x(1) + 4(1)$
$$= 2x^3y + 2x^2 - 3x + 4$$

7. $g(p,q) = \sqrt{pq} = (pq)^{1/2}$.
$g_p(p,q) = \frac{1}{2}(pq)^{-1/2} \cdot q = \frac{q}{2\sqrt{pq}}$;
$g_q(p,q) = \frac{1}{2}(pq)^{-1/2} \cdot p = \frac{p}{2\sqrt{pq}}$

9. $h(s,t) = \frac{s^2+4}{t-3}$. $h_s(s,t) = \frac{1}{t-3}(2s)$.   Since $h(s,t) =$
$(s^2+4)(t-3)^{-1}$, then $h_t(s,t) = (s^2+4)[(-1)(t-3)^{-2}(1)]$.
<u>Ans.</u> $h_s(s,t) = 2s/(t-3)$; $h_t(s,t) = -(s^2+4)/(t-3)^2$

11. $u_{q_1}(q_1,q_2) = \frac{3}{4}\cdot\frac{1}{q_1} + 0 = \frac{3}{4q_1}$; $u_{q_2}(q_1,q_2) = 0 + \frac{1}{4}\cdot\frac{1}{q_2} = \frac{1}{4q_2}$

13. $h(x,y) = \frac{x^2+3xy+y^2}{\sqrt{x^2+y^2}}$.

$h_x(x,y) = \dfrac{(x^2+y^2)^{1/2}[2x+3y] - (x^2+3xy+y^2)\left[\frac{1}{2}(x^2+y^2)^{-1/2}(2x)\right]}{\left[(x^2+y^2)^{1/2}\right]^2}$

$= \dfrac{(x^2+y^2)^{-1/2}\left[(x^2+y^2)(2x+3y) - (x^2+3xy+y^2)x\right]}{x^2+y^2}$

$= \dfrac{2x^3+3x^2y+2xy^2+3y^3-x^3-3x^2y-xy^2}{(x^2+y^2)^{3/2}} = \dfrac{x^3+xy^2+3y^3}{(x^2+y^2)^{3/2}}$;

$h_y(x,y) = \dfrac{(x^2+y^2)^{1/2}[3x+2y] - (x^2+3xy+y^2)\left[\frac{1}{2}(x^2+y^2)^{-1/2}(2y)\right]}{\left[(x^2+y^2)^{1/2}\right]^2}$

$= \dfrac{(x^2+y^2)^{-1/2}\left[(x^2+y^2)(3x+2y) - (x^2+3xy+y^2)y\right]}{x^2+y^2}$

$= \dfrac{3x^3+2x^2y+3xy^2+2y^3-x^2y-3xy^2-y^3}{(x^2+y^2)^{3/2}} = \dfrac{3x^3+x^2y+y^3}{(x^2+y^2)^{3/2}}$.

15. $\frac{\partial z}{\partial x} = e^{5xy}(5y) = 5ye^{5xy}$; $\frac{\partial z}{\partial y} = e^{5xy}(5x) = 5xe^{5xy}$

17. $x = 5x\ \ln(x^2+y)$.

$\frac{\partial z}{\partial x} = 5\left\{x\left[\frac{1}{x^2+y}(2x)\right] + \ln(x^2+y)\ [1]\right\} = 5\left[\frac{2x^2}{x^2+y} + \ln(x^2+y)\right]$;

$\frac{\partial z}{\partial y} = 5x\left(\frac{1}{x^2+y}[1]\right) = \frac{5x}{x^2+y}$

19. $f(r,s) = (r+2s)^{1/2}(r^3-2rs+s^2)$.

$f_r(r,s) = (r+2s)^{1/2}[3r^2-2s] + (r^3-2rs+s^2)\left[\frac{1}{2}(r+2s)^{-1/2}(1)\right]$

$= \sqrt{r+2s}(3r^2-2s) + \frac{r^3-2rs+s^2}{2\sqrt{r+2s}}$;

$$f_s(r,s) = (r+2s)^{1/2}[-2r+2s] + (r^3-2rs+s^2)\left[\frac{1}{2}(r+2s)^{-1/2}(2)\right]$$

$$= 2(s-r)\sqrt{r+2s} + \frac{r^3-2rs+s^2}{\sqrt{r+2s}}$$

21. $f(r,s) = e^{3-r} \ln(7-s)$.

$f_r(r,s) = \ln(7-s)\left[e^{3-r}(-1)\right] = -e^{3-r} \ln(7-s)$;

$f_s(r,s) = e^{3-r}\left[\frac{1}{7-s}(-1)\right] = \frac{e^{3-r}}{s-7}$

23. $g_x(x,y,z) = 3y(2x) + 2y^2z(1) + 0 = 6xy + 2y^2z$;

$g_y(x,y,z) = 3x^2(1) + 2xz(2y) + 0 = 3x^2 + 4xyz$;

$g_z(x,y,z) = 0 + 2xy^2(1) + 3(3z^2) = 2xy^2 + 9z^2$

25. $g_r(r,s,t) = e^{s+t}[2r+0] = 2re^{s+t}$;

$g_s(r,s,t) = e^{s+t}[0+21s^2] + (r^2+7s^3)[e^{s+t}(1)]$

$\qquad\qquad = (7s^3+21s^2+r^2)e^{s+t}$;

$g_t(r,s,t) = (r^2+7s^3)[e^{s+t}(1)] = e^{s+t}(r^2+7s^3)$

27. $f(x,y) = x^3y+7x^2y^2$. $f_x(x,y) = 3x^2y+14xy^2$. $f_x(1,-2) = 50$.
<u>Ans.</u>   50

29. $g(x,y,z) = e^x\sqrt{y+2z}$. $g_z(x,y,z) = e^x\left[\frac{1}{2}(y+2z)^{-1/2}(2)\right] =$
$e^x/\sqrt{y+2z}$. $g_z(0,1,4) = 1/\sqrt{1+8} = 1/3$.   <u>Ans.</u>   1/3

31. $h(r,s,t,u) = (s^2+tu) \ln(2r+7st)$.

$h_s(r,s,t,u) = \frac{7t(s^2+tu)}{2r+7st} + 2s \ln(2r+7st)$. $h_s(1,0,0,1) = 0$.
<u>Ans.</u>   0

33. $f(r,s,t) = rst(r^2+s^3+t^4) = r^3st + rs^4t + rst^5$.

$f_s(r,s,t) = r^3(1)t + r(4s^3)t + r(1)t^5 = r^3t + 4rs^3t + rt^5$.

$f_s(1,-1,2) = 2 + (-8) + 32 = 26$.   <u>Ans.</u>   26

35. $z = xe^{x-y} - ye^{y-x}$. $\frac{\partial z}{\partial x} = [xe^{x-y} + e^{x-y}] - [ye^{y-x}(-1)]$.

$\frac{\partial z}{\partial y} = [xe^{x-y}(-1)] - [ye^{y-x} + e^{y-x}]$.

Thus $\frac{\partial z}{\partial x} + \frac{\partial z}{\partial y} = e^{x-y} - e^{y-x}$, as was to be shown.

37. $F(b,C,T,i) = \frac{bT}{C} + \frac{iC}{2}$. $\frac{\partial F}{\partial C} = \frac{\partial}{\partial C}\left[\frac{bT}{C}\right] + \frac{\partial}{\partial C}\left[\frac{iC}{2}\right] = -\frac{bT}{C^2} + \frac{i}{2}$.

39. $R = f(r,a,n) = \frac{r}{1 + a^{\frac{n-1}{2}}} = r\left[1 + a^{\frac{n-1}{2}}\right]^{-1}$.

$\frac{\partial R}{\partial n} = r(-1)\left[1 + a^{\frac{n-1}{2}}\right]^{-2} \cdot \frac{a}{2} = -\frac{ra}{2\left[1 + a^{\frac{n-1}{2}}\right]^2}$.

EXERCISE 19.3

1. $c = 4x + 0.3y^2 + 2y + 500$. $\partial c/\partial y = 0.6y + 2$. When $x = 20$ and $y = 30$, then $\partial c/\partial y = 0.6(30) + 2 = 20$. <u>Ans.</u> 20

3. $c = 0.03(x+y)^3 - 0.6(x+y)^2 + 4.5(x+y) + 7700$. $\partial c/\partial x = 0.09(x+y)^2 - 1.2(x+y) + 4.5$. When $x = 50$ and $y = 50$, then $\partial c/\partial x = 784.5$. <u>Ans.</u> 784.5

5. To avoid confusion of the letter "l" with the number "1", we shall use "L" to denote the letter "l".

$\partial P/\partial k = 1.582(0.764)L^{0.192}k^{-0.236} = 1.208648L^{0.192}k^{-0.236}$

$\partial P/\partial L = 1.582(0.192)L^{-0.808}k^{0.764} = 0.303744L^{-0.808}k^{0.764}$

7. $\partial q_A/\partial p_A = -50$, $\partial q_A/\partial p_B = 2$, $\partial q_B/\partial p_A = 4$, $\partial q_B/\partial p_B = -20$. Since $\partial q_A/\partial q_B > 0$ and $\partial q_B/\partial q_A > 0$ the products are competitive.

9. $q_A = 100p_A^{-1}p_B^{-1/2}$; $q_B = 500p_B^{-1}p_A^{-1/3}$.

$\partial q_A/\partial p_A = 100(-1)p_A^{-2}p_B^{-1/2} = -100/(p_A^2 p_B^{1/2})$,

$\partial q_A/\partial p_B = 100(-1/2)p_A^{-1}p_B^{-3/2} = -50/(p_A p_B^{3/2})$,

$\partial q_B/\partial p_A = 500(-1/3)p_B^{-1}p_A^{-4/3} = -500/(3p_B p_A^{4/3})$,

$\partial q_B/\partial p_B = 500(-1)p_B^{-2}p_A^{-1/3} = -500/(p_B^2 p_A^{1/3})$.

Since $\partial q_A/\partial p_B < 0$ and $\partial q_B/\partial p_A < 0$, the products are complementary.

11. $\partial P/\partial B = 0.01A^{0.27}B^{-0.99}C^{0.01}D^{0.23}E^{0.09}F^{0.27}$;

    $\partial P/\partial C = 0.01A^{0.27}B^{0.01}C^{-0.99}D^{0.23}E^{0.09}F^{0.27}$

13. $\partial z/\partial x = 1120$. If a staff manager with an MBA degree had an extra year of work experience before the degree, the manager would receive $1120 per year in extra compensation.

15. (a) $\frac{\partial R}{\partial w} = -1.015$; $\frac{\partial R}{\partial s} = -0.846$

    (b) One for which $w = w_0$ and $s = s_0$ since increasing w by while holding s fixed decreases the reading ease score

17. $\frac{\partial g}{\partial x} = \frac{1}{V_F} > 0$ for $V_F > 0$. Thus if x increases and V_F and V_S are fixed, then g increases.

19. (a) $\frac{\partial q_A}{\partial p_A} = 30\sqrt{p_B}\left(-\frac{2}{3}p_A^{-5/3}\right)$; $\frac{\partial q_A}{\partial p_B} = \frac{30}{p_A^{2/3}}\left(\frac{1}{2}p_B^{-1/2}\right)$.

    When $p_A = 8$ and $p_B = 64$, then

    $\frac{\partial q_A}{\partial p_A} = 30(8)\left(-\frac{2}{3}\cdot\frac{1}{32}\right) = -5$, $\frac{\partial q_A}{\partial p_B} = \frac{30}{4}\left(\frac{1}{2}\cdot\frac{1}{8}\right) = \frac{15}{32}$.

    (b) From (a), when $p_A = 8$ and $p_B = 64$, then $\frac{\partial q_A}{\partial p_B} = \frac{15}{32}$.

    Hence each $1 *reduction* in p_B *decreases* q_A by approximately $\frac{15}{32}$ unit. Thus a $4 reduction in p_B (from $64 to $60) decreases the demand for A by approximately $\frac{15}{32}(4) = \frac{15}{8}$ units.

21. (a) $\frac{\partial R}{\partial E_r}$ = 2.5945 - 0.1608E_r - 0.0277I_r.

If E_r = 18.8 and I_r = 10, then $\partial R/\partial E_r$ = -0.70554.

Since $\partial R/\partial E_r < 0$, such a candidate should not be so advised.

(b) $\partial R/\partial N$ = 0.8579 - 0.0122N. If $\partial R/\partial N < 0$, then N > 70.3.

<u>Ans.</u> (a) no; (b) 70%

23. q_A = 1000-50p_A+2p_B.

$$\eta_{p_A} = \left(\frac{p_A}{q_A}\right)\frac{\partial q_A}{\partial p_A} = \left(\frac{p_A}{q_A}\right)(-50). \qquad \eta_{p_B} = \left(\frac{p_B}{q_A}\right)\frac{\partial q_A}{\partial p_B} = \left(\frac{p_B}{q_A}\right)(2).$$

When p_A = 2 and p_B = 10, then q_A = 920. Thus

<u>Ans.</u> η_{p_A} = -5/46, η_{p_B} = 1/46

25. q_A = 100/(p_A\sqrt{p_B}). $\eta_{p_A} = \left(\frac{p_A}{q_A}\right)\frac{\partial q_A}{\partial p_A} = \left(\frac{p_A}{q_A}\right)\left(\frac{-100}{p_A^2\sqrt{p_B}}\right).$

$\eta_{p_B} = \left(\frac{p_B}{q_A}\right)\frac{\partial q_A}{\partial p_B} = \left(\frac{p_B}{q_A}\right)\left(\frac{-50}{p_A\sqrt{p_B^3}}\right).$ When p_A = 1 and p_B = 4,

then q_A = 50. <u>Ans.</u> η_{p_A} = -1, η_{p_B} = -1/2

EXERCISE 19.4

1. $2x + 0 + 2z\frac{\partial z}{\partial x} = 0.$ $2z\frac{\partial z}{\partial x} = -2x,$ $\frac{\partial z}{\partial x} = -\frac{2x}{2z} = -\frac{x}{z}.$

3. $6z^2\frac{\partial z}{\partial y} - 0 - 8y = 0.$ $\frac{\partial z}{\partial y} = \frac{8y}{6z^2} = \frac{4y}{3z^2}.$

5. $x^2 - 2y - z^2 + y\left(x^2z^2\right) = 20.$

$2x - 0 - 2z\frac{\partial z}{\partial x} + y\left[x^2 \cdot 2z\frac{\partial z}{\partial x} + z^2 \cdot 2x\right] = 0,$

$\left(2x^2yz - 2z\right)\frac{\partial z}{\partial x} = -2x - 2xyz^2,$ $\frac{\partial z}{\partial x} = \frac{-2x(1+yz^2)}{2z(x^2y-1)} = \frac{x(yz^2+1)}{z(1-x^2y)}.$

7. $0 + e^y + e^z\frac{\partial z}{\partial y} = 0.$ $\frac{\partial z}{\partial y} = -\frac{e^y}{e^z} = -e^{y-z}.$

9. $\frac{1}{z}\frac{\partial z}{\partial x} + \frac{\partial z}{\partial x} - y = 0.$ $\left(\frac{1}{z} + 1\right)\frac{\partial z}{\partial x} = y,$ $\left(\frac{1+z}{z}\right)\frac{\partial z}{\partial x} = y,$
$\frac{\partial z}{\partial x} = \frac{yz}{1+z}.$

11. $\left(2z\frac{\partial z}{\partial y} + 6x\right)\sqrt{x^3+5} = 0.$ $2z\frac{\partial z}{\partial y} + 6x = 0,$ $\frac{\partial z}{\partial y} = \frac{-6x}{2z} = -\frac{3x}{z}.$

13. $\left[x\cdot 2z\frac{\partial z}{\partial x} + z^2\cdot 1\right] + y\frac{\partial z}{\partial x} - 0 = 0.$ $(2xz+y)\frac{\partial z}{\partial x} = -z^2,$
$\frac{\partial z}{\partial x} = -\frac{z^2}{2xz+y}.$ If $x = 2, y = -2, z = 3,$ then $\frac{\partial z}{\partial x} = -\frac{9}{12-2} = -\frac{9}{10}.$ **Ans.** $-\frac{9}{10}$

15. $e^{yz}\cdot y\frac{\partial z}{\partial x} = -y\left[x\frac{\partial z}{\partial x} + z\cdot 1\right].$ $(ye^{yz} + xy)\frac{\partial z}{\partial x} = -yz,$
$\frac{\partial z}{\partial x} = -\frac{yz}{y(e^{yz}+x)},$ $\frac{\partial z}{\partial x} = -\frac{z}{e^{yz}+x}.$ If $x = -e^2/2, y = 1,$
$z = 2,$ then $\frac{\partial z}{\partial x} = -\frac{2}{e^2+(-e^2/2)} = -\frac{2}{e^2/2} = -\frac{4}{e^2}.$
Ans. $-4/e^2$

17. $\frac{1}{z}\frac{\partial z}{\partial x} = 1 + 0.$ $\frac{\partial z}{\partial x} = z.$ If $x = 5, y = -5, z = 1,$ then
$\frac{\partial z}{\partial x} = 1.$ **Ans.** 1

19. $\frac{(rs)\left[2t\frac{\partial t}{\partial r}\right] - (s^2+t^2)[s]}{(rs)^2} = 0.$ $2rst\frac{\partial t}{\partial r} - s(s^2+t^2) = 0,$
$2rst\frac{\partial t}{\partial r} = s(s^2+t^2),$ $\frac{\partial t}{\partial r} = \frac{s(s^2+t^2)}{2rst} = \frac{s^2+t^2}{2rt}.$ If $r = 1,$
$s = 2, t = 4,$ then $\frac{\partial t}{\partial r} = \frac{4+16}{2\cdot 1\cdot 4} = \frac{20}{8} = \frac{5}{2}.$ **Ans.** $\frac{5}{2}$

21. $c + \sqrt{c} = 12 + q_A\sqrt{9 + q_B^2}.$
   (a) If $q_A = 6$ and $q_B = 4,$ then $c + \sqrt{c} = 12 + 6(5) = 42,$
   $\sqrt{c} = 42 - c,$ $c = (42 - c)^2 = 42^2 - 84c + c^2,$

$$c^2 - 85c + 1764 = 0,$$

$$c = \frac{85 \pm \sqrt{(-85)^2 - 4(1)(1764)}}{2} = \frac{85 \pm \sqrt{169}}{2} = \frac{85 \pm 13}{2}.$$

Thus c = 49 or c = 36. However, c = 49 is extraneous but c = 36 is not. <u>Ans.</u> 36

(b) Differentiating with respect to q_A:

$$\frac{\partial c}{\partial q_A} + \frac{1}{2\sqrt{c}} \cdot \frac{\partial c}{\partial q_A} = \sqrt{9 + q_B^2}. \qquad \left(1 + \frac{1}{2\sqrt{c}}\right)\frac{\partial c}{\partial q_A} = \sqrt{9 + q_B^2}.$$

When $q_A = 6$ and $q_B = 4$, then c = 36 and

$$\left(1 + \frac{1}{12}\right)\frac{\partial c}{\partial q_A} = 5, \quad \frac{13}{12} \cdot \frac{\partial c}{\partial q_A} = 5, \quad \text{or} \quad \frac{\partial c}{\partial q_A} = \frac{60}{13}.$$

Differentiating with respect to q_B:

$$\frac{\partial c}{\partial q_B} + \frac{1}{2\sqrt{c}} \cdot \frac{\partial c}{\partial q_B} = q_A \cdot \frac{q_B}{\sqrt{9 + q_B^2}}. \qquad \left(1 + \frac{1}{2\sqrt{c}}\right)\frac{\partial c}{\partial q_B} = \frac{q_A q_B}{\sqrt{9 + q_B^2}}.$$

When $q_A = 6$ amd $q_B = 4$, then c = 36 and

$$\left(1 + \frac{1}{12}\right)\frac{\partial c}{\partial q_B} = \frac{24}{5}, \quad \frac{13}{12} \cdot \frac{\partial c}{\partial q_B} = \frac{24}{5}, \quad \text{or} \quad \frac{\partial c}{\partial q_B} = \frac{288}{65}.$$

<u>Ans.</u> $\dfrac{\partial c}{\partial q_A} = \dfrac{60}{13}, \dfrac{\partial c}{\partial q_B} = \dfrac{288}{65}$

EXERCISE 19.5

1. $f_x(x,y) = 4(2x)y = 8xy;$ $f_{xy}(x,y) = 8x$

3. $f_y(x,y) = 3;$ $f_{yy}(x,y) = 0;$ $f_{yyx}(x,y) = 0$

5. $f_y(x,y) = 4[e^{2xy}(2x)] = 8xe^{2xy}$

$$f_{yx}(x,y) = 8[x(e^{2xy} \cdot 2y) + e^{2xy}(1)] = 8e^{2xy}(2xy+1)$$

$$f_{yxy}(x,y) = 8[e^{2xy}(2x) + (2xy+1)(e^{2xy} \cdot 2x)]$$

$$= 8e^{2xy}(2x)[1 + (2xy+1)] = 8e^{2xy}(2x)[2 + 2xy]$$

$$= 32x(1 + xy)e^{2xy}$$

7. $f(x,y) = (x+y)^2(xy) = (x^2+2xy+y^2)(xy)$
$= x^3y+2x^2y^2+xy^3$.

$f_x(x,y) = 3x^2y+4xy^2+y^3$. $f_y(x,y) = x^3+4x^2y+3xy^2$.

$f_{xx}(x,y) = 6xy+4y^2$. $f_{yy}(x,y) = 4x^2+6xy$.

9. $\frac{\partial z}{\partial x} = \frac{1}{2}(x^2+y^2)^{-1/2}[2x] = x(x^2+y^2)^{-1/2}$

$\frac{\partial^2 z}{\partial x^2} = x\left[-\frac{1}{2}(x^2+y^2)^{-3/2}[2x]\right] + (x^2+y^2)^{-1/2}[1]$

$= (x^2+y^2)^{-3/2}\left[-x^2 + (x^2+y^2)\right] = y^2(x^2+y^2)^{-3/2}$

11. $f_y(x,y,z) = 0$. $f_{yx}(x,y,z) = 0$. $f_{yxx}(x,y,z) = 0$.
Thus $f_{yxx}(4,3,-2) = 0$. <u>Ans.</u> 0 .

13. $f_k(L,k) = 30L^3k^5-7Lk^6$. $f_{kk}(L,k) = 150L^3k^4-42Lk^5$.

$f_{kkL}(L,k) = 450L^2k^4-42k^5$. Thus $f_{kkL}(2,1) =$
$450(4)(1)-(42)(1) = 1758$. <u>Ans.</u> 1758

15. $f_x(x,y) = y^2e^x + \frac{1}{x}$. $f_{xy}(x,y) = 2ye^x$.

$f_{xyy}(x,y) = 2e^x$. Thus $f_{xyy}(1,1) = 2e$.   <u>Ans.</u> 2e

17. $\frac{\partial c}{\partial q_B} = \frac{1}{3}(3q_A^2+q_B^3+4)^{-2/3}(3q_B^2) = q_B^2(3q_A^2+q_B^3+4)^{-2/3}$.

$\frac{\partial^2 c}{\partial q_A \partial q_B} = -\frac{2}{3}q_B^2(3q_A^2+q_B^3+4)^{-5/3}(6q_A) = -4q_Aq_B^2(3q_A^2+q_B^3+4)^{-5/3}$.

When $p_A = 25$ and $p_B = 4$, then $q_A = 10 - 25 + 16 = 1$ and

$q_B = 20 + 25 - 44 = 1$, and $\frac{\partial^2 c}{\partial q_A \partial q_B} = -4(8)^{-5/3} = -\frac{4}{32} =$

$-\frac{1}{8}$.   <u>Ans.</u> $-\frac{1}{8}$

19. $f_x(x,y) = 24x^2+4xy^2$. $f_y(x,y) = 4x^2y+20y^3$.
$f_{xy}(x,y) = 8xy$. $f_{yx}(x,y) = 8xy$.
Thus $f_{xy}(x,y) = f_{yx}(x,y)$.

21. $\frac{\partial z}{\partial x} = \frac{2x}{x^2+y^2}$. $\frac{\partial^2 z}{\partial x^2} = \frac{(x^2+y^2)(2) - (2x)(2x)}{(x^2+y^2)^2} = \frac{2(y^2-x^2)}{(x^2+y^2)^2}$.

$\frac{\partial z}{\partial y} = \frac{2y}{x^2+y^2}$. $\frac{\partial^2 z}{\partial y^2} = \frac{(x^2+y^2)(2) - (2y)(2y)}{(x^2+y^2)^2} = \frac{2(x^2-y^2)}{(x^2+y^2)^2}$.

$\frac{\partial^2 z}{\partial x^2} + \frac{\partial^2 z}{\partial y^2} = \frac{2(y^2-x^2)}{(x^2+y^2)^2} + \frac{2(x^2-y^2)}{(x^2+y^2)^2} = 0$.

23. $2z\frac{\partial z}{\partial y} + 2y = 0$. $\frac{\partial z}{\partial y} = -\frac{2y}{2z} = -\frac{y}{z}$. $\frac{\partial^2 z}{\partial y^2} = -\frac{z(1) - y \cdot \frac{\partial z}{\partial y}}{z^2} =$

$-\frac{z - y\left(-\frac{y}{z}\right)}{z^2} = -\frac{z^2+y^2}{z^3}$. From original equation,

$z^2+y^2 = 3x^2$. Thus $\frac{\partial^2 z}{\partial y^2} = -\frac{3x^2}{z^3}$. <u>Ans.</u> $-\frac{z^2+y^2}{z^3} = -\frac{3x^2}{z^3}$

EXERCISE 19.6

1. $z = 5x+3y$, $x = 2r+3s$, $y = r-2s$.

$\frac{\partial z}{\partial r} = \frac{\partial z}{\partial x}\frac{\partial x}{\partial r} + \frac{\partial z}{\partial y}\frac{\partial y}{\partial r} = (5)(2) + (3)(1) = 13$

$\frac{\partial z}{\partial s} = \frac{\partial z}{\partial x}\frac{\partial x}{\partial s} + \frac{\partial z}{\partial y}\frac{\partial y}{\partial s} = (5)(3) + (3)(-2) = 9$

3. $z = e^{x+y}$, $x = t^2+3$, $y = \sqrt{t^3}$.

$\frac{dz}{dt} = \frac{\partial z}{\partial x}\frac{dx}{dt} + \frac{\partial z}{\partial y}\frac{dy}{dt} = e^{x+y}(2t) + e^{x+y}\left(\frac{3}{2}t^{1/2}\right)$

$= \left(2t + \frac{3}{2}\sqrt{t}\right)e^{x+y}$

5. $w = x^2z^2+xyz+yz^2$, $x = 5t$, $y = 2t+3$, $z = 6-t$.

$\frac{dw}{dt} = \frac{\partial w}{\partial x}\frac{dx}{dt} + \frac{\partial w}{\partial y}\frac{dy}{dt} + \frac{\partial w}{\partial t}\frac{dz}{dt}$

$= (2xz^2+yz)(5) + (xz+z^2)(2) + (2x^2z+xy+2yz)(-1)$

$= 5(2xz^2+yz) + 2(xz+z^2) - (2x^2z+xy+2yz)$

7. $z = (x^2+xy^2)^3$, $x = r+s+t$, $y = 2r-3s+t$.

$\frac{\partial z}{\partial t} = \frac{\partial z}{\partial x}\frac{\partial x}{\partial t} + \frac{\partial z}{\partial y}\frac{\partial y}{\partial t}$

$= 3(x^2+xy^2)^2(2x+y^2)[1] + 3(x^2+xy^2)^2(2xy)[1]$

$= 3(x^2+xy^2)^2(2x+y^2+2xy)$

9. $w = x^2 + xyz + y^3z^2$, $x = r-s^2$, $y = rs$, $z = 2r-5s$.

$$\frac{\partial w}{\partial s} = \frac{\partial w}{\partial x}\frac{\partial x}{\partial s} + \frac{\partial w}{\partial y}\frac{\partial y}{\partial s} + \frac{\partial w}{\partial z}\frac{\partial z}{\partial s}$$

$$= (2x+yz)(-2s) + (xz+3y^2z^2)(r) + (xy+2y^3z)(-5)$$

$$= -2s(2x+yz) + r(xz+3y^2z^2) - 5(xy+2y^3z)$$

11. $y = x^2-7x+5$, $x = 15rs+2s^2t^2$.

$$\frac{\partial y}{\partial r} = \frac{dy}{dx}\frac{\partial x}{\partial r} = (2x-7)(15s) = 15s(2x-7)$$

13. $z = (4x+3y)^3$, $x = r^2s$, $y = r-2s$; $r = 0$, $s = 1$.

$$\frac{\partial z}{\partial r} = \frac{\partial z}{\partial x}\frac{\partial x}{\partial r} + \frac{\partial z}{\partial y}\frac{\partial y}{\partial r} = 12(4x+3y)^2(2rs) + 9(4x+3y)^2(1) =$$

$3(4x+3y)^2(8rs+3)$. When $r = 0$, $s = 1$, then $x = 0$, $y = -2$, and $\partial z/\partial r = 324$.   <u>Ans.</u> 324

15. $w = e^{3x-y}(x^2+4z^3)$, $x = rs$, $y = 2s-r$, $z = r+s$; $r = 1$, $s = -1$.

$$\frac{\partial w}{\partial s} = \frac{\partial w}{\partial x}\frac{\partial x}{\partial s} + \frac{\partial w}{\partial y}\frac{\partial y}{\partial s} + \frac{\partial w}{\partial z}\frac{\partial z}{\partial s}$$

$$= \left[e^{3x-y}(2x) + (x^2+4z^3)e^{3x-y}(3)\right](r) +$$

$$\left[e^{3x-y}(-1)(x^2+4z^3)\right](2) + \left[e^{3x-y}(12z^2)\right](1)$$

$$= e^{3x-y}[(3x^2+2x+12z^3)r - 2(x^2+4z^3) + 12z^2].$$

When $r = 1$ and $s = -1$, then $x = -1$, $y = -3$, $z = 0$, and $\partial w/\partial s = -1$.   <u>Ans.</u> -1

17. $\dfrac{\partial c}{\partial p_A} = \dfrac{\partial c}{\partial q_A}\dfrac{\partial q_A}{\partial p_A} + \dfrac{\partial c}{\partial q_B}\dfrac{\partial q_B}{\partial p_A}$

$$= \left[\tfrac{1}{3}(3q_A^2+q_B^3+4)^{-2/3}(6q_A)\right](-1) + \left[\tfrac{1}{3}(3q_A^2+q_B^3+4)^{-2/3}(3q_B^2)\right]$$

$$= (3q_A^2+q_B^3+4)^{-2/3}(-2q_A+q_B^2).$$

$$\frac{\partial c}{\partial p_B} = \frac{\partial c}{\partial q_A}\frac{\partial q_A}{\partial p_B} + \frac{\partial c}{\partial q_B}\frac{\partial q_B}{\partial p_B}$$

$$= \left[\tfrac{1}{3}(3q_A^2+q_B^3+4)^{-2/3}(6q_A)\right](2p_B) +$$

$$\left[\tfrac{1}{3}(3q_A^2+q_B^3+4)^{-2/3}(3q_B^2)\right](-11)$$

$$= (3q_A^2+q_B^3+4)^{-2/3}(4q_Ap_B-11q_B^2).$$

When $p_A = 25$ and $p_B = 4$, then $q_A = 10 - 25 + 16 = 1$,

$q_B = 20 + 25 - 44 = 1$, and

$\frac{\partial c}{\partial p_A} = (8)^{-2/3}(-1) = -\frac{1}{4}$ and $\frac{\partial c}{\partial p_B} = (8)^{-2/3}(5) = \frac{5}{4}$.

<u>Ans.</u> $\partial c/\partial p_A = -1/4$, $\partial c/\partial p_B = 5/4$

19. (a) $\frac{\partial w}{\partial s} = \frac{\partial w}{\partial x}\frac{\partial x}{\partial s} + \frac{\partial w}{\partial y}\frac{\partial y}{\partial s}$

(b) $\frac{\partial w}{\partial s} = \left[3x^2 \cdot \frac{1}{x-2y}(1) + \ln(x-2y)[6x]\right](\sqrt{t-2}) +$

$$\left[3x^2 \cdot \frac{1}{x-2y}(-2)\right][-3e^{1-s}(-1)]$$

$$= \left[\frac{3x^2}{x-2y} + 6x \ln(x-2y)\right]\sqrt{t-2} - \frac{18x^2 e^{1-s}}{x-2y}.$$

When $s = 1$ and $t = 3$, then $x = 1$, $y = 0$, and

$\frac{\partial w}{\partial s} = [3 + 0](1) - \frac{18}{1} = -15$.   <u>Ans.</u> -15

<u>EXERCISE 19.7</u>

1. $f(x,y) = x^2 + y^2 - 5x + 4y + xy$.

$\begin{cases} f_x(x,y) = 2x + y - 5 = 0 \\ f_y(x,y) = x + 2y + 4 = 0. \end{cases}$

Solving the system gives the critical point $(14/3, -13/3)$.

3. $f(x,y) = 2x^3 + y^3 - 3x^2 + 1.5y^2 - 12x - 90y$.

$\begin{cases} f_x(x,y) = 6x^2 - 6x - 12 = 0 \\ f_y(x,y) = 3y^2 + 3y - 90 = 0. \end{cases}$

Both equations are easily solved by factoring. The critical points are $(2, 5)$, $(2, -6)$, $(-1, 5)$, $(-1, -6)$.

5. $f(x,y,z) = 2x^2 + xy + y^2 + 100 - z(x + y - 200)$.

$\begin{cases} f_x(x,y,z) = 4x + y - z = 0 \\ f_y(x,y,z) = x + 2y - z = 0 \\ f_z(x,y,z) = -x - y + 200 = 0. \end{cases}$

Solving gives the critical point $(50, 150, 350)$.

7. $f(x,y) = x^2 + 3y^2 + 4x - 9y + 3$.

$$\begin{cases} f_x(x,y) = 2x + 4 = 0 \\ f_y(x,y) = 6y - 9 = 0. \end{cases}$$ Critical point $(-2, 3/2)$.

Second-Derivative Test

$f_{xx}(x,y) = 2$, $f_{yy}(x,y) = 6$, $f_{xy}(x,y) = 0$. At $(-2, 3/2)$,

$D = (2)(6) - 0^2 = 12 > 0$ and $f_{xx}(x,y) = 2 > 0$. Thus at

$(-2, 3/2)$ there is a relative minimum.

9. $f(x,y) = y - y^2 - 3x - 6x^2$.

$$\begin{cases} f_x(x,y) = -3 - 12x = 0 \\ f_y(x,y) = 1 - 2y = 0. \end{cases}$$ Critical point $(-1/4, 1/2)$.

Second-Derivative Test

$f_{xx}(x,y) = -12$, $f_{yy}(x,y) = -2$, $f_{xy}(x,y) = 0$.

At $(-1/4, 1/2)$, $D = (-12)(-2) - 0^2 = 24 > 0$ and $f_{xx}(x,y) =$

$-12 < 0$. Thus at $(-1/4, 1/2)$ there is a relative maximum.

11. $f(x,y) = x^3 - 3xy + y^2 + y - 5$.

$$\begin{cases} f_x(x,y) = 3x^2 - 3y = 0 \\ f_y(x,y) = -3x + 2y + 1 = 0. \end{cases}$$ Crit. pt.: $(1,1)$, $(1/2,1/4)$.

Second-Derivative Test

$f_{xx}(x,y) = 6x$, $f_{yy}(x,y) = 2$, $f_{xy}(x,y) = -3$. At $(1,1)$,

$D = (6)(2) - (-3)^2 = 3 > 0$ and $f_{xx}(x,y) = 6 > 0$; thus

there is a relative minimum at $(1, 1)$. At $(1/2, 1/4)$,

$D = (3)(2)-(-3)^2 = -3 < 0$; thus there is no relative

extremum at $(1/2, 1/4)$.

13. $f(x,y) = (1/3)(x^3 + 8y^3) - 2(x^2 + y^2) + 1$.

$$\begin{cases} f_x(x,y) = x^2 - 4x = 0 \\ f_y(x,y) = 8y^2 - 4y = 0. \end{cases}$$

Critical points $(0, 0)$, $(4, 1/2)$, $(0, 1/2)$, $(4, 0)$.

Second-Derivative Test

$f_{xx}(x,y) = 2x - 4$, $f_{yy}(x,y) = 16y - 4$, $f_{xy}(x,y) = 0$.

At $(0, 0)$, $D = (-4)(-4)-0^2 = 16 > 0$ and $f_{xx}(x,y) = -4 < 0$;

thus a relative maximum.

At $(4, 1/2)$, $D = (4)(4)-0^2 = 16 > 0$ and $f_{xx}(x,y) = 4 > 0$;
thus a relative minimum.

At $(0,1/2)$, $D = (-4)(4)-0^2 = -16 < 0$; thus neither.

At $(4,0)$, $D = (4)(-4)-0^2 = -16 < 0$; thus neither.

15. To avoid confusion of the letter "l" with the number "1",
we shall use "L" for the letter "l".

$f(L,k) = 2Lk - L^2 + 264k - 10L - 2k^2$.

$$\begin{cases} f_L(L,k) = 2k-2L-10 = 0 \\ f_k(L,k) = -4k+2L+264 = 0. \end{cases} \quad \text{Critical point } (122, 127).$$

Second-Derivative Test

$f_{LL}(L,k) = -2$, $f_{kk}(L,k) = -4$, $f_{Lk}(L,k) = 2$. At $(122, 127)$,

$D = (-2)(-4)-2^2 = 4 > 0$ and $f_{LL}(L,k) = -2 < 0$; thus there
is a relative maximum at $(122, 127)$.

17. $f(p,q) = pq - (1/p) - (1/q)$.

$$\begin{cases} f_p(p,q) = q + (1/p^2) = 0 \\ f_q(p,q) = p + (1/q^2) = 0. \end{cases} \quad \text{Critical point } (-1, -1).$$

Second-Derivative Test

$f_{pp}(p,q) = -2/p^3$, $f_{qq}(p,q) = -2/q^3$, $f_{pq}(p,q) = 1$. At

$(-1, -1)$, $D = (2)(2)-1^2 = 3 > 0$ and $f_{pp}(p,q) = 2 > 0$; thus
there is a relative minimum at $(-1, -1)$.

19. $f(x, y) = (y^2 - 4)(e^x - 1)$.

$$\begin{cases} f_x(x,y) = e^x(y^2 - 4) = 0 \quad (1) \\ f_y(x,y) = 2y(e^x - 1) = 0. \quad (2) \end{cases}$$

From (1), $y = \pm 2$. From (2), $x = 0$, $y = 0$. The only
critical points are $(0, -2)$ and $(0, 2)$. [$y = 0$ does not
give rise to a common solution of (1) and (2).]

Second-Derivative Test

$f_{xx}(x,y) = e^x(y^2 - 4)$, $f_{yy}(x,y) = 2(e^x - 1)$, $f_{xy}(x,y) =$

$2ye^x$. At $(0, -2)$, $D = (0)(0)-(-4)^2 = -16 < 0$; thus

neither. At $(0, 2)$, $D = (0)(0)-(4)^2 = -16 < 0$; thus

neither. Summary: neither point corresponds to a relative extremum.

21. To avoid confusion of the letter "l" with the number "1", we shall use "L" for the letter "l".

    $P = f(L,k) = 1.08L^2 - 0.03L^3 + 1.68k^2 - 0.08k^3.$

    $\begin{cases} P_L = 2.16L - 0.09L^2 = 0 \\ P_k = 3.36k - 0.24k^2 = 0. \end{cases}$

    Critical points $(0, 0)$, $(0, 14)$, $(24, 0)$, $(24, 14)$.
    Second-Derivative Test
    $P_{LL} = 2.16 - 0.18L$, $P_{kk} = 3.36 - 0.48k$, $P_{LK} = 0$. At $(0,0)$,
    $D = (2.16)(3.36) - 0^2 > 0$ and $P_{LL} = 2.16 > 0$; thus rel. min.
    At $(0,14)$, $D = (2.16)(-3.36) - 0^2 < 0$; thus no extremum. At
    $(24,0)$, $D = (-2.16)(3.36) - 0^2 < 0$; thus no extremum. At
    $(24,14)$, $D = (-2.16)(-3.36) - 0^2 > 0$ and $P_{LL} = -2.6 < 0$, so
    rel. max. Thus $L = 24$, $k = 14$ gives a relative maximum.

23. Profit per lb for A = $p_A - 60$.
    Profit per lb for B = $p_B - 70$.
    Total Profit = $P = (p_A - 60)q_A + (p_B - 70)q_B$.

    $P = (p_A - 60)[5(p_B - p_A)] + (p_B - 70)[500 + 5(p_A - 2p_B)]$.

    $\begin{cases} \partial P / \partial p_A = -10(p_A - p_B + 5) = 0 \\ \partial P / \partial p_B = 10(p_A - 2p_B + 90) = 0. \end{cases}$ Crit. pt.: $p_A = 80$, $p_B = 85$.

    Second-Derivative Test

    $\dfrac{\partial^2 P}{\partial p_A^2} = -10$, $\dfrac{\partial^2 P}{\partial p_B^2} = -20$, $\dfrac{\partial^2 P}{\partial p_B \partial p_A} = 10$. When $p_A = 80$ and

    $p_B = 85$, then $D = (-10)(-20) - (10)^2 = 100 > 0$ and $\dfrac{\partial^2 P}{\partial p_A^2} =$

    $-10 < 0$; thus rel. max. at $p_A = 80$, $p_B = 85$.

25. $p_A = 100 - q_A$, $p_B = 84 - q_B$, $c = 600 + 4(q_A + q_B)$.
    Revenue from market A = $r_A = p_A q_A = (100 - q_A)q_A$.
    Revenue from market B = $r_B = p_B q_B = (84 - q_B)q_B$.

Total Profit = Total Revenue - Total Cost
$$P = (100 - q_A)q_A + (84 - q_B)q_B - [600 + 4(q_A + q_B)].$$

$$\begin{cases} \partial P/\partial q_A = 96 - 2q_A = 0 \\ \partial P/\partial q_B = 80 - 2q_B = 0. \end{cases}$$ Crit. pt.: $q_A = 48$, $q_B = 40$

$\dfrac{\partial^2 P}{\partial q_A^2} = -2$, $\dfrac{\partial^2 P}{\partial q_B^2} = -2$, $\dfrac{\partial^2 P}{\partial q_B \partial q_A} = 0$. At $q_A = 48$ and $q_B = 40$,

then $D = (-2)(-2) - 0^2 = 4 > 0$ and $\dfrac{\partial^2 P}{\partial q_A^2} = -2 < 0$; thus rel.

max. at $q_A = 48$, $q_B = 40$. When $q_A = 48$ and $q_B = 40$, then
selling prices are $p_A = 52$, $p_B = 44$, and profit = 3304.

27. $c = 1.5q_A^2 + 4.5q_B^2$, $p_A = 36 - q_A^2$, $p_B = 30 - q_B^2$.

Total Profit = Total Revenue - Total Cost
$$P = (p_A q_A + p_B q_B) - c$$

$$P = 36q_A - q_A^3 + 30q_B - q_B^3 - (1.5q_A^2 + 4.5q_B^2)$$

$$\begin{cases} \partial P/\partial q_A = 36 - 3q_A - 3q_A^2 = 3(4 + q_A)(3 - q_A) \\ \partial P/\partial q_B = 30 - 9q_B - 3q_B^2 = 3(5 + q_B)(2 - q_B). \end{cases}$$

Since we want $q_A \geqq 0$ and $q_B \geqq 0$, the crit. pt. occurs when
$q_A = 3$ and $q_B = 2$.

$\dfrac{\partial^2 P}{\partial q_A^2} = -3 - 6q_A$, $\dfrac{\partial^2 P}{\partial q_B^2} = -9 - 6q_B$, $\dfrac{\partial^2 P}{\partial q_B \partial q_A} = 0$. When $q_A = 3$ and

$q_B = 2$, then $D = (-21)(-21) - 0^2 > 0$ and $\dfrac{\partial^2 P}{\partial q_A^2} = -21 < 0$; thus

rel. max. at $q_A = 3$, $q_B = 2$.

29. Refer to the diagram in the text.
$xyz = 6$. $C = 3xy + 2[1(xz)] + 2[0.5(yz)]$.
Note that $z = 6/(xy)$.   Thus
$$C = 3xy + 2xz + yz = 3xy + 2x[6/(xy)] + y[6/(xy)]$$
$$= 3xy + (12/y) + (6/x).$$

$$\begin{cases} \dfrac{\partial C}{\partial x} = 3y - (6/x^2) = 0 \\ \dfrac{\partial C}{\partial y} = 3x - (12/y^2) = 0. \end{cases}$$

A critical point occurs at x = 1 and y = 2. Thus z = 3.
$\frac{\partial^2 c}{\partial x^2} = \frac{12}{x^3}$, $\frac{\partial^2 c}{\partial y^2} = \frac{24}{y^3}$, $\frac{\partial^2 c}{\partial x \partial y} = 3$. When x = 1 and y = 2, the
D = (12)(3)-(3)^2 = 27 > 0 and $\partial^2 c/\partial x^2$ = 12 > 0. Thus we
have a minimum. The dimensions should be 1 ft by 2 ft by
3 ft.

31. y = (3x - 7)/2. f(x,y) = $-2x^2 + 5\left(\frac{3x - 7}{2}\right)^2$ + 7. Setting
the derivative equal to 0 gives $-4x + 5(2)\left(\frac{3x - 7}{2}\right)\left(\frac{3}{2}\right)$ = 0,
$-4x + \frac{15}{2}(3x - 7)$ = 0, -8x + 15(3x - 7) = 0, 37x = 105, or
x = 105/37. The second-derivative is 37/2 > 0, so we have
a relative minimum. If x = 105/37, then y = 28/37. Thus
there is a relative minimum at (105/37, 28/37).

33. c = $q_A^2 + 3q_B^2 + 2q_A q_B + aq_A + bq_B$ + d.
We are given that (q_A, q_B) = (3, 1) is a critical point.
$$\begin{cases} \partial c/\partial q_A = 2q_A + 2q_B + a = 0 \\ \partial c/\partial q_B = 6q_B + 2q_A + b = 0 \end{cases}$$
Substituting the given values for q_A and q_B into both
equations gives a = -8 and b = -12. Since c = 15 when
q_A = 3 and q_B = 1, from the joint-cost function we have
15 = $3^2 + 3(1^2)$ + 2(3)(1) + (-8)(3) + (-12) + d,
15 = -18 + d, 33 = d. Thus a = -8, b = -12, d = 33.

35. (a) Profit = total revenue - total cost
P = $p_A q_A + p_B q_B$ - total cost
$$= (35 - 2q_A^2 + q_B)q_A + (20 - q_B + q_A)q_B - \left(-8 - 2q_A^3 + 3q_A q_B + 30q_A + 12q_B + \frac{1}{2}q_A^2\right).$$
P = $5q_A - \frac{1}{2}q_A^2 - q_A q_B + 8q_B - q_B^2$ + 8.
$$\begin{cases} \partial P/\partial q_A = 5 - q_A - q_B = 0 \\ \partial P/\partial q_B = -q_A + 8 - 2q_B = 0 \end{cases}$$ Crit. pt.: q_A = 2, q_B = 3.
$\frac{\partial^2 P}{\partial q_A^2}$ = -1, $\frac{\partial^2 P}{\partial q_B^2}$ = -2, $\frac{\partial^2 P}{\partial q_B \partial q_A}$ = -1. At q_A = 2 and q_B = 3

then $D = (-1)(-2)-(-1)^2 = 1 > 0$ and $\frac{\partial^2 P}{\partial q_A^2} = -1 < 0$; thus

there is a relative maximum profit for 2 units of A
and 3 units of B.

(b) Substituting $q_A = 2$ and $q_B = 3$ into the formulas for
p_A, p_B, and P gives a selling price for A of 30, a
selling price for B of 19, and a relative maximum
profit of 25.

37. (a) $P = 5T(1 - e^{-x}) - 20x - 0.1T^2$.

(b) $\frac{\partial P}{\partial T} = 5(1 - e^{-x}) - 0.2T$. $\frac{\partial P}{\partial x} = 5Te^{-x} - 20$.

At the point $(T, x) = (20, \ln 5)$,

$\frac{\partial P}{\partial T} = 5(1 - e^{-\ln 5}) - 0.2(20) = 5\left(1 - \frac{1}{5}\right) - 4 = 0$

$\frac{\partial P}{\partial x} = 5(20)e^{-\ln 5} - 20 = 100\left(\frac{1}{5} - 20\right) = 0$.

Thus $(20, \ln 5)$ is a critical point. In a similar

fashion we verify that $\left(5, \ln \frac{5}{4}\right)$ is a critical point.

(c) $\frac{\partial^2 P}{\partial T^2} = -0.2$, $\frac{\partial^2 P}{\partial x^2} = -5Te^{-x}$, $\frac{\partial^2 P}{\partial T\partial x} = 5e^{-x}$.

At $(20, \ln 5)$,

$$D = (-0.2)[-5(20)e^{-\ln 5}] - (5e^{-\ln 5})^2$$

$$= 20(1/5) - [5(1/5)]^2 = 3 > 0,$$

and $\frac{\partial^2 P}{\partial T^2} = -0.2 < 0$.

Thus we get a relative maximum at $(20, \ln 5)$.

At $\left(5, \ln \frac{5}{4}\right)$,

$$D = (-0.2)[-5(5)e^{-\ln(5/4)}] - [5e^{-\ln(5/4)}]^2$$

$$= 5(4/5) - [5(4/5)]^2 = -12 < 0,$$

so there is no relative extremum at $\left(5, \ln \frac{5}{4}\right)$.

EXERCISE 19.8

1. $f(x,y) = x^2 + 4y^2 + 6$, $2x - 8y = 20$.

$F(x,y,\lambda) = x^2 + 4y^2 + 6 - \lambda(2x - 8y - 20)$.

$$\begin{cases} F_x = 2x - 2\lambda = 0 & (1) \\ F_y = 8y + 8\lambda = 0 & (2) \\ F_\lambda = -2x + 8y + 20 = 0. & (3) \end{cases}$$

From (1), $x = \lambda$; from (2), $y = -\lambda$. Substituting $x = \lambda$ and $y = -\lambda$ into (3) gives $-2\lambda - 8\lambda + 20 = 0$, $-10\lambda = -20$, so $\lambda = 2$. Thus $x = 2$ and $y = -2$. Critical point of F: $(2, -2, 2)$. Critical point of f: $(2, -2)$.

3. $f(x,y,z) = x^2 + y^2 + z^2$, $2x + y - z = 9$.

$F(x,y,z,\lambda) = x^2 + y^2 + z^2 - \lambda(2x + y - z - 9)$.

$$\begin{cases} F_x = 2x - 2\lambda = 0 & (1) \\ F_y = 2y - \lambda = 0 & (2) \\ F_z = 2z + \lambda = 0 & (3) \\ F_\lambda = -2x - y + z + 9 = 0. & (4) \end{cases}$$

From (1), $x = \lambda$; from (2), $y = \lambda/2$; from (3), $z = -\lambda/2$. Substituting into (4) gives $-2\lambda - (\lambda/2) + (-\lambda/2) + 9 = 0$, $-6\lambda + 18 = 0$, so $\lambda = 3$. Thus $x = 3$, $y = 3/2$, $z = -3/2$. Critical point of F: $(3, 3/2, -3/2, 3)$. Critical point of f: $(3, 3/2, -3/2)$.

5. $f(x,y,z) = x^2 + xy + 2y^2 + z^2$, $x - 3y - 4z = 16$.

$F(x,y,z,\lambda) = x^2 + xy + 2y^2 + z^2 - \lambda(x - 3y - 4z - 16)$.

$$\begin{cases} F_x = 2x+y-\lambda = 0 & (1) \\ F_y = x+4y+3\lambda = 0 & (2) \\ F_z = 2z+4\lambda = 0 & (3) \\ F_\lambda = -x+3y+4z+16 = 0. & (4) \end{cases}$$

Eliminating x from (1) and (2) yields $y = -\lambda$. Substituting $y = -\lambda$ into (2) yields $x = \lambda$. From (3), $z = -2\lambda$. Substituting $x = \lambda$, $y = -\lambda$, and $z = -2\lambda$ into (4) yields $\lambda = 4/3$. Crit. pt. of F: $(4/3, -4/3, -8/3, 4/3)$. Crit. pt. of f: $(4/3, -4/3, -8/3)$.

7. $f(x,y,z) = xyz$, $x + 2y + 3z = 18$ $(xyz \neq 0)$.

$F(x,y,z,\lambda) = xyz - \lambda(x+2y+3z-18)$.

$$\begin{cases} F_x = yz - \lambda = 0 & (1) \\ F_y = xz - 2\lambda = 0 & (2) \\ F_z = xy - 3\lambda = 0 & (3) \\ F_\lambda = -x - 2y - 3z + 18 = 0. & (4) \end{cases}$$

From (1) and (2), $y = x/2$. From (1) and (3), $z = x/3$.
Hence from (4), $x = 6$, so $y = 3$ and $z = 2$. Critical point
of f is (6, 3, 2). Note that it is not necessary to
determine λ.

9. $f(x,y,z) = x^2 + 2y - z^2$, $2x - y = 0$, $y + z = 0$. Since
there are two constraints, two Lagrange multipliers are
used.

$F(x,y,z,\lambda_1 \lambda_2) = x^2 + 2y - z^2 - \lambda_1(2x - y) - \lambda_2(y + z)$.

$$\begin{cases} F_x = 2x - 2\lambda_1 = 0 & (1) \\ F_y = 2 + \lambda_1 - \lambda_2 = 0 & (2) \\ F_z = -2z - \lambda_2 = 0 & (3) \\ F_{\lambda_1} = -2x + y = 0 & (4) \\ F_{\lambda_2} = -y - z = 0. & (5) \end{cases}$$

From (1), $x = \lambda_1$. From (3), $z = -\lambda_2/2$. From (4) and (5),
$2x = -z$, so $\lambda_1 = \lambda_2/4$. Substituting $\lambda_1 = \lambda_2/4$ into (2)
yields $\lambda_2 = 8/3$. Thus $\lambda_1 = 2/3$, $x = 2/3$, and $z = -4/3$.
From (5), $y = -z$ and hence $y = 4/3$.
Crit. pt. of f: (2/3, 4/3, -4/3).

11. $f(x,y,z) = xyz$, $x+y+z = 12$, $x+y-z = 0$ $(xyz \neq 0)$. Since
there are two constraints, we use two Lagrange multipliers.

$F(x,y,z,\lambda_1,\lambda_2) = xyz - \lambda_1(x+y+z-12) - \lambda_2(x+y-z)$.

$$\begin{cases} F_x = yz - \lambda_1 - \lambda_2 = 0 & (1) \\ F_y = xz - \lambda_1 - \lambda_2 = 0 & (2) \\ F_z = xy - \lambda_1 + \lambda_2 = 0 & (3) \\ F_{\lambda_1} = -x - y - z + 12 = 0 & (4) \\ F_{\lambda_2} = -x - y + z = 0. & (5) \end{cases}$$

Subtracting (2) from (1) gives $yz - xz = 0$, so $y = x$.
Subtracting (5) from (4) gives $-2z + 12 = 0$, so $z = 6$.
Substituting $y = x$ and $z = 6$ into (5) gives $-2x + 6 = 0$,
so $x = 3$. Thus $y = 3$. Critical point of f: (3, 3, 6).

13. We minimize $c = f(q_1, q_2) = 0.1q_1^2 + 7q_1 + 15q_2 + 1000$
subject to the constraint $q_1 + q_2 = 100$.

$F(q_1, q_2, \lambda) = 0.1q_1^2 + 7q_1 + 15q_2 + 1000 - \lambda(q_1 + q_2 - 100)$.

$$\begin{cases} F_{q_1} = 0.2q_1 + 7 - \lambda = 0 & (1) \\ F_{q_2} = 15 - \lambda = 0 & (2) \\ F_\lambda = -q_1 - q_2 + 100 = 0. & (3) \end{cases}$$

From (2), $\lambda = 15$. Substituting $\lambda = 15$ into (1) gives
$0.2q_1 + 7 - 15 = 0$, so $q_1 = 40$. Substituting $q_1 = 40$ into (3)
gives $-40 - q_2 + 100 = 0$, so $q_2 = 60$. Thus $\lambda = 15$, $q_1 = 40$
and $q = 60$. Thus Plant 1 should produce 40 units and
Plant 2 should produce 60 units.

15. To avoid confusion of the letter "l" with the number "1",
we shall use "L" for the letter "l".
We maximize $f(L, k) = 12L + 20k - L^2 - 2k^2$ subject to the
constraint $4L + 8k = 88$.

$F(L, k, \lambda) = 12L + 20k - L^2 - 2k^2 - \lambda(4L + 8k - 88)$.

$$\begin{cases} F_L = 12 - 2L - 4\lambda = 0 & (1) \\ F_k = 20 - 4k - 8\lambda = 0 & (2) \\ F_\lambda = -4L - 8k + 88 = 0. & (3) \end{cases}$$

Eliminating λ from (1) and (2) yields $k = L - 1$. Substituting
$k = L - 1$ into (3) yields $L = 8$, so $k = 7$. Therefore the
greatest output is $f(8, 7) = 74$ units (when $L = 8$, $k = 7$).

17. We maximize $P(x, y) = 9x^{1/4}y^{3/4} - x - y$ subject to the
constraint $x + y = 60,000$.

$F(x, y, \lambda) = 9x^{1/4}y^{3/4} - x - y - \lambda(x + y - 60,000)$

$$\begin{cases} F_x = \frac{9}{4}x^{-3/4}y^{3/4} - 1 - \lambda = 0 & (1) \\ F_y = \frac{27}{4}x^{1/4}y^{-1/4} - 1 - \lambda = 0 & (2) \\ F_\lambda = -x - y + 60,000 = 0. & (3) \end{cases}$$

Solving (2) for λ and substituting in (1) gives
$\frac{9}{4}x^{-3/4}y^{3/4} - \frac{27}{4}x^{1/4}y^{-1/4} = 0$, $\frac{9}{4}x^{-3/4}y^{3/4} = \frac{27}{4}x^{1/4}y^{-1/4}$,
$y = 3x$. Substituting for y in (3) $\Rightarrow -4x - 60,000 = 0$, so
$x = 15,000$, from which $y = 45,000$. Thus each month
\$15,000 should be spent on newspaper advertising and
\$45,000 on TV advertising.

19. We minimize $B(x,y,z) = x^2 + y^2 + 2z^2$ subject to $x + y = 20$ and $y + z = 20$. Since there are two constraints, we use two Lagrange multipliers.

$$F(x,y,z,\lambda_1,\lambda_2) = x^2+y^2+2z^2-\lambda_1(x+y-20)-\lambda_2(y+z-20).$$

$$\begin{cases} F_x = 2x-\lambda_1 = 0 & (1) \\ F_y = 2y-\lambda_1-\lambda_2 = 0 & (2) \\ F_z = 4z-\lambda_2 = 0 & (3) \\ F_{\lambda_1} = -x-y+20 = 0 & (4) \\ F_{\lambda_2} = -y-z+20 = 0. & (5) \end{cases}$$

Eliminating y from (4) and (5) gives $x = z$. From (1) and (3), $\lambda_1 = 2x$ and $\lambda_2 = 4z$. Substituting in (2) we have $2y - 2x - 4z = 0$, $2y - 2x - 4x = 0$, $2y - 6x = 0$, $y = 3x$. Substituting in (5) gives $-(3x) - x + 20 = 0$, so $x = 5$. Thus $z = 5$ and $y = 15$. Therefore, $x = 5$, $y = 15$, $z = 5$.

21. $U = x^3y^3$, $p_x = 2$, $p_y = 3$, $I = 48$ $(x^3y^3 \neq 0)$.

We want to maximize $U = x^3y^3$ subject to $2x + 3y = 48$.

$$F(x,y,\lambda) = x^3y^3 - \lambda(2x + 3y - 48).$$

$$\begin{cases} F_x = 3x^2y^3 - 2\lambda = 0 & (1) \\ F_y = 3x^3y^2 - 3\lambda = 0 & (2) \\ F_\lambda = -2x - 3y + 48 = 0. & (3) \end{cases}$$

From (1), $\lambda = (3/2)x^2y^3$ and from (2), $\lambda = x^3y^2$. Thus $\frac{3}{2}x^2y^3 = x^3y^2$, so $x = \frac{3}{2}y$. Substituting this expression for x into (3) yields $y = 8$. Hence $x = (3/2)8 = 12$.

23. $U = f(x,y,z) = xyz$. $p_x = 2$, $p_y = 1$, $p_z = 4$, $I = 60$ $(xyz \neq 0)$.

$$F(x,y,z,\lambda) = xyz - \lambda(2x + y + 4z - 60).$$

$$\begin{cases} F_x = yz - 2\lambda = 0 & (1) \\ F_y = xz - \lambda = 0 & (2) \\ F_z = xy - 4\lambda = 0 & (3) \\ F_\lambda = -2x - y - 4z + 60 = 0. & (4) \end{cases}$$

From (1) and (2), $(1/2)yz = xz$, so $y = 2x$. Similarly, from (1) and (3), $z = x/2$. Substituting $y = 2x$ and $z = x/2$ into (4) yields $x = 10$. Thus $y = 20$ and $z = 5$.

EXERCISE 19.9

1. $n = 6$, $\Sigma x_1 = 21$, $\Sigma y_i = 18.6$, $\Sigma x_i y_i = 75.7$, $\Sigma x_i^2 = 91$.

$$\hat{a} = \frac{(\Sigma x_i^2)(\Sigma y_i) - (\Sigma x_i)(\Sigma x_i y_i)}{n\Sigma x_i^2 - (\Sigma x_i)^2} = \frac{91(18.6) - 21(75.7)}{6(91) - (21)^2} = 0.98.$$

$$\hat{b} = \frac{n\Sigma x_i y_i - (\Sigma x_i)(\Sigma y_i)}{n\Sigma x_i^2 - (\Sigma x_i)^2} = \frac{6(75.7) - 21(18.6)}{6(91) - (21)^2} = 0.61.$$

Thus $\hat{y} = 0.98 + 0.61x$. When $x = 3.5$, then $\hat{y} = 3.12$.

3. $n = 5$, $\Sigma x_i = 22$, $\Sigma y_i = 37$, $\Sigma x_i y_i = 189$, $\Sigma x_i^2 = 112.5$.

$\hat{a} = 0.057$, $\hat{b} = 1.67$. Thus $\hat{y} = 0.057 + 1.67x$. When $x = 3.5$, then $\hat{y} = 5.90$.

5. $n = 6$, $\Sigma p_i = 260$, $\Sigma q_i = 329$, $\Sigma p_i q_i = 12,760$, $\Sigma p_i^2 = 13,600$.

$$\hat{a} = \frac{(\Sigma p_i^2)(\Sigma q_i) - (\Sigma p_i)(\Sigma p_i q_i)}{n\Sigma p_i^2 - (\Sigma p_i)^2}$$

$$= \frac{(13,600)(329) - (260)(12,760)}{6(13,600) - (260)^2} = 82.6;$$

$$\hat{b} = \frac{n\Sigma p_i q_i - (\Sigma p_i)(\Sigma q_i)}{n\Sigma p_i^2 - (\Sigma p_i)^2}$$

$$= \frac{6(12,760) - (260)(329)}{6(13,600) - (260)^2} = -0.641.$$

Thus $\hat{q} = 82.6 - 0.641p$.

7. $n = 4$, $\Sigma x_i = 160$, $\Sigma y_i = 420.8$, $\Sigma x_i y_i = 16,915.2$, $\Sigma x_i^2 = 7040$. $\hat{a} = 100$, $\hat{b} = 0.13$. Thus $\hat{y} = 100 + 0.13x$. When $x = 40$, then $\hat{y} = 105.2$.

9.

Year (x)	1	2	3	4	5
Production (y)	10	15	16	18	21

$n = 5$, $\Sigma x_i = 15$, $\Sigma y_i = 80$, $\Sigma x_i y_i = 265$, $\Sigma x_i^2 = 55$.

$$\hat{a} = \frac{(\Sigma x_i^2)(\Sigma y_i) - (\Sigma x_i)(\Sigma x_i y_i)}{n\Sigma x_i^2 - (\Sigma x_i)^2} = \frac{55(80) - 15(265)}{5(55) - (15)^2} = 8.5;$$

$$\hat{b} = \frac{n\Sigma x_i y_i - (\Sigma x_i)(\Sigma y_i)}{n\Sigma x_i^2 - (\Sigma x_i)^2} = \frac{5(265) - 15(80)}{5(55) - (15)^2} = 2.5.$$

Thus $\hat{y} = 8.5 + 2.5x$.

1. a.

Year (x)	1	2	3	4	5
Quantity (y)	35	31	26	24	26

   $n = 5$, $\Sigma x_i = 15$, $\Sigma y_i = 142$, $\Sigma x_i y_i = 401$, $\Sigma x_i^2 = 55$.

   $\hat{a} = 35.9$, $\hat{b} = -2.5$. Thus $\hat{y} = 35.9 - 2.5x$.

   b.

Year (x)	-2	-1	0	1	2
Quantity (y)	35	31	26	24	26

   $n = 5$, $\Sigma x_i = 0$, $\Sigma y_i = 142$, $\Sigma x_i y_i = -25$, $\Sigma x_i^2 = 10$.

   $\hat{a} = \dfrac{\Sigma y_i}{n} = 28.4$ and $\hat{b} = \dfrac{\Sigma x_i y_i}{\Sigma x_i^2} = -2.5$. Thus

   $\hat{y} = 28.4 - 2.5x$.

EXERCISE 19.11

1. $\displaystyle\int_0^3 \int_0^4 x \, dy \, dx = \int_0^3 xy \Big|_0^4 \, dx = \int_0^3 4x \, dx = 2x^2 \Big|_0^3 = 18$

3. $\displaystyle\int_0^1 \int_0^1 xy \, dx \, dy = \int_0^1 \frac{x^2 y}{2} \Big|_0^1 \, dy = \int_0^1 \frac{y}{2} \, dy = \frac{y^2}{4} \Big|_0^1 = \frac{1}{4}$

5. $\displaystyle\int_1^3 \int_1^2 (x^2 - y) \, dx \, dy = \int_1^3 \left(\frac{x^3}{3} - xy\right) \Big|_1^2 \, dy =$

   $\displaystyle\int_1^3 \left[\left(\frac{8}{3} - 2y\right) - \left(\frac{1}{3} - y\right)\right] \, dy = \int_1^3 \left(\frac{7}{3} - y\right) \, dy =$

   $\left(\dfrac{7}{3}y - \dfrac{y^2}{2}\right) \Big|_1^3 = \left(7 - \dfrac{9}{2}\right) - \left(\dfrac{7}{3} - \dfrac{1}{2}\right) = \dfrac{2}{3}$

7. $\displaystyle\int_0^1 \int_0^2 (x + y) \, dy \, dx = \int_0^1 \left(xy + \frac{y^2}{2}\right) \Big|_0^2 \, dx = \int_0^1 (2x + 2) \, dx =$

   $(x^2 + 2x) \Big|_0^1 = 3$

9. $\int_0^6 \int_0^{3x} y \, dy \, dx = \int_0^6 \frac{y^2}{2}\Big|_0^{3x} dx = \int_0^6 \frac{9x^2}{2} dx = \frac{3x^3}{2}\Big|_0^6 = 324$

11. $\int_0^1 \int_{3x}^{x^2} 2x^2 y \, dy \, dx = \int_0^1 (x^2 y^2)\Big|_{3x}^{x^2} dx = \int_0^1 (x^6 - 9x^4) \, dx =$

$\left(\frac{x^7}{7} - \frac{9x^5}{5}\right)\Big|_0^1 = -\frac{58}{35}$

13. $\int_0^2 \int_0^{\sqrt{4-y^2}} x \, dx \, dy = \int_0^2 \frac{x^2}{2}\Big|_0^{\sqrt{4-y^2}} dy = \int_0^2 \frac{4-y^2}{2} \, dy =$

$\int_0^2 \left(2 - \frac{y^2}{2}\right) dy = \left(2y - \frac{y^3}{6}\right)\Big|_0^2 = \left(4 - \frac{4}{3}\right) - 0 = \frac{8}{3}$

15. $\int_{-1}^1 \int_x^{1-x} (x+y) \, dy \, dx = \int_{-1}^1 \left(xy + \frac{y^2}{2}\right)\Big|_x^{1-x} dx =$

$\int_{-1}^1 \left[x(1-x) + \frac{(1-x)^2}{2} - \left(x^2 + \frac{x^2}{2}\right)\right] dx =$

$\int_{-1}^1 \left[x - \frac{5x^2}{2} + \frac{(1-x)^2}{2}\right] dx = \left[\frac{x^2}{2} - \frac{5x^3}{6} - \frac{(1-x)^3}{6}\right]\Big|_{-1}^1 =$

$\left[\frac{1}{2} - \frac{5}{6} - 0\right] - \left[\frac{1}{2} + \frac{5}{6} - \frac{4}{3}\right] = -\frac{1}{3}$

17. $\int_0^1 \int_0^y e^{x+y} \, dx \, dy = \int_0^1 e^{x+y}\Big|_0^y dy = \int_0^1 (e^{2y} - e^y) \, dy =$

$\left[\frac{e^{2y}}{2} - e^y\right]\Big|_0^1 = \frac{e^2}{2} - e - \left(\frac{1}{2} - 1\right) = \frac{e^2}{2} - e + \frac{1}{2}$

19. $\int_{-1}^0 \int_{-1}^2 \int_1^2 6xy^2 z^3 \, dx \, dy \, dz = \int_{-1}^0 \int_{-1}^2 3x^2 y^2 z^3\Big|_1^2 dy \, dz =$

$\int_{-1}^0 \int_{-1}^2 9y^2 z^3 \, dy \, dz = \int_{-1}^0 3y^3 z^3\Big|_{-1}^2 dz =$

$\int_{-1}^0 27z^3 \, dz = \frac{27z^4}{4}\Big|_{-1}^0 = -\frac{27}{4}$

21. $\int_0^1 \int_{x^2}^x \int_0^{xy} dz\ dy\ dx = \int_0^1 \int_{x^2}^x z\Big|_0^{xy}\ dy\ dx =$

$\int_0^1 \int_{x^2}^x xy\ dy\ dx = \int_0^1 \frac{xy^2}{2}\Big|_{x^2}^x\ dx = \int_0^1 \left[\frac{x^3}{2} - \frac{x^5}{2}\right]\ dx =$

$\left[\frac{x^4}{8} - \frac{x^6}{12}\right]\Big|_0^1 = \frac{1}{24}$

23. $P(0 \leq x \leq 2,\ 1 \leq y \leq 2) = \int_1^2 \int_0^2 e^{-(x+y)}\ dx\ dy = \int_1^2 -e^{-(x+y)}\Big|_0^2\ dy =$

$\int_1^2 [-e^{-(2+y)} + e^{-y}]\ dy = [e^{-(2+y)} - e^{-y}]\Big|_1^2 = e^{-4} - e^{-2} - e^{-3} + e^{-1}.$

<u>Ans.</u> $e^{-4} - e^{-2} - e^{-3} + e^{-1}$

25. $P(x \geq 1,\ y \geq 2) = \int_2^4 \int_1^2 \frac{x}{8}\ dx\ dy = \int_2^4 \frac{x^2}{16}\Big|_1^2\ dy =$

$\int_2^4 \left(\frac{4}{16} - \frac{1}{16}\right)\ dy = \int_2^4 \frac{3}{16}\ dy = \frac{3y}{16}\Big|_2^4 = \frac{12}{16} - \frac{6}{16} = \frac{3}{8}.$

<u>Ans.</u> 3/8

<u>CHAPTER 19 - REVIEW PROBLEMS</u>

1. $2x+3y+z = 9$ can be put in
   the form $Ax+By+Cz+D = 0$, so
   the graph is a plane. The
   intercepts are (9/2,0,0),
   (0,3,0), and (0,0,9).

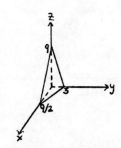

3. $z = y^2$. The y,z-trace is
   $z = y^2$, which is a parabola.
   For any fixed value of x,
   we obtain the parabola $z = y^2$

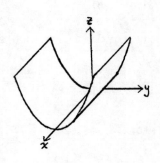

5. $f_x(x,y) = 2(2x)+3(1)y+0-0 = 4x+3y$
   $f_y(x,y) = 0+3x(y)+2y-0 = 3x+2y$

7. $\dfrac{\partial z}{\partial x} = \dfrac{(x+y)(1) - x(1)}{(x+y)^2} = \dfrac{y}{(x+y)^2}$.
   Because $z = x(x+y)^{-1}$, $\dfrac{\partial z}{\partial y} = x[(-1)(x+y)^{-2}(1)] = -\dfrac{x}{(x+y)^2}$.

9. $f(x,y) = \ln\sqrt{x^2+y^2} = \frac{1}{2} \ln(x^2+y^2)$.
   $\dfrac{\partial}{\partial y}[f(x,y)] = \frac{1}{2}\cdot\dfrac{1}{x^2+y^2}(2y) = \dfrac{y}{x^2+y^2}$.   <u>Ans.</u> $\dfrac{y}{x^2+y^2}$

11. $w_x(x,y) = 2xyze^{x^2yz}$;
    $w_{xy}(x,y) = 2xz\left[y(e^{x^2yz}\cdot x^2z) + e^{x^2yz}\cdot 1\right] = 2xze^{x^2yz}(x^2yz+1)$.
    <u>Ans.</u> $2xze^{x^2yz}(x^2yz+1)$

13. $\dfrac{\partial}{\partial z}[f(x,y,z)] = (x+y)[2z]$; $\dfrac{\partial^2}{\partial z^2}[f(x,y,z)] = (x+y)[2]$.
    <u>Ans.</u> $2(x+y)$

15. $\dfrac{\partial w}{\partial y} = (x \ln z)[e^{yz}\cdot z] = xze^{yz} \ln z$;
    $\dfrac{\partial w}{\partial z} = x\left[e^{yz}\cdot\frac{1}{z} + (\ln z)(e^{yz}\cdot y)\right] = xe^{yz}\left[\frac{1}{z} + y \ln z\right]$;

$$\frac{\partial^2 w}{\partial x \partial z} = e^{yz}\left[\frac{1}{z} + y \ln z\right].$$

Ans. $xze^{yz} \ln z$; $e^{yz}[(1/z) + y \ln z]$

17. $f(x,y,z) = \frac{x + y}{xz} = \frac{1}{z} + \frac{y}{xz}$.

$f_x(x,y,z) = -\frac{y}{x^2 z}$. $f_{xy}(x,y,z) = -\frac{1}{x^2 z}$.

$f_{xyz}(x,y,z) = \frac{1}{x^2 z^2}$. $f_{xyz}(2,3,4) = \frac{1}{2^2 \cdot 4^2} = \frac{1}{64}$.   Ans. $\frac{1}{64}$

19. $\frac{\partial w}{\partial r} = \frac{\partial w}{\partial x} \frac{\partial x}{\partial r} + \frac{\partial w}{\partial y} \frac{\partial y}{\partial r} = (2x+2y)(e^r) + (2x+6y)\left(\frac{1}{r+s}\right)$

$= 2(x+y)e^r + \frac{2(x+3y)}{r+s}$

$\frac{\partial w}{\partial s} = \frac{\partial w}{\partial x} \frac{\partial x}{\partial s} + \frac{\partial w}{\partial y} \frac{\partial y}{\partial s} = (2x+2y)(0) + (2x+6y)\left(\frac{1}{r+s}\right)$

$= \frac{2(x+3y)}{r+s}$

21. $2x + 2y - 4z\frac{\partial z}{\partial x} + \left[x\frac{\partial z}{\partial x} + z(1)\right] + 0 = 0,$

$(-4z+x)\frac{\partial z}{\partial x} = -(2x+2y+z), \quad \frac{\partial z}{\partial x} = \frac{-(2x+2y+z)}{-4z+x}.$

Ans. $(2x+2y+z)/(4z-x)$

23. In order to avoid confusion of the letter "l" with the number "1", we shall use "L" for the letter "l".

$P = 20L^{0.7}k^{0.3}$. Marginal productivity functions are given by $\frac{\partial P}{\partial L} = 20(0.7)L^{-0.3}k^{0.3}$ and $\frac{\partial P}{\partial k} = 20(0.3)L^{0.7}k^{-0.7}$.

Ans. $\partial P/\partial L = 14L^{-0.3}k^{0.3}$; $\partial P/\partial k = 6L^{0.7}k^{-0.7}$

25. $q_A = 200-3p_A+p_B$, $q_B = 50-5p_B+p_A$. Since $\partial q_A/\partial p_B = 1 > 0$ and $\partial q_B/\partial p_A = 1 > 0$, A and B are competitive products.

Ans. competitive

27. $f(x,y) = x^2+2y^2-2xy-4y+3$.

$\begin{cases} f_x(x,y) = 2x-2y = 0 \\ f_y(x,y) = 4y-2x-4 = 0. \end{cases}$ Crit. pt.: $(2,2)$.

$f_{xx}(x,y) = 2$, $f_{yy}(x,y) = 4$, $f_{xy}(x,y) = -2$. At $(2,2)$, D = $(2)(4)-(-2)^2 = 4 > 0$ and $f_{xx}(x,y) = 2 > 0$; thus rel. min.
<u>Ans.</u> $(2,2)$, rel. min.

29.

xyz = 32. Let S be the amount of cardboard used.
$$S = xy + 2yz + 2xz$$
$$= xy + 2y[32/(xy)] + 2x[32/(xy)]$$
$$= xy + (64/x) + (64/y).$$
$$\frac{\partial S}{\partial x} = y - \frac{64}{x^2}, \quad \frac{\partial S}{\partial y} = x - \frac{64}{y^2}.$$

The critical point occurs when $x = 4$, $y = 4$, and $z = 2$, which gives a minimum.   <u>Ans.</u> 4 ft by 4 ft by 2 ft

31. Profit = $P = (p_A-50)q_A + (p_B-60)q_B$.

$P = (p_A-50)[250(p_B-p_A)] + (p_B-60)[32,000+250(p_A-2p_B)]$.

$\frac{\partial P}{\partial p_A} = (p_A-50)(-250) + [250(p_B-p_A)](1) + 250(p_B-60)$

$= -500p_A + 500p_B - 250(10) = 500(-p_A + p_B - 5)$.

$\frac{\partial P}{\partial p_B} = (p_A-50)(250) + (p_B-60)(-500) + [32,000+250(p_A-2p_B)]$

$= 500p_A - 1000p_B + 49,500 = 500(p_A - 2p_B + 99)$.

Setting $\partial P/\partial p_A = 0$ and $\partial P/\partial p_B = 0$ gives

$$-p_A + p_B - 5 = 0 \qquad (1)$$
$$\text{and} \quad p_A - 2p_B + 99 = 0. \qquad (2)$$

Adding Eqs. (1) and (2) gives $-p_B + 94 = 0$. So $p_B = 94$. From Eq. (1), $p_A = p_B - 5$, so $p_A = 94 - 5 = 89$.

At $p_A = 89$ and $p_B = 94$, D = $\frac{\partial^2 P}{\partial p_A^2} \frac{\partial^2 P}{\partial p_B^2} - \frac{\partial^2 P}{\partial p_B \partial p_A} =$

$(-500)(-1000) - (500)^2 > 0$ and $\frac{\partial^2 P}{\partial p_A^2} = -500 < 0$. Thus there

is a relative maximum profit when $p_A = 89$ and $p_B = 94$.
<u>Ans.</u> A, 89 cents per pound; B, 94 cents per pound

3. $f(x,y,z) = x^2+y^2+z^2$, $3x+2y+z = 14$.

$F(x,y,z,\lambda) = x^2+y^2+z^2 - \lambda(3x+2y+z-14)$.

$$\begin{cases} F_x = 2x-3\lambda = 0 & (1) \\ F_y = 2y-2\lambda = 0 & (2) \\ F_z = 2z-\lambda = 0 & (3) \\ F_\lambda = -3x-2y-z+14 = 0. & (4) \end{cases}$$

From (1), $x = \frac{3\lambda}{2}$; from (2), $y = \lambda$; from (3), $z = \frac{\lambda}{2}$.

Substituting into (4) gives $-3\left(\frac{3\lambda}{2}\right) - 2\lambda - \frac{\lambda}{2} + 14 = 0$,

from which $\lambda = 2$. Thus $x = 3$, $y = 2$, and $z = 1$.

Crit. pt. of F: $(3,2,1,2)$. Crit. pt. of f: $(3,2,1)$.

<u>Ans.</u> $(3,2,1)$

5.

Year (x)	1	2	3	4	5	6
Expenditures (y)	15	22	21	27	26	34

$n = 6$, $\Sigma x_i = 21$, $\Sigma y_i = 145$, $\Sigma x_i y_i = 564$, $\Sigma x_i^2 = 91$.

$$\hat{a} = \frac{(\Sigma x_i^2)(\Sigma y_i) - (\Sigma x_i)(\Sigma x_i y_i)}{n\Sigma x_i^2 - (\Sigma x_i)^2} = 12.87,$$

$$\hat{b} = \frac{n\Sigma x_i y_i - (\Sigma x_i)(\Sigma y_i)}{n\Sigma x_i^2 - (\Sigma x_i)^2} = 3.23.$$

Thus $\hat{y} = 12.87 + 3.23x$. <u>Ans.</u> $\hat{y} = 12.87 + 3.23x$

37. $\int_0^4 \int_{y/2}^2 xy\, dx\, dy = \int_0^4 \frac{x^2 y}{2}\bigg|_{y/2}^2 dy = \int_0^4 \left(2y - \frac{y^3}{8}\right) dy =$

$\left(y^2 - \frac{y^4}{32}\right)\bigg|_0^4 = (16 - 8) - 0 = 8$

39. $\int_0^1 \int_{\sqrt{x}}^{x^2} (x^2+2xy-3y^2)\, dy\, dx = \int_0^1 (x^2 y+xy^2-y^3)\bigg|_{\sqrt{x}}^{x^2} dx =$

$\int_0^1 [(x^4+x^5-x^6) - (x^{5/2}+x^2-x^{3/2})]\, dx =$

$\left[\frac{x^5}{5} + \frac{x^6}{6} - \frac{x^7}{7} - \frac{2x^{7/2}}{7} - \frac{x^3}{3} + \frac{2x^{5/2}}{5}\right]\bigg|_0^1 =$

$\left[\frac{1}{5} + \frac{1}{6} - \frac{1}{7} - \frac{2}{7} - \frac{1}{3} + \frac{2}{5}\right] - 0 = \frac{1}{210}$

CHAPTER 19 — MATHEMATICAL SNAPSHOT

1. $y = Ce^{ax} + 5$, $y-5 = Ce^{ax}$, $\ln(y-5) = ax + \ln C$.

x	y	y - 5	ln(y - 5)
0	15	10	2.30259
1	12	7	1.94591
4	9	4	1.38629
7	7	2	0.69315
10	6	1	0.00000

$n = 5$, $\Sigma x_i = 22$, $\Sigma \ln(y_i-5) = 6.32794$,

$\Sigma[x_i \ln(y_i-5)] = 12.34312$, $\Sigma x_i^2 = 166$.

$$a = \frac{n\Sigma[x_i \ln(y_i-5)] - (\Sigma x_1)[\Sigma \ln(y_i-5)]}{n(\Sigma x_i^2) - (\Sigma x_i)^2}$$

$$= \frac{5(12.34312) - 22(6.32794)}{5(166) - (22)^2} \approx -0.22399;$$

$$\ln C = \frac{\Sigma(x_i^2)[\Sigma \ln(y_i-5)] - (\Sigma x_i)\{\Sigma[x_i \ln(y_i-5)]\}}{n(\Sigma x_i^2) - (\Sigma x_i)^2}$$

$$= \frac{166(6.32794) - 22(12.34312)}{5(166) - (22)^2} \approx 2.25113.$$

$C \approx e^{2.25113} \approx 9.50$.   <u>Ans.</u> $y = 9.50e^{-0.22399x} + 5$

3. Newton's law of cooling: $\frac{dT}{dt} = k(T-a)$, where $a = 45$. Thus

$\frac{dT}{dt} = k(T-45)$, $\frac{dT}{T-45} = k\,dt$, $\int \frac{dT}{T-45} = \int k\,dt$,

$\ln|T-45| = kt + C$. Because $T-45 > 0$, $\ln(T-45) = kt + C$.

Thus $T-45 = e^{kt+C}$, or $T = e^{kt+C} + 45 = e^C e^{kt} + 45 =$

$C_1 e^{kt} + 45$, where $C_1 = e^C$. So

$$T = C_1 e^{kt} + 45.$$

When $t = 0$, then $T = 124$. Hence $124 = C_1 + 45$, or $C_1 = 79$

Thus $T = 79e^{kt} + 45$. When $t = 128$, then $T = 64$, so

$64 = 79e^{128k} + 45$, $19 = 79e^{128k}$, $e^{128k} = \frac{19}{79}$,

$128k = \ln\frac{19}{79}$, $k = \frac{\ln(19/79)}{128} \approx -0.01113$.

<u>Ans.</u> $T = 79e^{-0.01113t} + 45$